ELASTOKINETICS

An Introduction to the Dynamics of Elastic Systems—presents the fundamental concepts and ideas used in formulating mathematical models of dynamic physical systems. Both discrete and continuous models are treated.

A variety of techniques are presented for the solution of dynamics problems. Emphasis is on classical techniques, with Hamilton's principle occupying a central position.

Also, thorough coverage of modal analysis is given, with emphasis upon the general case of the forced motion problem. The authors develop the intimate relationship between wave motion and normal modes.

Elastokinetics is intended for beginning graduate students in the areas of mechanical, civil, and aeronautical engineering, as well as students in engineering science.

ELASTOKINETICS

An Introduction
to the
Dynamics of
Elastic Systems

ELASTOKINETICS

An Introduction to the Dynamics of Elastic Systems

Herbert Reismann
State University of New York at Buffalo

Peter S. Pawlik
State University College at Buffalo

WEST PUBLISHING CO.
St. Paul • New York • Boston
Los Angeles • San Francisco

Library of Congress Catalog Number: 74–4510
ISBN: 0–8299–0016–0

To

The State University of New York at Buffalo

On the occasion of

its 125th anniversary

1846-1971

PREFACE

> The scientist must put things in order; science is made of facts as a house is made of stones, but science is no more an accumulation of facts than a house is a pile of stones.
>
> *Henri Poincare'*

Predicting the dynamic response of solids and structures subjected to time-dependent loads and/or boundary conditions is a task which occurs in many branches of modern engineering and applied science. Such problems arise from considerations as diverse as gust response of aircraft, earthquake response of tall buildings, vibration of machine parts, and shock response of electron tube filaments. The mathematical models used most often in such analyses assume a linearly elastic material behavior and characterize the mass distribution of the structure in either a discrete or continuous fashion. The present book was written to present the fundamentals of both approaches, and this is reflected by its subdivision into Part I (Discrete Systems) and Part II (Continuous Systems). Equal emphasis is placed upon both a rational derivation of the pertinent equations and the techniques for their solution. With regard to this distribution of emphasis, there is a uniformity of the presentation throughout the book: With the exception of Chapter 1, the derivation of the pertinent equations is always based upon Hamilton's principle and the dynamic response of a solid or structure of bounded extent is always characterized by an eigenfunction or normal mode expansion. Furthermore, the intimate relationship which exists between the modal response and wave motion in the solid is emphasized. We have endeavored to clearly indicate the roots from which the various mathematical models grow, viz.: particle mechanics in the case of discrete representations and continuum mechanics in the case of continuous representations. Although our ultimate aim is to present a number of useful, linear models, we have shown their nonlinear origins and present rational linearization procedures which should leave little doubt about their range of validity.

Various parts of the material contained in this book have been presented to students from a number of engineering disciplines (engineering science, mechanical engineering, civil engineering, and electrical engineering). The presentation is directed to beginning graduate students, although some parts have been presented, with considerable success, to engineering seniors. It was the interest and enthusiasm of the audience which provided the initial impetus to expand a set of lecture notes into a book. The level of mathematical maturity expected of the reader is consistent with that of a senior or first year graduate student in engineering. However, we have deliberately attempted to keep the required mathematical background to a minimum in order to reach as wide an audience as possible. Thus, a chapter on Cartesian tensors is included for the benefit of those readers who are not familiar with this subject or who wish a brief review of it. A brief introduction to the basic concepts of continuum mechanics is also included for the convenience of those readers who have not had a formal course in that subject. Our aim is to present several methods which a practicing engineer or engineering student can easily learn and apply to the dynamics problems which confront him. Consequently, we have not included any existence, completeness or convergence theorems and in this respect we make no claim to rigor in the modern mathematical sense. However, we do believe that our conclusions are correct and we do cite the pertinent references where such questions are treated.

The present treatment provides considerable flexibility, and it is possible to adapt the material presented to a variety of courses. For instance, Part I (Chapters 1 through 7) can be used for a one semester course in Vibration Theory in a science oriented program, or it can be used to succeed a more earthy, introductory course in Mechanical Vibrations. Part II (Chapters 8 through 14), being almost entirely self-contained, can be used for either a one or a two-semester graduate course in Dynamical Elasticity, Vibrations of Continua, Vibrations of Solids, or whatever designation is in vogue at the time and place the course is offered. Selected portions of Parts I and II are readily adapted to provide the material for a comprehensive two-semester graduate course in Elastokinetics. Although intended for engineers, we even venture the thought that the material might be of interest to the community of (classical) physics and (applied) mathematics students who have a taste for applications. We have endeavored to make the text suitable for self-study for those who have left the university and wish to acquire the material without the assistance of an instructor. All readers who desire to derive the maximum benefit are urged to enhance and extend their understanding of the textual material by solving the exercises at the ends of the various chapters. Further elucidation can be obtained by consulting the references interspersed in the text or listed at the ends of Parts I and II. The references cited are not complete and certainly not exhaustive, but they are the ones which we consulted. An accurate and complete, critical, annotated bibliography of all research in such an active field would require a separate monograph. We hope that we shall be forgiven any particular omission deemed important.

At this point, it is our pleasant task to acknowledge the assistance of Dr. Paul Culkowski, Dr. Yu Chung Lee, and Mr. James Greene. Their formal as well as informal contributions to some of the detailed examples have helped to clarify many a point which otherwise might have been less clear. We also express our appreciation for the patience and diligence displayed by Mrs. Ruth Fenton, Mrs. Donna Miller and Miss Ginger Moronski in typing the manuscript and Mr. Takashi Yamaguchi for his careful preparation of some of the illustrations. We also owe thanks to the Air Force Office of Scientific Research of the U.S. Air Force who provided financial support for the research effort during the course of which we recognized the need for a book such as this one. We thank Professor I.H. Shames for providing encouragement. Finally, one of us (H.R.) would like to express his appreciation to the State University of New York at Buffalo for the grant of a sabbatical leave during which a major part of the manuscript took shape.

Buffalo, New York

Herbert Reismann
Peter Pawlik

CONTENTS

I **DISCRETE SYSTEMS** 1

1 **THE LINEAR OSCILLATOR (VIBRATIONS OF A ONE-DEGREE OF FREEDOM SYSTEM)** 2

mathematical models, *3*
the laplace transform, *5*
free vibrations, *7*
forced motion, *12*

2 **MECHANICS OF DISCRETE SYSTEMS** 24

some general observations, *25*
hamilton's principle, *25*
generalized coordinates and lagrange's equations, *35*
applications of lagrange's equations, *43*

3 **SMALL OSCILLATIONS OF DISCRETE SYSTEMS (THEORY)** 53

linearization of the equations of motion, *54*
free vibrations, *56*
normal coordinates, *62*
forced motion, *64*

4 SMALL OSCILLATIONS OF DISCRETE SYSTEMS (APPLICATIONS) 68

the linear oscillator in a gravitational field, *69*
free and forced motion of a two-degree of freedom system, *71*
free vibration and forced motion of an iterated structure, *79*
transition to a continuous structure, *86*

5 SMALL OSCILLATIONS OF A DISCRETE SYSTEM (FURTHER DEVELOPMENTS) 91

the energy equation of motion, *94*
uniqueness of solution, *94*
the reciprocal theorem, *95*
a stationary property of the natural frequencies, *97*
the effect of a constraint on the natural frequencies, *102*

6 DAMPING 108

introduction, *109*
the dissipation function, *109*
free damped vibrations, *112*
forced, harmonic excitation of damped systems, *117*
classical normal modes in the presence of damping, *120*

7 WAVE PROPAGATION IN ITERATED STRUCTURES 124

introduction, *125*
energy and momentum flux in a simple iterated structure, *126*
harmonic wave propagation in an unbounded, simple, periodic structure (a mechanical low pass filter), *128*
transmission and reflection of harmonic waves in an iterated structure, *134*
transient wave motion in a simple periodic structure, *137*
the mechanical band-pass filter, *142*

REFERENCES: PART I 152

II CONTINUOUS SYSTEMS
155

8 CARTESIAN TENSORS
156

introduction, *157*
index notation, *158*
vector algebra, *160*
coordinate transformations, *166*
tensor algebra, *172*
tensor calculus, *178*
curvilinear coordinates, *182*

9 MECHANICS OF CONTINUA
203

introduction, *204*
kinematics of continua, *205*
dynamics of continua, *224*
theorems on work and energy in elastic solids, *235*

10 DYNAMICS OF ELASTIC BODIES (LINEAR THEORY)
260

linearization of the basic equations, *261*
f. neumann's uniqueness theorem, *271*
reciprocal relations, *274*
free vibrations of bounded elastic bodies, *277*
forced motions of bounded elastic bodies, *284*
forced motion of a spherical shell, *292*

11 WAVE MOTION IN ELASTIC MEDIA
315

introduction, *316*
waves in unbounded media, *316*
wave reflection from a plane boundary, *328*
rayleigh surface waves, *335*
progressive waves in plates (the rayleigh-lamb problem), *340*
progressive waves in cylinders, *347*

12 DYNAMICS OF ELASTIC RODS
360

torsional motion of rods, *361*
longitudinal motion of rods, *366*
the rod of finite length, *368*
the rod of unbounded length, *381*

13 DYNAMICS OF ELASTIC BEAMS 399

equations of motion, *400*
forced motion of euler-bernoulli beams, *408*
forced motion of timoshenko beams, *418*
wave motion in beams, *431*

14 DYNAMICS OF ELASTIC PLATES 456

preliminaries, *457*
equations of motion, *462*
forced motion, *472*
classical plate theory, *496*
wave motion in plates, *514*

REFERENCES: PART II 549

APPENDICES 553

A THE CALCULUS OF VARIATIONS 554

B KINEMATICS OF WAVE MOTION 558

INDEX 576

Part I

DISCRETE SYSTEMS

Chapter

1

THE LINEAR OSCILLATOR (VIBRATIONS OF A ONE-DEGREE OF FREEDOM SYSTEM)

First of all one must observe that each pendulum has its own time of vibration, so definite and determinate that it is not possible to make it move with any other period than that which nature has given it. On the other hand one can confer motion upon even a heavy pendulum which is at rest by simply blowing against it. By repeating these blasts with a frequency which is the same as that of the pendulum one can impart considerable motion.

G. Galilei, Discorsi a Due Nuove Scienze (1638).

1.1 MATHEMATICAL MODELS

A large number of mechanical as well as other physical systems may be described, to a first approximation, by a single degree of freedom model. In such a model, a single quantity (the coordinate) completely describes the configuration of the system at any given instant. Fig. 1.1 shows a mass attached to a linearly elastic, massless spring. The mass is constrained to move (in pure translation) along a straight line, and its motion is retarded by a (massless) viscous damping device which is often used to symbolize the damping in the spring to a first approximation. We denote by m (lbs-sec²/ft) the magnitude of the mass, k (lbs/ft) the spring constant, and c (lbs-sec/ft) the viscous damping coefficient. At any instant of time, the spring force $-kx$ and the force due to the damper $-c\dot{x}$, both act on the mass m. The quantity F(t)(lbs) is an applied force which is specified as a function of time. If Newton's second law of motion is applied to the system shown in Fig. 1.1, the following linear, ordinary differential equation with constant coefficients is obtained:

$$m\ddot{x} = -kx-c\dot{x}+F(t) \tag{1.1}$$

where x (ft) denotes the displacement of the mass from its equilibrium position, i.e., that position for which the spring force vanishes. If, in addition, we specify the initial displacement of the mass $x(0)=x_0$ and its initial velocity $\dot{x}(0)=\dot{x}_0$, then the theory of linear differential equations guarantees us a unique solution of the problem posed, i.e., we can obtain $x = x(t)$ for $t > 0$.

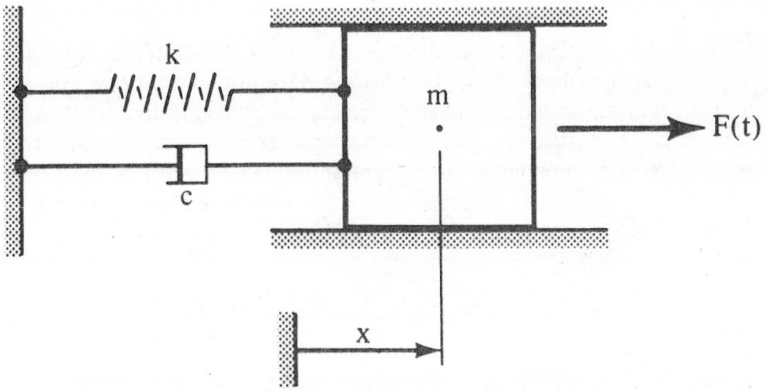

Figure 1.1

To insure that the viscous damping force is dissipative, it is necessary that the viscous damping coefficient c be a positive quantity.

To demonstrate this we set $F(t)\equiv 0$ and write (1.1) in the form

$$\frac{d}{dt}(T+V) = -c\dot{x}^2 \tag{1.2}$$

where $T = \frac{1}{2}m\dot{x}^2$ and $V = \frac{1}{2}kx^2$ are the kinetic and potential energy, respectively, of the linear oscillator. Since there is no external energy supplied to the oscillating system, its energy can not increase as time increases. Hence,

$$\frac{d}{dt}(T+V) = -c\dot{x}^2 \leq 0 \tag{1.3}$$

and c is necessarily non-negative.

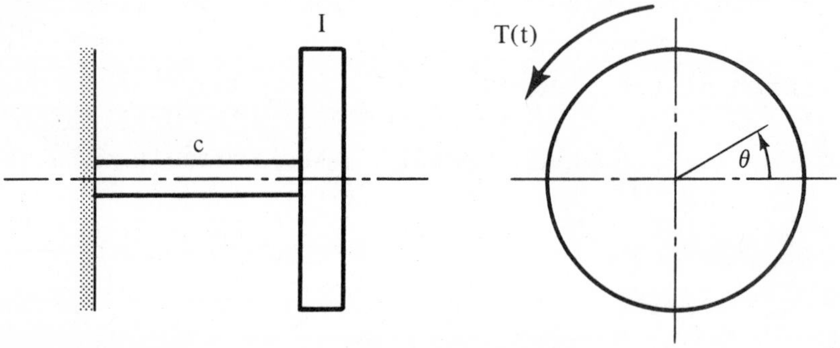

Figure 1.2

Another mechanical model which can be characterized by an equation which is equivalent to (1.1) is shown in Fig. 1.2. In this case a disc with polar moment of inertia I(ft-lbs-sec^2) is attached to a massless shaft with torsional stiffness c (ft-lbs) and viscous damping coefficient γ (ft-lbs-sec). An externally applied torque T (ft-lbs) is applied to the disc. Euler's equations, applied to the disc, require that the rate of change of angular momentum $I\ddot{\theta}$ of the disc must be set equal to the sum of the elastic restoring torque $-c\theta$, the viscous damping torque $-\gamma\dot{\theta}$, and the applied torque $T(t)$. Hence

$$I\ddot{\theta} = -c\theta-\gamma\dot{\theta}+T(t) \tag{1.4}$$

where the reference $\theta=0$ defines that position of the disc for which the restoring torque vanishes. The appropriate initial conditions, in this case, are given by $\theta(0)=\theta_0$, $\dot{\theta}(0)=\dot{\theta}_0$.

The concept of a linear oscillator with one degree of freedom is not restricted to mechanical systems. If we consider a series circuit as shown in Fig. 1.3, with inductance L (henries), capacitance C (farads), and resistance R (ohms), and if we equate the sum of the voltage drops around the loop to the applied electro-motive force e(t) (volts), we obtain

$$L\frac{di}{dt} + Ri + \frac{q}{C} = e(t) \tag{1.5}$$

where i (amperes) is the loop current and q (coulombs) is the charge on the capacitor. Since $i = \frac{dq}{dt}$, (1.5) may be rewritten as

$$L\frac{d^2q}{dt^2} + R\frac{dq}{dt} + \frac{1}{C}q = e(t) \tag{1.6}$$

The appropriate initial conditions associated with (1.6) are $q(0)=q_0$ and $\dot{q}(0) = i_0$, i.e., the loop current and charge on the capacitor must be specified at t=0.

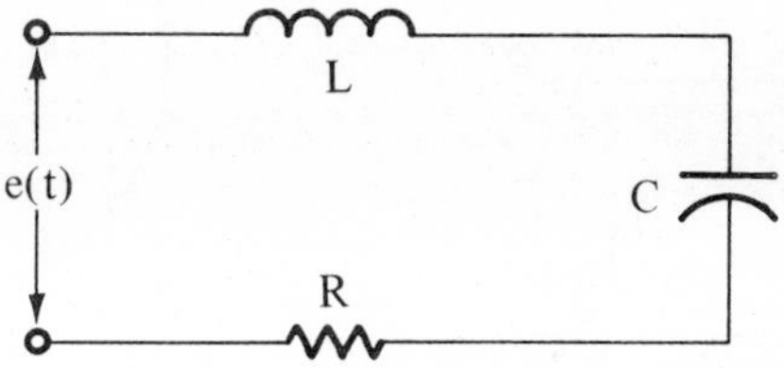

Figure 1.3

From the point of view of mathematics, equations (1.1), (1.4), and (1.6) are equivalent, although the physical systems which they characterize are entirely different. The solution of the linear oscillator equation for arbitrary initial conditions and arbitrary forcing functions of physical interest has intrinsic value. It should be noted, however, that techniques and insights which accrue from such an effort carry over to the most complex linear vibrating systems whether they be discrete or continuous. The techniques which are developed in the present chapter to analyze this system will be used extensively in succeeding chapters.

1.2 THE LAPLACE TRANSFORM

Linear differential equations can be solved by a variety of methods. In the present and succeeding chapters we shall be confronted by initial value problems and in this case the method of Laplace transforms provides a useful and efficient means to obtain solutions. The present development of the method of Laplace transforms is in the form of an outline or review, and concerns itself only with the formal, manipulative aspects of the problem. Readers who have had no previous exposure to this subject or who are concerned with questions of rigor are urged to consult the standard references.

If $x(t)$ is a known function of the variable t for $t>0$, its Laplace transform $\bar{x}(p)$ is defined by

$$\bar{x}(p) = L\{x(t)\} = \int_{0-}^{\infty} e^{-pt}x(t)dt \tag{1.7}$$

where p may be real or complex. If

$$|x(t)| \leq Me^{\alpha t} \tag{1.8}$$

for all $t>0$ and some constants M and α, then $\bar{x}(p)$ exists for all p such that $Re(p)>\alpha$. As an example, consider the Heaviside unit step function

$$\left.\begin{aligned} H(t) &= 0, \quad t<0 \\ &= 1, \quad t>0 \end{aligned}\right\} \tag{1.9}$$

The Laplace transform of this function is

$$\bar{H}(p) = \int_{0+}^{\infty} e^{-pt}dt = \left[-\frac{1}{p}\, e^{-pt}\right]_{0+}^{\infty} = \frac{1}{p} \tag{1.10}$$

According to (1.8), any real $p>0$ will insure the existence of the integral in (1.7) for this case. We conclude that the Laplace transform of the Heaviside unit step function is $\frac{1}{p}$, and this result is recorded as entry no. 8 in Table 1.1 at the end of this chapter. In the process of solving differential equations, we shall require the Laplace transform of $\dot{x}(t)$ given that $x(0^-)=x_0$. Here we have

$$L\{\dot{x}(t)\} = \int_{0-}^{\infty} e^{-pt}\,\frac{dx}{dt}\,dt = \left(xe^{-pt}\right)_{0-}^{\infty} + p\int_{0-}^{\infty} e^{-pt}x\,dt = -x_0 + p\bar{x}(p) \tag{1.11}$$

and this result forms entry no. 3 in Table 1.1.

It is obviously possible to compile a "dictionary" of Laplace transforms, and Table 1.1 at the end of this chapter provides us with such a compilation. The choice of entries in Table 1.1 was dictated by the needs of the present and succeeding chapters. The principal effort in effecting a solution of a differential equation consists of finding the x(t) corresponding to an $\bar{x}(p)$. This can often be accomplished by looking up $\bar{x}(p)$ in Table 1.1. For the class of functions which are considered in this book, it can be shown that this process of "inversion" is unique.

1.3 FREE VIBRATIONS

We now return to the problem of the linear oscillator, and consider its motion when external excitation is absent, i.e., with reference to (1.1) we set $F(t)\equiv 0$. If we divide (1.1) by m and set $\frac{c}{m} = 2\beta\omega$, $\frac{k}{m} = \omega^2$, we obtain the following equation of motion and associated initial conditions.

$$\left.\begin{aligned} &\ddot{x} + 2\beta\omega\dot{x} + \omega^2 x = 0 \\ &x(0) = x_0,\ \dot{x}(0) = \dot{x}_0 \end{aligned}\right\} \tag{1.12}$$

With the assistance of Table 1.1 we now take Laplace transforms of (1.12):

$$p^2\bar{x} - (px_0 + \dot{x}_0) + 2\beta\omega(p\bar{x} - x_0) + \omega^2\bar{x} = 0 \tag{1.13}$$

and solving for $\bar{x}$, we obtain

$$\bar{x}(p) = \frac{x_0 p + (\dot{x}_0 + 2\beta\omega x_0)}{p^2 + 2\beta\omega p + \omega^2} \tag{1.14}$$

When inverting (1.14), it is necessary to distinguish between four physically distinct situations, depending upon the character of the roots of the denominator of the right hand side of (1.14). These roots are negative and unequal, negative and equal, conjugate complex, and conjugate imaginary, corresponding to the case of supercritical, critical, subcritical and zero damping, respectively.

Case (a): $\beta > 1$, Supercritical Damping

Factoring the denominator and subsequently decomposing (1.14) into partial fractions, we obtain

$$\bar{x}(p) = \frac{x_0 p + (\dot{x}_0 + 2\beta\omega x_0)}{(p+\lambda_1)(p+\lambda_2)} = \frac{C_1}{p+\lambda_1} + \frac{C_2}{p+\lambda_2} \tag{1.15}$$

where

$$\left.\begin{aligned} \lambda_1 &= \omega(\beta + \sqrt{\beta^2-1}) > 0 \\ \lambda_2 &= \omega(\beta - \sqrt{\beta^2-1}) > 0 \end{aligned}\right\} \tag{1.16}$$

and

$$\left.\begin{aligned} C_1 &= \frac{-\dot{x}_o - \lambda_2 x_o}{2\omega\sqrt{\beta^2-1}} \\ C_2 &= \frac{\dot{x}_0 + \lambda_1 x_0}{2\omega\sqrt{\beta^2-1}} \end{aligned}\right\} \tag{1.17}$$

Equation (1.15) is now inverted with the help of Table 1.1:

$$x(t) = C_1 e^{-\lambda_1 t} + C_2 e^{-\lambda_2 t} \tag{1.18}$$

and we note that the constants C_1 and C_2 are functions of the initial conditions. Inspection of the solution (1.18) reveals that the response is non-oscillatory and subsident, i.e., $x \to 0$ as $t \to \infty$ for arbitrary initial conditions. Moreover, x passes through its zero value at most once for $t>0$:

$$C_1 e^{-\lambda_1 t} + C_2 e^{-\lambda_2 t} = 0 \text{ for } t>0 \tag{1.19}$$

provided

$$t = t^* = \frac{1}{2\omega\sqrt{\beta^2-1}} \log_e \frac{\dot{x}_o + \lambda_2 x_o}{\dot{x}_o + \lambda_1 x_o} \tag{1.20}$$

Equation (1.19) has at most one positive real root.

Case (b): $\beta = 1$, **Critical Damping**

Equation (1.14) assumes the form

$$\bar{x}(p) = \frac{x_o p + (\dot{x}_o + 2\omega x_o)}{(p+\omega)^2} = \frac{x_o}{p+\omega} + \frac{\dot{x}_o + \omega x_o}{(p+\omega)^2} \tag{1.21}$$

and its inverse is

$$x(t) = x_o e^{-\omega t} + (\dot{x}_o + \omega x_o) t e^{-\omega t} \tag{1.22}$$

We note that (1.22) has at most one positive real root $t=t^*$ for $x(t)=0$ given by

$$t^* = \frac{-x_o}{\dot{x}_o+\omega x_o} \tag{1.23}$$

and that $x(t)\to 0$ for $t\to\infty$. As in Case (a), the motion is non-oscillatory and subsident.

Case (c): $\beta < 1$, Subcritical Damping

Equation (1.14) is written as

$$\bar{x}(p)= \frac{x_o(p+\beta\omega)}{p^2+2\beta\omega p+\omega^2} + \frac{\dot{x}_o+\beta\omega x_o}{p^2+2\beta\omega p+\omega^2} \tag{1.24}$$

Equation (1.24) may be inverted with the help of Table 1.1:

$$\left.\begin{aligned} x(t)&=e^{-\beta\omega t}\left(x_o\cos\,\omega\sqrt{1-\beta^2}\;t+\frac{\dot{x}_o+\beta\omega x_o}{\omega\sqrt{1-\beta^2}}\sin\,\omega\sqrt{1-\beta^2}\;t\right)\\ &=\sqrt{\frac{\dot{x}_o^2+2\beta\omega\dot{x}_o x_o+\omega^2x_o^2}{\omega^2(1-\beta)^2}}\;e^{-\beta\omega t}\cos\,(\omega\sqrt{1-\beta^2}\;t-\psi)\end{aligned}\right\} \tag{1.25}$$

where $\tan\psi= \dfrac{\dot{x}_o+\beta\omega x_o}{x_o\omega\sqrt{1-\beta^2}}$

A function x(t) is said to be periodic with period T if T is the smallest number such that $f(t+nT)=f(t)$, $n=1,2,3,\ldots$ In vibration theory, this definition is often applied to expressions of the type (1.25), where the decaying exponential which modulates the periodic function is simply ignored. Hence with reference to (1.25) and to Fig. 1.4, the "period" T of the oscillatory, subsident motion is the time elapsed between two passages in the same direction. This turns out to be

$$T = \frac{2\pi}{\omega\sqrt{1-\beta^2}} \text{ (seconds)} \tag{1.26}$$

and the frequency is defined to be

$$f = \frac{1}{T} = \frac{\omega\sqrt{1-\beta^2}}{2\pi} \text{ (cycles per second)} \tag{1.27}$$

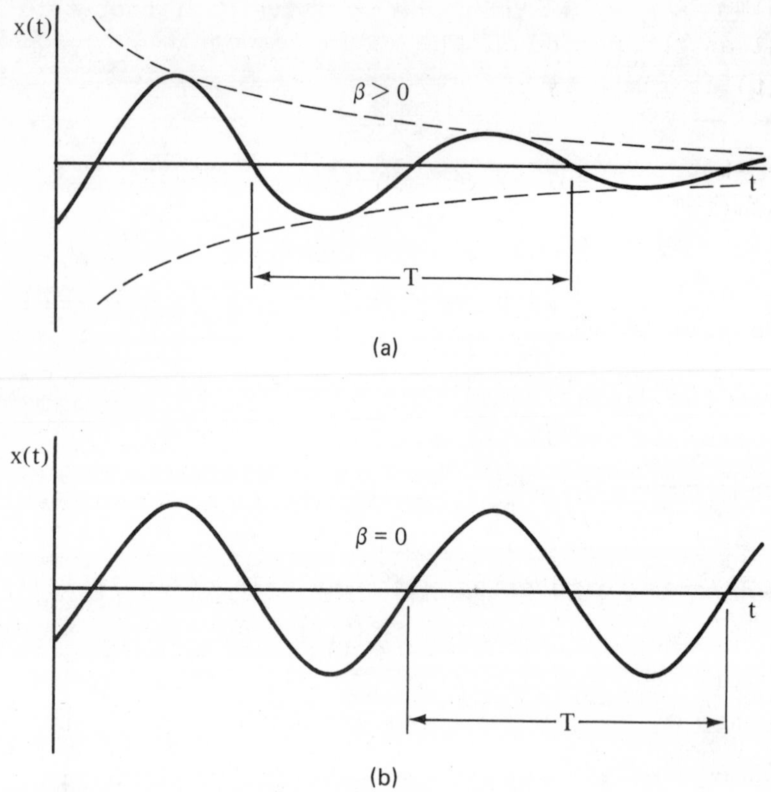

Figure 1.4

A (dimensionless) plot of frequency vs. damping factor is shown in Fig. 1.5. We note that $f_0 = \frac{1}{T_0} = \frac{\omega}{2\pi}$ is the undamped, natural frequency of the oscillator.

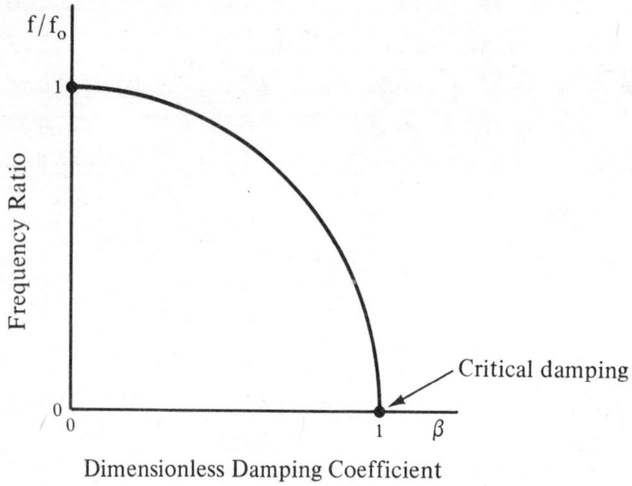

Figure 1.5

The maxima of x(t) occur at equidistant values of t having the same interval as the period T. The ratio between two consecutive maxima of x(t) is given by

$$\frac{x_1}{x_2} = \frac{e^{-\beta\omega t}}{e^{-\beta\omega(t+T)}} = e^{\beta\omega T} \qquad (1.28)$$

The quantity

$$\delta = \log_e \frac{x_1}{x_2} = \beta\omega T = \frac{2\pi\beta}{\sqrt{1-\beta^2}} \qquad (1.29)$$

has been named the "logarithmic decrement." It is a useful and alternative means to specify or describe the degree of system damping. Equation (1.29) can be expanded in a power series in (x_1-x_2):

$$\delta = \log_e\left[\frac{x_2+(x_1-x_2)}{x_2}\right] = \log_e\left(1 + \frac{x_1-x_2}{x_2}\right)$$

$$= \frac{x_1-x_2}{x_2} - \frac{1}{2}\left(\frac{x_1-x_2}{x_2}\right)^2 + \frac{1}{3}\left(\frac{x_1-x_2}{x_2}\right)^3 - \ldots$$

and therefore $\delta \approx \dfrac{x_1-x_2}{x_2}$, i.e., for sufficiently small damping β the logarithmic decrement δ is approximately equal to the fractional decrease in amplitude during one cycle of the vibration.

Case (d): $\beta = 0$**, Zero Damping**

In this case (1.25) becomes

$$x(t) = x_o \cos \omega t + \frac{\dot{x}_o}{\omega} \sin \omega t = \sqrt{\frac{\dot{x}_o^2+\omega^2 x_o^2}{\omega^2}} \cos (\omega t-\psi) \qquad (1.30)$$

where

$$\tan \psi = \frac{\dot{x}_o}{\omega x_o}$$

The motion characterized by (1.30) is oscillatory and undamped, with period $T = \frac{2\pi}{\omega}$ (seconds) and frequency $f = \frac{\omega}{2\pi}$ (cycles per second). The amplitude of the motion $\sqrt{\dfrac{\dot{x}_o^2+\omega^2 x_o^2}{\omega^2}}$ remains constant for all $t>0$ as shown in Fig. 1.4b.

1.4 FORCED MOTION

The present section is concerned with the dynamic response of the linear oscillator to a prescribed forcing function F(t). If (1.1) is Laplace transformed with respect to t, we obtain

$$\bar{x}(p) = \frac{x_o p+(\dot{x}_o+2\beta\omega x_o)}{p^2+2\beta\omega p+\omega^2} + \frac{\bar{F}(p)}{m(p^2+2\beta\omega p+\omega^2)} \tag{1.31}$$

In most mechanical and structural considerations, damping effects arise either because of internal friction or external interactions with the surrounding medium. Generally, in these and similar applications $\beta<1$, i.e., the damping of the system is subcritical. In the present section we shall assume the latter.

(a) Response to a Suddenly Applied Load.

Let $F(t)=F_o H(t)$, where H(t) is the Heaviside unit step function as defined by (1.9). If quiescent initial conditions $x(0^-)=x_o=0$, $\dot{x}(0^-)=\dot{x}_o=0$ are assumed, then (1.31) becomes

$$\bar{x}(p) = \frac{F_o}{m}\,\frac{1}{p(p^2+2\beta\omega p+\omega^2)} = \frac{F_o}{k}\left(\frac{1}{p} - \frac{p+2\beta\omega}{p^2+2\beta\omega p+\omega^2}\right) \tag{1.32}$$

Inverting (1.32) with the help of Table 1.1, we obtain

$$x(t) = \frac{F_o}{k}\left[1-e^{-\beta\omega t}\left(\cos\sqrt{1-\beta^2}\,\omega t + \frac{\beta}{\sqrt{1-\beta^2}}\sin\sqrt{1-\beta^2}\,\omega t\right)\right] \tag{1.33}$$

In the undamped case, $\beta=0$, and

$$x(t) = \frac{F_o}{k}(1-\cos\omega t) \tag{1.34}$$

The static displacement of the mass m under the load F_o is $x_{st}=\frac{F_o}{k}$. Hence

$$\frac{x(t)}{x_{st}} = (1-\cos\omega t) \tag{1.35}$$

and therefore the dynamic amplification factor $\left[\frac{x(t)}{x_{st}}\right]_{max}=2$, i.e., the maximum displacement is twice the static displacement, as shown in Fig. 1.6.

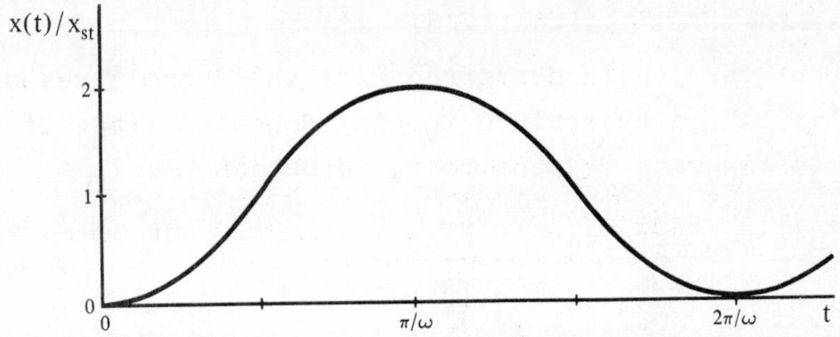

Figure 1.6

(b) Response to a Pure Impulse I_o Applied at t = 0.

We assume that the oscillator is initially at rest and that the applied force $F(t)=I_o\delta(t)$, where $\delta(t)$ is the Dirac delta function. Thus

$$\text{Impulse} = \int_{0^-}^{t} F(t)dt = I_o \int_{0^-}^{t} \delta(t)dt = I_o$$

and $x_o=\dot{x}_o=0$. In this case (1.31) assumes the form

$$\bar{x}(p) = \frac{I_o}{m} \frac{1}{p^2+2\beta\omega p+\omega^2} \tag{1.36}$$

Inverting (1.36),

$$x(t) = \frac{I_o}{m} \frac{e^{-\beta\omega t} \sin \omega\sqrt{1-\beta^2}\, t}{\omega\sqrt{1-\beta^2}} \quad t>0 \tag{1.37}$$

We note that $\dot{x}(t)$ is discontinuous at x=0, i.e., $\dot{x}(0^-)=0$, $\dot{x}(0^+)=\frac{I_o}{m}$.
The present problem is completely equivalent to the following:
No external forces are acting on the oscillator $(F(t)\equiv 0)$ and the mass is started from its equilibrium position $x(0^-)=x_o=0$ with an initial velocity $\dot{x}(0^-)=\dot{x}_o$. In this case

$$\bar{x}(p) = \frac{\dot{x}_o}{p^2+2\beta\omega p+\omega^2} \tag{1.38}$$

and upon comparing (1.38) with (1.36) we note that

$$\dot{x}_o = \frac{I_o}{m}, \text{ or } I_o = m\dot{x}_o$$

When $\beta=0$, (1.37) becomes

$$x(t) = \frac{I_o}{m}\frac{1}{\omega} \sin \omega t, \quad t>0. \tag{1.39}$$

(c) Harmonic Forcing Function.

The oscillator is subjected to the harmonic forcing function $F(t)=F_o \cos \omega_o t$ and initial conditions $x(0^-)=x_o$, $\dot{x}(0^-)=\dot{x}_o$. Equation (1.31) assumes the form

$$\bar{x}(p) = \frac{x_o p+(\dot{x}_o+2\beta\omega x_o)}{(p^2+2\beta\omega p+\omega^2)} + \frac{F_o}{m}\frac{p}{(p^2+\omega_o^2)(p^2+2\beta\omega p+\omega^2)} \tag{1.40}$$

Straightforward but somewhat lengthy algebraic manipulation will reveal that

$$\frac{p}{(p^2+\omega_o^2)(p^2+2\beta\omega p+\omega^2)} = \frac{p\frac{A}{\omega^2}\cos\psi + \frac{\omega_o}{\omega^2} A \sin\psi}{(p^2+\omega_o^2)} - \frac{p\frac{A}{\omega^2}\cos\psi + \frac{A}{\omega_o}\sin\psi}{(p^2+2\beta\omega p+\omega^2)} \tag{1.41a}$$

where

$$\left.\begin{aligned} A &= \left[\left(1-\frac{\omega_o^2}{\omega^2}\right)^2 + \left(2\beta\frac{\omega_o}{\omega}\right)^2\right]^{-\frac{1}{2}} > 0 \\ \sin\psi &= A\left(2\beta\frac{\omega_o}{\omega}\right); \quad \cos\psi = A\left(1-\frac{\omega_o^2}{\omega^2}\right) \end{aligned}\right\} \tag{1.41b}$$

With the help of relations (1.41) and Table 1.1, equation (1.40) is readily inverted:

$$x = x_T + x_{SS} \tag{1.42}$$

where

$$x_T(t) = \left(x_o - \frac{F_o}{k} A \cos\psi\right) e^{-\beta\omega t}\cos \omega\sqrt{1-\beta^2}\, t \tag{1.43}$$

$$+\left[\frac{\dot{x}_o}{\omega}+\beta x_o+\frac{F_o}{k}A\left(\beta\cos\psi-\frac{\omega}{\omega_o}\sin\psi\right)\right]\frac{e^{-\beta\omega t}}{\sqrt{1-\beta^2}}\sin\omega\sqrt{1-\beta^2}\,t$$

$$x_{SS}(t)=\frac{F_o}{k}A\cos(\omega_o t-\psi) \tag{1.44}$$

We note that the transient part of the solution, (1.43), approaches zero as $t\to\infty$. In actual applications, x_T becomes negligibly small compared to the steady state part x_{SS} after a few cycles of oscillation, and the steady state part x_{SS} will predominate. A plot of the amplification factor A as a function of frequency ratio ω_o/ω is shown in Fig. 1.7a. We note that for all values of $\beta\geq 0$, $A\to 0$ as $\omega_o\to\infty$, and that A=1 for $\omega_o=0$. To find the maximum of A, we set $dA/d\omega_o=0$, and obtain $\omega_o=\omega\sqrt{1-2\beta^2}$ for the "resonant frequency," i.e., the forcing function $F(t)=F_o\cos\omega\sqrt{1-2\beta^2}\,t$ maximizes the amplitude of the steady

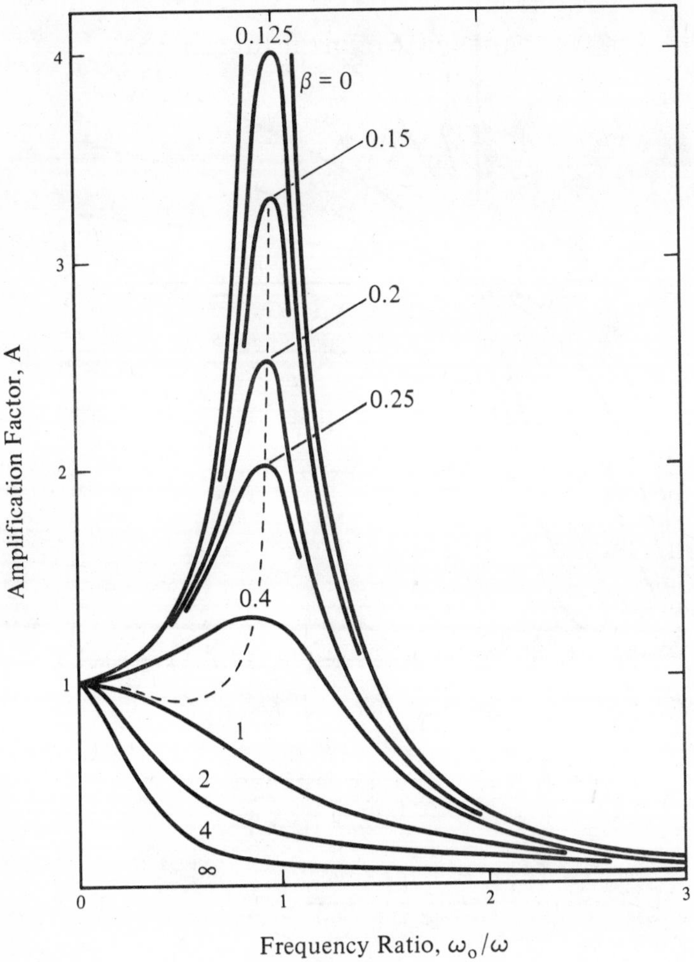

Fig. 1.7
(a)

state part of the response for all $\beta < 1/\sqrt{2}$. The dynamic amplification A reaches a maximum value $A_{max} = 1/2\beta\sqrt{1-\beta^2}$ at the excitation frequency $\omega_o = \omega\sqrt{1-2\beta^2} > 0$ provided $\beta < 1/\sqrt{2}$; when $\beta \geq 1/\sqrt{2}$, $A_{max}=1$ and this maximum occurs for $\omega_o=0$. The phase angle ψ of the steady state part of the solution as a function of frequency is shown in Fig. 1.7b. We note that for the case $\beta=0$, the applied harmonic force is in phase with the displacement below the resonant frequency, whereas it is 180° out of phase when the forcing frequency is larger than the resonant frequency $\omega = \sqrt{k/m}$. In the latter case the direction of the excitation force vector is opposite to that of the velocity vector.

Figure 1.7

(b)

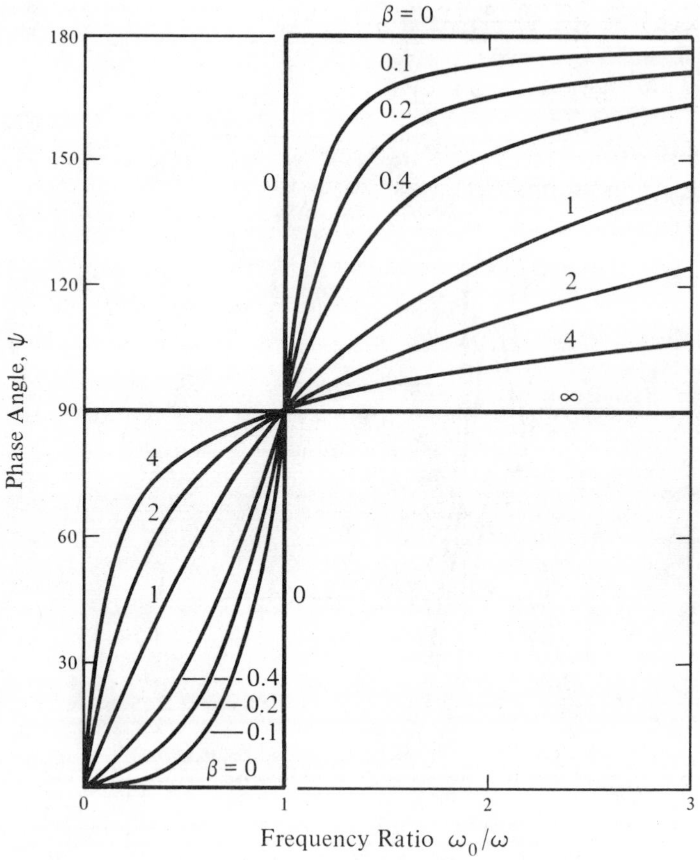

If damping is neglected in the motion characterized by (1.42), the solution assumes the form

$$x(t) = x_o \cos \omega t + \frac{\dot{x}_o}{\omega} \sin \omega t + \frac{F_o}{k} \frac{(\cos \omega_o t - \cos \omega t)}{\left(1 - \frac{\omega_o^2}{\omega^2}\right)} \tag{1.45}$$

and if the motion starts with zero initial conditions $x_0 = \dot{x}_0 = 0$, then

$$x(t) = \frac{F_0}{k} \frac{(\cos \omega_0 t - \cos \omega t)}{\left(1 - \frac{\omega_0^2}{\omega^2}\right)} \tag{1.46}$$

Let us consider the case of a forcing frequency which is close to the natural frequency of the oscillator, but not equal to it, and let the frequency difference be $\Delta\omega = \frac{1}{2}(\omega - \omega_0)$. Then $\cos \omega_0 t - \cos \omega t = 2 \sin \frac{1}{2}(\omega + \omega_0)t \cdot \sin \Delta\omega t$ and (1.46) assumes the form

$$x(t) = \frac{F_0}{k} \frac{\omega^2}{\Delta\omega \cdot (\omega + \omega_0)} \sin \tfrac{1}{2}(\omega + \omega_0)t \cdot \sin \Delta\omega t \tag{1.47}$$

The response, as characterized by (1.47), is a harmonic motion with angular frequency $\frac{1}{2}(\omega + \omega_0)$ modulated by an amplitude which varies periodically with angular frequency $\Delta\omega$. The phenomenon, as shown in Fig. 1.8, is called beats, and there are $\Delta\omega/\pi$ beats per unit of time.

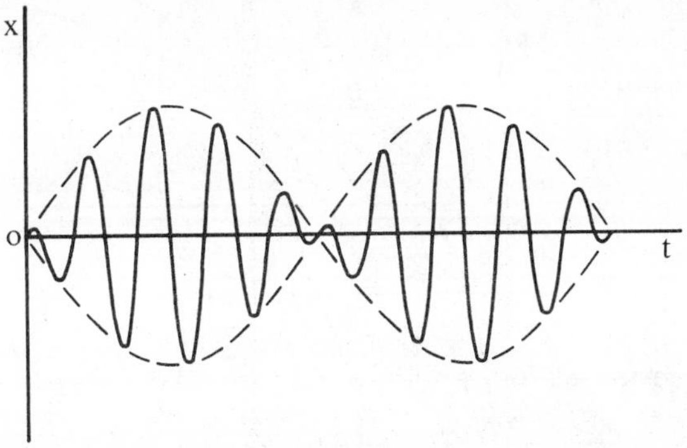

Figure 1.8

To obtain the case of resonance, we write (1.47) as

$$x(t) = \frac{F_0}{k} \frac{\omega^2 t}{\omega + \omega_0} \cdot \frac{\sin \Delta\omega \cdot t}{\Delta\omega \cdot t} \cdot \sin \tfrac{1}{2}(\omega + \omega_0)t \tag{1.48}$$

and take the limit of (1.48) as $\omega_0 \to \omega$. The result is

$$x(t) = \tfrac{1}{2} \frac{F_0}{k} (\omega_0 t) \sin \omega_0 t \tag{1.49}$$

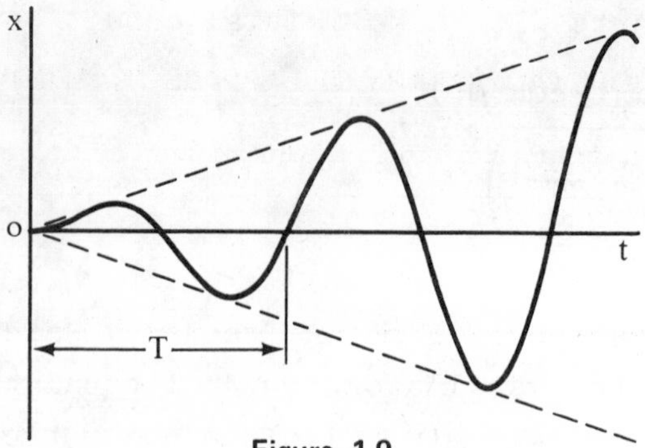

Figure 1.9

A plot of (1.49) is shown in Fig. 1.9. As time increases, the amplitude of displacement becomes unbounded, and the basic physical assumptions of our mathematical model are violated. However, it does explain how the amplitude grows with time for sufficiently small displacements when the forcing frequency coincides with the natural frequency of oscillation and $\beta \equiv 0$.

(d) Arbitrary Forcing Function F(t).

Equation (1.31) can be inverted by application of the convolution theorem (see entry no. 6 in Table 1.1). The solution may be written in two different ways:

$$x(t) = x_C + x_P \tag{1.50a}$$

where

$$x_C(t)=e^{-\beta\omega t}\left(x_o\cos\sqrt{1-\beta^2}\,\omega t + \frac{\dot{x}_o+\beta\omega x_o}{\omega\sqrt{1-\beta^2}}\sin\sqrt{1-\beta^2}\,\omega t\right) \tag{1.50b}$$

and

$$x_P(t) = \frac{1}{\omega m\sqrt{1-\beta^2}}\int_o^t F(\tau)e^{-\beta\omega(t-\tau)}\sin[\sqrt{1-\beta^2}\,\omega(t-\tau)]d\tau \tag{1.50c}$$

or

$$x_P(t) = \frac{1}{\omega m\sqrt{1-\beta^2}}\int_o^t F(t-\tau)\,e^{-\beta\omega\tau}\sin\left(\sqrt{1-\beta^2}\,\omega\tau\right)d\tau \tag{1.50d}$$

The integrals in (1.50) can be evaluated for most forcing functions F(t) which appear in practical applications, if not analytically, then certainly by standard numerical methods.

TABLE 1.1

A SHORT DICTIONARY OF LAPLACE TRASFORMS

$$\bar{x}(p) = \int_{0^-}^{\infty} e^{-pt} x(t)\, dt$$

$\bar{x}(p)$	$x(t)$
(1) $k\bar{x}(p)$	$kx(t)$
(2) $\bar{x}_1(p)+\bar{x}_2(p)$	$x_1(t)+x_2(t)$
(3) $p\bar{x}(p)-x(0^-)$	$\frac{d}{dt}[x(t)]$
(4) $p^2\bar{x}(p)-[px(0^-)+\dot{x}(0^-)]$	$d^2/dt^2[x(t)]$
(5) $\frac{1}{p}\bar{x}(p)$	$\int_0^t x(\tau)d\tau$
(6) $\bar{x}_1(p)\cdot\bar{x}_2(p)$	$\int_0^t x_1(\tau)\cdot x_2(t-\tau)d\tau=$ $\int_0^t x_1(t-\tau)\cdot x_2(\tau)d\tau$
(7) 1	$\delta(t)$
(8) $\frac{1}{p}$	$H(t)$
(9) $\frac{1}{p+\lambda}$	$e^{-\lambda t}$
(10) $\frac{1}{(p+\lambda)^2}$	$te^{-\lambda t}$
(11) $\frac{\omega}{p^2+\omega^2}$	$\sin \omega t$
(12) $\frac{p}{p^2+\omega^2}$	$\cos \omega t$
(13) $\frac{p}{(p^2+\omega^2)^2}$	$\frac{t}{2\omega}\sin \omega t$
(14) $\frac{p^2-\omega^2}{(p^2+\omega^2)^2}$	$t \cos \omega t$
(15) $\frac{1}{p^2+2\beta\omega p+\omega^2}$	$\frac{1}{\omega\sqrt{1-\beta^2}} e^{-\beta\omega t} \sin \sqrt{1-\beta^2}\, \omega t;\ \beta<1$
(16) $\frac{p+\beta\omega}{p^2+2\beta\omega p+\omega^2}$	$e^{-\beta\omega t}\cos \sqrt{1-\beta^2}\, \omega t;\ \beta<1$

TABLE 1.1, Continued

(17) $\frac{(\sqrt{p^2+a^2}-p)^n}{na^n}$	$\frac{J_n(at)}{t}$, $n>0$
(18) $\frac{(\sqrt{p^2+a^2}-p)^n}{a^n\sqrt{p^2+a^2}}$	$J_n(at)$, $n>-1$
(19) $\frac{e^{-\lvert b\rvert\sqrt{p^2+a^2}}}{\sqrt{p^2+a^2}}$	$\begin{cases} 0 \text{ when } 0<t<\lvert b\rvert \\ J_0(a\sqrt{t^2-b^2}) \text{ when } t>\lvert b\rvert \end{cases}$
(20) $\frac{1}{p}e^{-\frac{a}{p}}$	$J_0(2\sqrt{at})$

EXERCISES

1.1 Two properly meshed gears with moments of inertia I_1 and I_2 are elastically restrained as shown in the figure. Assuming that the shafts are massless and neglecting friction, find the natural frequency of oscillation of the system if the gear ratio $r_1/r_2 = n$.

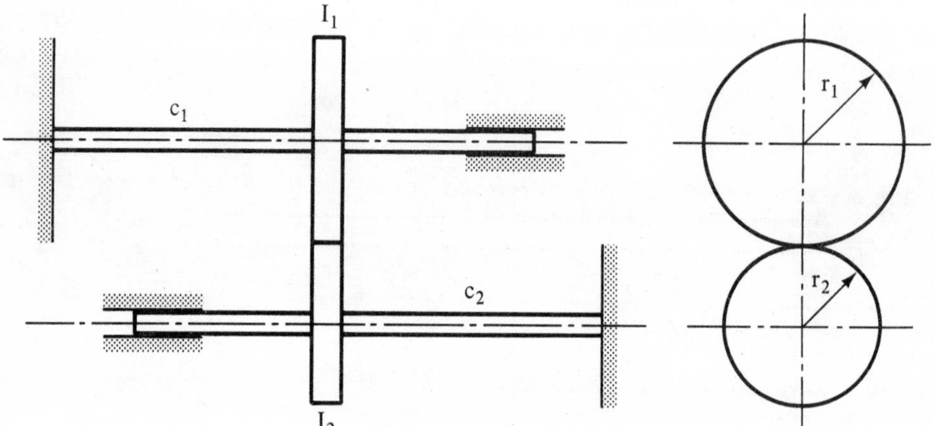

Figure: Exercise 1.1

1.2 The solution (1.50) of equation (1.1) can also be derived by the use of Lagrange's method of the variation of parameters. Obtain the solution of $m\ddot{x}+kx=F(t)$, $x(0)=x_o$, $\dot{x}(0)=\dot{x}_o$, by this method.

1.3 A single degree of freedom ocsillator having an amount of viscous damping just equal to the critical damping value is initially at rest. At time $t=0$ a velocity v_o is given to the mass. Find the maximum deflection of the mass in the direction of the initial velocity.

1.4 Solve the problem $m\ddot{x}+kx=F(t)$, $x(0)=0$, $\dot{x}(0)=0$, where $F(t)=F_o$ for $0<t<T$, $F(t)=0$ for $t>T$. Evaluate the solution for the limiting case $T\to 0$, $F_oT=I_o$. Show that this result is the solution of (a) $m\ddot{x}+kx=I_o\delta(t)$, $x(0)=\dot{x}(0)=0$, and (b) $m\ddot{x}+kx=0$, $x(0)=0$, $\dot{x}(0)=I_o/m$.

1.5 Let $h(t)$ be the response of an undamped linear oscillator to a unit impulse at $t=0$. Let $A(t)$ be the response of the same undamped oscillator to a unit step function applied at $t=0$. In both cases $x(0)=\dot{x}(0)=0$. Show that $dA/dt = h(t)$.

1.6 A weight of magnitude W is moving along a rigid beam with constant speed v. The beam rests on an elastic support at B and is supported by a hinge at A. Assuming that the moving weight does not change the moment of inertia and the natural frequency of the beam appreciably, find the equation for the (small) deflection of the beam under the weight of the moving load.

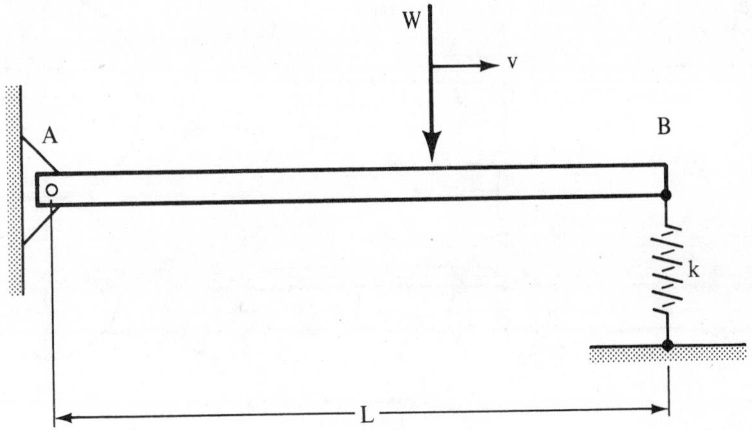

Figure: Exercise 1.6

1.7 As a result of unbalanced rotating masses, a machine exerts an alternating force $A\cos\omega_o t$ upon the foundation on which it rests. The machine is isolated from the foundation by a device characterized by a spring and a dashpot. Discuss the effectiveness of the spring and the dashpot in reducing the force transmitted to the foundation.

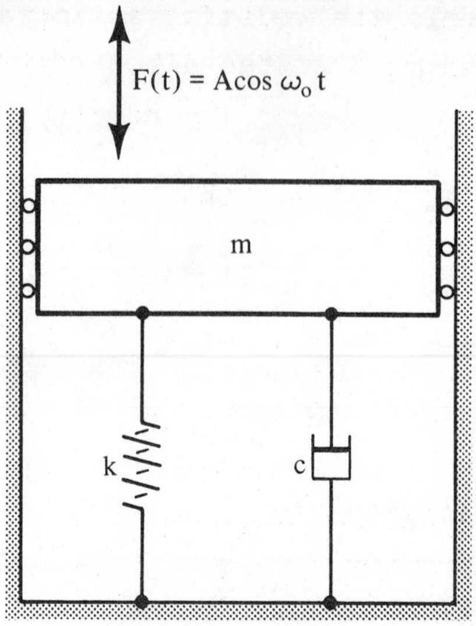

Figure: Exercise 1.7

1.8 An idealized model of a spring supported vehicle travelling over a rough road is shown in the figure. Determine the equation for the amplitude of the weight W as a function of the constant speed, and determine the most unfavorable speed.

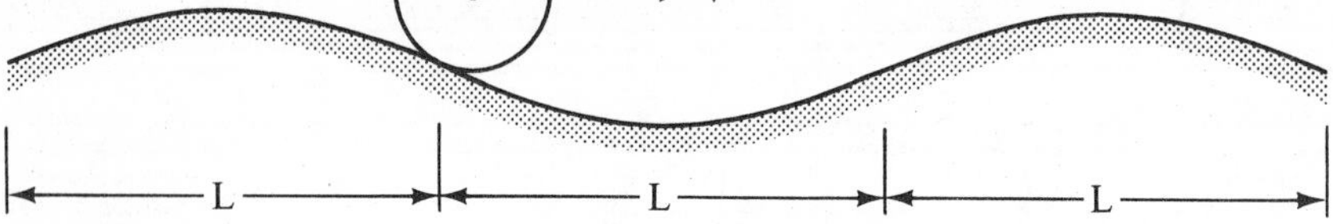

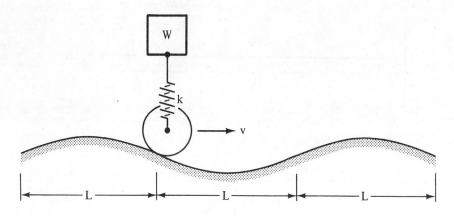

Figure: Exercise 1.8

1.9 A typical seismic instrument is shown in the figure. The motion of the translating seismic mass is restrained by a spring and a damping device. Assume that the instrument case is in steady, harmonic motion $y=A \cos \omega_0 t$. Discuss the relative motion of the seismic mass with particular reference to the special cases $\omega=\sqrt{k/m} >> \omega_0$, $\omega << \omega_0$, and $\omega \approx \omega_0$.

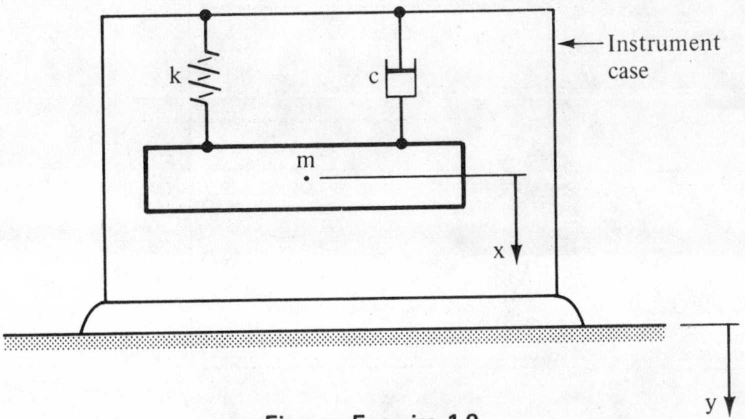

Figure: Exercise 1.9

1.10 Consider the seismic instrument of Exercise 1.9. What will be the relative motion of the seismic mass if the motion of the case is given by $y=0$ for $t<0$, $y=A \sin \pi t/T$ for $0<t<T$, $y=0$ for $t>T$?

Chapter 2

MECHANICS OF DISCRETE SYSTEMS

An intelligent being who knew for a given instant all the forces by which nature is animated and possessed complete information on the state of matter of which nature consists—providing his mind were powerful enough to analyze these data—could express in the same equation the motions of the largest bodies of the universe and the motion of the smallest atoms. Nothing would be uncertain for him, and he would see the future as well as the past at one glance.

Marquis de Laplace, Théorie Analytique des Probabilités (1820).

2.1 SOME GENERAL OBSERVATIONS

In many complex physical and engineering situations it is possible and desirable to model the deformable mechanical system in a discrete fashion for dynamical purposes. When this is the case, those portions of the system with distinct inertial characteristics are isolated from those which display predominantly deformational characteristics. For example, we may consider the torsional motion of a shaft with attached discs in such a way that the shaft segments between discs are characterized by massless torsional springs while the discs are taken to be rigid and supply the inertial elements which are characterized by their polar moment of inertia. It should be noted that the lumping process (i.e., separation of inertial from elastic elements of the structure) is not unique and requires a certain amount of judgement, experience, as well as common sense.

Once the structure has been characterized as a discrete, mechanical system, it is possible to apply Newton's second law of motion to each inertial element, and to express the interaction of these elements with the help of Newton's third law. In general, this procedure results in a system of simultaneous, ordinary, differential equations of the second order. Although straightforward, the results are generally so complex as to make it useful to seek other, more sophisticated approaches resulting in a simpler problem characterization as well as a unified point of view.

2.2 HAMILTON'S PRINCIPLE

Newton's laws are inherently vectorial in character, but the integration of the equations of motion is usually accomplished after their scalar components have been referred to a particular coordinate system. The structure of the scalar components of the vector equations of motion, physical insight obtainable, as well as the ability to integrate the equations of motion all depend strongly upon the choice of the particular coordinate system. To facilitate the selection of a convenient coordinate system, and to circumvent the necessity for transformations which are often awkward and unwieldy, we shall seek a formulation of the basic laws of motion which is independent of the particular choice of the coordinates used. Such a formulation is provided by Hamilton's principle which we shall now proceed to derive, using Newton's second law of motion as a point of departure.

Consider a dynamical system consisting of N mass particles, each of mass m_i, $i=1,2,...,N$. The position of each mass particle m_i is specified by the position vector $\vec{r}_i$ with rectangular, Cartesian components (x_i,y_i,z_i). The reference axis system is assumed to be inertial. For any complete set of prescribed initial conditions, each of the coordinates must be a single valued function of time. If it is assumed, for the present, that these functions are known and have the form $\vec{r}_i=\vec{r}_i(t)$, $i=1,2,...,N$, then these 3N scalar equations may be regarded as the parametric equations of a path in a configuration space of 3N dimensions. Accordingly, the motion of the dynamical system can be correlated with that of a point which moves along this path. It is noted that, once the initial conditions are prescribed, the future motion of a given dynamical system is determined by Newton's laws of motion. There is, therefore, a unique path in the 3N dimensional configuration space resulting from a given set of initial conditions. We intend to compare this (actual) path with another, varied path in the configuration space. The two paths will have the same end points such that both paths will be traversed in the same time interval t_2-t_1. It is clear that the varied path corresponds to a motion of the dynamical system which is not in accord with Newton's laws of motion. We assume that at any instant of time the "distance" between the two points in the configuration space is infinitesimally small, in a sense to be defined below.

Newton's second law of motion applied to the i'th mass particle is

$$m_i\dot{\vec{v}}_i = \vec{F}^*_i \tag{2.1}$$

where $\vec{v}_i=\dot{\vec{r}}_i$ is the velocity vector and $\vec{F}^*_i$ is the sum total of all forces acting upon the i'th mass particle. We now suppose that the motion of the i'th particle does not take place in accordance with (2.1), but instead moves along a varied (hypothetical) path. The position vector of the i'th particle in the varied path is $\vec{r}_i'$. The variation (difference) between the particle position on the actual and varied path is designated by

$$\delta\vec{r}_i = \vec{r}_i'-\vec{r}_i \tag{2.2}$$

where $\vec{r}_i'$ and $\vec{r}_i$ are measured at the same instant of time. In like manner, we designate by

$$\delta\vec{v}_i = \vec{v}_i'-\vec{v}_i = \frac{d}{dt}(\vec{r}_i'-\vec{r}_i) \tag{2.3}$$

the variation of the velocity vector, where $\vec{v}_i'$ and $\vec{v}_i$ are to be evaluated at the same instant of time. If we take the scalar product of (2.1) with $\delta\vec{r}_i$ we obtain

$$m_i\dot{\vec{v}}_i\cdot\delta\vec{r}_i = \vec{F}^*{}_i\cdot\delta\vec{r}_i \tag{2.4}$$

But

$$\begin{aligned}\dot{\vec{v}}_i\cdot\delta\vec{r}_i &= \frac{d}{dt}(\vec{v}_i\cdot\delta\vec{r}_i)-\vec{v}_i\cdot\frac{d}{dt}(\delta\vec{r}_i) = \\ &= \frac{d}{dt}(\vec{v}_i\cdot\delta\vec{r}_i)-\vec{v}_i\cdot\frac{d}{dt}(\vec{r}_i'-\vec{r}_i) = \\ &= \frac{d}{dt}(\vec{v}_i\cdot\delta\vec{r}_i)-\vec{v}_i\cdot(\vec{v}_i'-\vec{v}_i) = \\ &= \frac{d}{dt}(\vec{v}_i\cdot\delta\vec{r}_i)-\vec{v}_i\cdot\delta\vec{v}_i\end{aligned} \tag{2.5}$$

Combination of (2.4) and (2.5) results in

$$m_i\left[\frac{d}{dt}(\vec{v}_i\cdot\delta\vec{r}_i)-\vec{v}_i\cdot\delta\vec{v}_i\right] = \vec{F}^*{}_i\cdot\delta\vec{r}_i \tag{2.6}$$

If the system under consideration consists of N mass particles, then there will be N equations of the type (2.6). Addition of these N equations results in

$$\sum_{i=1}^{N} m_i\left[\frac{d}{dt}(\vec{v}_i\cdot\delta\vec{r}_i)-\vec{v}_i\cdot\delta\vec{v}_i\right] = \sum_{i=1}^{N}\vec{F}_i^*\cdot\delta\vec{r}_i \tag{2.7}$$

We now define the kinetic energy of the N mass particles in the actual, dynamical path as

$$T = \frac{1}{2}\sum_{i=1}^{N} m_i\vec{v}_i\cdot\vec{v}_i \tag{2.8}$$

and in the varied, hypothetical path as

$$T' = \frac{1}{2}\sum_{i=1}^{N} m_i\vec{v}_i'\cdot\vec{v}_i' \tag{2.9}$$

Hence the variation in kinetic energy is

$$\delta T = T'-T = \frac{1}{2}\sum_{i=1}^{N} m_i(\vec{v}_i'\cdot\vec{v}_i'-\vec{v}_i\cdot\vec{v}_i) \tag{2.10}$$

At this point we make the assumption that, at any instant of time, the variation of the velocity vector (i.e., the difference of the velocities in the varied and actual paths) is negligible compared to the velocity vector in the actual path. Consequently,

$$\vec{v}_i' \cdot \vec{v}_i' - \vec{v}_i \cdot \vec{v}_i = (\vec{v}_i + \delta\vec{v}_i)\cdot(\vec{v}_i + \delta\vec{v}_i) - \vec{v}_i \cdot \vec{v}_i = \delta\vec{v}_i \cdot (2\vec{v}_i + \delta\vec{v}_i) \approx 2\vec{v}_i \cdot \delta\vec{v}_i$$

in the sense that $|2\vec{v}_i| >> |\delta\vec{v}_i|$. Therefore

$$\delta T = \sum_{i=1}^{N} m_i \vec{v}_i \cdot \delta\vec{v}_i \tag{2.11}$$

Combining (2.11) with (2.7) results in

$$\frac{d}{dt}\{\sum_{i=1}^{N} m_i(\vec{v}_i \cdot \delta\vec{r}_i)\} = \delta T + \sum_{i=1}^{N} \vec{F}^*_i \cdot \delta\vec{r}_i \tag{2.12}$$

Integrating (2.12) with respect to time, we obtain

$$\left\{\sum_{i=1}^{N} m_i(\vec{v}_i \cdot \delta\vec{r}_i)\right\}_{t_1}^{t_2} = \int_{t_1}^{t_2} \left[\delta T + \sum_{i=1}^{N} \vec{F}^*_i \cdot \delta\vec{r}_i\right] dt \tag{2.13}$$

As the variations $\delta\vec{r}_i$ vanish at $t=t_1$ and $t=t_2$, the left hand side of (2.13) vanishes and

$$\int_{t_1}^{t_2} \delta T \, dt = -\int_{t_1}^{t_2} \left[\sum_{i=1}^{N} \vec{F}^*_i \cdot \delta\vec{r}_i\right] dt \tag{2.14}$$

It is now assumed that the forces $\vec{F}^*_i$ acting on the particles can be divided into two classes: conservative forces $\vec{F}_i^{(c)}$ and non-conservative forces $\vec{F}_i$. The conservative forces can be derived from a potential function. Let

$$\vec{F}_i^{(c)} = \hat{i}X_i^{(c)} + \hat{j}Y_i^{(c)} + \hat{k}Z_i^{(c)} \tag{2.15}$$

where the unit vectors $\hat{i}$, $\hat{j}$, $\hat{k}$ are parallel to the x, y, z axes, respectively, of the rectangular, Cartesian, inertial axis system. The quantities $X_i^{(c)}$, $Y_i^{(c)}$, $Z_i^{(c)}$ are the Cartesian components of the conservative forces $\vec{F}_i^{(c)}$. Since these forces are conservative, we may write

$$X_i^{(c)} = -\frac{\partial V}{\partial x_i}, \quad Y_i^{(c)} = -\frac{\partial V}{\partial y_i}, \quad Z_i^{(c)} = -\frac{\partial V}{\partial z_i} \tag{2.16}$$

where the potential V is a function of the position of the system in the actual dynamical path in the 3N- dimensional configuration space, i.e.,

$$V = V(x_1,y_1,z_1;\ x_2,y_2,z_2;\ \ldots;\ x_N,y_N,z_N) \tag{2.17}$$

In the varied path,

$$V' = V(x_1+\delta x_1,y_1+\delta y_1,z_1+\delta z_1;x_2+\delta x_2,y_2+\delta y_2,z_2+\delta z_2;\ \ldots$$
$$\ldots;\ x_N+\delta x_N,y_N+\delta y_N,z_N+\delta z_N) \tag{2.18}$$

If we expand V' in a Taylor series about the point $(x_1,y_1,z_1;\ x_2,y_2,z_2;\ \ldots;\ x_N,y_N,z_N)$ we obtain

$$V' = V + \sum_{i=1}^{N} \left(\frac{\partial V}{\partial x_i}\right)\delta x_i+\left(\frac{\partial V}{\partial y_i}\right)\delta y_i+\left(\frac{\partial V}{\partial z_i}\right)\delta z_i+ O(\delta x_i^2,\delta x_i\cdot\delta y_i,\ \text{etc.}) \tag{2.19}$$

Using (2.16) and (2.19), and neglecting $O(\delta x_i^2,\ \delta x_i\cdot\delta y_i,\ \text{etc.})$

$$\delta V=V'-V=-\sum_{i=1}^{N}(X_i^{(c)}\delta x_i+Y_i^{(c)}\delta y_i+Z_i^{(c)}\delta z_i) = -\sum_{i=1}^{N}\vec{F}_i^{(c)}\cdot\delta\vec{r}_i \tag{2.20}$$

We note that the variations $\delta\vec{r}_i$ may be identified as virtual displacements of the particle coordinates since no variation of time is involved. The varied path in configuration space can therefore be thought of as built up by a succession of virtual displacements from the actual path. Each virtual displacement occurs at some given instant of time, and at that instant the forces acting on the system have definite values. Thus the quantity

$$\delta W = \sum_{i=1}^{N}\vec{F}_i\cdot\delta r_i \tag{2.21}$$

may be identified as the work done by the non-conservative forces on the system during the virtual displacement from the actual to the varied path. If we now combine (2.14), (2.20), and (2.21) we obtain

$$\int_{t_1}^{t_2} (\delta T-\delta V)dt = \int_{t_1}^{t_2} \delta(T-V)dt = -\int_{t_1}^{t_2}\delta W dt \tag{2.22}$$

To obtain the conventional form of Hamilton's principle, we set L=T-V, where L is the Lagrangian function. Then

$$\int_{t_1}^{t_2}\delta(T-V)dt = \int_{t_1}^{t_2}\delta L dt = \int_{t_1}^{t_2}(L'-L)dt = \int_{t_1}^{t_2}L'dt - \int_{t_1}^{t_2}L dt = \delta\int_{t_1}^{t_2}L dt$$

so that

$$\delta\int_{t_1}^{t_2}L dt = -\int_{t_1}^{t_2}\delta W dt \tag{2.23}$$

Equation (2.23) is Hamilton's principle for a discrete mechanical system subjected to non-conservative forces. If there are no non-conservative forces acting on the system, $\delta W \equiv 0$ and (2.23) reduces to $\delta I \equiv 0$, where

$$I = \int_{t_1}^{t_2}L dt \tag{2.24}$$

Equation (2.24) implies that the motion of a discrete, conservative, mechanical system from time t_1 to time t_2 is such that the line integral I is an extremum for the path of motion, i.e., from among all possible paths by which the system point in configuration space could travel from its position at time t_1 to its position at time t_2, it will actually travel along the path for which the integral (2.24) has a stationary value. In this form Hamilton's principle for conservative mechanical systems indicates that the motion according to Newton's laws is distinguished from all other types of motion by possessing the property that I, for any given interval of time, has a stationary value. In addition, we note that L and δW in (2.23) are scalar quantities and therefore are invariant with respect to coordinate transformations. For these reasons, Hamilton's principle emerges as a fundamental principle of classical mechanics which may be used in place of Newton's laws to deduce all important principles of that branch of physics.

In order to illustrate the methodology employed in the application of Hamilton's principle, we shall consider four illustrative problems.

Example 1: Newton's second law of motion for a particle in free space.

We have

$$T= \frac{1}{2}m(\dot{x}^2+\dot{y}^2+\dot{z}^2);\quad V=0,\ \text{and}\ \delta W=\vec{F}\cdot\delta\vec{r}=X\delta x + Y\delta y + Z\delta z.$$

In this case L=T and $\delta L=m(\dot{x}\delta\dot{x} + \dot{y}\delta\dot{y} + \dot{z}\delta\dot{z})$ in accordance with (2.11). Upon substitution in (2.23) we obtain

$$-\int_{t_1}^{t_2} m(\dot{x}\delta\dot{x}+\dot{y}\delta\dot{y}+\dot{z}\delta\dot{z})dt = \int_{t_1}^{t_2}(X\delta x+Y\delta y+Z\delta z)dt$$

But

$$\delta\dot{x} = \delta\left(\frac{dx}{dt}\right) = \frac{dx'}{dt} - \frac{dx}{dt} = \frac{d}{dt}(x'-x) = \frac{d}{dt}(\delta x)$$

and

$$\int_{t_1}^{t_2}\dot{x}\delta\dot{x}dt = \int_{t_1}^{t_2}\dot{x}\,\frac{d}{dt}(\delta x)dt = (\dot{x}\delta x)_{t_1}^{t_2} - \int_{t_1}^{t_2}\ddot{x}\delta x dt = -\int_{t_1}^{t_2}\ddot{x}\delta x dt$$

because $\delta x=0$ for $t=t_1$ and $t=t_2$. Thus

$$\int_{t_1}^{t_2}\left[(-m\ddot{x}+X)\delta x + (-m\ddot{y}+Y)\delta y + (-m\ddot{z}+Z)\delta z\right]dt = 0$$

The interval t_2-t_1 is arbitrary. Hence the integrand vanishes:

$$(-m\ddot{x}+X)\delta x + (-m\ddot{y}+Y)\delta y + (-m\ddot{z}+Z)\delta z = 0$$

The variations δx, δy, δz are independent. Consequently, $m\ddot{x}=X$, $m\ddot{y}=Y$, $m\ddot{z}=Z$. These are the equations of motion for the particle.

Example 2: The linear oscillator (Fig. 2.1).

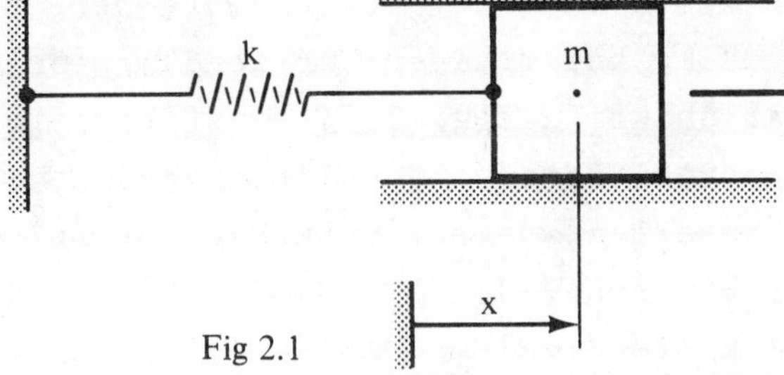

Figure 2.1

The mass of magnitude m is constrained by frictionless guides to move along the x-axis. A linear spring with spring constant k provides the restoring force. A force of magnitude F acts on the mass.

If $x=0$ denotes that position of the mass for which the spring force vanishes, then $T=\frac{1}{2}m\dot{x}^2$, $V=\frac{1}{2}kx^2$, and $\delta W=F\delta x$. Therefore $L=T-V=\frac{1}{2}(m\dot{x}^2-kx^2)$ and $\delta L=m\dot{x}\delta\dot{x}-kx\delta x$. According to (2.23),

$$\int_{t_1}^{t_2}(m\dot{x}\delta\dot{x}-kx\delta x)dt = -\int_{t_1}^{t_2}F\ \delta x dt$$

But

$$\int_{t_1}^{t_2}m\dot{x}\delta\dot{x}dt = (m\dot{x}\delta x)_{t_1}^{t_2} - \int_{t_1}^{t_2}m\ddot{x}\delta x dt$$

and $(m\dot{x}\delta x)_{t_1}^{t_2}= 0$. Therefore

$$\int_{t_1}^{t_2}(m\ddot{x}+kx-F)\delta x dt = 0$$

and because t_2-t_1 and δx are arbitrary, we obtain the equation of motion $m\ddot{x}+kx=F$. This is of course, the same equation which we obtained by the application of Newton's second law of motion in Chapter 1.

Example 3: A spring-connected, two-mass system (Fig. 2.2).

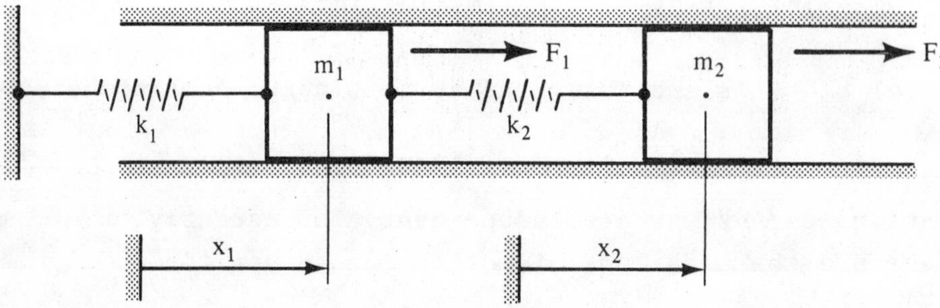

Figure 2.2

Two masses of magnitude m_1 and m_2 are connected by springs with spring constants k_1 and k_2 as shown in Fig. 2.2. The forces in the springs are zero when $x_1=x_2=0$. External forces F_1 and F_2 act on the masses m_1 and m_2, respectively. We have $T = \frac{1}{2}m_1\dot{x}_1^2 + \frac{1}{2}m_2\dot{x}_2^2$, $V = \frac{1}{2}k_1x_1^2 + \frac{1}{2}k_2(x_2-x_1)^2$ and $\delta W = F_1\delta x_1+F_2\delta x_2$. Since $L=T-V$, $\delta L=m_1\dot{x}_1\delta\dot{x}_1 + m_2\dot{x}_2\delta\dot{x}_2$ $-(k_1x_1+k_2x_1-k_2x_2)\delta x_1-(k_2x_2-k_2x_1)\delta x_2$.
Using (2.23), and integrating δT by parts with respect to time, we obtain

$$\int_{t_1}^{t_2}[(-m_1\ddot{x}_1-k_1x_1-k_2x_1+k_2x_2+F_1)\delta x_1+(-m_2\ddot{x}_2-k_2x_2+k_2x_1+F_2)\delta x_2]dt=0$$

Because t_2-t_1 is arbitrary, and since the variations δx_1 and δx_2 are independent, we obtain the equations of motion

$$m_1\ddot{x}_1+k_1x_1-k_2(x_2-x_1) = F_1$$

$$m_2\ddot{x}_2+k_2(x_2-x_1) = F_2$$

Example 4: The pendulum (Fig. 2.3)

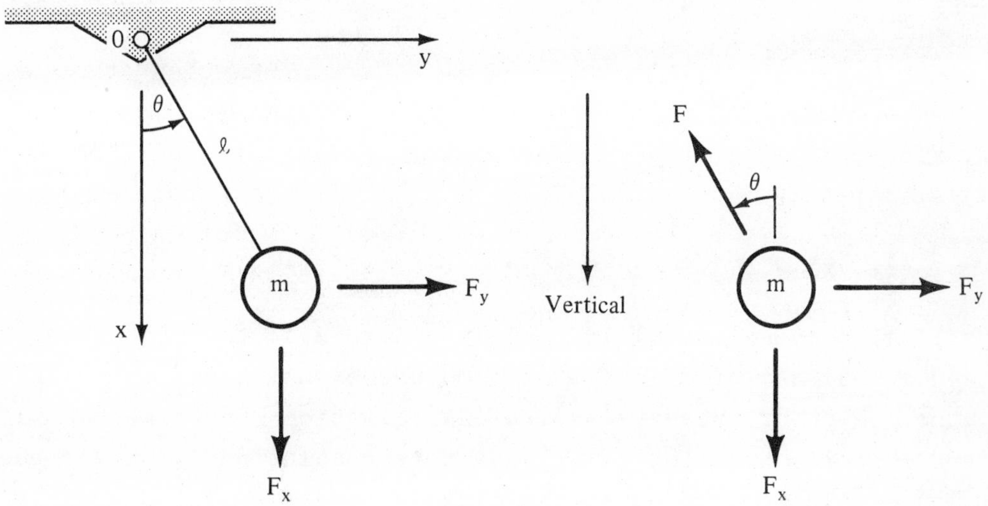

Figure 2.3

A mathematical pendulum is constrained to move in the vertical x-y plane as shown. The pendulum mass is concentrated at a distance ℓ from the frictionless point of suspension 0. The rigid arm connecting the pendulum mass to the hinge is massless. We have

$$T = \frac{1}{2} m(\dot{x}^2+\dot{y}^2), \quad V=mg(\ell-x)$$

$$\delta W = (F_x-F \cos \theta)\delta x + (F_y-F \sin \theta)\delta y$$

where F_x, F_y are the Cartesian components of the external force and F is the force exerted by the pendulum arm on the mass (constraint force). Upon substitution in (2.23) we obtain

$$\int_{t_1}^{t_2}(m\dot{x}\delta\dot{x} + m\dot{y}\delta\dot{y} + mg\delta x)dt = -\int_{t_1}^{t_2}[(F_x-F \cos \theta)\delta x+(F_y-F \sin \theta)\delta y]\, dt$$

Upon integration of $\int_{t_1}^{t_2}\delta T\, dt$ by parts,

$$\int_{t_1}^{t_2}[(m\ddot{x}-F_x+F \cos \theta- mg)\delta x + (m\ddot{y}-F_y+F \sin \theta)\delta y]dt = 0$$

Because the interval t_2-t_1 is arbitrary,

$$(m\ddot{x}-F_x+F\cos\theta-mg)\delta x+(m\ddot{y}-F_y+F\sin\theta)\delta y=0$$

In this case the variations δx and δy are not independent. The equation of constraint is

$$x^2+y^2=\ell^2 \qquad \text{(a)}$$

and therefore

$$x\delta x+y\delta y=0$$

Consequently

$$(m\ddot{x}-F_x+F\cos\theta-mg)y-(m\ddot{y}-F_y+F\sin\theta)x=0$$

or

$$m\frac{d}{dt}(\dot{x}y-\dot{y}x)=(F_x-F\cos\theta+mg)y-(F_y-F\sin\theta)x \qquad \text{(b)}$$

The physical meaning of (b) is readily recognized: the left side represents the rate of change of the moment of momentum; the right side, the moment of the external forces. Equation (b), together with the equation of constraint (a), leads to the solution of the problem. It is interesting to note that, in this case, a single and greatly simplified equation of motion can be obtained by a simple transformation. Let

$$x=\ell\cos\theta,\ y=\ell\sin\theta$$

These transformation equations satisfy (a) and transform (b) into

$$m\ell\ddot{\theta}+mg\sin\theta=F_y\cos\theta-F_x\sin\theta=F_\theta \qquad \text{(c)}$$

Equation (c) characterizes the motion of the pendulum with a single dependent variable θ. Moreover, the force of constraint F does not appear. We also note that the virtual work can be written

$$\delta W=F_x\delta x+F_y\delta y-F(\cos\theta\delta x+\sin\theta\delta y)$$

But

$$\delta x=x'-x=\ell\cos(\theta+\delta\theta)-\ell\cos\theta\simeq-\delta\theta\ell\sin\theta$$

$$\delta y=y'-y=\ell\sin(\theta+\delta\theta)-\ell\sin\theta\simeq\delta\theta\ell\cos\theta$$

and therefore

$$\cos\theta\delta x+\sin\theta\delta y=0$$

We thus conclude that the virtual work of the force of constraint F vanishes, and

$$\delta W = F_x \delta x + F_y \delta y$$

2.3 GENERALIZED COORDINATES AND LAGRANGE'S EQUATIONS

Although Cartesian coordinates provide a useful vehicle for the derivation of many formulae and principles (such as Hamilton's principle in Section 2), they are frequently not convenient for the solution of specific problems. Consider, for example, a mathematical pendulum in the vertical plane (see Example 4 of Section 2). There are two Cartesian coordinates required to locate the pendulum mass, but a single angular coordinate suffices to locate the pendulum in a unique manner. In this case the two Cartesian coordinates x,y are not independent, and they are related by the equation $x^2+y^2=\ell^2$. Similarly, even though a rigid body is composed of a very large number of particles, only six independent quantities are required to uniquely locate it in space. For instance, we can specify the location of three non-collinear points of the rigid body. If (x_1, y_1, z_1), (x_2, y_2, z_2), and (x_3, y_3, z_3) are the coordinates of these points, they will uniquely determine the position of the body. However, they are not independent. They are connected by the relations

$$(x_1-x_2)^2 + (y_1-y_2)^2 + (z_1-z_2)^2 = c_1^2$$

$$(x_2-x_3)^2 + (y_2-y_3)^2 + (z_2-z_3)^2 = c_2^2$$

$$(x_3-x_1)^2 + (y_3-y_1)^2 + (z_3-z_1)^2 = c_3^2$$

which express the invariance of the distances c_1, c_2, and c_3 between the three points. Consequently, there are only 9-3=6 independent coordinates required to specify the position of the rigid body in space. An entirely different way to locate a rigid body in space is to specify the three coordinates of a point in the rigid body with respect to a rectangular Cartesian axis system, along with the three angles which a straight, directed line segment in the rigid body subtends with respect to the same axis system.

The number of independent quantities necessary to specify uniquely the configuration of a mechanical system is known as the number of degrees of freedom of the system. Aggregates or systems of mass particles are often constrained in a variety of ways. The number of degrees of freedom n for a system of N mass particles subjected to k independent constraints is given by $n=3N-k$. Throughout the present and succeeding developments, we shall assume that the equations of constraint, if any, can always be written as

$$f_i = f_i(\vec{r}_1,\vec{r}_2,\ldots\vec{r}_n);\ i=1,2,\ldots k$$

Constraints of this type are called holonomic and scleronomous, i.e., the constraints are expressed in terms of particle coordinates only (and not in terms of velocities), and the constraint equations do not contain the time variable t explicitly.

When the constraints are holonomic and scleronomous, it is usually possible to find a set of n generalized coordinates which are independent and which, in number, are equal to the number n of degrees of freedom. These coordinates are unrestricted by constraints (i.e., they take into account whatever constraints have been imposed upon the mechanical system) and they are called a proper set of generalized coordinates. With reference to Example 4 of Section 2, the angle θ is a proper coordinate, while the Cartesian coordinates x and y cannot be called by that name because an equation of constraint $x^2+y^2=\ell^2$ renders x and y interdependent. The specific choice of a set of proper generalized coordinates is not unique, and, in general, many different sets will completely specify the configuration of a given system. The ultimate test of the suitability of a particular set of generalized coordinates is the structure of the resulting equations of motion. If these equations permit a simple physical interpretation and/or yield to a straightforward solution, then the choice of generalized coordinates is indeed a good one. It is obvious that skill, experience, as well as insight enter into the process of selection of the appropriate generalized coordinates.

Let us assume that we are dealing with a holonomic, scleronomous, mechanical system, and let us select a set of proper, generalized coordinates $q_1,q_2,q_3,\ldots,q_n$. In this case the equations of transformation from generalized coordinates to Cartesian coordinates are

$$x_i = x_i(q_1,q_2,\ldots q_n) \tag{2.25a}$$

$$y_i = y_i(q_1,q_2,\ldots q_n) \tag{2.25b}$$

$$z_i = z_i(q_1,q_2,\ldots q_n) \tag{2.25c}$$

where $i=1,2,\ldots N$; and $n \leq 3N$.
The components of velocity are given by

$$\dot{x}_i = \sum_{j=1}^{n} \frac{\partial x_i}{\partial q_j} \dot{q}_j \tag{2.26a}$$

$$\dot{y}_i = \sum_{j=1}^{n} \frac{\partial y_i}{\partial q_j} \dot{q}_j \tag{2.26b}$$

$$\dot{z}_i = \sum_{j=1}^{n} \frac{\partial z_i}{\partial q_j} \dot{q}_j \tag{2.26c}$$

In view of (2.26), the expression for kinetic energy of the system of N mass particles is transformed to

$$T = \frac{1}{2} \sum_{i=1}^{N} m_i(\dot{x}_i^2 + \dot{y}_i^2 + \dot{z}_i^2) = \frac{1}{2} \sum_{r,s=1}^{n} t_{rs}\dot{q}_r\dot{q}_s \tag{2.27}$$

where

$$t_{rs} = \sum_{i=1}^{N} m_i \left(\frac{\partial x_i}{\partial q_r}\frac{\partial x_i}{\partial q_s} + \frac{\partial y_i}{\partial q_r}\frac{\partial y_i}{\partial q_s} + \frac{\partial z_i}{\partial q_r}\frac{\partial z_i}{\partial q_s} \right) \tag{2.28}$$

and with reference to (2.25) it is clear that, in general $t_{rs}=t_{rs}(q_1,q_2,\ldots,q_n)$, and that the kinetic energy expression is a function of the generalized coordinates q_r as well as of the generalized velocities $\dot{q}_r$, $r=1,2,\ldots,n$, i.e.,

$$T = T(q_1,q_2,\ldots,q_n;\ \dot{q}_1,\dot{q}_2,\ldots,\dot{q}_n) \tag{2.29}$$

A strict interpretation limits the above arguments to systems consisting of discrete mass particles. However, they are readily extended to the case of rigid bodies (which are collections of mass particles), provided the summations are replaced by suitable integrations. For example, consider the case of a rigid disc of thickness h, radius a, and uniformly distributed mass-density ρ, rotating about a fixed z axis as shown in Fig. 2.4.

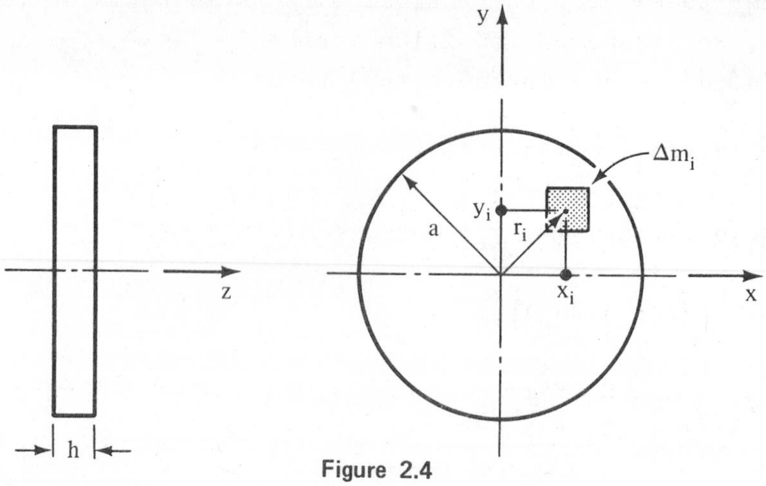

Figure 2.4

According to (2.27)

$$T = \lim_{\substack{N\to\infty \\ \Delta m_i \to 0}} \frac{1}{2} \sum_{i=1}^{N} (\dot{x}_i^2 + \dot{y}_i^2)\Delta m_i$$

But

$$x_i = r_i \cos\theta, \quad \dot{x}_i = -r_i\dot{\theta}\sin\theta$$

$$y_i = r_i \sin\theta, \quad \dot{y}_i = r_i\dot{\theta}\cos\theta$$

$$\dot{x}_i^2 + \dot{y}_i^2 = r_i^2\dot{\theta}^2$$

and

$$T = \lim_{\substack{N\to\infty \\ \Delta m_i \to 0}} \frac{1}{2}\dot{\theta}^2 \sum_{i=1}^{N} r_i^2\Delta m_i = \frac{1}{2}\dot{\theta}^2 \int_M r^2 dm = \frac{1}{2}\dot{\theta}^2 \int_S r^2\rho h ds$$

where S is the area of the circle of radius a. If we take $dm = \rho h\, 2\pi r\, dr$ then

$$\int_M r^2 dm = 2\pi\rho h \int_0^a r^3 dr = \frac{1}{2}\rho h\pi a^4 = \frac{1}{2}Ma^2 = I$$

where $M = h\pi a^2\rho$ denotes the mass of the entire disc. Thus, in this instance we take $T = 1/2\, I\dot{\theta}^2$.

In order to express the virtual work δW in terms of generalized coordinates, we expand the position vector in the varied path $\vec{r}'$ about the point $(q_1,q_2,\ldots q_n)$ in the actual path:

$$\vec{r}_i' = \vec{r}_i(q_1+\delta q_1,\ q_2+\delta q_2,\ +\ldots,+\ q_n+\delta q_n) =$$

$$= \vec{r}_i(q_1,q_2,\ldots q_n) + \sum_{r=1}^{n} \frac{\partial \vec{r}_i}{\partial q_r}\delta q_r + 0[(\delta q_r)(\delta q_s)]$$

If we neglect $0[(\delta q_r)(\delta q_s)]$,

$$\delta\vec{r}_i = \vec{r}_i'-\vec{r}_i = \sum_{r=1}^{n} \frac{\partial \vec{r}_i}{\partial q_r}\delta q_r \tag{2.30}$$

so that

$$\delta W = \sum_{i=1}^{N} \vec{F}^*_i\cdot\delta\vec{r}_i = \sum_{i=1}^{n} \Theta_i \delta q_i \tag{2.31}$$

where

$$\Theta_i = \sum_{j=1}^{N} \vec{F}^*_j \cdot \frac{\partial \vec{r}_j}{\partial q_i};\ i=1,2,\ldots n. \tag{2.32}$$

When treating a constrained system, it is convenient to divide the forces $\vec{F}^*_i$ and the related generalized forces Θ_i into two classes: applied forces which arise from causes external to the system, and forces which arise due to internal constraints. When the system is subjected to ideal constraints, the virtual work of the forces of constraint vanishes. This has already been observed in the special case depicted by Example 4 of Section 2. As an additional illustration of this phenomenon, consider the case of any two particles in a rigid body. Denote by $\vec{F}^*_{ij}$, $i\neq j$, the force which the particle j exerts on particle i. Let $\vec{F}^*_{ii} \equiv \vec{F}^*_i$ be the applied force acting on particle i. If we denote by $\vec{r}_i$ and $\vec{r}_j$ the position vectors of the particles i and j, respectively, we have

$$(\vec{r}_i-\vec{r}_j)\cdot(\vec{r}_i-\vec{r}_j) = \text{constant}$$

and

$$(\vec{r}_i-\vec{r}_j)\cdot(\delta\vec{r}_i-\delta\vec{r}_j) = 0$$

According to Newton's third law,

$$\vec{F}^*_{ij} = -\vec{F}^*_{ji} = \lambda_{ij} \frac{(\vec{r}_i - \vec{r}_j)}{|\vec{r}_i - \vec{r}_j|}, \quad i \neq j$$

where λ_{ij} is a scalar and we have

$$\vec{F}^*_{ij} \cdot \delta\vec{r}_i + \vec{F}^*_{ji} \cdot \delta\vec{r}_j = \frac{\lambda_{ij} (\vec{r}_i - \vec{r}_j) \cdot (\delta\vec{r}_i - \delta\vec{r}_j)}{|\vec{r}_i - \vec{r}_j|} = 0$$

Consequently the virtual work of all forces, applied as well as forces due to constraints is given by

$$\delta W = \sum_{i,j=1}^{N} \vec{F}^*_{ij} \cdot \delta\vec{r}_i = \sum_{i=1}^{N} \vec{F}^*_i \cdot \delta r_i$$

and we must conclude that the virtual work due to constraining forces in the rigid body vanishes. Other forces arising because of constraints which do not contribute to the virtual work term are:

(a) The normal reaction which a fixed, smooth surface exerts on a sliding body, the force being perpendicular to the displacement.

(b) The reactive force acting on a non-slipping, rolling body, the point of contact being the instantaneous center of rotation, and the reacting force experiencing no displacement.

(c) A force of constraint acting at a fixed point, such as the force at a frictionless hinge.

(d) The forces arising because of the interaction of two bodies connected by a frictionless hinge, the forces being equal and opposite, and acting at the same point.

(e) Forces transmitted by rigid bars, and forces transmitted by taut, inextensible strings.

In the following we shall assume that our system has ideal constraints and, as a consequence, that the virtual work of all forces of constraint vanishes. Thus $\vec{F}^*_i$ and the resulting generalized forces Θ_i will be entirely due to causes which are independent of the manner of constraint of our mechanical system.

With reference to (2.31), the quantity Θ_i is the generalized force which corresponds to the generalized displacement q_i. We note that the virtual work expression δW always has dimensional units of work (ft-lbs); however, the generalized force Θ_i does not necessarily have units of force just as the generalized coordinates are not restricted to units of displacement.

To express the variation of kinetic energy δT in terms of generalized coordinates, we expand the kinetic energy T' in the varied path about the point $(q_1,q_2,\ldots,q_n;\ \dot{q}_1,\dot{q}_2,\ldots,\dot{q}_n)$ in the actual dynamical path. By Taylor's theorem,

$$T' = T(q_1,q_2,\ldots q_n;\ \dot{q}_1,\dot{q}_2,\ldots\dot{q}_n)+\sum_{i=1}^{n}\frac{\partial T}{\partial q_i}\delta q_i+\sum_{i=1}^{n}\frac{\partial T}{\partial \dot{q}_i}\delta\dot{q}_i$$
$$+\ O[\delta q_i\cdot\delta q_j] + O[\delta\dot{q}_i\cdot\delta\dot{q}_j]$$

Neglecting higher order terms,

$$\delta T = T'-T = \sum_{i=1}^{n}\frac{\partial T}{\partial q_i}\delta q_i + \sum_{i=1}^{n}\frac{\partial T}{\partial \dot{q}_i}\delta\dot{q}_i \tag{2.33}$$

and upon substitution of (2.31) and (2.33) into (2.14) we obtain

$$\int_{t_1}^{t_2}\left\{\sum_{i=1}^{n}\left(\frac{\partial T}{\partial \dot{q}_i}\delta\dot{q}_i + \frac{\partial T}{\partial q_i}\delta q_i+\Theta_i\delta q_i\right)\right\}dt=0 \tag{2.34}$$

But

$$\int_{t_1}^{t_2}\frac{\partial T}{\partial \dot{q}_i}\delta\dot{q}_i dt = \int_{t_1}^{t_2}\frac{\partial T}{\partial \dot{q}_i}\frac{d}{dt}(\delta q_i)dt= \left(\frac{\partial T}{\partial \dot{q}_i}\delta q_i\right)_{t_1}^{t_2} -\int_{t_1}^{t_2}\frac{d}{dt}\left(\frac{\partial T}{\partial \dot{q}_i}\right)\delta q_i dt$$

and $\delta q_i=0$ at $t=t_1$ and $t=t_2$, because the varied and the actual dynamical path coincide at $t=t_1$ and $t=t_2$. Consequently (2.34) can be written as

$$\int_{t_1}^{t_2}\left\{\left[\sum_{i=1}^{n}\frac{d}{dt}\left(\frac{\partial T}{\partial \dot{q}_i}\right) - \frac{\partial T}{\partial q_i} - \Theta_i\right]\delta q_i\right\} dt=0 \tag{2.35}$$

The time interval t_2-t_1 is arbitrary. Hence at every instant $t_1<t<t_2$ we have

$$\sum_{i=1}^{n}\left[\frac{d}{dt}\left(\frac{\partial T}{\partial \dot{q}_i}\right) - \frac{\partial T}{\partial q_i} - \Theta_i\right]\delta q_i = 0 \tag{2.36}$$

We have pre-supposed that the holonomic system under consideration has n degrees of freedom, i.e., whatever values are assigned to the variations of the generalized coordinates δq_i, the new configuration $q_i+\delta q_i$ is a possible configuration, one that the system can assume without violating the constraints. Consequently

$$\frac{d}{dt}\left(\frac{\partial T}{\partial \dot{q}_i}\right) - \frac{\partial T}{\partial q_i} = \Theta_i, \quad i=1,2,\ldots n \tag{2.37}$$

It is now convenient to divide the generalized forces Θ_i into those which are conservative, $Q_i^{(c)}$, and those which are non-conservative, Q_i, such that

$$\Theta_i = Q_i^{(c)} + Q_i \tag{2.38}$$

With the help of (2.32) and (2.16) we obtain

$$Q_i^{(c)} = -\sum_{j=1}^{N}\left(\frac{\partial V}{\partial x_j}\frac{\partial x_j}{\partial q_i} + \frac{\partial V}{\partial y_j}\frac{\partial y_j}{\partial q_i} + \frac{\partial V}{\partial z_j}\frac{\partial z_j}{\partial q_i}\right) = -\frac{\partial V}{\partial q_i}, \quad i=1,2,\ldots n.$$

In this case (2.37) assumes the form

$$\frac{d}{dt}\left(\frac{\partial T}{\partial \dot{q}_i}\right) - \frac{\partial T}{\partial q_i} + \frac{\partial V}{\partial q_i} = Q_i \tag{2.39}$$

If we define the Lagrangian function $L=T-V$, and recognize the fact that the potential function V is independent of the generalized velocities, i.e., $V=V(q_1,q_2,\ldots q_n)$, then (2.39) assumes the form

$$\frac{d}{dt}\left(\frac{\partial L}{\partial \dot{q}_i}\right) - \frac{\partial L}{\partial q_i} = Q_i \tag{2.40}$$

Equations (2.37), (2.39), and (2.40) are different forms of Lagrange's equations which provide us with an unusually convenient means for deriving equations of motion of complex mechanical systems.

If all forces are derivable from a potential function, $Q_i\equiv 0$, and the motion of the holonomic, scleronomous mechanical system is characterized by

$$\frac{d}{dt}\left(\frac{\partial L}{\partial \dot{q}_i}\right) - \frac{\partial L}{\partial q_i} = 0 \tag{2.41}$$

At this point it is possible to establish a formal connection between Lagrange's equation in the form (2.41) and Hamilton's principle via the calculus of variations. Within the present context, the principle of Hamilton states:

The actual motion of a holonomic, scleronomous mechanical system, whose Lagrangian is given by $L=T-V=L(q_1,q_2,\ldots,q_n;\ \dot{q}_1,\dot{q}_2,\ldots,\dot{q}_n)$ is such as to render the integral

$$I = \int_{t_1}^{t_2} L\,dt \tag{2.42}$$

an extremum with respect to the functions $q_i(t)$, $i=1,2,\ldots,n$, where t_1 and t_2 are two arbitrary instants of time. The necessary condition for I to be an extremum is that the Lagrangian function L satisfy the Euler-Lagrange differential equations as provided by the calculus of variations, and (2.41) are, in fact, those equations, as shown in Appendix A.

2.4 APPLICATIONS OF LAGRANGE'S EQUATIONS

We next consider four typical applications of the theory developed in Section 3.

(a) Particle Motion in Cylindrical Coordinates.

A particle with mass m moves in free space. We are to find the equations of motion in cylindrical coordinates. If the coordinates of the particle with respect to a cylindrical coordinate system are given by $(r,\ \theta,\ z)$, its position in the Cartesian frame is

$$x = r\cos\theta,\ y = r\sin\theta,\ z = z$$

and the square of its speed is

$$v^2 = \dot{x}^2+\dot{y}^2+\dot{z}^2 = \dot{r}^2+r^2\dot{\theta}^2+\dot{z}^2$$

Therefore the kinetic energy of the particle is

$$T = \frac{1}{2}mv^2 = \frac{1}{2}m(\dot{r}^2+r^2\dot{\theta}^2+\dot{z}^2)$$

Let the components of the force acting on the particle be F_x, F_y, F_z when referred to Cartesian axes, and let Q_r, Q_θ, Q_z be the generalized forces corresponding to the generalized coordinates r,θ,z, respectively. Then according to (2.32),

$$Q_r = F_x \frac{\partial x}{\partial r} + F_y \frac{\partial y}{\partial r} + F_z \frac{\partial z}{\partial r} = F_x \cos\theta + F_y \sin\theta$$

$$Q_\theta = F_x \frac{\partial x}{\partial \theta} + F_y \frac{\partial y}{\partial \theta} + F_z \frac{\partial z}{\partial \theta} = -F_x r \sin\theta + F_y r \cos\theta$$

$$Q_z = F_x \frac{\partial x}{\partial z} + F_y \frac{\partial y}{\partial z} + F_z \frac{\partial z}{\partial z} = F_z$$

In this case T≡L, and upon substitution in (2.40) we obtain the desired equations of motion:

$$m\ddot{r} - mr\dot{\theta}^2 = Q_r = F_x \cos\theta + F_y \sin\theta$$

$$m \frac{d}{dt} (r^2\dot{\theta}) = Q_\theta = r(F_y \cos\theta - F_x \sin\theta)$$

$$m\ddot{z} = F_z$$

(b) A Two-mass, Single Degree of Freedom System (Fig. 2.5).

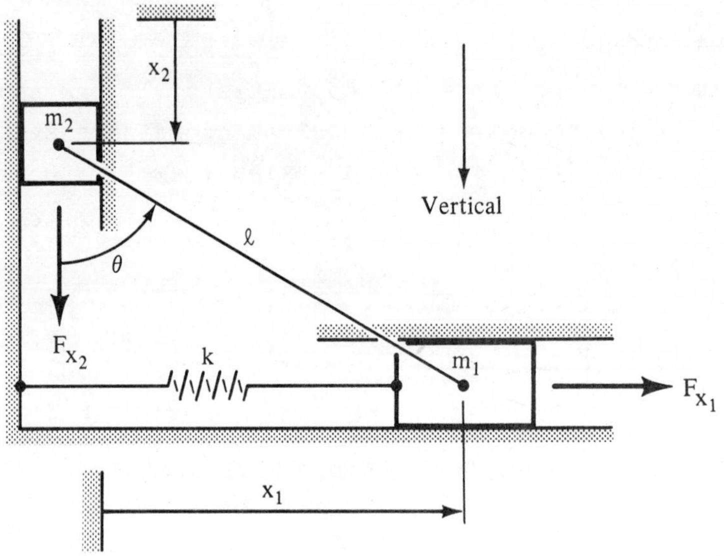

Figure 2.5

The masses m_1 and m_2 translate over frictionless surfaces. The spring is elastic with spring constant k, and the rigid connecting rod of length ℓ is mass-less. When the connecting rod is vertical, $x_1 = x_2 = \theta = 0$ and the force in the spring is zero. We have

$$x_1 = \ell \sin\theta, \quad x_2 = \ell(1-\cos\theta), \text{ and}$$

$$T = \tfrac{1}{2}m_1\dot{x}_1^2 + \tfrac{1}{2}m_2\dot{x}_2^2 = \tfrac{1}{2}\ell^2\dot{\theta}^2(m_1\cos^2\theta + m_2\sin^2\theta)$$

$$V = \tfrac{1}{2}kx_1^2 - m_2 g x_2 = \tfrac{1}{2}k\ell^2\sin^2\theta - m_2 g\ell(1-\cos\theta)$$

$$Q_\theta = F_{x_1}\frac{\partial x_1}{\partial\theta} + F_{x_2}\frac{\partial x_2}{\partial\theta} = F_{x_1}\ell\cos\theta + F_{x_2}\ell\sin\theta$$

Upon substitution in (2.39) we obtain the equation of motion

$$\ell\frac{d}{dt}\left[(m_1\cos^2\theta + m_2\sin^2\theta)\dot{\theta}\right] + \left[\dot{\theta}^2\ell(m_1-m_2)+k\ell\right]\sin\theta\cos\theta - m_2 g\sin\theta$$

$$= F_{x_1}\cos\theta + F_{x_2}\sin\theta$$

(c) Pendulum with Moving, Spring-restrained Point of Suspension (Fig. 2.6).

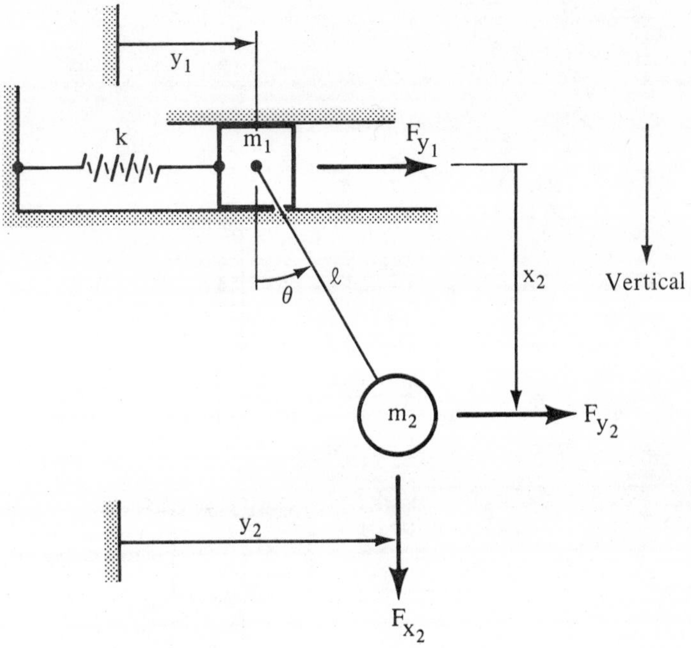

Figure 2.6

The mass m_1 is restrained by an elastic spring with modulus k, and slides, without friction, on the horizontal guide. The pendulum mass m_2 is attached to the sliding mass m_1 by means of a rigid, massless arm of length ℓ. When the system is in static equilibrium, $y_1=y_2=\theta=0$ and the force in the spring vanishes. Thus

$$x_2 = \ell \cos\theta, \; y_2 = y_1 + \ell \sin\theta, \text{ and}$$

$$v_2^2 = \dot{x}_2^2 + \dot{y}_2^2 = \ell^2\dot{\theta}^2 + \dot{y}_1^2 + 2\ell\dot{\theta}\dot{y}_1\cos\theta$$

$$T = \tfrac{1}{2}m_1\dot{y}_1^2 + \tfrac{1}{2}m_2(\ell^2\dot{\theta}^2 + \dot{y}_1^2 + 2\ell\dot{\theta}\dot{y}_1 \cos\theta)$$

$$V = \tfrac{1}{2}ky_1^2 + m_2 g\ell(1-\cos\theta)$$

With the aid of (2.32) we have

$$Q_{y_1} = F_{y_1}\frac{\partial y_1}{\partial y_1} + F_{y_2}\frac{\partial y_2}{\partial y_1} + F_{x_2}\frac{\partial x_2}{\partial y_1} = F_{y_2} + F_{y_1}$$

$$Q_\theta = F_{y_1}\frac{\partial y_1}{\partial \theta} + F_{y_2}\frac{\partial y_2}{\partial \theta} + F_{x_2}\frac{\partial x_2}{\partial \theta} = F_{y_2}\ell\cos\theta - F_{x_2}\ell\sin\theta$$

Substituting in (2.39), we obtain the equations of motion

$$(m_1+m_2)\ddot{y}_1 + ky_1 + m_2\ell\frac{d}{dt}(\dot{\theta}\cos\theta) = F_{y_1} + F_{y_2}$$

$$m_2\ell\ddot{\theta} + m_2(g+\dot{\theta}\dot{y}_1)\sin\theta + m_2\frac{d}{dt}(\dot{y}_1\cos\theta) = F_{y_2}\cos\theta - F_{x_2}\sin\theta$$

(d) The Spring-restrained Fly-ball Governor (Fig. 2.7).

The Cartesian x, y, z system is stationary. Fig. 2.7a shows the fly-ball governor at an instant when $\psi = 0$. We assume the entire mechanism to be mass-less with the exception of the two balls, each of which has mass m/2. All hinges, and the vertical slide are

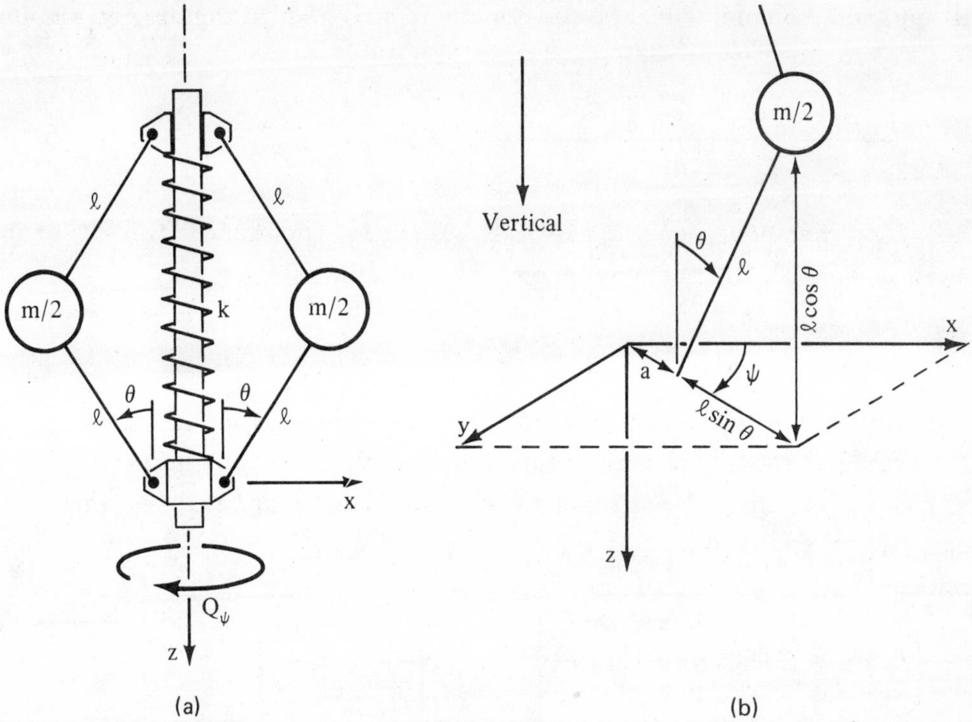

Figure 2.7

frictionless. When θ = 0, the force in the elastic spring is zero. The torque Q_ψ causes an angular acceleration $\ddot{\psi}$ by rotating the entire system about the z-axis. We have

$$x = (a + \ell \sin\theta)\cos\psi$$

$$y = (a + \ell \sin\theta)\sin\psi$$

$$z = -\ell \cos\theta$$

$$v^2 = \dot{x}^2 + \dot{y}^2 + \dot{z}^2 = \ell^2\dot{\theta}^2 + (a + \ell \sin\theta)^2\dot{\psi}^2$$

and

$$T = \tfrac{1}{2} m[\ell^2\dot{\theta}^2 + (a+\ell \sin\theta)^2\dot{\psi}^2]$$

$$V = -mg\ell(1-\cos\theta) + 2k\ell^2(1-\cos\theta)^2$$

Upon substitution in (2.39), we obtain the equations of motion

$$m\ell^2\ddot{\theta} - m\ell\dot{\psi}^2(a+\ell \sin\theta)\cos\theta + 4k\ell^2\sin\theta(1-\cos\theta) - mg\ell \sin\theta = 0$$

$$m\frac{d}{dt}[\dot{\psi}(a+\ell \sin\theta)^2] = Q_\psi$$

If the applied torque $Q_\psi=0$, $\dot{\psi}=\Omega=$ constant and $\theta=\theta_0=$ constant. In this case

$$\Omega^2 = \frac{\left[4\,\frac{k}{m}\,(1-\cos\,\theta_0)-\frac{g}{\ell}\right]\tan\,\theta_0}{\frac{a}{\ell}+\sin\,\theta_0}$$

We note that when $\Omega=0$, $\cos\,\theta_0=1-\frac{1}{4}\,\frac{mg}{k\ell}$. This is the condition of stable, static equilibrium of the governor.

EXERCISES

2.1 Three spring connected masses are constrained to move retilinearly as shown in the figure. Neglecting frictional forces, derive the equations of motion by (a) Hamilton's principle, and (b) Lagrange's equations.

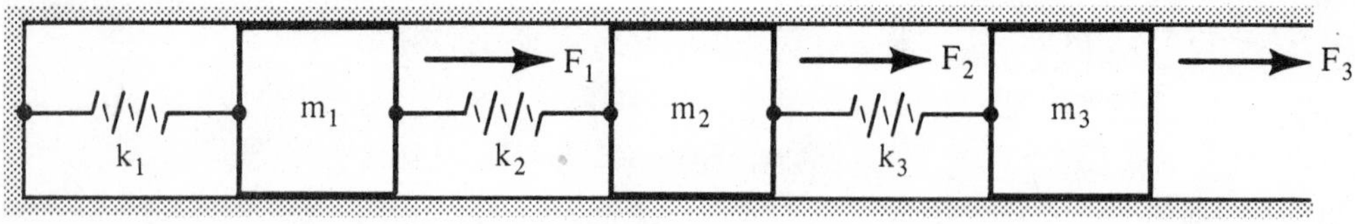

Figure: Exercise 2.1

2.2 By means of Lagrange's equations of motion, derive the differential equations of motion for a particle in spherical polar coordinates r, θ, ψ, where $x=r\,\sin\,\psi\,\cos\,\theta$, $y=r\,\sin\,\psi\,\sin\,\theta$, $z=r\,\cos\,\psi$.

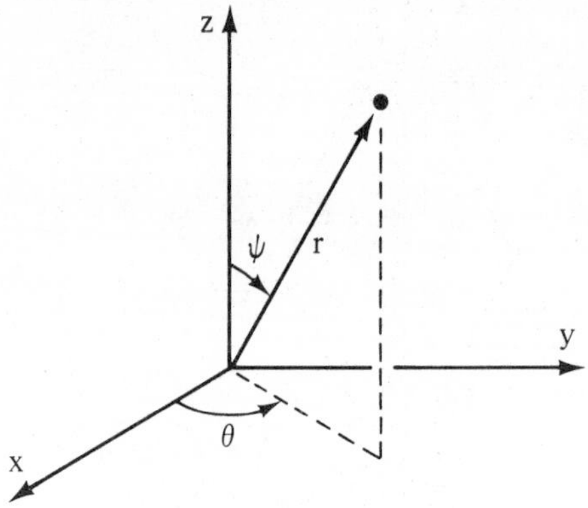

Figure: Exercise 2.2

2.3 Using Lagrange's equations of motion, write the equations of motion of the double pendulum subject to external forces F_{x_1}, F_{y_1}, F_{x_2}, F_{y_2}.

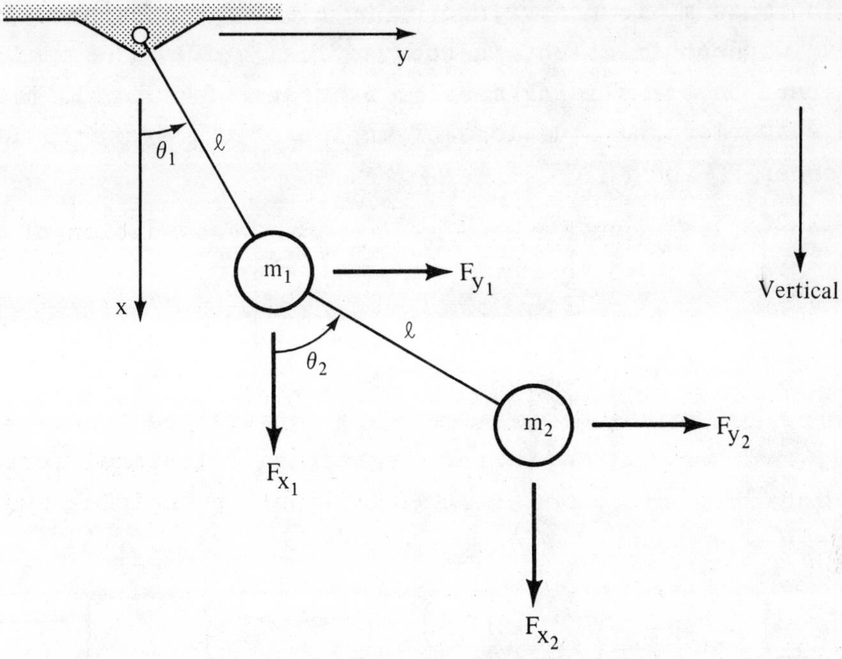

Figure: Exercise 2.3

2.4 Write the equations of motion of the fly-ball governor shown in the figure. Assume that the shaft and connecting linkages are massless.

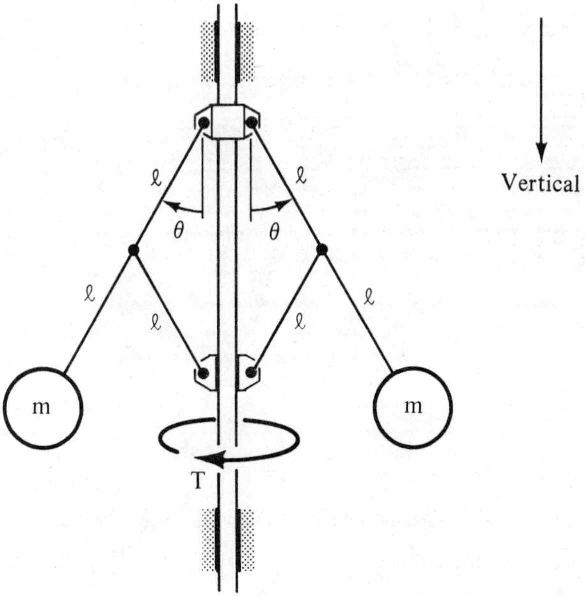

Figure: Exercise 2.4

2.5 The mass m_1 is restrained by an elastic spring with modulus k and slides, without friction, in the vertical guide. The pendulum mass m_2 is attached to the sliding mass m_1 by means of a rigid, massless arm of length ℓ. Derive the equations of motion of the system under the applied forces F_x and F_y.

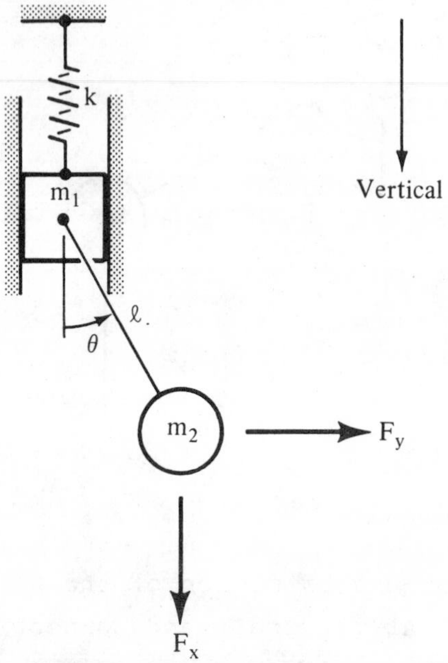

Figure: Exercise 2.5

2.6 Two spring connected masses m_1 and m_2 are constrained to move rectilinearly as shown in the figure. A pendulum mass m_3 is hinged to mass m_2 by means of a massless rigid rod of length ℓ. Write the equations of motion of the system if a force F is acting on the pendulum mass.

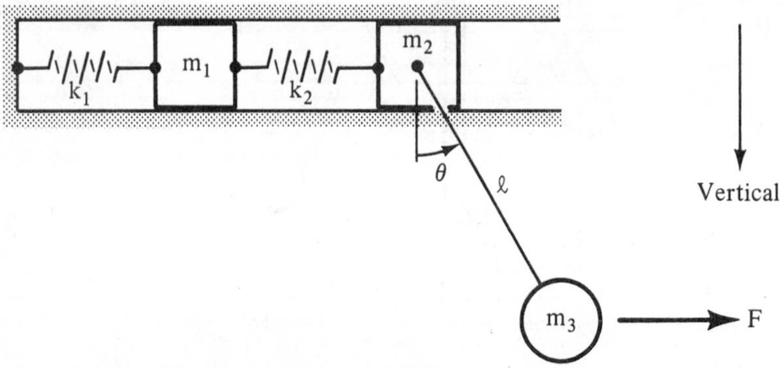

Figure: Exercise 2.6

2.7 A pendulum of length ℓ and mass m can pivot freely about a horizontal axis which, in turn, rotates about a vertical axis. Assuming that the frame has a moment of inertia I, set up the equations of motion.

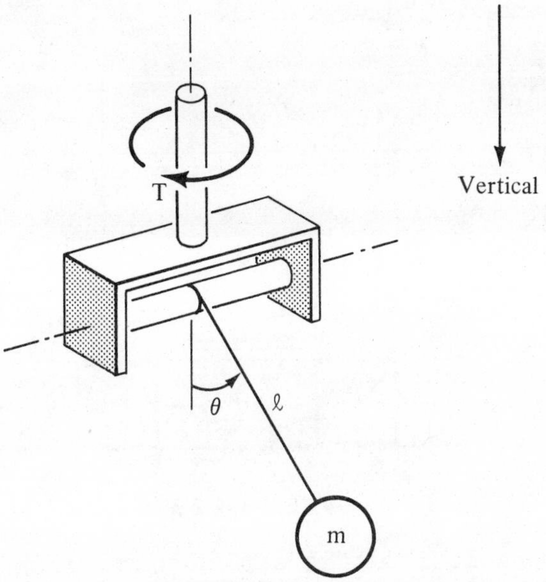

Figure: Exercise 2.7

2.8 A slender rod of length ℓ and mass M is attached to a spring and constrained to move as shown in the figure. Neglecting friction, write the equation of motion if $0 < \theta < \frac{\pi}{2}$.

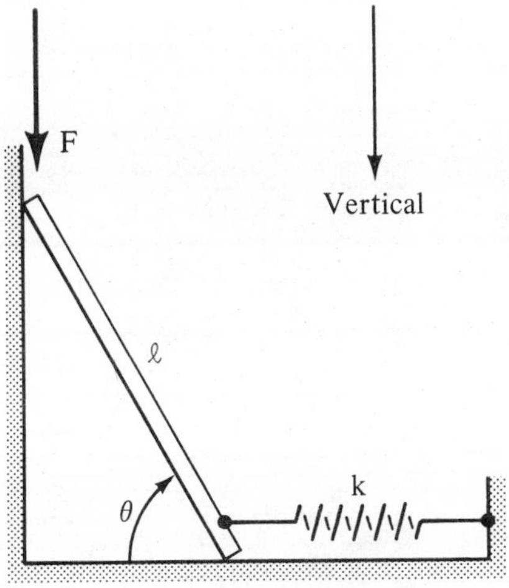

Figure: Exercise 2.8

2.9 Three masses are constrained and connected as shown in the figure. Neglecting friction and gravity, write the equations of motion. Assume that the rod connecting m_1 and m_2 is rigid and massless.

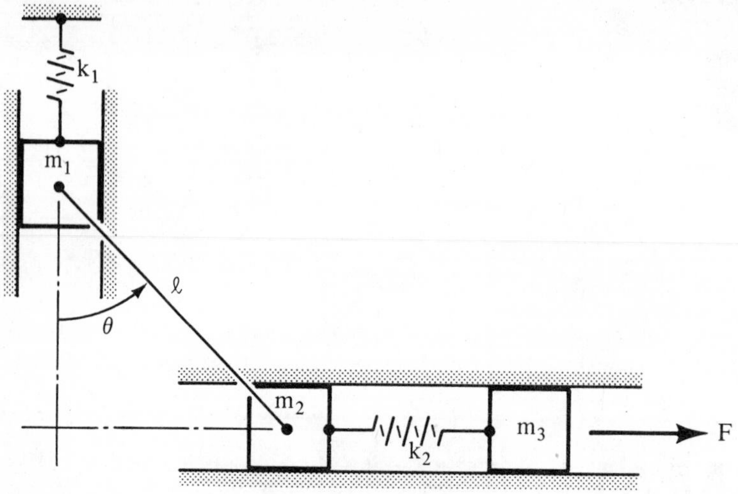

Figure: Exercise 2.9

2.10 Consider the motion of a set of n coupled linear oscillators shown in the figure. In the equilibrium position, the distances between the identical masses are equal. Write the equations of motion of this system.

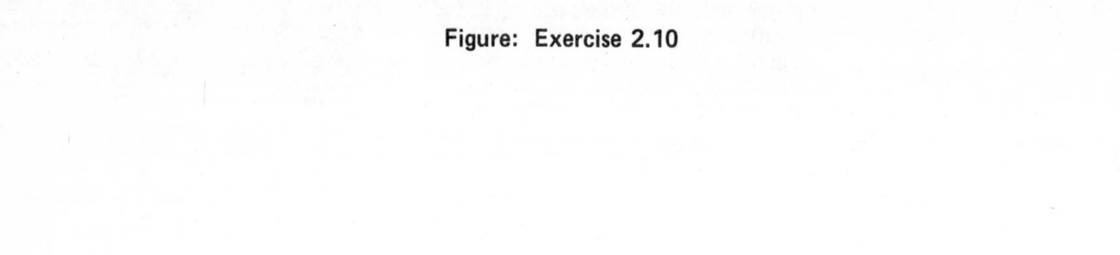

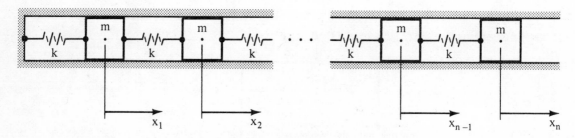

Figure: Exercise 2.10

Chapter

3

SMALL OSCILLATIONS OF DISCRETE SYSTEMS (THEORY)

The methods which I present here do not require either constructions or reasonings of geometrical or mechanical nature, but only algebraic operations proceeding after a regular and uniform plan. Those who love the Analysis will see with pleasure Mechanics made a branch of it, and will be grateful to me for having thus extended its domain.

J.L. Lagrange, Mécanique Analytique (1788).

3.1 LINEARIZATION OF THE EQUATIONS OF MOTION

Equations of motion of discrete mechanical systems are usually non-linear, a fact which is readily verified by an inspection of the equations of motion derived at the end of Chapter 2. As a rule, the explicit solution of non-linear equations of motion gives rise to major difficulties, and a general, unifying theory is not available at this time. For a large class of important problems which arise in technical and scientific applications, a major simplification is possible: the equations can be linearized by the introduction of a consistent and uniform set of assumptions. When these simplifying assumptions are introduced, a complete and explicit solution of the problems of free and forced motion with appropriate initial conditions becomes possible. In addition, we can obtain insight into many aspects of the behavior of a system which, at first sight, appears to present a problem of hopeless complexity.

We shall assume that our discrete system is holonomic and scleronomous, and that its configuration at any instant of time is described by a set of proper generalized coordinates q_i, $i=1,2,\ldots,n$ in the sense of Chapter 2. As a matter of convenience, and without loss of generality, we select the q_i in such a manner that

$$V_o = V(0,0,\ldots,0) = 0 \tag{3.1}$$

i.e., the vanishing of the potential energy coincides with the zero (reference) values of the generalized coordinates. In addition, we shall assume that our system is conservative, and that it is in a state of stable, static equilibrium when the generalized forces and generalized velocities vanish. The kinetic energy T vanishes for $\dot{q}_i \equiv 0$. Thus 2.39 of Chapter 2 provides us with the following condition for static equilibrium:

$$Q_i = \left(\frac{\partial V}{\partial q_i}\right)_o = 0, \quad i = 1,2,\ldots,n. \tag{3.2}$$

The equilibrium position characterized by (3.2) is stable if and only if V has a minimum at $q_i=0$. A sufficient condition which insures such a minimum is provided by (see, for instance, W.F. Osgood, Advanced Calculus, The Macmillan Co., New York, 1925).

$$k_{11}>0, \quad \begin{vmatrix} k_{11} & k_{12} \\ k_{21} & k_{22} \end{vmatrix} >0, \ldots, \begin{vmatrix} k_{11} & k_{12} & \cdots & k_{1n} \\ k_{21} & k_{22} & \cdots & k_{2n} \\ \vdots & \vdots & & \vdots \\ k_{n1} & k_{n2} & \cdots & k_{nn} \end{vmatrix} >0 \tag{3.3}$$

where

$$k_{ij} = k_{ji} = \left(\frac{\partial^2 V}{\partial q_i \partial q_j}\right)_o \tag{3.4}$$

If we now expand the potential energy function V in a Taylor series about the position of stable equilibrium, we obtain

$$V = V_o + \sum_i^n \left(\frac{\partial V}{\partial q_i}\right)_o q_i + \frac{1}{2} \sum_{i,j}^n \left(\frac{\partial^2 V}{\partial q_i \partial q_j}\right)_o q_i q_j + O(q_i q_j q_k) \tag{3.5}$$

If we neglect $O(q_i q_j q_k)$ in the generalized coordinates, then in view of (3.1), (3.2) and (3.5), we obtain

$$V = \frac{1}{2} \sum_{i,j}^n k_{ij} q_i q_j \tag{3.6}$$

Equation (3.6) characterizes the potential energy as a quadratic form in the generalized coordinates which is non-negative for all real values of the variables q_i. Hence it is a positive, definite quadratic form.

With reference to (2.27), the kinetic energy, expressed in terms of generalized coordinates, is given by

$$T = \frac{1}{2} \sum_{i,j}^n t_{ij} \dot{q}_i \dot{q}_j \tag{3.7}$$

where

$$t_{ij} = t_{ji} = t_{ij}(q_1, q_2, \ldots, q_n) \tag{3.8}$$

If we expand t_{ij} in a Taylor series about the equilibrium position,

$$t_{ij}(q_1,q_2,\ldots,q_n) = t_{ij}(0,0,\ldots,0) + \sum_{k=1}^{n} \left(\frac{\partial t_{ij}}{\partial q_k}\right)_0 q_k + O(q_k q_\ell) \qquad (3.9)$$

To obtain an approximation for T which is of the same order as that obtained for V in (3.6), we must retain only the first term of the expansion in (3.9). For notational reasons, we also set

$$t_{ij}(0,0,\ldots,0) = m_{ij} \qquad (3.10)$$

and in view of (3.8),

$$m_{ij} = m_{ji} \qquad (3.11)$$

If we now drop terms $O(q_k)$ in (3.9) and utilize (3.11), then (3.7) becomes

$$T = \frac{1}{2} \sum_{i,j}^{n} m_{ij}\dot{q}_i\dot{q}_j \qquad (3.12)$$

where the m_{ij} are n^2 constants. With reference to its original definition (Section 3 of Chapter 2), the kinetic energy is necessarily a non-negative quantity. Consequently (3.12) is a positive, definite quadratic form in the generalized velocities $\dot{q}_i$, and the m_{ij} satisfy n inequalities of the form (3.3).

If we now substitute (3.6) and (3.12) into Lagrange's equations (2.39) we obtain the linearized equations of motion

$$\sum_{i}^{n} (m_{ij}\ddot{q}_i + k_{ij}q_i) = Q_j;\ j=1,2,\ldots,n \qquad (3.13)$$

Further consideration of (3.13) will be divided into two parts: free vibrations and forced motion.

3.2 FREE VIBRATIONS

By the term "free vibrations" we mean the possible oscillations of our conservative system about its position of stable equilibrium in the absence of external forces, In this case $Q_i \equiv 0$ in (3.13), and our study of free vibrations begins with

$$\sum_i^n (m_{ij}\ddot{q}_i + k_{ij}q_i) = 0; \; j=1,2,\ldots,n \tag{3.14}$$

Equation (3.14) constitutes a system of n simultaneous, linear, ordinary differential equations with constant coefficients. To obtain a solution, we assume harmonic vibrations

$$q_i(t) = a_i \begin{matrix}\cos \omega t\\ \sin \omega t\end{matrix}; \; i=1,2,\ldots,n \tag{3.15}$$

where the amplitudes a_i and the (circular) frequency ω are constants. Substitution of (3.15) into (3.14) results in

$$\sum_i^n (k_{ij}-\omega^2 m_{ij})a_i = 0; \; j=1,2,\ldots,n. \tag{3.16a}$$

or, in the notation of matrix algebra,

$$\begin{bmatrix} k_{11}-\omega^2 m_{11} & k_{12}-\omega^2 m_{12} & \cdots\cdots & k_{1n}-\omega^2 m_{1n} \\ k_{21}-\omega^2 m_{21} & k_{22}-\omega^2 m_{22} & \cdots\cdots & k_{2n}-\omega^2 m_{2n} \\ \vdots & \vdots & & \vdots \\ k_{n1}-\omega^2 m_{n1} & k_{n2}-\omega^2 m_{n2} & \cdots\cdots & k_{nn}-\omega^2 m_{nn} \end{bmatrix} \begin{bmatrix} a_1 \\ a_2 \\ \vdots \\ a_n \end{bmatrix} = 0 \tag{3.16b}$$

Equations (3.16) pose the classical eigenvalue problem: Determine the eigenvalues ω^2 for which a non-trivial solution exists. The solution (a_i; i=1,2,...,n) corresponding to a particular eigenvalue is called an eigenvector. Each eigenvector characterizes what is commonly called a normal mode of vibration and is therefore also referred to as a modal vector. Equations (3.16) are a homogeneous system of n linear algebraic equations in the n unknown modal vector components a_i. For a non-trivial solution we require that the determinant of the (square) coefficient matrix in (3.16b) vanish, i.e.,

$$|k_{ij}-\omega^2 m_{ij}| = 0; \; i=1,2,\ldots,n; \; j=1,2,\ldots,n \tag{3.17}$$

Upon expansion of (3.17), we obtain a polynomial equation of degree n in the unknown ω^2. A solution of this equation will result in n values of the square of the natural frequency ω_r^2, r=1,2,...,n. In non-degenerate cases the n values of ω_r^2 will be distinct, special considerations being required when this is not the case. For each value of ω_r^2, there exists a modal vector (a_{ir}; i=1,2,...,n) obtainable from (3.16). Because (3.16) is homogeneous, the a_{ir} cannot be uniquely defined without the aid of additional relations. Equations (3.16) only fix their ratios, i.e., only the orientation of the r modal vectors in the n-dimensional space is defined at this point. It is customary, and subsequently convenient, to adjoin the (arbitrary) normalization conditions

$$\sum_{i,j}^{n} m_{ij} a_{ir} a_{jr} = 1; \; r = 1,2,\ldots,n. \tag{3.18}$$

to equations (3.16). Equation (3.18) fixes the length of each of the r modal vectors. If we divide both sides of (3.18) by the square of the k'th component of the r'th modal vector a_{kr}, and then solve for a_{kr} we obtain

$$a_{kr} = \pm\left[\sum_{i,j}^{n} m_{ij} \left(\frac{a_{ir}}{a_{kr}}\right)\left(\frac{a_{jr}}{a_{kr}}\right)\right]^{-\frac{1}{2}} \tag{3.19}$$

Equation (3.19) together with the preceding discussion indicate that the r modal vectors are now defined with respect to magnitude and orientation, but that their sense is still ambiguous. However, if consistency is maintained (we choose either plus or minus in (3.19)), the assumed sense of the vector will in no way influence the validity of the subsequent analysis.

We shall now show that the frequencies ω_r^2 and the associated modal vector components a_{ir} are real. Consider the possibility of a complex frequency ω_r^2. With reference to (3.16a) we have

$$\omega_r^2 \sum_{i}^{n} m_{ij} a_{ir} = \sum_{i}^{n} k_{ij} a_{ir} \tag{3.20a}$$

If we take the complex conjugate of (3.20a), and interchange the indices i and j, we obtain

$$(\omega_r^2)^* \sum_j^n m_{ji}(a_{jr})^* = \sum_j^n k_{ji}(a_{jr})^* \tag{3.20b}$$

where the star, in this instance, indicates the operation of taking the conjugate complex quantity.

Multiply (3.20a) by $(a_{jr})^*$ and sum over the index j; similarly, multiply (3.20b) by a_{ir} and sum over the index i. If we take the difference of the sums of the products so obtained, and also observe (3.4) and (3.11), we obtain

$$\left[\omega_r^2 - \left(\omega_r^2\right)^*\right] \sum_{i,j}^n m_{ij} a_{ir}(a_{jr})^* = 0 \tag{3.21}$$

Let $a_{ir} = \alpha_{ir} + i\beta_{ir}$, where α_{ir} and β_{ir} are real. Then

$$\sum_{i,j}^n m_{ij} a_{ir}(a_{jr})^* = \sum_{i,j}^n m_{ij}(\alpha_{ir}+i\beta_{ir})(\alpha_{jr}-i\beta_{jr}) =$$

$$\sum_{i,j}^n m_{ij}\alpha_{ir}\alpha_{jr} + \sum_{i,j}^n m_{ij}\beta_{ir}\beta_{jr} + i \sum_{i,j}^n (m_{ij}\alpha_{jr}\beta_{ir} - m_{ij}\alpha_{ir}\beta_{jr})$$

By an interchange of indices, and in view of (3.11),

$$\sum_{i,j}^n m_{ij}\alpha_{jr}\beta_{ir} = \sum_{i,j}^n m_{ji}\alpha_{ir}\beta_{jr} = \sum_{i,j}^n m_{ij}\alpha_{ir}\beta_{jr}$$

and therefore

$$\sum_{i,j}^n m_{ij} a_{ir}(a_{jr})^* = \sum_{i,j}^n m_{ij}\alpha_{ir}\alpha_{jr} + \sum_{i,j}^n m_{ij}\beta_{ir}\beta_{jr} \tag{3.22}$$

Each of the two sums on the right-hand side of (3.22) is twice the kinetic energy where the velocities $\dot{q}_i$ have the values α_{ir} and β_{ir}, respectively. The kinetic energy is a positive, definite quadratic function in the generalized velocities $\dot{q}_i$, hence it is non-negative. Moreover, we require a non-trivial solution of (3.16), one for which not all of the a_{ir} vanish. Consequently the left-hand side of (3.22) is positive, and with reference to (3.21), this requires that $\omega_r^2 = (\omega_r^2)^*$. We thus conclude that our original supposition was erroneous, and that the ω_r^2 are real. If we multiply (3.20a) by a_{jr}, sum over j, and solve for ω_r^2, we obtain

$$\omega_r^2 = \frac{\sum_{i,j}^{n} k_{ij} a_{ir} a_{jr}}{\sum_{i,j}^{n} m_{ij} a_{ir} a_{jr}} \tag{3.23}$$

In view of (3.18), the denominator in (3.23) is equal to one. The numerator in (3.23) is twice the potential energy where the generalized coordinates are equal to a_{ir}. Moreover, not all a_{ir} vanish. Since the potential energy is a positive definite form in the generalized displacements, we conclude that the numerator of (3.23) is positive. Thus we have shown that the square of the frequency ω_r^2 is a positive quantity.

We can now readily show that the modal vector components a_{ir} are real. Since the ω_r^2 have been shown to be real, all elements of the square matrix in (3.16b) are real. The n-1 ratios of the a_{ir} obtained from (3.16) will be real, because the only algebraic operations used in their solution are addition, subtraction, multiplication, and division. With reference to (3.19), a_{kr} is necessarily real because the sum on the right-hand side, as noted earlier, is a positive definite form. Hence we conclude that all a_{ir}; i,r=1,2,...,n are real.

We shall now show that the r modal vectors are orthogonal in an n-dimensional space. If we replace r by s, and interchange i and j in (3.20b), we obtain

$$\omega_s^2 \sum_{j}^{n} m_{ji} a_{js} = \sum_{j}^{n} k_{ji} a_{js} \tag{3.24}$$

Multiply (3.20a) by a_{js} and sum over j. Similarly, multiply (3.24) by a_{ir} and sum over i. We thus obtain

$$\omega_r^2 \sum_{i,j}^{n} m_{ij} a_{ir} a_{js} = \sum_{i,j}^{n} k_{ij} a_{ir} a_{js} \tag{3.25a}$$

$$\omega_s^2 \sum_{i,j}^{n} m_{ji} a_{ir} a_{js} = \sum_{i,j}^{n} k_{ji} a_{ir} a_{js} \tag{3.25b}$$

We now subtract (3.25b) from (3.25a). In view of (3.4) and (3.11), we obtain

$$(\omega_r^2-\omega_s^2)\sum_{i,j}^{n} m_{ij}a_{ir}a_{js} = 0 \tag{3.26}$$

If $\omega_r^2 \neq \omega_s^2$, we obtain the orthogonality condition

$$\sum_{i,j}^{n} m_{ij}a_{ir}a_{js} = 0 \tag{3.27}$$

It is convenient to combine (3.18) and (3.27) into a single orthonormality condition

$$\sum_{i,j}^{n} m_{ij}a_{ir}a_{js} = \delta_{rs} \tag{3.28}$$

where $\delta_{rs}=1$ for $r=s$, and $\delta_{rs}=0$ for $r\neq s$, is the Kronecker delta symbol.

Because of the linearity of the system (3.14), and in view of the assumed solution (3.15), it is now clear that the complete solution of (3.14) is given by

$$\left.\begin{aligned} q_i(t) &= \sum_{r=1}^{n} a_{ir}(A_r\cos\,\omega_r t + B_r\,\sin\,\omega_r t) \\ & \qquad\qquad i = 1,2,\ldots,n \end{aligned}\right\} \tag{3.29}$$

The 2n constants A_r, B_r are determined by application of the 2n initial conditions

$$q_i(0) = q_{i_o}\,,\quad \dot{q}_i(0) = \dot{q}_{i_o} \tag{3.30}$$

Substitution of (3.29) into (3.30) results in

$$q_{i_o} = \sum_{r}^{n} a_{ir}A_r \tag{3.31a}$$

$$\dot{q}_{i_o} = \sum_{r}^{n} \omega_r a_{ir}B_r \tag{3.31b}$$

To determine A_r, multiply both sides of (3.31a) by $m_{ij}a_{js}$ and sum over i and j:

$$\sum_{i,j}^{n} q_{i_0} m_{ij} a_{js} = \sum_{r}^{n} A_r \sum_{i,j}^{n} m_{ij} a_{ir} a_{js} = \sum_{r}^{n} A_r \delta_{rs} = A_s$$

Therefore

$$A_r = \sum_{i,j}^{n} q_{i_0} m_{ij} a_{jr} \tag{3.32a}$$

and in an entirely analogous manner, it can be shown that

$$\omega_r B_r = \sum_{i,j}^{n} \dot{q}_{i_0} m_{ij} a_{jr} \tag{3.32b}$$

Thus the constants A_r and B_r, r=1,2,...,n are fully determined by (3.32). This completes the problem of free vibrations, and we can now proceed to the study of forced motion.

3.3 NORMAL COORDINATES

Inspection of the solution (3.29) of the free vibration problem reveals that the motion resulting from initial conditions (3.30) is a superposition of n simple harmonic motions, each with a different frequency. The phase of any one simple oscillation of a particular coordinate depends on initial conditions, but the ratios of the amplitudes of the different coordinates for a particular frequency are determined by the physical characteristics of the system, i.e., by its inertia and stiffness parameters. If $\omega_r/\omega_1 = K_r$ for r=2,3,...,n and some positive integers K_r, then the motion described by (3.29) is periodic. However, this is generally not the case.

There exists a special set of coordinates which decouples the motion, i.e., one for which each generalized coordinate responds in simple harmonic motion with a single frequency ω_r. Such coordinates are called normal (or principal) coordinates, and the (linear) transformation between our present generalized coordinates q_i and the normal coordinates ξ_i is given by

$$q_i = \sum_{r}^{n} a_{ir} \xi_r, \quad i = 1,2,\ldots,n. \tag{3.33}$$

where the n^2 coefficients a_{ir} of the transformation are the components of the modal vectors. To invert (3.33), multiply it by $m_{ij}a_{js}$ and sum over i and j. We obtain

$$\sum_{i,j}^{n} q_i m_{ij} a_{js} = \sum_{r}^{n} \xi_r \sum_{i,j}^{n} m_{ij} a_{ir} a_{js} = \sum_{r}^{n} \xi_r \delta_{rs} = \xi_s$$

where we have used (3.33) and (3.28). Thus we may express the normal coordinates as a linear function of the original generalized coordinates by

$$\xi_r = \sum_{i}^{n} q_i \sum_{j}^{n} m_{ij} a_{jr} \tag{3.34}$$

To express the kinetic energy in terms of the normal coordinates, we substitute (3.34) into (3.12):

$$T = \frac{1}{2}\sum_{i,j}^{n} m_{ij}\dot{q}_i\dot{q}_j = \frac{1}{2}\sum_{i,j}^{n}\sum_{r,s}^{n} m_{ij}a_{ir}a_{js}\dot{\xi}_r\dot{\xi}_s = \frac{1}{2}\sum_{r,s}^{n}\dot{\xi}_r\dot{\xi}_s\sum_{i,j}^{n} m_{ij}a_{ir}a_{js}$$

$$= \frac{1}{2}\sum_{r,s}^{n}\dot{\xi}_r\dot{\xi}_s\delta_{rs} = \frac{1}{2}\sum_{r}^{n}\dot{\xi}_r^2 \tag{3.35}$$

where we have used (3.28). To express the potential energy in terms of the normal coordinates, we substitute (3.34) into (3.36):

$$V = \frac{1}{2}\sum_{i,j}^{n} k_{ij}q_iq_j = \frac{1}{2}\sum_{r,s}^{n}\xi_r\xi_s\sum_{i,j}^{n} k_{ij}a_{ir}a_{js} = \frac{1}{2}\sum_{r,s}^{n}\xi_r\xi_s\omega_r^2\delta_{rs} = \frac{1}{2}\sum_{r}^{n}\omega_r^2\xi_r^2 \tag{3.36}$$

where we have used (3.25a) and (3.28). It is interesting to note that cross-products of velocities and coordinates disappear from the kinetic and potential energy functions, respectively, when these are expressed in normal coordinates, as in (3.35) and (3.36). The Lagrangian function when expressed in normal coordinates, assumes the form

$$L = T-V = \frac{1}{2}\sum_{i}^{n}(\dot{\xi}_i^2 - \omega_i^2\xi_i^2) \tag{3.37}$$

The applicable form of Lagrange's equations is (see (2.41))

$$\frac{d}{dt}\left(\frac{\partial L}{\partial \dot{\xi}_r}\right) - \frac{\partial L}{\partial \xi_r} = 0,\ r=1,2,\ldots,n. \tag{3.38}$$

and upon substitution of (3.37) into (3.38) we obtain

$$\ddot{\xi}_r + \omega_r^2 \xi_r = 0 \tag{3.39}$$

The solution of (3.39) is

$$\xi_r(t) = A_r \cos \omega_r t + B_r \sin \omega_r t \tag{3.40}$$

where the constants A_r and B_r can be determined from initial conditions $\xi_r(0)$ and $\dot{\xi}_r(0)$.

With reference to (3.39) and (3.40), it is now clear that when our mechanical system is described by normal coordinates, each normal coordinate acts as an independent linear oscillator (see Chapter 1). Additional physical insight can be obtained from the following considerations. Let us assume that we encounter a motion wherein only one normal coordinate ξ_k is excited. Then, with reference to (3.33) and (3.40) our system response is characterized by

$$\left.\begin{aligned} q_i = a_{ik}\xi_k = a_{ik}(A_k \cos \omega_k t + B_k \sin \omega_k t) \\ i = 1,2,\ldots,n. \end{aligned}\right\} \tag{3.41}$$

Equations (3.41) show that each generalized displacement coordinate q_i is proportional to the components of the k'th modal vector a_{ik}; $i=1,2,\ldots,n$. The motion characterized by (3.41) is called the normal mode oscillation corresponding to the frequency ω_k, i.e., the system responds in its k'th normal mode of oscillation. It is for this reason that the set of modal vector components a_{ik}; $i=1,2,\ldots,n$, is referred to as the k'th normal mode.

3.4 FORCED MOTION

We now return to the complete equations of motion (3.13) which include the applied forces Q_j. If we transform (3.13) to normal coordinates defined by (3.33), we obtain,

$$\sum_{i,r}^{n} m_{ij} a_{ir} \ddot{\xi}_r + \sum_r^n \xi_r \sum_i^n k_{ij} a_{ir} = Q_j \tag{3.42}$$

Upon utilization of (3.20a), equation (3.42) simplifies to

$$\sum_{i,r}^{n} m_{ij} a_{ir} (\ddot{\xi}_r + \omega_r^2 \xi_r) = Q_j \tag{3.43}$$

If we multiply (3.43) by a_{js} and sum over j, we obtain

$$\sum_r^n (\ddot{\xi}_r + \omega_r^2 \xi_r) \sum_{i,j}^n m_{ij} a_{ir} a_{js} = \sum_j^n Q_j a_{js} \tag{3.44}$$

and upon application of (3.28), equation (3.44) reduces to

$$\ddot{\xi}_r + \omega_r^2 \xi_r = P_r; \; r=1,2,\ldots,n \tag{3.45}$$

where

$$P_r = \sum_j^n Q_j a_{jr} \tag{3.46}$$

As shown in Section 4 of Chapter 1, the solution of (3.45) is given by

$$\xi_r(t) = A_r \cos \omega_r t + B_r \sin \omega_r t + \frac{1}{\omega_r} \int_o^t P_r(\tau) \sin \omega_r (t-\tau) d\tau \qquad r = 1,2,\ldots,n. \tag{3.47}$$

It now becomes necessary to express the initial conditions on the normal coordinates ξ_r in terms of the initial conditions on the generalized coordinates q_i. With reference to (3.33) and (3.47) we have

$$q_i(0) = \sum_r^n a_{ir} \xi_r(0) = \sum_r^n a_{ir} A_r \tag{3.48}$$

If we multiply (3.48) by $m_{ij} a_{js}$ and sum over the indices i and j, we obtain with the aid of (3.28)

$$\sum_{i,j}^n q_i(0) m_{ij} a_{js} = \sum_r^n A_r \sum_{i,j}^n m_{ij} a_{ir} a_{js} = \sum_r^n A_r \delta_{rs} = A_s$$

Therefore

$$A_r = \sum_j^n a_{jr} \sum_i^n m_{ij} q_i(0) \tag{3.49a}$$

By an entirely analogous procedure it can be shown that

$$\omega_r B_r = \sum_j^n a_{jr} \sum_i^n m_{ij} \dot{q}_i(0) \tag{3.49b}$$

Equations (3.47) in conjunction with (3.46) constitute a complete solution of the forced motion problem expressed in normal coordinates

ξ_r. Transformation to the originally selected generalized coordinates q_i is readily accomplished with the aid of (3.33) and (3.49). We also note that the basic building blocks of the present method of solution are the modal vector components a_{ir} and associated frequencies ω_r obtained from a free vibration analysis of our system.

We may also utilize the natural frequencies ω_r and associated modal vectors a_{ir}, $i=1,2,\ldots,n$; $r=1,2,\ldots,n$, to obtain a static solution of the problem characterized by (3.13) with inertia terms deleted:

$$\sum_{i=1}^{n} k_{ij} q_i = Q_j \tag{3.50}$$

In the static case, $\ddot{\xi}_r=0$, and (3.45) reduces to

$$\xi_r = \frac{P_r}{\omega_r^2} \tag{3.51}$$

where P_r is given by (3.46). Upon substitution of (3.51) into (3.33), we obtain

$$q_i = \sum_{r}^{n} a_{ir} \frac{P_r}{\omega_r^2} \tag{3.52}$$

and further substitution of (3.46) into (3.52) results in

$$q_i = \sum_{r}^{n} \sum_{j}^{n} \frac{a_{ir} a_{jr}}{\omega_r^2} Q_j \tag{3.53}$$

which is the solution of the static problem characterized by (3.50) in terms of the free vibration characteristics of our mechanical system.

If we set $Q_j = \delta_{jk}$, then (3.53) reduces to

$$q_i \equiv q_{ik} \equiv \sum_{r}^{n} \frac{a_{ir} a_{kr}}{\omega_r^2} \tag{3.54}$$

The symbol q_{ik} in (3.54) denotes the generalized displacement q_i caused by a unit generalized force $Q_j=\delta_{jk}$, and in the linear theory of static structures, q_{ik} is called the influence coefficient. With reference to (3.54), we note that $q_{ik}=q_{ki}$, a result which is in accordance with the well-known reciprocal theorem (Maxwell) of the theory of structures.

EXERCISES

3.1 Consider the quadratic function

$$V = \frac{1}{2} k_{11}q_1q_1 + \frac{1}{2} k_{12}q_1q_2 + \frac{1}{2} k_{21}q_2q_1 + \frac{1}{2} k_{22}q_2q_2$$

where $k_{12}=k_{21}$ and q_1 and q_2 do not vanish simultaneously.

(a) Given that $k_{11}>0$, $k_{11}k_{22}-k_{12}^2>0$, show that $V>0$.

(b) Given that $V>0$, shown that $k_{11}>0$, $k_{11}k_{22}-k_{12}^2>0$.

3.2 Consider the vibratory motion

$$q_i(t) = a_{i1}A_1 \cos \omega_1 t + a_{i2}A_2 \cos 2\omega_1 t + a_{i3}A_3 \cos 3\omega_1 t$$

Show that $q_i(t)$ is periodic with respect to time with period $\frac{2\pi}{\omega_1}$.

3.3 Derive equation (3.49b).

3.4 Derive equation (3.32b).

Additional exercises can be found at the end of Chapter 4.

Chapter

4

SMALL OSCILLATIONS OF DISCRETE SYSTEMS (APPLICATIONS)

Practice is best served by a good theory.

L. Boltzmann.

4.1 THE LINEAR OSCILLATOR IN A GRAVITATIONAL FIELD

We consider the vertical (rectilinear) motion of a mass of magnitude m suspended by a spring in a gravitational field as shown in Fig. 4.1. Let us denote the potential energy stored in the spring by $V_S = V_S(q)$, where $V_S(q)$ is an arbitrary analytic function of the vertical displacement q of the mass in a finite interval which includes the interior point q=0. Then

$$V_S(q) = V_S(0)+qV_S'(0) + \frac{1}{2} q^2 V_S''(0)+O(q^3) \tag{4.1}$$

where primes denote differentiation with respect to q. The potential energy of the mass m is

$$V_G = c - mgq \tag{4.2}$$

where c is a constant. Hence the total potential energy function is

$$V = V_S + V_G \tag{4.3}$$

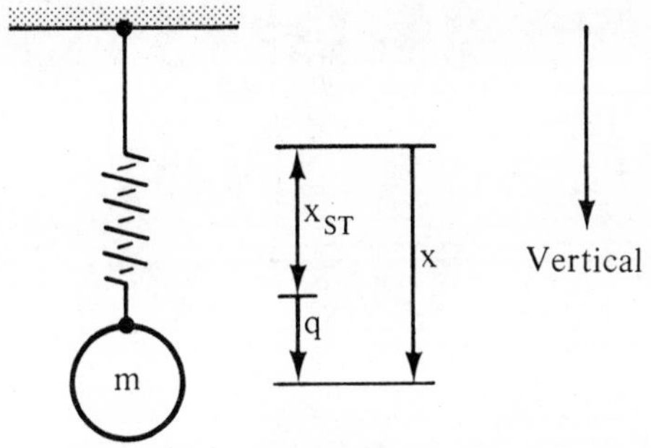

Figure 4.1

In line with the conditions stipulated by (3.1) and (3.2), we require

$$V(0) = 0 \tag{4.4a}$$

$$\left(\frac{dV}{dq}\right)_{q=0} = Q = 0 \tag{4.4b}$$

Application of (4.4) to (4.3) in conjunction with (4.1) and (4.2) results in

$$V_S(0) = -c, \quad V_S'(0) = mg$$

and if we set $V_S''(0)=k$, V(q) assumes the form

$$V = \frac{1}{2} kq^2 + O(q^3) \tag{4.5}$$

To insure that the motion takes place about a position of stable equilibrium, we apply (3.3) to (4.5):

$$\left(\frac{d^2V}{dq^2}\right)_{q=0} = k>0.$$

Let the force which the mass exerts upon the spring be $-F_S$. Then

$$-F_S = \frac{dV_S}{dq} = mg + kq + O(q^2)$$

The force applied to the spring vanishes when $F_S=0$, i.e., when q is a root of the equation

$$mg + kq + O(q^2) = 0$$

We shall call this root $q = -x_{ST}$. Thus, with reference to Fig. 4.1, we define $x=x_{ST}+q$, and therefore

$$k = \left[\frac{d(-F_S)}{dq}\right]_{q=0} = \left[\frac{d(-F_S)}{dx} \cdot \frac{dx}{dq}\right]_{x=x_{ST}} = \left[\frac{d(-F_S)}{dx}\right]_{x=x_{ST}} \quad (4.6)$$

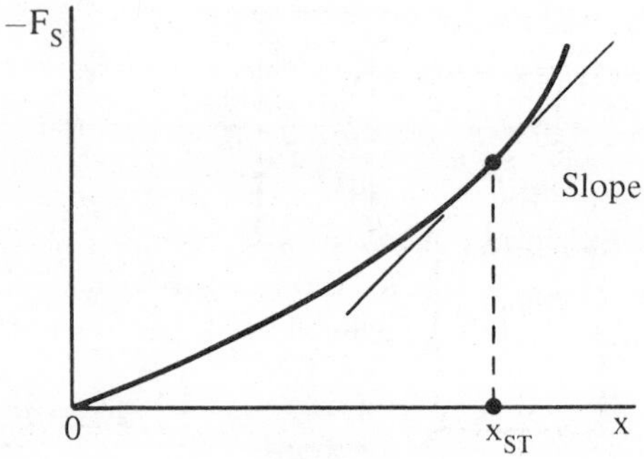

Figure 4.2

The force-deflection curve of our (non-linear) spring is shown in Fig. 4.2. With reference to that figure and to (4.6) it is now clear that k is the slope of the (non-linear) force-deflection curve measured at the point $x=x_{ST}$. Moreover, with reference to Fig. 4.1, x_{ST} is seen to be the vertical displacement of the mass from a position of vanishing spring force to the position of stable, static equilibrium (when $Q\equiv 0$). Thus q is the displacement of the mass measured from its static equilibrium position $x=x_{ST}$ or $q=0$. In accordance with (3.6) we now drop terms of $O(q^3)$ in 4.5 and obtain $V = \frac{1}{2} kq^2$. The kinetic energy of the mass is $T = \frac{1}{2} m\dot{q}^2$, and the Lagrangian function is $L=T-V= \frac{1}{2} (m\dot{q}^2-kq^2)$. Upon substitution in Lagrange's equation (2.39), we obtain the linearized equation of motion for sufficiently small q:

$$m\ddot{q} + kq = Q(t) \tag{4.7}$$

This is the equation of the linear oscillator treated in Chapter 1. We conclude that the effect of the gravitational field results in a simple shift of the reference about which the mass oscillates. In the case of free vibrations, $Q(t)=0$, and the mass will vibrate about its stable, static equilibrium position $x=x_{ST}$ or $q=0$. A complete solution of (4.7) for arbitrary forcing functions $Q(t)$ and arbitrary initial conditions $q(0)$ and $\dot{q}(0)$ is readily obtained by the techniques developed in Chapter 1.

4.2 FREE VIBRATIONS AND FORCED MOTION OF A TWO-DEGREE OF FREEDOM SYSTEM

Two particles with mass m_{11} and m_{22} are connected by two linearly elastic, mass-less springs with spring constants k_1 and k_2 as shown in Fig. 4.3. The forces in the springs vanish when $q_1=q_2=0$. The masses are forced to move rectinlinearly by frictionless constraints. We want to find the motion caused by an applied force $Q_2(t)$ for initial conditions $q_1(0)=q_2(0)=\dot{q}_1(0)=\dot{q}_2(0)=0$.

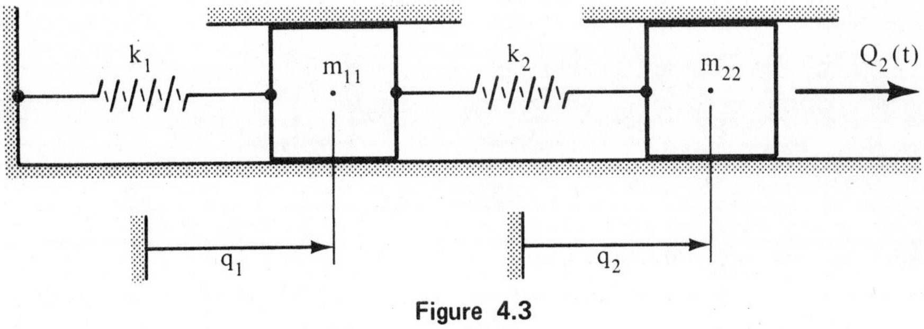

Figure 4.3

The expressions for kinetic and potential energy are

$$T = \frac{1}{2} m_{11}\dot{q}_1^2 + \frac{1}{2} m_{22}\dot{q}_2^2 \tag{4.8}$$

$$V = \frac{1}{2} k_1 q_1^2 + \frac{1}{2} k_2(q_2-q_1)^2 \tag{4.9}$$

To bring the present problem into the notational framework of Chapter 3, we set

$$k_1 = k_{11}-k_{22},\ k_2 = -k_{12} = -k_{21} = k_{22}$$

or

$$k_{11} = k_1+k_2 > 0,\ k_{22} = k_2 > 0,\ k_{12} = k_{21} = -k_2 < 0$$

so that (4.9) becomes

$$V = \frac{1}{2} k_{11}q_1q_1 + \frac{1}{2} k_{12}q_1q_2 + \frac{1}{2} k_{21}q_2q_1 + \frac{1}{2} k_{22}q_2q_2 \tag{4.11}$$

Upon substitution of (4.8) and (4.11) into Lagrange's equations (2.39), we obtain the equations of motion

$$\begin{aligned} m_{11}\ddot{q}_1 + k_{11}q_1 + k_{12}q_2 &= 0 \\ m_{22}\ddot{q}_2 + k_{22}q_2 + k_{21}q_1 &= Q_2 \end{aligned} \tag{4.12}$$

(a) Free Vibrations.

We assume a solution of (4.12) in the form

$$\begin{aligned} q_1(t) &= Ca_1 \cos \omega t \\ q_2(t) &= Ca_2 \cos \omega t \end{aligned} \tag{4.13}$$

and substitute (4.13) into (4.12) (with $Q_2 \equiv 0$). The result is

$$\begin{aligned} (k_{11}-\omega^2 m_{11})a_1 + k_{12}a_2 &= 0 \\ k_{12}a_1 + (k_{22}-\omega^2 m_{22})a_2 &= 0 \end{aligned} \tag{4.14}$$

In order to obtain a non-trivial solution for the modal vector components a_1 and a_2, we require that

$$\begin{vmatrix} (k_{11}-\omega^2 m_{11}) & k_{12} \\ k_{12} & (k_{22}-\omega^2 m_{22}) \end{vmatrix} = 0 \tag{4.15}$$

Upon expanding (4.15), we obtain

$$\omega^4 - (\omega_a^2+\omega_b^2)\omega^2 + (\omega_a^2\omega_b^2-\omega_{ab}^4) = 0 \tag{4.16}$$

where

$$\omega_a^2 = \frac{k_{11}}{m_{11}} = \frac{k_1+k_2}{m_{11}} > 0$$

$$\omega_b^2 = \frac{k_{22}}{m_{22}} = \frac{k_2}{m_{22}} > 0 \qquad (4.17)$$

$$\omega_{ab}^2 = \frac{k_{12}}{\sqrt{m_{11}m_{22}}} = -\frac{k_2}{\sqrt{m_{11}m_{22}}} < 0$$

The solution of the biquadratic equation (4.16) is

$$\omega_1^2 = \frac{\omega_a^2+\omega_b^2}{2} - \sqrt{\left(\frac{\omega_a^2-\omega_b^2}{2}\right)^2 + (\omega_{ab}^2)^2} \qquad (4.18a)$$

$$\omega_2^2 = \frac{\omega_a^2+\omega_b^2}{2} + \sqrt{\left(\frac{\omega_a^2-\omega_b^2}{2}\right)^2 + (\omega_{ab}^2)^2} \qquad (4.18b)$$

We also note that

$$\omega_1^2 + \omega_2^2 = \omega_a^2 + \omega_b^2$$

$$\omega_1^2\omega_2^2 = \omega_a^2\omega_b^2 - \omega_{ab}^4 \qquad (4.19)$$

For the present problem we assume that $\omega_a^2 < \omega_b^2$. Then

$$\omega_1^2 < \omega_a^2 < \omega_b^2 < \omega_2^2 \qquad (4.20)$$

A convenient way to visualize the relations (4.18) through (4.20) is made possible by the Mohr circle construction shown in Fig. 4.4. The ratios of the modal vector components are obtained from (4.14):

$$\frac{a_{1r}}{a_{2r}} = -\frac{k_{12}}{k_{11}-\omega_r^2 m_{11}}, \quad r=1,2 \qquad (4.21)$$

or, utilizing the notations of (4.17),

$$\frac{\sqrt{m_{11}}a_{1r}}{\sqrt{m_{22}}a_{2r}} = \frac{\omega_{ab}^2}{\omega_r^2-\omega_a^2}; \quad r=1,2 \qquad (4.22)$$

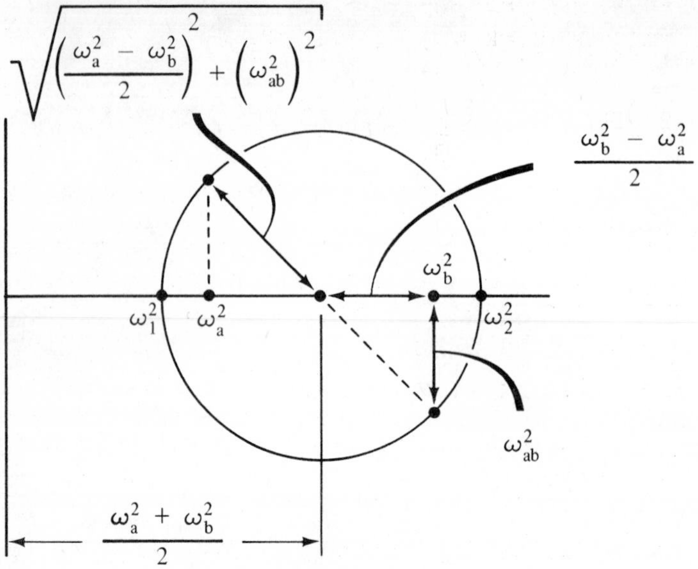

Figure 4.4

In the present case we have $\omega_{ab}^2<0$, $\omega_1^2-\omega_a^2<0$, and $\omega_2^2-\omega_a^2>0$. Therefore

$$\frac{a_{11}}{a_{21}} > 0 \text{ and } \frac{a_{12}}{a_{22}} < 0 \tag{4.23}$$

i.e., a_{11} and a_{21} must have like signs, while a_{12} and a_{22} must have opposite signs. We now invoke the condition of orthogonality (3.27) which in this particular instance reduces to

$$m_{11}a_{1r}^2 + m_{22}a_{2r}^2 = 1, \; r=1,2 \tag{4.24}$$

Equations (4.22) and (4.24) are now solved for the squares of the modal vector components. With the aid of (4.19), we obtain

$$\begin{aligned} m_{11}a_{11}^2 &= \frac{\omega_b^2-\omega_1^2}{\omega_2^2-\omega_1^2}; \quad & m_{11}a_{12}^2 &= \frac{\omega_2^2-\omega_b^2}{\omega_2^2-\omega_1^2} \\ m_{22}a_{21}^2 &= \frac{\omega_a^2-\omega_1^2}{\omega_2^2-\omega_1^2}; \quad & m_{22}a_{22}^2 &= \frac{\omega_2^2-\omega_a^2}{\omega_2^2-\omega_1^2} \end{aligned} \tag{4.25}$$

Let us consider the special case of equal masses and equal springs. In this case $m_{11}=m_{22}=m$, $k_1=k_2=k$, $k_{11}=2k$, $k_{22}=k$, $k_{12}=k_{21}=-k$. Using (4.17) we have

$$\omega_a^2 = 2\,\frac{k}{m},\quad \omega_b^2 = \frac{k}{m},\quad \omega_{ab}^2 = -\,\frac{k}{m}$$

and from (4.18) we obtain

$$\omega_1^2 = \frac{3-\sqrt{5}}{2}\,\frac{k}{m} = 0.382\,\frac{k}{m};\quad \omega_1 = 0.618\sqrt{\frac{k}{m}}$$

$$\omega_2^2 = \frac{3+\sqrt{5}}{2}\,\frac{k}{m} = 2.618\,\frac{k}{m};\quad \omega_2 = 1.618\sqrt{\frac{k}{m}} \qquad (4.26)$$

$$\omega_2^2-\omega_1^2 = \sqrt{5}\,\frac{k}{m} = 2.236\,\frac{k}{m}$$

The squares of the modal vector components are obtained from (4.25):

$$ma_{11}^2 = \frac{5-\sqrt{5}}{10} = 0.276 \qquad ma_{12}^2 = \frac{5+\sqrt{5}}{10} = 0.724$$

$$ma_{21}^2 = \frac{5+\sqrt{5}}{10} = 0.724 \qquad ma_{22}^2 = \frac{5-\sqrt{5}}{10} = 0.276$$

Observing the relations (4.23), we obtain

$$\sqrt{m}\,a_{11} = 0.525, \qquad \sqrt{m}\,a_{12} = 0.851$$

$$\sqrt{m}\,a_{21} = 0.851, \qquad \sqrt{m}\,a_{22} = -0.525 \qquad (4.27)$$

(b) Forced Motion.

We now return to the general case. With reference to (3.45) and (3.46), the equations of forced motion in normal coordinates are

$$\ddot{\xi}_1 + \omega_1^2\xi_1 = P_1 = a_{21}Q_2$$

$$\ddot{\xi}_2 + \omega_2^2\xi_2 = P_2 = a_{22}Q_2 \qquad (4.28)$$

and the relations between our generalized and normal coordinates are (see (3.33))

$$q_1 = a_{11}\xi_1 + a_{12}\xi_2$$

$$q_2 = a_{21}\xi_1 + a_{22}\xi_2 \qquad (4.29)$$

Also, since $q_r(0)=\dot{q}_r(0)=0$, $r=1,2$, the appropriate initial conditions in normal coordinates are simply

$$\xi_r(0) = \dot{\xi}_r(0) = 0, \quad r=1,2 \tag{4.30}$$

Let us first assume that the load $Q_2(t)$ is applied suddenly and is thereafter maintained, i.e.,

$$Q_2(t) = QH(t) \tag{4.31}$$

where H(t) is the unit step function applied at t=0, and Q is the magnitude of the constant force. In this case solutions of (4.28) subject to (4.30) are (see 1.34)

$$\begin{aligned} \xi_1(t) &= \frac{a_{21}}{\omega_1^2} Q(1-\cos \omega_1 t) \\ \xi_2(t) &= \frac{a_{22}}{\omega_2^2} Q(1-\cos \omega_2 t) \end{aligned} \tag{4.32}$$

and upon substitution of (4.32) into (4.29) and a certain amount of simplification, we obtain the desired forced motion solution of our problem:

$$\frac{q_1(t)}{(q_1)_{st}} = 1 + \frac{\omega_1^2 \cos \omega_2 t - \omega_2^2 \cos \omega_1 t}{\omega_2^2 - \omega_1^2} \tag{4.33a}$$

$$\frac{q_2(t)}{(q_2)_{st}} = 1 - \frac{\omega_2^2 \left(\omega_a^2-\omega_1^2\right)}{\omega_a^2 \left(\omega_2^2-\omega_1^2\right)} \cos \omega_1 t - \frac{\omega_1^2 \left(\omega_2^2-\omega_a^2\right)}{\omega_a^2 \left(\omega_2^2-\omega_1^2\right)} \cos \omega_2 t \tag{4.33b}$$

The quantities

$$(q_1)_{st} = \frac{Q}{k_1} \; ; \; (q_2)_{st} = Q \frac{k_1+k_2}{k_1 k_2} \tag{4.34}$$

are obtained by solving (4.12) with inertia terms deleted. Thus $(q_1)_{st}$ and $(q_2)_{st}$ characterize the static displacement of our system when subjected to the load $Q_2=Q$. The right-hand sides of (4.33) are the ratios of dynamical to static displacements, and when their magnitudes are maximized with respect to time t, they become dynamic amplification factors. Upper bounds for (4.33) can be determined (see J.S. Pistiner and H. Reismann, "Dynamic Amplification of a Two-Degree-of-Freedom

System," The Journal of Environmental Sciences, Vol. 3, No. 5, October 1960, pp. 4-8), and they are shown in Fig. 4.5 as a function of the system parameter ω_1/ω_2 where

$$\frac{\omega_1^2}{\omega_2^2} = \frac{1-\lambda}{1+\lambda} \tag{4.35a}$$

and where

$$\lambda = \sqrt{\left(\frac{\omega_a^2-\omega_b^2}{\omega_a^2+\omega_b^2}\right)^2 + \left(\frac{2\omega_{ab}^2}{\omega_a^2+\omega_b^2}\right)^2} \tag{4.35b}$$

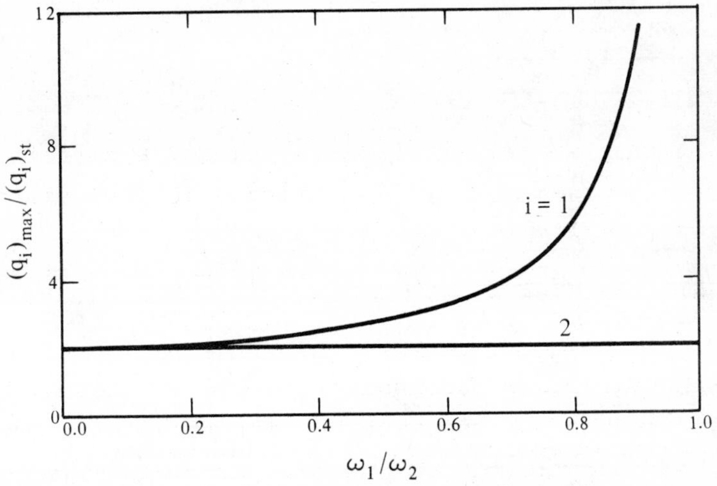

Figure 4.5

We note that the particle with mass m_1 has a maximum dynamic displacement which is always greater than twice its corresponding static displacement. Conversely, the particle with mass m_2 has a maximum dynamic displacement which is always less than or at most equal to twice its corresponding static displacement.

Let us next consider the steady state vibrations of our two-mass system under the action of a periodic forcing function

$$Q_2(t) = Q \cos \omega t \tag{4.36}$$

where Q is the force amplitude and ω is the forcing frequency.

Substituting (4.36) into (4.28), we obtain

$$\left.\begin{aligned} \ddot{\xi}_1 + \omega_1^2\xi_1 &= a_{21}\ Q\cos\omega t \\ \ddot{\xi}_2 + \omega_2^2\xi_2 &= a_{22}\ Q\cos\omega t \end{aligned}\right\} \tag{4.37}$$

and the steady state part of the solution of (4.37) are given by (see 1.44)

$$\left.\begin{aligned} \xi_1(t) &= \frac{a_{21}Q}{\omega_1^2-\omega^2}\cos\omega t \\ \xi_2(t) &= \frac{a_{22}Q}{\omega_2^2-\omega^2}\cos\omega t \end{aligned}\right\} \tag{4.38}$$

By substituting (4.38) into (4.29) and performing a certain amount of rearranging, we obtain the desired steady state part of the solution:

$$\frac{q_1(t)}{(q_1)_{st}} = \frac{\cos\omega t}{\left(1-\frac{\omega^2}{\omega_1^2}\right)\left(1-\frac{\omega^2}{\omega_2^2}\right)} \tag{4.39a}$$

$$\frac{q_2(t)}{(q_2)_{st}} = \frac{\left(1-\frac{\omega^2}{\omega_a^2}\right)\cos\omega t}{\left(1-\frac{\omega^2}{\omega_1^2}\right)\left(1-\frac{\omega^2}{\omega_2^2}\right)} \tag{4.39b}$$

where $(q_1)_{st}$ and $(q_2)_{st}$ are defined by (4.34). A plot of (4.39) is shown in Fig. 4.6. We note the two resonant frequencies $\omega=\omega_1$ and $\omega=\omega_2$.

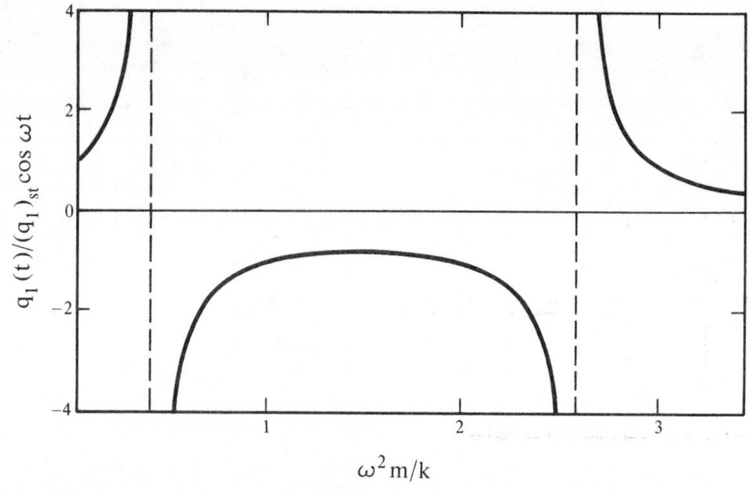

(a)

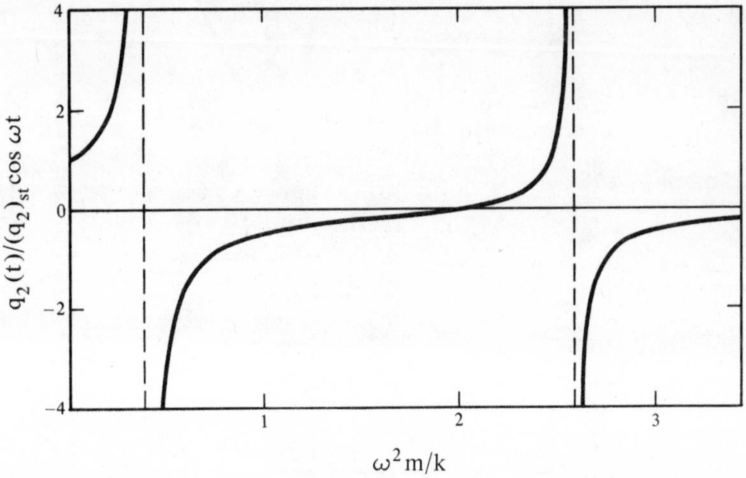

(b)

Figure 4.6

4.3 FREE VIBRATIONS AND FORCED MOTION OF AN ITERATED STRUCTURE

We shall consider the torsional motion of the iterated structure shown in Fig. 4.7. The collinear, massless shaft segments of length a with torsional spring constant c provide the connection between n rigid circular discs each with moment of inertia I about the z-axis. A time dependent, applied twisting moment $Q_i(t)$ acts on each disc. The internal twisting moment in any shaft segment is given by

$$M_i = c(\theta_{i+1} - \theta_i) \tag{4.40}$$

and expressions for kinetic and potential energy are

$$T = \frac{1}{2}\sum_i^n I\dot{\theta}_i^2, \qquad V = \frac{1}{2}\sum_i^n c(\theta_{i+1} - \theta_i)^2 \tag{4.41}$$

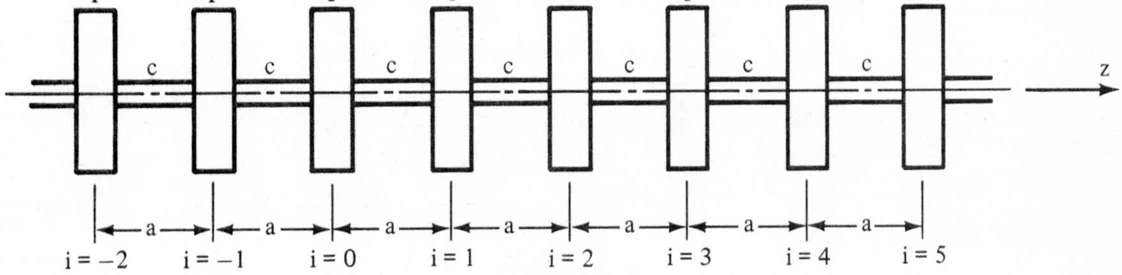

Figure 4.7

Substitution of (4.41) into Lagrange's equations (2.39) results in the equations of motion

$$I\ddot{\theta}_i - c\theta_{i+1} + 2c\theta_i - c\theta_{i-1} = Q_i \tag{4.42}$$

or, introducing (4.40),

$$I\ddot{\theta}_i + M_{i-1} - M_i = Q_i \tag{4.43}$$

Free Vibrations.

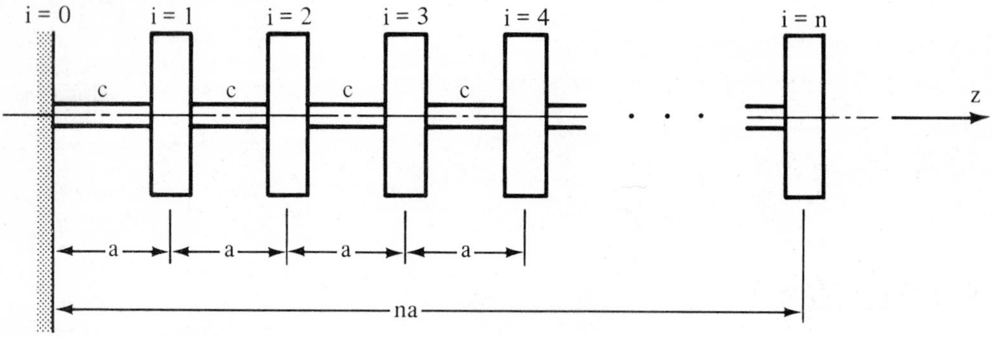

Figure 4.8

Let us consider the case of free vibrations for the iterated structure shown in Fig. 4.8. The shaft end at i=0 is fixed, and the disc at i=n is free. In this case we shall seek a solution of

$$I\ddot{\theta}_i - c\theta_{i+1} + 2c\theta_i - c\theta_{i-1} = 0; \qquad i=1,2,\ldots,n \tag{4.44}$$

subject to the boundary conditions

$$\theta_o = 0, \quad \theta_{n+1} - \theta_n = 0 \tag{4.45}$$

We assume harmonic vibrations of the type

$$\theta_i = Ca_i \cos \omega t \tag{4.46}$$

and upon substitution of (4.46) into (4.44) we obtain

$$a_{i+1} + (\Omega^2 - 2)a_i + a_{i-1} = 0 \tag{4.47}$$

where $\Omega^2 = \frac{I\omega^2}{c}$.

Equation (4.47) is a linear difference equation with constant coefficients. In many ways, it is similar to a linear differential equation with constant coefficients, and in line with this observation we assume a solution of (4.47) in the form

$$Ca_i = \sin i\lambda \tag{4.48}$$

where λ is an as yet undetermined parameter.
Upon substitution of (4.48) into (4.47), we obtain

$$\sin(i+1)\lambda + (\Omega^2-2)\sin i\lambda + \sin(i-1)\lambda = 0 \tag{4.49}$$

By the application of trigonometric identities, (4.49) reduces to

$$\Omega^2 = \frac{I\omega^2}{c} = 2(1-\cos\lambda) = 4\sin^2\frac{\lambda}{2} \tag{4.50}$$

or

$$\omega = 2\sqrt{\frac{c}{I}}\sin\frac{\lambda}{2} \tag{4.51}$$

The parameter λ can be determined by the application of boundary conditions (4.45). We note that the assumed solution satisfies $\theta_o=0$. To satisfy $\theta_{n+1}=\theta_n$ we must set

$$\sin(n+1)\lambda = \sin n\lambda \tag{4.52}$$

Application of trigonometric identities to (4.52) reveals that

$$\tan n\lambda = \frac{\sin\lambda}{1-\cos\lambda} = \frac{1}{\tan\frac{\lambda}{2}}$$

or

$$1-\tan\frac{\lambda}{2}\tan n\lambda = 0$$

Consequently

$$\frac{1}{\tan(n\lambda+\frac{\lambda}{2})} = \frac{1-\tan\frac{\lambda}{2}\tan n\lambda}{\tan\frac{\lambda}{2}+\tan n\lambda} = 0$$

and

$$\lambda(n+\frac{1}{2}) = (2r-1)\frac{\pi}{2}, \quad r=1,2,\ldots,n.$$

Therefore

$$\lambda_r = \frac{(2r-1)\pi}{(2n+1)} \tag{4.53a}$$

$$\omega_r = 2\sqrt{\frac{c}{I}}\,\sin\frac{(2r-1)\pi}{(2n+1)2} \tag{4.53b}$$

When r=1,2,...,n, (4.53) yields distinct roots. If r=n+1, $\lambda_r=\pi$ and $a_{n+1}=0$. For the cases r=n+2, n+3,..., etc., the values of ω_r, r=1,2,...,n are repeated. We thus conclude that there are n distinct frequencies ω_r. Moreover, it is seen that $\omega = 2\sqrt{c/I}$ is an upper bound for the frequencies of our system which, we note, is independent of n. The spectrum of natural frequencies for the case of 6 discs is shown in Fig. 4.9. When the number of discs n is increased, the frequencies move closer together and especially crowd together near the highest frequency.

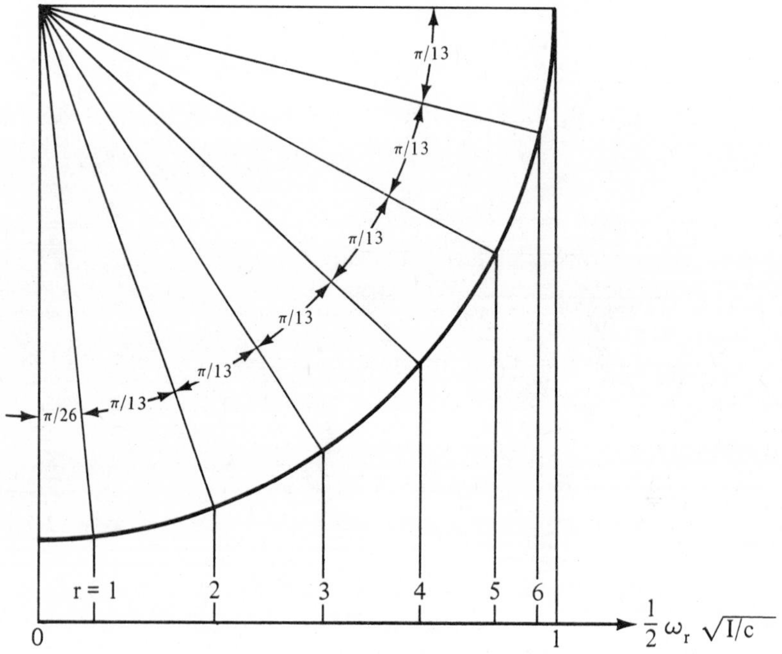

Figure 4.9

We shall now normalize the modal vector components in accordance with (3.18). In the present case it is required that

$$\sum_{i=1}^{n} Ia_{ir}^2 = 1 \tag{4.54}$$

With reference to (4.48), the component of the modal vector which corresponds to the i'th generalized coordinate vibrating in the r'th mode is

$$a_{ir} = \frac{A_r}{\sqrt{I}} \sin i\lambda_r, \quad \begin{array}{l} i = 1,2,\ldots,n \\ r = 1,2,\ldots,n \end{array} \tag{4.55}$$

Upon substitution of (4.55) into (4.54), we obtain

$$\frac{1}{A_r^2} = \sum_{i=1}^{n} \sin^2 i\lambda_r \tag{4.56}$$

The series in (4.56) can be summed (see E.P. Adams, Smithsonian Mathematical Formulae, The Smithsonian Institution, Washington, 1947, p. 82.) We have

$$\sum_{i=1}^{n} \sin^2 i\lambda_r = \frac{n}{2} - \frac{\cos(n+1)\lambda_r \cdot \sin n\lambda_r}{2 \sin \lambda_r} = \frac{2n \sin\lambda_r - \sin 2(n+1)\lambda_r}{4 \sin \lambda_r}$$

Consequently

$$A_r = 2\sqrt{\frac{\sin \lambda_r}{2n \sin \lambda_r - \sin 2(n+1)\lambda_r}} \tag{4.57}$$

and

$$\sqrt{I}\, a_{ir} = 2\sqrt{\frac{\sin \lambda_r}{2n \sin \lambda_r - \sin 2(n+1)\lambda_r}} \sin i\lambda_r \tag{4.58}$$

Let us apply these results to the special case of three discs (n=3). With the aid of (4.53a), (4.53b) and (4.58) we obtain the following values for frequencies and modal vector components:

$$\omega_1 = 0.445 \sqrt{\frac{c}{I}}$$

$$\sqrt{I}\, a_{11} = 0.328, \quad \sqrt{I}\, a_{21} = 0.591, \quad \sqrt{I}\, a_{31} = 0.736 \tag{4.59a}$$

$$\omega_2 = 1.247 \sqrt{\frac{c}{I}}$$

$$\sqrt{I}\, a_{12} = -0.736, \quad \sqrt{I}\, a_{22} = -0.328, \quad \sqrt{I}\, a_{32} = 0.591 \tag{4.59b}$$

$$\omega_3 = 1.801 \sqrt{\frac{c}{I}}$$

$$\sqrt{I}\, a_{13} = 0.591, \quad \sqrt{I}\, a_{23} = -0.736, \quad \sqrt{I}\, a_{33} = 0.328 \tag{4.59c}$$

The amplitudes of the three normal modes are shown in Fig. 4.10.

Forced Motion.

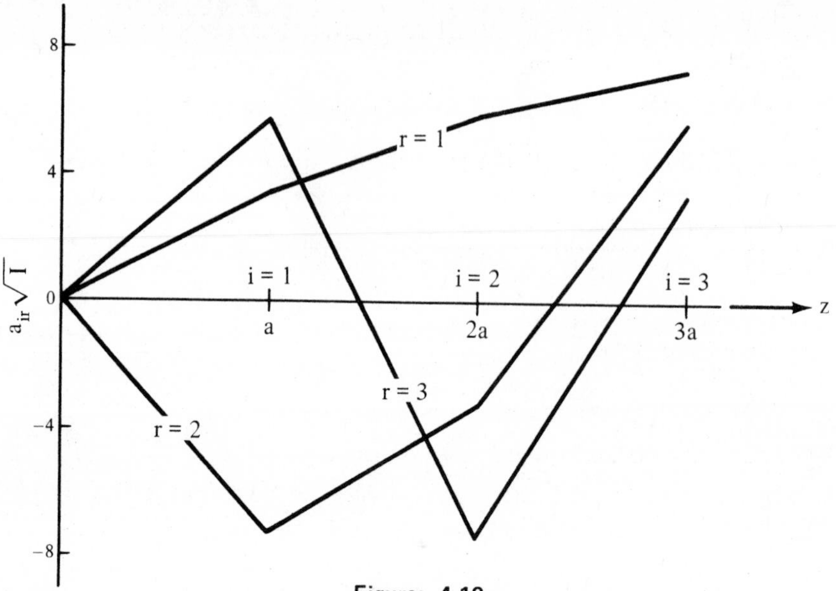

Figure: 4.10

Let us consider the response of the iterated structure to a torque $Q_n(t)$ applied to the disc i=n. In accordance with (3.45) and (3.46), the forced motion in normal coordinates is characterized by

$$\ddot{\xi}_r + \omega_r^2 \xi_r = P_r = \sum_j^n Q_j a_{jr} = a_{nr} Q_n \tag{4.60}$$

If quiescent conditions exist at t=0, we have

$$\xi_r(0) = \dot{\xi}_r(0) = 0 \tag{4.61}$$

We now assume that a constant torque is suddenly applied to the disc i=n at t=0, and that this torque continues to act for t>0. Hence

$$Q_n(t) = Q_n^o H(t) \tag{4.62}$$

where H(t) is the unit step function and Q_n^o is the magnitude of the applied torque. In this case the solution of (4.60), subject to conditions (4.61) is (see (1.34))

$$\xi_r(t) = \frac{Q_n^o a_{nr}}{\omega_r^2} (1-\cos \omega_r t) \tag{4.63}$$

and in view of (3.33),

$$\theta_i(t) = \sum_{r=1}^{n} a_{ir}\xi_r = Q_n^o \sum_{r=1}^{n} \frac{a_{nr}a_{ir}}{\omega_r^2}(1-\cos\omega_r t) \tag{4.64}$$

If we utilize (4.58), the solution (4.64) assumes the form

$$\theta_i(t) = \frac{4Q_n^o}{I}\sum_{r=1}^{n} \frac{\sin\lambda_r\cdot\sin n\lambda_r\cdot\sin i\lambda_r}{2n\sin\lambda_r - \sin 2(n+1)\lambda_r}\cdot\frac{1-\cos\omega_r t}{\omega_r^2} \tag{4.65}$$

where λ_r and ω_r^2 are defined by (4.53a) and (4.53b), respectively. For the special case of three discs, n=3, and with the aid of (4.59) and (4.64) the explicit solution of our problem is

$$\theta_1 = \frac{Q_3^o}{c}\Bigg[\frac{(0.736)(0.328)}{(0.445)^2}(1-\cos 0.445\sqrt{c/I}\,t)$$

$$+\frac{(0.591)(-0.736)}{(1.247)^2}(1-\cos 1.247\sqrt{c/I}\,t)$$

$$+\frac{(0.328)(0.591)}{(1.801)^2}(1-\cos 1.801\sqrt{c/I}\,t\Bigg] \tag{4.66}$$

and similar expressions for $\theta_2(t)$ and $\theta_3(t)$.

Let us now consider the case of a harmonically varying torque with frequency ω and amplitude Q_n^o applied to the disc i=n:

$$Q_n(t) = Q_n^o\cos\omega t \tag{4.67}$$

In accordance with (3.45) and (3.46), the forced motion in normal coordinates is characterized by

$$\ddot{\xi}_r + \omega_r^2\xi_r = P_r = \sum_j Q_j a_{jr} = Q_n^o a_{nr}\cos\omega t \tag{4.68}$$

Seeking only the steady state part of the solution of (4.68) (see (1.44), we obtain

$$\xi_r = \frac{Q_n^o a_{nr}}{\omega_r^2-\omega^2}\cos\omega t \tag{4.69}$$

and in view of (3.33), the solution expressed in generalized coordinates is

$$\theta_i(t) = \sum_{r=1}^{n} a_{ir}\xi_r = Q_n^o \sum_{r=1}^{n} \frac{a_{nr}a_{ir}}{\omega_r^2 - \omega^2} \cos \omega t$$

$$= \frac{4Q_n^o}{I} \sum_r^n \frac{\sin \lambda_r \cdot \sin n\lambda_r \cdot \sin i\lambda_r}{2n \sin \lambda_r - \sin 2(n+1)\lambda_r} \cdot \frac{\cos \omega t}{(\omega_r^2 - \omega^2)} \tag{4.70}$$

Inspection of (4.70) reveals that the solution becomes unbounded for $\omega^2=\omega_r^2$, $r=1,2,\ldots,n$. We thus conclude that our iterated structure has n resonant frequencies.

4.4 TRANSITION TO A CONTINUOUS STRUCTURE

It is intuitively obvious that we may use our iterated structural model to approximate, in some sense, an equivalent continuous model. Thus, if we increase the number of discs n indefinitely but keep the length of the shaft constant, then the system of ordinary differential equations of motion associated with the iterated structure should convert to a single partial differential equation characterizing the motion of the equivalent continuous structure. If we divide (4.43) by the distance a we obtain

$$\frac{I\ddot{\theta}_i}{a} = \frac{M_i - M_{i-1}}{a} + \frac{Q_i}{a} \tag{4.71}$$

$M_{i-1}=M(z,t)$, $\theta_i=\theta(z,t)$ and $Q_i = \int_{z-a}^{z} q(z,t)dz$, where $q(z,t)$ is the intensity of the distributed torque along the z-axis. Thus (4.71) becomes

$$\frac{I}{a}\frac{\partial^2\theta}{\partial t^2} = \frac{M(z+a,t)-M(z,t)}{a} + \frac{1}{a}\int_{z-a}^{z} q(z,t)dz \tag{4.72}$$

and in the limit as $a\to 0$ and $\frac{I}{a} \to J\rho$, we obtain

$$J\rho\,\frac{\partial^2\theta}{\partial t^2} = \frac{\partial M}{\partial z} + q(z,t) \tag{4.73}$$

where J is the polar moment of inertia of the circular cross-section of a continuous rod and ρ is its mass per unit volume. Similarly, if we divide (4.40) by ac we obtain

$$\frac{M_i}{ac} = \frac{\theta_{i+1}-\theta_i}{a} \tag{4.74}$$

But $\theta_{i+1}=\theta(z+a,t)$; $\theta_i=\theta(z,t)$; $M_i=M(z+a,t)$, and $\lim_{\substack{a\to 0\\ c\to\infty}} ac = GJ$, where G is the shear modulus of the continuous shaft and GJ is its torsional stiffness. Thus

$$\frac{M(z+a,t)}{ac} = \frac{\theta(z+a,t)-\theta(z,t)}{a} \tag{4.75}$$

and taking the limit of (4.75) as $a\to 0$ and $c\to\infty$, we obtain

$$\frac{M}{GJ} = \frac{\partial\theta}{\partial z} \tag{4.76}$$

Upon substitution of (4.76) into (4.73), we obtain

$$\rho\,\frac{\partial^2\theta}{\partial t^2} = G\,\frac{\partial^2\theta}{\partial z^2} + \frac{q}{J} \tag{4.77}$$

Equations (4.73), (4.76), and (4.77) are the well-known stress equation of motion, stress-displacement relation, and displacement equation of motion, respectively, of a continuous, elastic rod of circular cross-section in torsion, and an entirely different derivation of these equations is given in Chapter 12.

If we now seek a solution for the free vibrations of a continuous rod which is clamped at one end and free at the other end, we require a solution of

$$\frac{\partial^2\theta}{\partial t^2} = \frac{G}{\rho}\,\frac{\partial^2\theta}{\partial z^2} \tag{4.78a}$$

$$\theta(0) = 0,\ M(\ell) = 0 \tag{4.78b}$$

The problem characterized by (4.78) is the continuous equivalent of the free vibration problem of our iterated, discrete structure. A solution of (4.78) is readily obtained by the method of separation of variables (see Chapter 12):

$$\theta_r(z,t) = C \sin \omega_r \sqrt{\rho/G}\, z \cdot \cos \omega_r t$$

where the frequencies are now given by

$$\omega_r = \frac{1}{\ell}\sqrt{G/\rho}\,\frac{(2r-1)\pi}{2}\,,\quad r=1,2,\ldots \tag{4.79}$$

We shall now show that the frequency equation of the discrete structure (4.53b) approaches (4.79) as the number of discs n increases indefinitely while the length of the shaft ℓ is held constant. Let us expand (4.53b) in a power series:

$$\omega_r = a\sqrt{c/I}\,\frac{(2r-1)\pi}{2a(n+\frac{1}{2})} + \sqrt{c/I}\;O\left\{\left[\frac{(2r-1)\pi}{2(n+\frac{1}{2})}\right]^3\right\} \tag{4.80}$$

We now hold r and ℓ=na constant and let $n\to\infty$ and $a\to 0$. Then $(n+\frac{1}{2})\to n$, $ac\to GJ$, and $I/a\to J\rho$. Therefore in the limit as $n\to\infty$, (4.80) becomes (4.79). Graphs of (4.53b) and (4.79) plotted to the same scale offer an interesting and useful comparison for frequencies of the discrete and equivalent continuous structure (see Fig. 4.11). The frequencies of

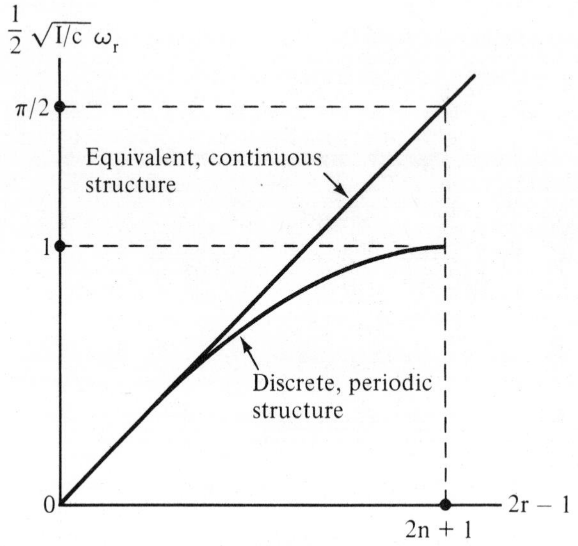

Figure 4.11

the lowest modes of the discrete model agree will with those of the continuous model. The higher modes of the discrete model are crowded together and they approach the limiting value $\omega=2\sqrt{c/I}$. We also note that the conitnuous model has an infinite number of frequencies corresponding to its infinite number of degrees of freedom.

EXERCISES

4.1 Consider the vibrations of the two degree of freedom system shown in the figure about a position of static equilibrium. Neglect gravity forces, and assume linear, elastic springs.

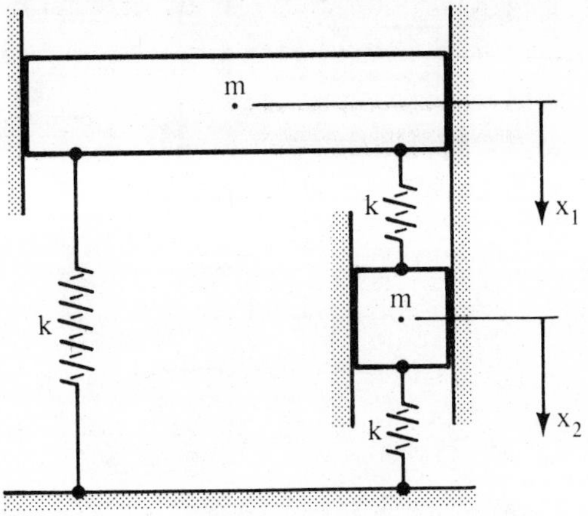

Figure: Exercise 4.1

(a) Write expressions for kinetic and potential energy.

(b) Use Lagrange's equations of motion to arrive at the proper equations of motion.

(c) Assume harmonic motion (free vibration), and find the natural frequencies.

(d) Find the (normalized) components of the eigenvectors, and check them by using the orthogonality relations.

(e) Find the normal coordinates.

(f) Given the initial conditions

$$x_1(0) = x_2(0) = x_0, \dot{x}_1(0) = \dot{x}_2(0) = 0$$

determine the motion $x_1(t)$ and $x_2(t)$ for $t>0$.

4.2 The figure shows the main chassis (weight W_1) of an instrument that has been cushioned by springs for shipment within the rigid container. A subchassis of weight W_2 is mounted on springs within the main chassis. The system is completely symmetrical. Assume that W_1 = 100 lbs, W_2=25 lbs, and that the equivalent spring constants in the vertical direction are k_1=250 lbs/in and k_2=100 lbs/in.

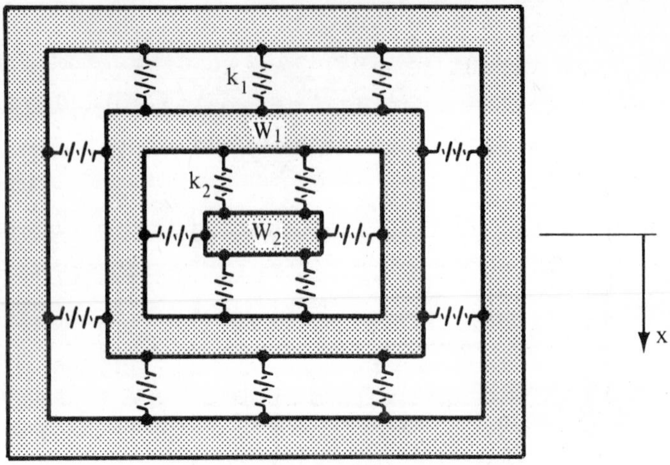

Figure: Exercise 4.2

(a) Calculate the natural frequencies and associated modal vectors.
(b) Find the motion of W_2 if the container motion is given by

$$x(t) = 0 \text{ for } t<0, \quad x(t) = A \sin \frac{\pi t}{T} \text{ for } 0<t<T$$

$$x(t) = 0 \text{ for } t>T.$$

(c) Find the steady state part of the motion of W_2 if $x(t)=A \sin \omega_0 t$ and discuss the response of W_2 with particular reference to the excitation frequency ω_0.

4.3 Two identical pendulums are arranged to swing freely in parallel planes about the same horizontal shaft. They are coupled together by a slender helical spring mounted on the shaft. The spring is unstrained when the pendulums are parallel. Initially both pendulums are at rest and hang vertically, and one of them is given an initial angular velocity. Give a complete discussion of the ensuing motion.

4.4 A uniform horizontal, rectangular plate rests on four similar springs at the corners. Find the natural frequencies and associated modal vectors assuming the plate to be thin and rigid. Neglect any horizontal motion of the plate.

4.5 A slender bar of mass M is supported by two springs of modulus k and 2k as shown in the figure. If the motion of the bar is restrained to the plane of the paper, find (a) the linearized equations of motion, neglecting any horizontal motion, (b) the natural frequencies and corresponding modal vectors, (c) the response of the bar to the force $F(t)=F_o H(t)$ for quiescent initial conditions, (d) the steady state part of the response if $F(t)=A \sin \omega_o t$. Neglect the effects of gravity and any horizontal motion of the bar.

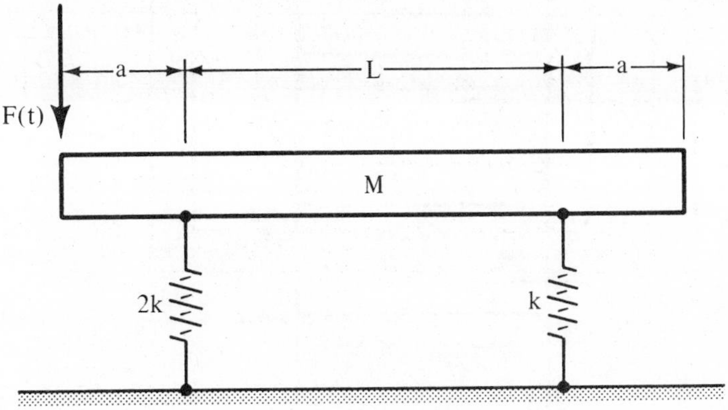

Figure: Exercise 4.5

4.6 Discuss the linearized dynamic response of the governor shown in Fig. 2.7 if $\dot{\psi} = \Omega$ = constant for $t<0$ and $\dot{\psi} = \Omega + p$ for $t>0$, where p is a constant such that $p<<\Omega$.

4.7 Four equal spring connected masses are constrained to move rectilinearly as shown in the figure. What will be the response of each mass if a constant force $F(t)=F_o H(t)$ is suddenly applied to the system. Neglect friction and assume that the system is at rest and in equilibrium at $t=0$.

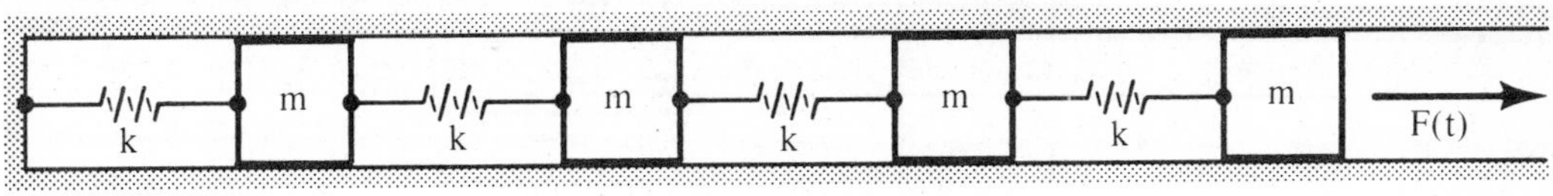

Figure: Exercise 4.7

4.8 Consider the 12 story building shown in the figure where the mass of each floor is m and the lateral (or shear) stiffness of each section between floors is k lbs/in. Assume that the column masses are negligible compared to the floor masses and that the floors remain horizontal when the building vibrates. (a) Find the natural frequencies and the associated modal vectors of the building. (b) Find the response of the building to horizontal motion of the ground caused by an earthquake: $x_o(t)=0$ for $t<0$, $x_o(t)=A \sin \frac{\pi t}{T}$ for $0<t<T$ and $x_o(t)=0$ for $t>T$. Neglect gravity effects.

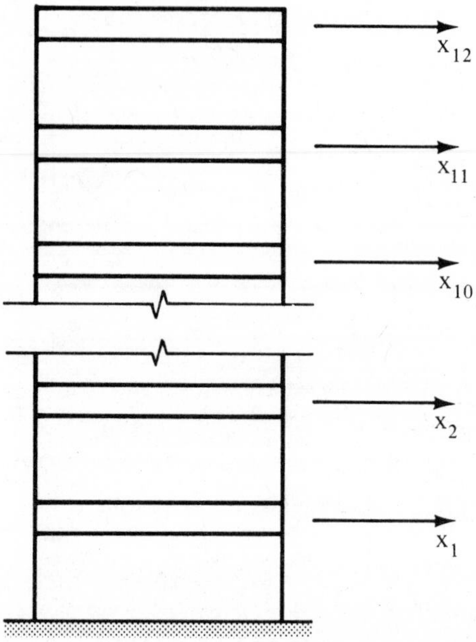

Figure: Exercise 4.8

4.9 The device shown in the figure is sometimes called a "torsional vibration absorber." If the steady state torsional excitation of the main disc I_1 is $T_o \cos \omega_o t$, and I_1 and c_1 are fixed, find the values of I_2 and c_2 so as to eliminate completely the motion of I_1.

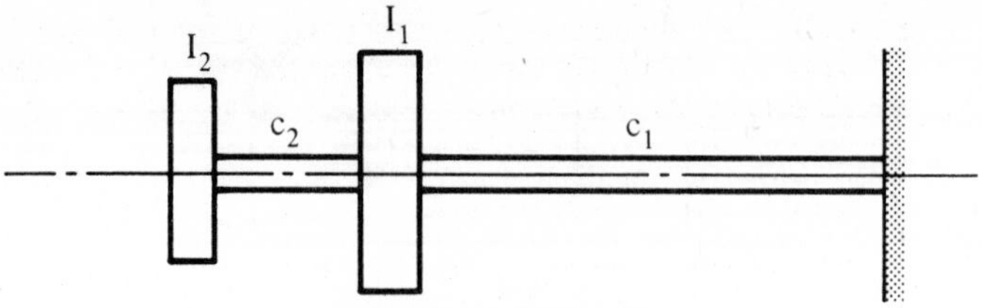

Figure: Exercise 4.9

Chapter

5

SMALL OSCILLATIONS OF A DISCRETE SYSTEM (FURTHER DEVELOPMENTS)

In many directions in engineering science, that vague commodity known as common sense will lead you correctly for quite a long way, but common sense applied to vibration problems may be an untrustworthy guide.

Sir Charles Inglis, Applied Mechanics for Engineers (1951).

5.1 THE ENERGY EQUATION OF MOTION

We shall now derive an equation which provides a relationship between the total energy contained in our linearized, vibrating conservative system and the applied forces which act on it.

If we multiply the equations of (small) motion (3.13) by $\dot{q}_j$ and sum over the j, we obtain

$$\sum_{i,j}^{n} (m_{ij}\ddot{q}_i\dot{q}_j + k_{ij}q_i\dot{q}_j) = \sum_{j}^{n} Q_j\dot{q}_j \tag{5.1}$$

But with reference to (3.6) and (3.12) we have

$$\frac{dV}{dt} = \sum_{i,j}^{n} k_{ij}q_i\dot{q}_j; \quad \frac{dT}{dt} = \sum_{i,j}^{n} m_{ij}\dot{q}_i\ddot{q}_j$$

and therefore (5.1) assumes the form

$$\frac{dE^*}{dt} = \frac{dV}{dt} + \frac{dT}{dt} = \sum_{j}^{n} Q_j\dot{q}_j \tag{5.2}$$

where E* characterizes the total internal energy of our mechanical system. Equation (5.2) expresses the fact that the time rate of change of the total energy of the conservative, linearized, mechanical system is equal to the power of all applied forces. Equation (5.2) is called the energy equation of motion. With reference to (5.2) we note that for a system in a state of free vibrations $Q_j \equiv 0$ and E^*=constant, i.e., the energy content of a linear, conservative vibrating system in a state of free vibrations is constant.

5.2 UNIQUENESS OF SOLUTION

Consider the forced motion problem as characterized by (3.13):

$$\sum_{i}^{n} (m_{ij}\ddot{q}_i + k_{ij}q_i) = Q_j, \; j=1,2,\ldots,n. \tag{5.3a}$$

and the associated (given) initial conditions in the form of 2n constants

$$q_i(0) = \alpha_i, \; \dot{q}_i(0) = \beta_i, \; i=1,2,\ldots,n \tag{5.3b}$$

Let us assume that it is possible to find two different solutions $q_i^{(1)}(t)$ and $q_i^{(2)}(t)$, both of which satisfy (5.3). We now form the difference solution

$$q_i^{(1)}-q_i^{(2)} = q_i \tag{5.4}$$

Because of the linearity of (5.3a), the difference solution satisfies the system of homogeneous differential equations

$$\sum_i^n (m_{ij}\ddot{q}_i + k_{ij}q_i) = 0, \; j=1,2,\ldots,n \tag{5.5a}$$

and associated initial conditions

$$q_i(0) = \dot{q}_i(0) = 0, \; i=1,2,\ldots,n. \tag{5.5b}$$

The problem characterized by (5.5) corresponds to the case of free vibrations, and with reference to Section 1, the total energy of our conservative mechanical system is constant in this case, i.e.,

$$T(t) + V(t) = C = \text{constant for } t\geq 0.$$

But in view of (5.5b), (3.6), and (3.7), $T(0) = V(0) = 0$, and therefore $C=0$. The kinetic energy T and the potential energy V as characterized by (3.7) and (3.6) are positive, definite, quadratic forms in the generalized velocities and displacements, respectively, i.e., $T(t)\geq 0$ and $V(t)\geq 0$. Consequently $T(t)\equiv 0$, $V(t)\equiv 0$ for all $t\geq 0$. But this result implies that $q_i(t)\equiv 0$, $\dot{q}_i(t)\equiv 0$ for all $t\geq 0$, and with reference to (5.4) it is clear that $q_i^{(1)}=q_i^{(2)}$ for all $t\geq 0$. We are thus forced to conclude that our original supposition was erroneous, and that the problem characterized by (5.3) does indeed have a unique solution.

5.3 THE RECIPROCAL THEOREM

Let us consider two separate and distinct sets of forces and the corresponding displacements caused by them. The set $(q_i^{(1)}, Q_j^{(1)})$ satisfies the equations

$$\sum_i^n (m_{ij}\ddot{q}_i^{(1)} + k_{ij}q_i^{(1)}) = Q_j^{(1)}, \tag{5.6a}$$

while the set $(q_i^{(2)}, Q_j^{(2)})$ satisfies the equations

$$\sum_i^n (m_{ij}\ddot{q}_i^{(2)} + k_{ij}q_i^{(2)}) = Q_j^{(2)} \tag{5.6b}$$

where $j=1,2,\ldots,n$ in (5.6). We now multiply (5.6a) by $q_j^{(2)}$ and sum over j. Similarly, we multiply (5.6b) by $q_j^{(1)}$ and sum over j. The results are

$$\sum_{i,j}^n (m_{ij}\ddot{q}_i^{(1)}q_j^{(2)} + k_{ij}q_i^{(1)}q_j^{(2)}) = \sum_j^n Q_j^{(1)}q_j^{(2)} \tag{5.7a}$$

$$\sum_{i,j}^n (m_{ij}\ddot{q}_i^{(2)}q_j^{(1)} + k_{ij}q_i^{(2)}q_j^{(1)}) = \sum_j^n Q_j^{(2)}q_j^{(1)} \tag{5.7b}$$

By an interchange of the indices i and j, and in view of (3.4) we have

$$\sum_{i,j}^n k_{ij}q_i^{(1)}q_j^{(2)} = \sum_{i,j}^n k_{ji}q_j^{(1)}q_i^{(2)} = \sum_{i,j}^n k_{ij}q_j^{(1)}q_i^{(2)} \tag{5.8}$$

With the aid of (5.8) we can combine (5.7a) and (5.7b) to read

$$\sum_j^n q_j^{(2)}(Q_j^{(1)} - \sum_i^n m_{ij}\ddot{q}_i^{(1)}) = \sum_j^n q_j^{(1)}(Q_j^{(2)} - \sum_i^n m_{ij}\ddot{q}_i^{(2)}) \tag{5.9}$$

Equation (5.9) is a statement of the reciprocal theorem applicable to linearized, conservative, discrete mechanical systems: The work performed by the first set of forces (including inertia forces) acting over the displacements caused by the second set is equal to the work performed by the second set of forces, acting over the displacements caused by the first set. We note that (5.9) reduces to the well-known reciprocal theorem for static systems if we delete the inertia terms.

5.4 A STATIONARY PROPERTY OF THE NATURAL FREQUENCIES

If we apply the rule for differentiation of the quotient of two functions to (3.23), we obtain

$$\frac{\partial(\omega_r^2)}{\partial a_{kr}} = \frac{\left(\sum_{i,j}^{n} m_{ij}a_{ir}a_{jr}\right)\left(\sum_{i}^{n} k_{ik}a_{ir} + \sum_{j}^{n} k_{kj}a_{jr}\right)}{\left(\sum_{i,j}^{n} m_{ij}a_{ir}a_{jr}\right)^2}$$

$$-\frac{\left(\sum_{i,j}^{n} k_{ij}a_{ir}a_{jr}\right)\left(\sum_{i}^{n} m_{ik}a_{ir} + \sum_{j}^{n} m_{kj}a_{jr}\right)}{\left(\sum_{i,j}^{n} m_{ij}a_{ir}a_{jr}\right)^2} \tag{5.10}$$

With the aid of (3.23), (3.28), (3.4) and (3.11), equation (5.10) readily simplifies to

$$\frac{\partial(\omega_r^2)}{\partial a_{ir}} = 2\sum_{j}^{n} k_{ji}a_{jr} - 2\omega_r^2\sum_{j}^{n} m_{ji}a_{jr} \tag{5.11}$$

and in view of (3.20a), the right-hand side of (5.11) vanishes. Consequently

$$\frac{\partial(\omega_r^2)}{\partial a_{ir}} = 0 \tag{5.12}$$

Equation (5.12) is often called Lagrange's theorem: If we consider the square of the circular frequency ω_r^2 as a function of the modal vector components a_{ir}, $i=1,2,\ldots,n$, then ω_r^2 is stationary with respect to the components of the modal vector, i.e., if

$$\omega_r^2=\omega_r^2(a_{1r},a_{2r},\ldots,a_{nr}),\ \text{then}\ \frac{\partial(\omega_r^2)}{\partial a_{ir}} = 0,\ i=1,2,\ldots,n.$$

Let us next consider the change in ω_r^2 when the modal vector components a_{ir} are changed to $a_{ir} + \delta a_{ir}$. By Taylor's theorem we have

$$\omega_r^2 + \delta(\omega_r^2) = \omega_r^2 + \sum_{i}^{n}\left[\frac{\partial(\omega_r^2)}{\partial a_{ir}}\right]\delta a_{ir} + \frac{1}{2}\sum_{i,j}^{n}\left[\frac{\partial^2(\omega_r^2)}{\partial a_{ir}\partial a_{jr}}\right]\delta a_{ir}\delta a_{jr}$$
$$+ 0\left[(\delta a_{ir})(\delta a_{jr})(\delta a_{kr})\right] \tag{5.13}$$

But because of (5.12) and in view of (5.13), the increment in ω_r^2 is given by

$$\delta(\omega_r^2) = \frac{1}{2} \sum_{i,j}^{n} \left[\frac{\partial^2(\omega_r^2)}{\partial a_{ir}\partial a_{jr}}\right] \delta a_{ir}\delta a_{jr} + O\,[(\delta a_{ir})(\delta a_{jr})(\delta a_{kr})] \tag{5.14}$$

Equation (5.14) may be interpreted as follows: A first order change in the modal vector components causes a second order change in the square of the frequency. In a qualitative sense, this simply means that the squares of the frequency ω_r^2 are insensitive to sufficiently small changes in the modal vector components a_{ir}, $i=1,2,\ldots,n$.

It is useful as well as instructive to re-derive Lagrange's theorem by utilizing normal coordinates. Let us assume that we constrain our dynamical system in such a manner that it has a single degree of freedom ξ by setting

$$\xi_i = \mu_i\xi, \quad i=1,2,\ldots,n. \tag{5.15}$$

where the ξ_i are normal coordinates and the μ_i are constants. In this characterization, the μ_i are the quantities which define the constrained mode shape, i.e., a particular choice of the μ_i fixes the ratios of the generalized coordinates q_i. In view of (3.35) and (3.36), the expressions for kinetic and potential energy now assume the form

$$T = \frac{1}{2}\dot{\xi}^2 \sum_{i}^{n} \mu_i^2 \tag{5.16}$$

$$V = \frac{1}{2}\xi^2 \sum_{i}^{n} \omega_i^2\mu_i^2 \tag{5.17}$$

and the Lagrangian function becomes

$$L = T-V = \frac{1}{2}\left(\dot{\xi}^2 \sum_{i}^{n} \mu_i^2 - \xi^2 \sum_{i}^{n} \omega_i^2\mu_i^2\right) \tag{5.18}$$

Upon substitution of (5.18) into the applicable form of Lagrange's equation (2.41), we obtain the equation of motion

$$\ddot{\xi}\left(\sum_{i}^{n} \mu_i^2\right) + \xi\left(\sum_{i}^{n} \omega_i^2\mu_i^2\right) = 0 \tag{5.19}$$

With reference to (5.19), it is clear that the square of the circular frequency of the constrained motion is given by

$$\omega^2 = \frac{\sum_i^n \omega_i^2 \mu_i^2}{\sum_i^n \mu_i^2} \tag{5.20}$$

and

$$\frac{\partial(\omega^2)}{\partial\mu_r} = \frac{\left(\sum_i^n \mu_i^2\right)\left(2\mu_r\omega_r^2\right) - \left(\sum_i^n \omega_i^2\mu_i^2\right)(2\mu_r)}{\left(\sum_i^n \mu_i^2\right)^2} \tag{5.21}$$

If we substitute (5.20) into (5.21) and simplify, we obtain

$$\frac{\partial(\omega^2)}{\partial\mu_r} = \frac{2\mu_r(\omega_r^2-\omega^2)}{\sum_i^n \mu_i^2} \tag{5.22}$$

The right-hand side vanishes when $\omega^2=\omega_r^2$. Hence

$$\frac{\partial(\omega_r^2)}{\partial\mu_r} = 0 \tag{5.23}$$

We also note that when $\omega^2=\omega_r^2$, $\mu_i=0$ for $i\neq r$ and $\mu_r=\pm 1$. Thus, within the present context, Lagrange's theorem may be stated as follows: The square of the circular frequency of a linear conservative system in free vibration about its equilibrium position is stationary when the system is vibrating in one of its normal modes.

In the following, let us assume that the squares of the free vibration frequencies are ordered in the sense that

$$\omega_1^2 \le \omega_2^2 \le \omega_3^2 \le \ldots\ldots \le \omega_n^2 \tag{5.24}$$

If we subtract ω_1^2 from both sides of (5.20), we obtain

$$\omega^2-\omega_1^2 = \frac{\sum_i^n (\omega_i^2-\omega_1^2)\mu_i^2}{\sum_i^n \mu_i^2} \ge 0 \tag{5.25}$$

If we subtract both sides of (5.20) from ω_n^2, we obtain

$$\omega_n^2-\omega^2 = \frac{\sum_i^n (\omega_n^2-\omega_i^2)\mu_i^2}{\sum_i^n \mu_i^2} \geq 0 \tag{5.26}$$

Equations (5.25) and (5.26) indicate that

$$\omega_n^2 \geq \omega^2 \geq \omega_1^2 \tag{5.27}$$

i.e., (5.20) provides an upper bound to the lowest frequency and a lower bound to the highest frequency.

The relations (5.27), in conjunction with (3.23), provide an interesting and useful means for the approximate determination of the free vibration frequencies. If we assume that our mechanical system is oscillating in a particular mode, and if we estimate a "mode shape" by a set of modal vector components a_{ir} which differ slightly from the actual values, then we shall obtain an approximate value of ω_r^2 with the aid of (3.23). Moreover, the relation (5.27) guarantees that the approximate frequency calculated in this manner is bounded by the least and greatest natural frequency of the mechanical system. The error in ω_r^2 obtained by this method will be considerably smaller than the errors in the assumed values of a_{ir} in view of (5.15). Since the first mode shape (i.e., the values a_{i1}, $i=1,2,\ldots,n$), corresponding to ω_1^2 can usually be estimated rather accurately, this procedure often results in surprisingly accurate values of the fundamental (lowest) frequency. This method, which is known as Rayleigh's principle, forms the basis of many useful numerical techniques employed to compute the frequencies of systems with a large number of degrees of freedom.

As an example we consider the free vibrations in torsion of three rigid discs attached to a mass-less, linearly elastic shaft as shown in Fig. 5.1. If the angles of rotation of the discs are chosen as the generalized coordinates, the kinetic and potential energy expressions are given by

$$T = \frac{1}{2}\, I\dot{\theta}_1^2 + \frac{1}{2}\, I\dot{\theta}_2^2 + \frac{1}{2}\, I\dot{\theta}_3^2 \tag{5.28}$$

$$V = \frac{1}{2}\, c(\theta_3-\theta_2)^2 + \frac{1}{2}\, c(\theta_2-\theta_1)^2 + \frac{1}{2}\, c\theta_1^2 \tag{5.29}$$

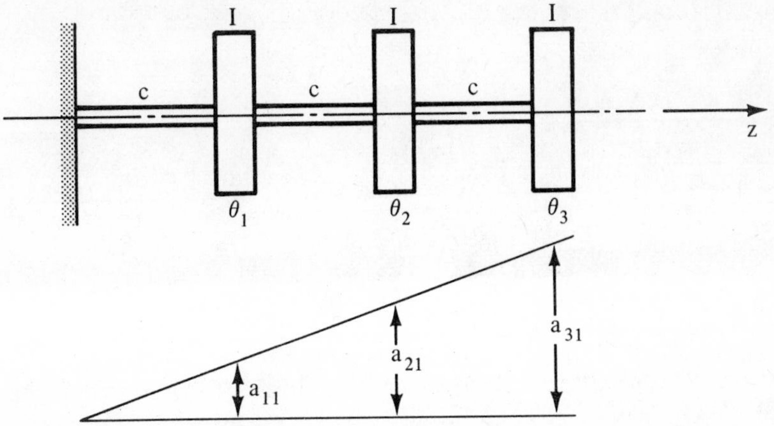

Figure 5.1

The moment of inertia of each disc with respect to the shaft axis is I and the torsional spring constant of each shaft segment is c. Comparing (5.28) with (3.12), and (5.29) with (3.6) results in the identification

$$\left.\begin{array}{l} m_{ij}=0 \text{ for } i \neq j,\ m_{ij}=I \text{ for } i=j. \\ k_{33} = c,\ k_{22}=k_{11}=2c,\ k_{12}=k_{21}=-c,\ k_{23}=k_{32}=-c \\ k_{13}=k_{31}=0. \end{array}\right\} \quad (5.30)$$

In view of (3.23) and (3.18),

$$\omega_r^2 = \sum_{i,j}^{3} k_{ij} a_{ir} a_{jr} \quad (5.31)$$

If we want to use (5.31) to obtain an approximation of the fundamental frequency ω_1, we have to "guess" the mode shape which corresponds to ω_1. We shall assume a linear mode shape as shown in Fig. 5.1 i.e.,

$$\frac{a_{21}}{a_{11}} = 2,\ \frac{a_{31}}{a_{11}} = 3. \quad (5.32)$$

Moreover, according to (3.18), the components of the modal vector satisfy the normalization condition

$$\left.\begin{array}{l} Ia_{11}^2 + Ia_{21}^2 + Ia_{31}^2 = 1, \\ \text{or} \\ 1 + \left(\frac{a_{21}}{a_{11}}\right)^2 + \left(\frac{a_{31}}{a_{11}}\right)^2 = \frac{1}{Ia_{11}^2} \end{array}\right\} \quad (5.33)$$

Upon substitution of (5.32) into (5.33), we obtain

$$Ia_{11}^2 = \frac{1}{14} \tag{5.34}$$

We now substitute (5.30) into (5.31) and divide by ca_{11}^2, obtaining

$$\frac{\omega_1^2}{ca_{11}^2} = 2 + 2\left(\frac{a_{21}}{a_{11}}\right)^2 + \left(\frac{a_{31}}{a_{11}}\right)^2 - 2\left(\frac{a_{21}}{a_{11}}\right) - 2\left(\frac{a_{21}}{a_{11}}\right)\left(\frac{a_{31}}{a_{11}}\right) \tag{5.35}$$

Further substitution of (5.32) and (5.34) into (5.35) results in

$$\omega_1^2 = \frac{3}{14}\,\frac{c}{I}, \text{ or } \omega_1 = \sqrt{3/14}\,\sqrt{c/I} \simeq 0.462\,\sqrt{c/I} \tag{5.36}$$

The actual value of $\omega_1 = 0.445\sqrt{c/I}$ as given by (4.59a). Thus the approximate value of ω_1 as obtained by the application of Rayleigh's principle is 3.82% greater than the exact value. Considering our crude estimate of the modal vector components, Rayleigh's principle resulted in a very good approximation of the actual frequency in this case.

5.5 THE EFFECT OF A CONSTRAINT ON THE NATURAL FREQUENCIES

Let us consider a linearized, conservative mechanical system with n degrees of freedom with proper generalized coordinates q_i; $i=1,2,\ldots,n$, and distinct natural frequencies ω_r; $r=1,2,\ldots,n$. A constraint is now imposed upon the system and the equation which characterizes the manner of constraint is

$$f(q_1,q_2,\ldots,q_n) = 0 \tag{5.37}$$

If the equilibrium configuration $q_i=0$, $i=1,2,\ldots,n$, is compatible with the imposed constraint,

$$f(0,0,\ldots,0) = 0 \tag{5.38}$$

We now expand (5.37) into a Taylor series about the equilibrium position,

$$f(q_1,q_2,\ldots,q_n)=f(0,0,\ldots,0) + \sum_{i=1}^{n}\left(\frac{\partial f}{\partial q_i}\right)_0 q_i + O(q_i\cdot q_j) \tag{5.39}$$

In the spirit of our linear model, we drop $O(q_i \cdot q_j)$. Then in view of (5.38) and (5.37), we obtain

$$\sum_{i=1}^{n} B_i q_i = 0 \tag{5.40}$$

where $B_i = (\partial f/\partial q_i)_0$. We have shown (see (3.33)) that the generalized coordinates q_i can always be expressed as a linear function of the normal coordinates ξ_r. Consequently the linearized equation of constraint assumes the form

$$\sum_{r=1}^{n} A_r \xi_r = 0 \tag{5.41}$$

where the A_r are constants. We shall use (5.41) to eliminate the n'th term from the expressions for kinetic and potential energy (in normal coordinates):

$$T = \tfrac{1}{2} \sum_{i=1}^{n} \dot{\xi}_i^2 = \tfrac{1}{2}\left[\sum_{i=1}^{n-1} \dot{\xi}_i^2 + \frac{1}{A_n^2}\left(\sum_{i=1}^{n-1} A_i \dot{\xi}_i\right)^2\right] \tag{5.42}$$

$$V = \tfrac{1}{2} \sum_{i=1}^{n} \omega_i^2 \xi_i^2 = \tfrac{1}{2}\left[\sum_{i=1}^{n-1} \omega_i^2 \xi_i^2 + \frac{\omega_n^2}{A_n^2}\left(\sum_{i=1}^{n-1} A_i \xi_i\right)^2\right] \tag{5.43}$$

Substituting (5.42) and (5.43) into the homogeneous form of Lagrange's equations, we obtain

$$\ddot{\xi}_r + \omega_r^2 \xi_r + \mu A_r = 0; \; r=1,2,\ldots,n-1. \tag{5.44}$$

where

$$\mu = \frac{1}{A_n^2} \sum_{i=1}^{n-1} A_i \ddot{\xi}_i + \frac{\omega_n^2}{A_n^2} \sum_{i=1}^{n-1} A_i \xi_i = -\frac{\ddot{\xi}_n}{A_n} - \frac{\omega_n^2}{A_n} \xi_n \tag{5.45}$$

In view of (5.44) and (5.45) we have

$$\ddot{\xi}_r + \omega_r^2 \xi_r + \mu A_r = 0; \; r=1,2,\ldots,n \tag{5.46}$$

Equation (5.46) characterizes the free vibrations of the constrained system. We therefore assume

$$\xi_r = \alpha_r \cos \omega t; \; r=1,2,\ldots,n \tag{5.47a}$$

$$\mu = \nu \cos \omega t \tag{5.47b}$$

where ω is the frequency of the constrained system. Substitution of (5.47) into (5.46) results in

$$\alpha_r = \frac{\nu A_r}{\omega^2-\omega_r^2} \tag{5.48}$$

and substitution of (5.47a) into (5.41) results in

$$\sum_{r=1}^{n} A_r \alpha_r = 0 \tag{5.49}$$

Upon substituting (5.48) into (5.49), we obtain

$$\sum_{r=1}^{n} \frac{A_r^2}{\omega_r^2-\omega^2} = 0 \tag{5.50}$$

Equation (5.50) has n-1 roots, and they will be designated by $\omega^2=\Omega_1^2,\ \Omega_2^2,\ \ldots,\ \Omega_{n-1}^2$. If (5.50) is multiplied by $(\omega_1^2-\omega^2)(\omega_2^2-\omega^2)\ldots\ldots(\omega_n^2-\omega^2)$, we obtain

$$\begin{aligned} F(\omega^2)=0=&A_1^2(1)(\omega_2^2-\omega^2)(\omega_3^2-\omega^2)\ldots\ldots(\omega_{n-1}^2-\omega^2)(\omega_n^2-\omega^2)\\ &+A_2^2(\omega_1^2-\omega^2)(1)(\omega_3^2-\omega^2)\ldots\ldots(\omega_{n-1}^2-\omega^2)(\omega_n^2-\omega^2)\\ &+\ldots\ldots\ldots\ldots\ldots\ldots\ldots\ldots\ldots\ldots\ldots\ldots\\ &+A_n^2(\omega_1^2-\omega^2)(\omega_2^2-\omega^2)(\omega_3^2-\omega^2)\ldots\ldots(\omega_{n-1}^2-\omega^2)(1) \end{aligned} \tag{5.51}$$

If the frequencies of the unconstrained system are ordered according to $\omega_1^2<\omega_2^2<\omega_3^2<\ldots\ldots<\omega_{n-1}^2<\omega_n^2$, then with reference to (5.51) we have $F(0)>0$, and $F(\omega_1^2)>0$, $F(\omega_2^2)<0$, $F(\omega_3^2)>0$, etc. so that (see Fig. 5.2)

$$\omega_1^2 \le \Omega_1^2 \le \omega_2^2 \le \Omega_2^2 \le\ldots\ldots\le \Omega_{n-1}^2 \le \omega_n^2.$$

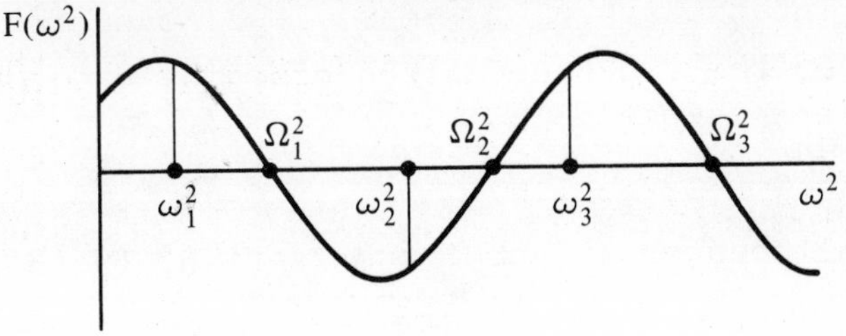

Figure 5.2

With reference to (5.51) we note that $\Omega_1^2=\omega_1^2$ if and only if $A_1=0$, and in general, $\Omega_r^2=\omega_r^2$ if and only if $A_r=0$. Additionally, $\Omega_1^2=\omega_2^2$ if and only if $A_2=0$ and then $\Omega_2^2=\omega_3^2$, $\Omega_3^2=\omega_4^2,\ldots\ldots,\Omega_{n-1}^2=\omega_n^2$. We thus conclude that in the non-trivial case of $A_r\neq 0$; $r=1,2,\ldots,n$, the frequencies of the constrained system are interdigitated with respect to the frequencies of the corresponding unconstrained system.

As a concrete demonstration of the above development, consider the torsional oscillations of two rigid discs about a mass-less shaft as shown in Fig. 5.3a. If the moment of inertia of each disc is I, and if the torsional stiffness of each shaft segment is c, then the natural frequencies of this system are given by (4.53b), with n=2:

$$\begin{aligned}\omega_1^2 &= 4\,\frac{c}{I}\sin^2\frac{\pi}{10} = \frac{3-\sqrt{5}}{2}\,\frac{c}{I}\\ \omega_2^2 &= 4\,\frac{c}{I}\sin^2\frac{3\pi}{10} = \frac{3+\sqrt{5}}{2}\,\frac{c}{I}\end{aligned} \tag{5.52}$$

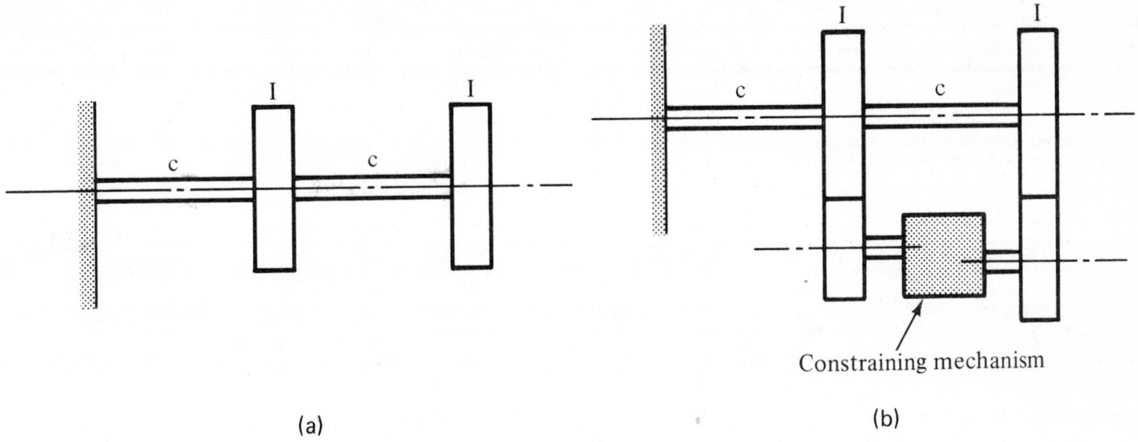

Figure 5.3

Let us now introduce a rigid, mass-less mechanism which provides a constraint between the two generalized coordinates q_1 and q_2 defined by

$$q_1 = \gamma \sin q_2 \tag{5.53}$$

where γ is a real constant. Upon linearization, (5.53) becomes

$$q_1 = \gamma q_2 \tag{5.54}$$

The expressions for kinetic and potential energy of the original system, Fig. 5.3a, are

$$T = \tfrac{1}{2}I\dot{q}_1^2 + \tfrac{1}{2}I\dot{q}_2^2 \tag{5.55}$$

$$V = \tfrac{1}{2}cq_1^2 + \tfrac{1}{2}c(q_2-q_1)^2 \tag{5.56}$$

The expressions for kinetic and potential energy of the constrained system are obtained by substituting (5.54) into (5.55) and (5.56):

$$T = \tfrac{1}{2}I(\gamma^2+1)\dot{q}_2^2 \tag{5.57}$$

$$V = \tfrac{1}{2}c(2\gamma^2-2\gamma+1)q_2^2 \tag{5.58}$$

Substitution of (5.57) and (5.58) into Lagrange's equations results in

$$I(\gamma^2+1)\ddot{q}_2+c(2\gamma^2-2\gamma+1)q_2 = 0 \tag{5.59}$$

and the natural frequency of the constrained system is

$$\omega^2 = \Omega_1^2 = g(\gamma)\ \frac{c}{I} \tag{5.60}$$

where

$$g(\gamma) = \frac{2\gamma^2-2\gamma+1}{\gamma^2+1}$$

It can be readily verified that

$$g\left(\frac{-1+\sqrt{5}}{2}\right) = \frac{3-\sqrt{5}}{2} \le g(\gamma) \le \frac{3+\sqrt{5}}{2} = g\left(\frac{-1-\sqrt{5}}{2}\right)$$

Thus in view of (5.52) and (5.60),

$$\frac{3-\sqrt{5}}{2}\ \frac{c}{I} = \omega_1^2 \le \Omega_1^2 \le \omega_2^2 = \frac{3+\sqrt{5}}{2}\ \frac{c}{I}$$

for all possible values of the constant γ. We also note that $\Omega_1^2=\omega_1^2=\frac{3-\sqrt{5}}{2}\frac{c}{I}$ only when $\gamma=\frac{-1+\sqrt{5}}{2}$ and $\Omega_1^2=\omega_2^2=\frac{3+\sqrt{5}}{2}\frac{c}{I}$ only when $\gamma=\frac{-1-\sqrt{5}}{2}$. The present results confirm the preceding general discussion.

EXERCISES

5.1 Estimate the components of the modal vector corresponding to the first mode of torsional oscillation of the structure shown in Fig. 4.8 when n=6. Use Rayleigh's principle to obtain an approximation of the fundamental natural frequency ω_1, and compare with the exact value calculated with the aid of (4.53b).

5.2 Calculate the 6 natural frequencies ω_r, r=1,2,3,4,5,6, of the torsional oscillations of the structure shown in Fig. 4.8 when n=6. Next, freeze the motion of the disc at r=1 and calculate the natural frequencies Ω_r, r=1,2,3,4,5, of the modified structure. Show that

$$\omega_1<\Omega_1<\omega_2<\Omega_2<\ldots\ldots<\omega_5<\Omega_5<\omega_6$$

and relate this result to the findings of Section 5.

5.3 Apply Rayleigh's principle to estimate the fundamental frequency of vibration of the building in Exercise 4.8 and compare to the exact value.

Chapter 6

DAMPING

For convenience of treatment, we have considered apart the two great classes of forced vibrations and free vibrations; but there is, of course, nothing to prevent their coexistence. After the lapse of a sufficient interval of time, the free vibrations always disappear, however small the friction may be. The case of absolutely no friction is purely ideal.

Lord Rayleigh, The Theory of Sound, Vol. I, (1894).

6.1 INTRODUCTION

In our development of Lagrange's equations and their application to the theory of small oscillations in Chapters 2 through 5, we have assumed that the internal forces of our mechanical system are conservative, i.e., we developed a mathematical model which neglects to account for damping and dissipative effects. The consequences of this omission are usually not serious, and the neglect of damping often does not alter the predicted motion of the system in an important way. However, there are definite indications that a dissipationless system has characteristics which are not in consonance with the behavior of real systems. For example, once started, free vibrations of a mechanical system in the sense of Chapters 3 through 5 will continue indefinitely, whereas a real system will ultimately come to rest. In the case of harmonically forced, steady state oscillations, the generalized displacements become unbounded when the forcing frequency coincides with one of the natural frequencies of the undamped system (see, for instance, (4.39) and Fig. 4.6). However, if damping is added to the system, the generalized displacements will remain bounded near resonance, although they may become large.

It is well known that all macroscopic mechanical systems experience structural damping, the ultimate cause of which is internal friction in materials and damping due to radiation of energy to the surrounding medium (such as air). A treatment of damping from first principles is beyond the scope of the present discussion. It is our intention to treat damping from a phenomenological viewpoint. Even within the framework of this self-imposed limitation, a variety of choices are possible, because there exist a number of ad hoc theories of damping. We shall limit ourselves to the choice of viscous damping primarily for two reasons: (a) it provides us with a reasonable approximation to the real world, and (b) it results in a mathematical description which is convenient and allows us to develop solutions which require only a minor extension of the treatment of the corresponding conservative systems of Chapters 2 through 4.

6.2 THE DISSIPATION FUNCTION

Let us re-consider the motion of a mechanical system consisting of N mass points. Then in the spirit of Chapter 2, the components of the frictional forces referred to a Newtonian, Cartesian axis system are given by

$$X_i^{(f)} = -\kappa_{x_i}\dot{x}_i,\quad Y_i^{(f)} = -\kappa_{y_i}\dot{y}_i,\quad Z_i^{(f)} = -\kappa_{z_i}\dot{z}_i \tag{6.1}$$

where κ_{x_i}, κ_{y_i} and κ_{z_i} are real, positive constants. If we define the dissipation function (first proposed by Lord Rayleigh) as

$$R = \tfrac{1}{2}\sum_{i=1}^{N}(\kappa_{x_i}\dot{x}_i^2 + \kappa_{y_i}\dot{y}_i^2 + \kappa_{z_i}\dot{z}_i^2) \tag{6.2}$$

then the components of the friction forces can be obtained from (6.2), because

$$X_i^{(f)} = -\frac{\partial R}{\partial \dot{x}_i},\quad Y_i^{(f)} = -\frac{\partial R}{\partial \dot{y}_i},\quad Z_i^{(f)} = -\frac{\partial R}{\partial \dot{z}_i} \tag{6.3}$$

It is possible and useful to obtain a simple physical interpretation of the quantity R. The time rate of work done by the system against friction is

$$\frac{dW^{(f)}}{dt} = -\sum_{i=1}^{N}(X_i^{(f)}\dot{x}_i + Y_i^{(f)}\dot{y}_i + Z_i^{(f)}\dot{z}_i)$$

$$= \sum_{i=1}^{N}(\kappa_{x_i}\dot{x}_i^2 + \kappa_{y_i}\dot{y}_i^2 + \kappa_{z_i}\dot{z}_i^2) = 2R \geq 0 \tag{6.4}$$

and in view of (6.4) it is clear that 2R is the rate of energy dissipation caused by friction in the system.

We shall now transform (6.2) and (6.3) to generalized coordinates. Direct substitution of (2.26) into (6.2) results in

$$R = \tfrac{1}{2}\sum_{r,s}^{n} b_{rs}\dot{q}_r\dot{q}_s \tag{6.5}$$

where

$$b_{rs} = b_{sr} = \sum_{i=1}^{N}\left(\kappa_{x_i}\frac{\partial x_i}{\partial q_r}\frac{\partial x_i}{\partial q_s} + \kappa_{y_i}\frac{\partial y_i}{\partial q_r}\frac{\partial y_i}{\partial q_s} + \kappa_{z_i}\frac{\partial z_i}{\partial q_r}\frac{\partial z_i}{\partial q_s}\right) \tag{6.6}$$

and with reference to (2.25) we note that, in general, $b_{rs} = b_{rs}(q_1, q_2, \ldots, q_n)$. Therefore, the Rayleigh dissipation function R is a function of the generalized coordinates q_r as well as of the generalized velocities $\dot{q}_r$, $r=1,2,\ldots,n$. The generalized friction forces $Q_i^{(f)}$ are now obtained with the aid of (6.3) and (2.32):

$$Q_i^{(f)} = \sum_{j=1}^{N} X_j^{(f)} \frac{\partial x_j}{\partial q_i} + Y_j^{(f)} \frac{\partial y_j}{\partial q_i} + Z_j^{(f)} \frac{\partial z_j}{\partial q_i}$$

$$= -\sum_{j=1}^{N} \left(\frac{\partial R}{\partial \dot{x}_j} \frac{\partial \dot{x}_j}{\partial \dot{q}_i} + \frac{\partial R}{\partial \dot{y}_j} \frac{\partial \dot{y}_j}{\partial \dot{q}_i} + \frac{\partial R}{\partial \dot{z}_j} \frac{\partial \dot{z}_j}{\partial \dot{q}_i} \right) = -\frac{\partial R}{\partial \dot{q}_i} \qquad (6.7)$$

where we have used the relations

$$\frac{\partial \dot{\vec{r}}_j}{\partial \dot{q}_i} = \frac{\partial \vec{r}_j}{\partial q_i}$$

which are readily obtained with the aid of (2.26). Equation (2.38) is presently replaced by

$$\Theta_i = Q_i^{(c)} + Q_i^{(f)} + Q_i \qquad (6.8)$$

i.e., the generalized forces are now divided into three distinct types: conservative, frictional, and applied (external). In view of (6.8) and (6.7), Lagrange's equations (2.37) now assume the form

$$\frac{d}{dt}\left(\frac{\partial T}{\partial \dot{q}_i}\right) - \frac{\partial T}{\partial q_i} + \frac{\partial V}{\partial q_i} + \frac{\partial R}{\partial \dot{q}_i} = Q_i \qquad (6.9)$$

We shall now linearize the dissipation function in the sense of the theory of small oscillations as presented in Chapter 3. If we take

$$R = \tfrac{1}{2} \sum_{i,j}^{n} b_{ij} \dot{q}_i \dot{q}_j \qquad (6.10)$$

and expand b_{ij} in a Taylor series about the stable equilibrium position $q_i=0$, $i=1,2,\ldots,n$, we obtain

$$b_{ij}(q_1,q_2,\ldots,q_n) = b_{ij}(0,0,\ldots,0) + \sum_{k=1}^{n} \left(\frac{\partial b_{ij}}{\partial q_k}\right) q_k + O(q_k^2) \qquad (6.11)$$

In order to obtain an approximation for R which is of the same order as that of T and V in (3.12) and (3.6), respectively, we must retain only the first term of the expansion in (6.11). For notational reasons, we also set

$$c_{ij} = b_{ij}(0,0,\ldots,0) \tag{6.12}$$

Consequently, the dissipation function (6.10) assumes the form

$$R = \tfrac{1}{2}\sum_{i,j}^{n} c_{ij}\dot{q}_i\dot{q}_j \tag{6.13}$$

where $c_{ij}=c_{ji}$ in view of (6.6) and (6.12). Equation (6.13) characterizes the dissipation function as a quadratic form in the generalized coordinates which, according to (6.4), is non-negative for all real values of the variable $\dot{q}_i$. Moreover, R vanishes only when all $\dot{q}_i=0$. Hence (6.13) is a positive definite quadratic form.

If we now substitute (6.13) as well as (3.6) and (3.12) into (6.9), we obtain the linearized system of equations of motion

$$\sum_{i=1}^{n}(m_{ij}\ddot{q}_i + c_{ij}\dot{q}_i + k_{ij}q_i) = Q_j(t) \qquad j=1,2,\ldots,n \tag{6.14}$$

6.3 FREE, DAMPED VIBRATIONS

We now set $Q_j\equiv 0$ in (6.14), and consider free vibrations of a damped system characterized by

$$\sum_{i=1}^{n}(m_{ij}\ddot{q}_i + c_{ij}\dot{q}_i + k_{ij}q_i) = 0 \tag{6.15}$$

If we multiply (6.15) by $\dot{q}_j$ and sum over the index j=1,2,...,n, then the resulting equation is readily cast in the form

$$\frac{d}{dt}(T+V) = -\sum_{i,j}^{n} c_{ij}\dot{q}_i\dot{q}_j = -2R \le 0 \tag{6.16}$$

Equation (6.16) indicates that the total energy of the mechanical system can not increase as time increases.

We now assume a solution of (6.15) in the form

$$q_i(t) = a_i \exp \lambda t \tag{6.17}$$

and substitute (6.17) into (6.15). The result is

$$\lambda^2 \sum_i^n m_{ij} a_i + \lambda \sum_i^n c_{ij} a_i + \sum_i^n k_{ij} a_i = 0 \tag{6.18}$$

If a non-trivial solution is required of (6.18), the determinant of its coefficient matrix must vanish:

$$|m_{ij}\lambda^2 + c_{ij}\lambda + k_{ij}| = 0 \qquad \begin{matrix} i=1,2,\ldots,n \\ j=1,2,\ldots,n \end{matrix} \tag{6.19}$$

Equation (6.19) has 2n roots which we shall assume to be distinct. By superposition, the general solution of the free vibration problem can now be written as

$$q_i(t) = \sum_{r=1}^{2n} a_{ir} \exp \lambda_r t, \qquad i=1,2,\ldots,n. \tag{6.20}$$

We note that there are $2n^2$ constants in (6.20), but it can be shown that only 2n of these constants are independent, and the latter are obtained by application of the 2n initial conditions.

We shall now prove that all the roots λ of (6.19) will have negative, real parts. Toward this end, let

$$\lambda = \kappa + i\omega, \qquad \lambda^* = \kappa - i\omega \tag{6.21}$$

and

$$a_i = \alpha_i + i\beta_i, \qquad a_i^* = \alpha_i - \beta_i \tag{6.22}$$

where κ, ω, α_i, and β_i are real numbers, and $i = \sqrt{-1}$ when it is not used as a subscript. We now multiply (6.18) by a_j^* and sum the products so obtained over the index $j=1,2,\ldots,n$:

$$\lambda^2 \sum_{i,j}^n m_{ij}(\alpha_i\alpha_j + \beta_i\beta_j) + \lambda \sum_{i,j}^n c_{ij}(\alpha_i\alpha_j + \beta_i\beta_j) + \sum_{i,j}^n k_{ij}(\alpha_i\alpha_j + \beta_i\beta_j) = 0 \tag{6.23}$$

Utilizing (6.21), the sum of the roots of (6.23) is

$$\lambda + \lambda^* = 2\kappa = -\frac{\sum_{i,j}^{n} c_{ij}(\alpha_i\alpha_j + \beta_i\beta_j)}{\sum_{i,j}^{n} m_{ij}(\alpha_i\alpha_j + \beta_i\beta_j)} \tag{6.24}$$

It is now assumed that the solution (6.17) is non-trivial, i.e., not all α_i and β_i will vanish. Moreover, both numerator and denominator of (6.24) are positive, definite quadratic forms. It can therefore be concluded that the right-hand side of (6.24) is a real negative number. Consequently the ensuing free vibration will tend to the stable equilibrium position of the mechanical system in either an oscillatory or entirely subsident manner, i.e., $q_i(t)\to 0$ as $t\to\infty$.

As a demonstration of the foregoing general discussion, let us consider the free vibrations of the two-degree of freedom system shown in Fig. 6.1. We denote by k_1 and k_2 the spring constants of the elastic springs, m_{11} and m_{22} the masses, c_1 and c_2 the coefficients of viscous friction of the springs, and c_3 and c_4 the coefficients of viscous friction characterizing the resistance to motion which the masses experience while sliding on their horizontal constraints. The expressions for kinetic and potential energy are

$$T = \tfrac{1}{2}\, m_{11}\dot{q}_1^2 + \tfrac{1}{2}\, m_{22}\dot{q}_2^2 \tag{6.25}$$

$$V = \tfrac{1}{2}\, k_1 q_1^2 + \tfrac{1}{2}\, k_2(q_2-q_1)^2 \tag{6.26}$$

The dissipation function is

$$R = \tfrac{1}{2}\, c_1\dot{q}_1^2 + \tfrac{1}{2}\, c_2(\dot{q}_2-\dot{q}_1)^2 + \tfrac{1}{2}\, c_3\dot{q}_1^2 + \tfrac{1}{2}\, c_4\dot{q}_2^2 \tag{6.27}$$

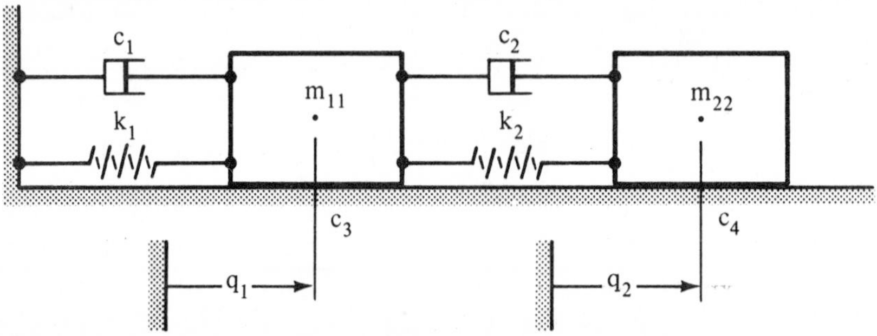

Figure 6.1

To bring (6.26) and (6.27) into the notational framework of the previous general discussion, we set

$$k_{11} = k_1+k_2,\ k_{12} = k_{21} = -k_2,\ k_{22} = k_2$$

$$c_{11} = c_1+c_2+c_3,\ c_{12} = c_{21} = -c_2,\ c_{22} = c_2+c_4$$

With these changes, (6.26) and (6.27) assume the form

$$V = \tfrac{1}{2}k_{11}q_1q_1 + \tfrac{1}{2}k_{12}q_1q_2 + \tfrac{1}{2}k_{21}q_2q_1 + \tfrac{1}{2}k_{22}q_2q_2 \tag{6.28}$$

$$R = \tfrac{1}{2}c_{11}\dot{q}_1\dot{q}_1 + \tfrac{1}{2}c_{12}\dot{q}_1\dot{q}_2 + \tfrac{1}{2}c_{21}\dot{q}_2\dot{q}_1 + \tfrac{1}{2}c_{22}\dot{q}_2\dot{q}_2 \tag{6.29}$$

If we now substitute (6.25), (6.28), and (6.29) into Lagrange's equations (6.9) (with $Q_i\equiv 0$), we obtain the equations of motion characterizing the free, damped vibrations of our mechanical system:

$$m_{11}\ddot{q}_1 + c_{11}\dot{q}_1 + c_{12}\dot{q}_2 + k_{11}q_1 + k_{12}q_2 = 0 \tag{6.30a}$$

$$m_{22}\ddot{q}_2 + c_{22}\dot{q}_2 + c_{12}\dot{q}_1 + k_{22}q_2 + k_{21}q_1 = 0 \tag{6.30b}$$

To find a solution of (6.30), we assume

$$q_1 = a_1 \exp \lambda t,\ q_2 = a_2 \exp \lambda t \tag{6.31}$$

where a_1, a_2, and λ are constants. Substitution of (6.31) into (6.30) results in

$$(m_{11}\lambda^2 + c_{11}\lambda + k_{11})a_1 + (k_{12} + c_{12}\lambda)\, a_2 = 0 \tag{6.32a}$$

$$(k_{12} + c_{12}\lambda)a_1 + (m_{22}\lambda^2 + c_{22}\lambda + k_{22})a_2 = 0 \tag{6.32b}$$

For a non-trivial solution, we require that the determinant of the coefficient matrix of (6.32) vanish, and upon expanding the determinant we obtain

$$(m_{11}\lambda^2+c_{11}\lambda+k_{11})(m_{22}\lambda^2+c_{22}\lambda+k_{22})-(k_{12}+c_{12}\lambda)^2 = 0 \tag{6.33}$$

Equation (6.33) is a quartic in λ. We shall distinguish between three distinct possibilities.

Case (a): (6.33) has four real, distinct roots $\lambda=\kappa_1,\kappa_2,\kappa_3,\kappa_4$. In this case the solution assumes the form

$$q_1(t) = a_{11} \exp \kappa_1 t+a_{12} \exp \kappa_2 t + a_{13} \exp \kappa_3 t+a_{14} \exp \kappa_4 t \tag{6.34a}$$

$$q_2(t) = a_{21} \exp \kappa_1 t+a_{22} \exp \kappa_2 t + a_{23} \exp \kappa_3 t+a_{24} \exp \kappa_4 t \tag{6.34b}$$

and with reference to (6.32), we note that the eight constants a_{ij}; $i=1,2$; $j=1,2,3,4$ are related by the four equations

$$\frac{a_{2j}}{a_{1j}} = \alpha_j;\ j=1,2,3,4 \tag{6.35}$$

If we substitute (6.35) into (6.34b), we obtain

$$q_2(t) = a_{11}\alpha_1 \exp \kappa_1 t + a_{12}\alpha_2 \exp \kappa_2 t + a_{13}\alpha_3 \exp \kappa_3 t + a_{14}\alpha_4 \exp \kappa_4 t \tag{6.36}$$

The present form of the solution contains four constants of integration and these can be obtained by application of the four initial conditions.

Case (b): (6.33) has two pairs of conjugate complex roots $\lambda_1=\kappa_1+i\omega_1$, $\lambda_2=\kappa_1-i\omega_1$, $\lambda_3=\kappa_2+i\omega_2$, $\lambda_4=\kappa_2-i\omega_2$.

In this case the solution is written as

$$q_1(t)=a_{11}\exp(\kappa_1+i\omega_1)t+a_{12}\exp(\kappa_1-i\omega_1)t+a_{13}\exp(\kappa_2+i\omega_2)t+a_{14}\exp(\kappa_2-i\omega_2)t \tag{6.37a}$$

$$q_2(t)=a_{21}\exp(\kappa_1+i\omega_1)t+a_{22}\exp(\kappa_1-i\omega_1)t+a_{23}\exp(\kappa_2+i\omega_2)t+a_{24}\exp(\kappa_2-i\omega_2)t \tag{6.37b}$$

The generalized displacements q_i are real. Consequently we set

$$\left.\begin{aligned} a_{11} &= \tfrac{1}{2}C_1 \exp i\psi_1, & a_{12} &= \tfrac{1}{2}C_1 \exp(-i\psi_1) \\ a_{13} &= \tfrac{1}{2}C_2 \exp i\psi_2, & a_{14} &= \tfrac{1}{2}C_2 \exp(-i\psi_2) \end{aligned}\right\} \tag{6.38a}$$

$$\left.\begin{aligned} \frac{a_{21}}{a_{11}} &= \mu_1 \exp i\Psi_1, & \frac{a_{22}}{a_{12}} &= \mu_1 \exp(-i\Psi_1) \\ \frac{a_{23}}{a_{13}} &= \mu_2 \exp i\Psi_2, & \frac{a_{24}}{a_{14}} &= \mu_2 \exp(-i\Psi_2) \end{aligned}\right\} \tag{6.38b}$$

where C_1, C_2, ψ_1, ψ_2 are real quantities and where μ_1, μ_2, Ψ_1, Ψ_2 are real quantities obtainable with the aid of (6.32). Thus the solution can now be written in the form

$$\begin{aligned} q_1(t) &= C_1\exp \kappa_1 t\cdot\cos(\omega_1 t+\psi_1)+C_2\exp \kappa_2 t\cdot\cos(\omega_2 t+\psi_2) \\ q_2(t) &= \mu_1 C_1\exp \kappa_1 t\cdot\cos(\omega_1 t+\psi_1+\Psi_1)+\mu_2 C_2\exp \kappa_2 t\cdot\cos(\omega_2 t+\psi_2+\Psi_2) \end{aligned} \tag{6.39}$$

The four constants of integration C_1, C_2, ψ_1, ψ_2, can now be determined by the application of the four initial conditions.

Case (c): (6.33) has two distinct real roots $\lambda_1=\kappa_1$, $\lambda_2=\kappa_2$, and a pair of conjugate complex roots $\lambda_3=\kappa_3+i\omega_3$, $\lambda_4=\kappa_3-i\omega_3$. The solution can be written as

$$q_1(t)=a_{11}\exp \kappa_1 t+a_{12}\exp \kappa_2 t+a_{13}\exp(\kappa_3+i\omega_3)t+a_{14}\exp(\kappa_3-i\omega_3)t \tag{6.40a}$$

$$q_2(t)=a_{21}\exp \kappa_1 t+a_{22}\exp \kappa_2 t+a_{23}\exp(\kappa_3+i\omega_3)t+a_{24}\exp(\kappa_3-i\omega_3)t \tag{6.40b}$$

We know that the displacements q_i are real. Hence we set

$$a_{13} = \tfrac{1}{2}C_2 \exp i\psi_3, \quad a_{14} = \tfrac{1}{2}C_2 \exp(-i\psi_3) \tag{6.41a}$$

$$\frac{a_{21}}{a_{11}} = \alpha_1, \quad \frac{a_{22}}{a_{12}} = \alpha_2, \tag{6.41b}$$

$$\frac{a_{23}}{a_{13}} = \mu \exp i\Psi_3, \quad \frac{a_{24}}{a_{14}} = \mu \exp(-i\Psi_3) \tag{6.41c}$$

We note that the real constants α_1, α_2, μ, Ψ_3 are readily obtained with the aid of (6.32). If equations (6.41) and (6.40) are combined, the result is

$$\begin{aligned} q_1(t) &= a_{11}\exp \kappa_1 t+a_{12}\exp \kappa_2 t+C\exp \kappa_3 t\cdot\cos(\omega_3 t+\psi_3) \\ q_2(t) &= a_{11}\alpha_1\exp \kappa_1 t+a_{12}\alpha_2\exp \kappa_2 t+\mu C\exp \kappa_3 t\cdot\cos(\omega_3 t+\psi_3+\Psi_3) \end{aligned} \tag{6.42}$$

The four constants a_{11}, a_{12}, C and ψ_3 can be determined by the application of initial conditions to (6.42).

6.4 FORCED, HARMONIC EXCITATION OF DAMPED SYSTEMS

The case of applied, harmonic generalized forces affords an interesting and useful comparison between the behavior of a damped system and the corresponding undamped (conservative) system. Let the generalized forces $Q_j(t)$ be periodic with frequency ω and complex amplitude C_j. Then (6.14) becomes

$$\sum_{i=1}^{n}(m_{ij}\ddot{q}_i + c_{ij}\dot{q}_i + k_{ij}q_i) = C_j \exp i\omega t \tag{6.43}$$

$$j=1,2,\ldots,n$$

Discarding the transient part of the solution which tends to zero as $t\to\infty$, we shall focus on the steady state solution of (6.43) which we characterize by

$$q_i(t) = A_i \exp i\omega t \tag{6.44}$$

Substitution of (6.44) into (6.43) results in the system of algebraic equations

$$\sum_{i=1}^{n} A_i(-\omega^2 m_{ij} + i\omega c_{ij} + k_{ij}) = C_j \qquad j=1,2,\ldots,n \tag{6.45}$$

The solution of (6.45) is readily obtained by the application of Cramer's rule:

$$A_i = \frac{D_i(\lambda)}{D(\lambda)}, \text{ where } \lambda = i\omega \tag{6.46}$$

The symbol $D(\lambda)$ stands for the determinant of the coefficients of A_i in (6.45), while $D_i(\lambda)$ is the same determinant as $D(\lambda)$ except that entries in the i'th column are replaced by $C_1, C_2, \ldots, C_n$. We note that the determinant $D(\lambda)$ is associated with the corresponding free vibration problem, and the 2n roots of $D(\lambda)=0$ play a key role in the solution of the latter (see (6.19)). For this reason $D(\lambda)$ can be written

$$D(\lambda) = B(\lambda_1-\lambda)(\lambda_2-\lambda)\ldots\ldots.(\lambda_{2n}-\lambda) \tag{6.47}$$

where B is a constant and λ_r, $r=1,2,\ldots,2n$ are the roots of (6.19). In most cases of practical interest, damping will be sufficiently small to cause (6.19) to have n pairs of conjugate complex roots, and therefore we set

$$\left.\begin{aligned} \lambda_r &= \kappa_r + i\omega_r \\ \lambda_{n+r} &= \kappa_r - i\omega_r \\ r&=1,2,\ldots,n \end{aligned}\right\} \tag{6.48}$$

If the product notation is employed, (6.47) in conjunction with (6.48) assumes the form

$$D(i\omega) = B\prod_{r=1}^{n} [\kappa_r - i(\omega-\omega_r)]\,[\kappa_r - i(\omega+\omega_r)] \tag{6.49}$$

In the solution of a problem of this type, it becomes necessary to separate A_i in (6.46) into real and imaginary parts. This, in turn, requires that the denominator of the right-hand side of (6.46) be rationalized. Multiplication of D by its complex conjugate D* results in

$$D(i\omega)\cdot D^*(i\omega) = B^2 \prod_{r=1}^{n} [\kappa_r^2+(\omega-\omega_r)^2][\kappa_r^2+(\omega+\omega_r)^2] \tag{6.50}$$

In view of (6.24), the quantity κ_r is related to the damping coefficients c_{ij} of the system. With reference to (6.44), (6.46), and (6.50), we note that in the presence of damping, the solution always remains bounded, but amplitudes can be made as large as desired by choosing the damping coefficients sufficiently small. When the damping coefficients c_{ij} vanish, $\kappa_r=0$, $r=1,2,\ldots,n$ and the forced vibration amplitude becomes unbounded at the natural frequencies of the undamped system, i.e., the forcing frequency coincides with one of the natural frequencies of the undamped system, and we have the condition of resonance. It is also interesting to note that if the damping coefficients c_{ij} (or κ_r) are sufficiently small, the resonant frequencies of the damped system will be very close to that of the corresponding undamped system.

As a specific application of the foregoing, we now consider the damped two-degree of freedom system shown in Fig. 6.1, and assume that a harmonic force $Q_2(t)=C_2 \exp i\omega t$ is applied to the mass m_{22}. In this case the appropriate equations of motion are

$$m_{11}\ddot{q}_1 + c_{11}\dot{q}_1 + c_{12}\dot{q}_2 + k_{11}q_1 + k_{12}q_2 = 0 \tag{6.51a}$$

$$m_{22}\ddot{q}_2 + c_{22}\dot{q}_2 + c_{12}\dot{q}_1 + k_{22}q_2 + k_{12}q_1 = C_2 \exp i\omega t \tag{6.51b}$$

Neglecting the transient part of the solution, we assume a steady state solution of the form

$$q_1(t) = A_1 \exp i\omega t, \quad q_2(t) = A_2 \exp i\omega t \tag{6.52}$$

Upon substitution of (6.52) into (6.51), we obtain

$$\begin{aligned} A_1(-m_{11}\omega^2+c_{11}i\omega+k_{11}) + A_2(c_{12}i\omega+k_{12}) &= 0 \\ A_1(c_{12}i\omega+k_{12}) + A_2(-m_{22}\omega^2+c_{22}i\omega+k_{22}) &= C_2 \end{aligned} \tag{6.53}$$

Solving A_1 and A_2, we obtain

$$A_1 = -\frac{C_2(c_{12}i\omega+k_{12})}{D(i\omega)} \tag{6.54a}$$

$$A_2 = \frac{C_2(-m_{11}\omega^2+c_{11}i\omega+k_{11})}{D(i\omega)} \tag{6.54b}$$

where

$$D(i\omega)=(-m_{11}\omega^2+c_{11}i\omega+k_{11})(-m_{22}\omega^2+c_{22}i\omega+k_{22})-(c_{12}i\omega+k_{12})^2$$
$$=m_{11}m_{22}(\lambda_1-i\omega)(\lambda_2-i\omega)(\lambda_3-i\omega)(\lambda_4-i\omega) \tag{6.55}$$

and where λ_1, λ_2, λ_3, λ_4 are the roots of the equation

$$D(\lambda) = 0 \tag{6.56}$$

Let us now assume that the damping coefficients c_{11}, c_{12}, c_{22} are sufficiently small so that the roots of (6.56) are given by

$$\begin{aligned}\lambda_1 &= \kappa_1+i\omega_1, \quad \lambda_2 = \kappa_1-i\omega_1\\ \lambda_3 &= \kappa_2+i\omega_2, \quad \lambda_4 = \kappa_2-i\omega_2\end{aligned} \tag{6.57}$$

Substitution of (6.57) into (6.56) results in

$$D(i\omega)=m_{11}m_{22}[\kappa_1-i(\omega-\omega_1)]\,[\kappa_1-i(\omega+\omega_1)][\kappa_2-i(\omega-\omega_2)][\kappa_2-i(\omega+\omega_2)]$$

and therefore

$$D(i\omega)\cdot D^*(i\omega) =$$

$$= m_{11}^2m_{22}^2\left[\kappa_1^2+(\omega-\omega_1)^2\right]\left[\kappa_1^2+(\omega+\omega_1)^2\right]\left[\kappa_2^2+(\omega-\omega_2)^2\right]\left[\kappa_2^2+(\omega+\omega_2)^2\right] \tag{6.58}$$

With reference to (6.52), (6.54), and (6.58), it is now clear that the displacement amplitudes will remain bounded as long as $\kappa_1\neq 0$ and $\kappa_2\neq 0$. However, these amplitudes can be made as large as desired by selecting a sufficiently small κ_1 for $\omega=\omega_1$ or a sufficiently small κ_2 for $\omega=\omega_2$.

6.5 CLASSICAL NORMAL MODES IN THE PRESENCE OF DAMPING

When an undamped linear dynamical system vibrates in a particular normal mode, the various parts of the system will all vibrate in the same phase (or 180° out of phase). In general, this property fails for damped systems, and classical normal modes in the sense of Chapter 3 do not exist. We shall now show that the concept of classical normal modes is applicable to damped systems, provided that the matrix of damping coefficients can be characterized as a linear combination of the mass and stiffness matrices, i.e.,

$$c_{ij} = \alpha m_{ij} + \beta k_{ij} \tag{6.59}$$

where α and β are real constants. Upon substitution of (6.59) into (6.14), we obtain

$$\sum_{i}^{n} m_{ij}\ddot{q}_i + (\alpha m_{ij}+\beta k_{ij})\dot{q}_i + k_{ij}q_i = Q_j \tag{6.60}$$

If we now consider the free vibrations of the associated undamped problem characterized by (6.60) with $Q_j=0$ and $\alpha=\beta=0$, then we can readily obtain a set of frequencies ω_r and modal vector components a_{ir} as described in Chapter 3. The eigenvector components of the reduced (undamped) problem satisfy the relations (see (3.28) and (3.25a))

$$\sum_{i,j}^{n} m_{ij}a_{ir}a_{js} = \delta_{rs} \tag{6.61a}$$

$$\sum_{i,j}^{n} k_{ij}a_{ir}a_{js} = \omega_r^2\delta_{rs} \tag{6.61b}$$

We are now ready to transform (6.60) to normal coordinates. Let

$$q_i = \sum_{r}^{n} a_{ir}\xi_r \tag{6.62}$$

Upon substitution of (6.62) into (6.60) we obtain

$$\sum_{i,r}^{n} m_{ij}a_{ir}\ddot{\xi}_r + (\alpha m_{ij}+\beta k_{ij})a_{ir}\dot{\xi}_r + k_{ij}a_{ir}\xi_r = Q_j \tag{6.63}$$

We now multiply (6.63) by a_{js} and sum over the index j. The result is

$$\sum_{r}^{n}\sum_{i,j}^{n} m_{ij}a_{ir}a_{js}\ddot{\xi}_r+(\alpha m_{ij}a_{ir}a_{js}+\beta k_{ij}a_{ir}a_{js})\dot{\xi}_r+k_{ij}a_{ir}a_{js}\xi_r$$

$$= \sum_{j}^{n} Q_j a_{js} = P_s \tag{6.64}$$

If relations (6.61) are now applied to (6.64), we obtain

$$\ddot{\xi}_r + 2\theta_r\dot{\xi}_r + \omega_r^2\xi_r = P_r \tag{6.65}$$

where

$$2\theta_r = \alpha + \beta\omega_r^2$$

With the aid of (1.50), the solution of (6.65) is given by

$$\xi_r(t) = \exp(-\theta_r t)\cdot\left(A_r\cos\sqrt{\omega_r^2-\theta_r^2}\; t + \frac{B_r+\theta_r A_r}{\sqrt{\omega_r^2-\theta_r^2}}\sin\sqrt{\omega_r^2-\theta_r^2}\; t\right)$$

$$+ \frac{1}{\sqrt{\omega_r^2-\theta_r^2}}\int_0^t P(t-\tau)\cdot\exp(-\theta_r\tau)\cdot\sin\sqrt{\omega_r^2-\theta_r^2}\;\tau\cdot d\tau \tag{6.66}$$

where $A_r=\xi_r(0)$, $B_r=\dot{\xi}_r(0)$, and where, with the aid of (3.49), we obtain

$$A_r = \xi_r(0) = \sum_j^n a_{jr}\sum_i^n m_{ij}q_i(0) \tag{6.67a}$$

$$\omega_r B_r = \omega_r\dot{\xi}_r(0) = \sum_j^n a_{jr}\sum_i^n m_{ij}\dot{q}_i(0) \tag{6.67b}$$

EXERCISES

6.1 Consider the damped, two-degree of freedom system shown in Fig. 6.1. The mass m_{22} is acted upon by the force $Q_2(t)=QH(t)$, where $H(t)$ is the Heaviside unit step function, and $q_1(0)=q_2(0)=\dot{q}_1(0)=\dot{q}_2(0)=0$. Let $c_1=\beta k_1$, $c_2=\beta k_2$, $c_3=\alpha m_{11}$, and $c_4=\alpha m_{22}$, where α and β are real constants. Find the dynamic response of the system and give a full discussion of the motion, with particular reference to the damping characteristics of the structure.

6.2 Consider two identical harmonic oscillators that are coupled via a force that is proportional to the relative velocity of the two masses. Discuss the normal modes of oscillation of this system.

6.3 Extend the uniqueness theorem in Section 2 of Chapter 4 to the problem characterized by (6.14) and appropriate initial conditions.

6.4 Modify Exercise 4.3 to include the effects of viscous damping in such a way as to preserve classical normal modes.

6.5 Modify all parts of Exercise 4.8 to include the effects of viscous damping in such a way as to preserve classical normal modes.

Chapter

7

WAVE PROPAGATION IN ITERATED STRUCTURES

The question of the vibration of connected particles is a peculiarly interesting and important problem . . . it is going to have many applications.

Lord Kelvin, Baltimore Lectures, 1884.

7.1 INTRODUCTION

In Chapters 3 through 6 we have considered the motion of bounded, discrete mechanical systems. For example, the iterated structure treated in Chapter 4 consists of n rigid discs connected by n mass-less shaft segments, each with torsional stiffness c as shown in Fig. 4.8. When n is a finite number, we showed that the free vibrations of the iterated structure gave rise to n modes of oscillations characterized by the equations

$$\theta_{ir} = \sin i\lambda_r \cos \omega_r t, \quad \begin{matrix} i=1,2,\ldots,n \\ r=1,2,\ldots,n \end{matrix} \tag{7.1}$$

where

$$\omega_r = 2\sqrt{\frac{c}{I}} \sin \frac{(2r-1)\pi}{(2n+1)2} \tag{7.2}$$

$$\lambda_r = \frac{(2r-1)\pi}{(2n+1)} \tag{7.3}$$

Equation (7.1) characterizes n different standing waves, i.e., any two θ_{jr} and θ_{kr}, $j \neq k$, will oscillate either in phase or exactly 180° out of phase. For the present purpose it will be instructive to re-write (7.1) in a somewhat different form. Toward this end we set

$$z=ia, \quad k_r = \frac{\lambda_r}{a}, \quad v_r = \frac{\omega_r}{k_r}$$

where $i=1,2,\ldots,n$ and a is the distance between discs. With these notations, (7.1) can be re-written to read

$$\theta_i(z,t) = \theta_{ir} = \tfrac{1}{2}\sin k_r(v_r t+z) - \tfrac{1}{2}\sin k_r(v_r t-z) \tag{7.4}$$

The representation (7.4) admits an alternate interpretation of a normal mode oscillation in this case. Each standing wave consists of the superposition of the two traveling waves $\tfrac{1}{2} \sin k_r(v_r t+z)$ and $\tfrac{1}{2} \sin k_r(v_r t-z)$, the former advancing in the negative z-direction, and the latter advancing in the positive z-direction. With reference to the nomenclature developed in Appendix B, each wave travels with phase velocity v_r and wave number k_r. The traveling waves are continuously reflected from the boundaries of the iterated structure at $z=0$ and $z=na$, and their superposition gives rise to a standing wave. Moreover, because of the boundary conditions (4.45), the values of v_r and k_r cannot be assigned arbitrarily, and each pair must assume the appropriate values

$$k_r = \frac{\lambda_r}{a}\,, \qquad v_r = \frac{\omega_r}{k_r}\,, \qquad r=1,2,\ldots,n.$$

The present chapter is concerned with the propagation of waves in unbounded, iterated structures, and is designed to explain the process of propagation of force (or moment), deformation, energy, and momentum in a discrete mechanical system. In addition, the study of harmonic wave motion in an iterated structure serves to develop the concept of a mechanical wave filter, whereas the consideration of transient wave motion in such a structure is useful in the assessment of initial structural response of a similarly constructed device of bounded extent before wave reflections have occurred.

7.2 ENERGY AND MOMENTUM FLUX IN A SIMPLE ITERATED STRUCTURE

Let us consider the iterated structure shown in Fig. 4.7. The appropriate equations of motion were established in Chapter 4 and are given by (4.43) and (4.40). For the present purpose, these equations are now rewritten as follows:

$$I\ddot{\theta}_{r+1} - M_{r+1} + M_r = 0 \tag{7.5}$$

$$M_r = c(\theta_{r+1} - \theta_r) \tag{7.6}$$

Upon substitution of (7.6) into (7.5), we obtain the displacement equation of motion

$$I\ddot{\theta}_{r+1} = c(\theta_{r+2} - 2\theta_{r+1} + \theta_r) \tag{7.7}$$

We shall first consider the flux of energy (energy per unit of time) at a generic point in our structure. Let us denote by E* the total energy at any instant of time t which is contained in a segment of the iterated structure extending from r=p to r=q:

$$E^* = T+V = \sum_{j=p}^{j=q-1} \tfrac{1}{2}I\dot{\theta}_{j+1}^2 + \tfrac{1}{2}c(\theta_{j+1} - \theta_j)^2 \tag{7.8}$$

If (7.8) is differentiated with respect to time, we obtain

$$\begin{aligned}\dot{E}^* &= \sum_{j=p}^{j=q-1}\left[\dot{\theta}_{j+1} I\ddot{\theta}_{j+1} + c(\theta_{j+1}-\theta_j)(\dot{\theta}_{j+1}-\dot{\theta}_j)\right] \\ &= \sum_{j=p}^{j=q-1}\left[\dot{\theta}_{j+1}(I\ddot{\theta}_{j+1} - M_{j+1} + M_j) + (\dot{\theta}_{j+1}M_{j+1} - \dot{\theta}_j M_j)\right] \\ &= \dot{\theta}_q M_q - \dot{\theta}_p M_p \qquad (7.9)\end{aligned}$$

where we have utilized (7.5) and (7.6).

If we now define the energy flux at r as

$$S_r = -M_r\dot{\theta}_r \qquad (7.10)$$

then

$$\dot{E}^* = S_p - S_q \qquad (7.11)$$

Equation (7.11) can be interpreted as an energy continuity equation. It expresses the fact that the instantaneous rate of change of energy contained in a segment of the lattice structure between $r=p$ and $r=q$, $p<q$, is equal to the energy flux entering at $r=p$ minus the energy flux leaving the segment at $r=q$. It is assumed, of course, that there are no energy sources located between $r=p$ and $r=q$.

We note that E^*_r is the total instantaneous energy contained between the points $r+1$ and r in the iterated structure. Therefore the appropriate definition of energy density (energy per unit of length, in this case) is given by ε_r, where

$$E^*_r = a\varepsilon_r = T_r + V_r \qquad (7.12)$$

and

$$T_r = \tfrac{1}{2}I\dot{\theta}^2_{r+1}; \qquad V_r = \tfrac{1}{2}c(\theta_{r+1}-\theta_r)^2 \qquad (7.13)$$

and where the symbol a denotes the characteristic length of the lattice i.e., a is the distance between any two adjacent discs. We now define the velocity of energy transport of a harmonic wave as the time average of energy flux divided by the time average of energy density, i.e.,

$$\text{velocity of energy transport} = \frac{\langle S_r\rangle}{\langle \varepsilon_r\rangle} \qquad (7.14)$$

It will be shown that in the present case, the velocity of energy transport as obtained from (7.14) is equal to the group velocity v_G as defined in Appendix B.

We shall next consider the momentum flux in the lattice. The instantaneous angular momentum of the lattice between points r=p and r=q is given by the expression

$$H^* = \sum_{j=p}^{j=q-1} I\dot{\theta}_{j+1} \tag{7.15}$$

Differentiating (7.15) with respect to time, we obtain

$$\dot{H}^* = \sum_{j=p}^{j=q-1} I\ddot{\theta}_{j+1} = \sum_{j=p}^{j=q-1} (M_{j+1}-M_j) = M_q-M_p \tag{7.16}$$

where we used (7.5). It is now appropriate to define the momentum flux at r by

$$J^*_r = -M_r \tag{7.17}$$

Consequently,

$$\dot{H}^* = J^*_p - J^*_q \tag{7.18}$$

Equation (7.18) may be interpreted as an (angular) momentum continuity equation. It expresses the fact that the rate of change of the instantaneous angular momentum contained in a segment of the lattice structure extending from r=p to r=q, p<q, is equal to the momentum flux entering at r=p minus the momentum flux leaving the structure at r=q.

7.3 HARMONIC WAVE PROPAGATION IN AN UNBOUNDED, SIMPLE, PERIODIC STRUCTURE (A Mechanical Low Pass Filter)

Let us consider the possibility of a traveling, harmonic wave progressing in the positive z-direction of an unbounded lattice structure characterized by

$$\theta_r = A \exp i(\omega t - \nu r) \tag{7.19}$$

where the amplitude A is a real constant, $\nu = ka$ is the dimensionless wave number and $i = \sqrt{-1}$. When working with the complex characterization (7.19), it is understood that we shall be interested in either the real or imaginary part of (7.19) in all physical applications. Upon substitution of (7.19) into (7.7), we readily obtain

$$\Omega^2 = 4\sin^2 \frac{\nu}{2} = 2(1-\cos \nu)$$

or

$$\Omega = 2 \sin \frac{\nu}{2}, \quad \cos \nu = 1- \frac{\Omega^2}{2}, \quad \sin \nu = \Omega \sqrt{1- \frac{\Omega^2}{4}} \tag{7.20}$$

where $\Omega=\sqrt{I/c}\ \omega>0$. To insure a real (dimensionless) wave number ν, it is necessary that $0<\sin^2 \nu/2<1$, and in view of (7.20) this will be the case if values of the frequency are restricted to $0<\Omega^2/4<1$. When $\Omega>2$, $\cos \nu<-1$, and therefore ν will be complex. For this reason a somewhat different form of the solution is indicated. Toward this end, let $\nu=\eta-i\mu$, where η and $\mu>0$ are real numbers. In this case

$$\cos \nu = \cos(\eta-i\mu) = \cos \eta \cosh \mu+i \sin \eta \sinh \mu<-1$$

and the inequality will be satisfied by setting $\eta=\pi$. In this case

$$\cos \nu = \cos(\pi-i\mu) = -\cosh \mu = 1- \frac{\Omega^2}{2}$$

or

$$\cosh \mu = \frac{\Omega^2}{2} -1 \tag{7.21}$$

and the solution corresponding to $\Omega>2$ assumes the form

$$\begin{aligned}\theta_r &= A \exp i[\omega t-(\pi-i\mu)r] \\ &= A(-1)^r \cdot \exp(-\mu r) \cdot \exp i\omega t\end{aligned} \tag{7.22}$$

where μ is defined by (7.21). It is now possible to draw the following conclusions from the preceding analysis: It is possible to propagate a sinusoidal wave (7.19) through the unbounded, iterated structure in the low frequency range $0\leq\Omega\leq2$. In this case, the dimensionless wave number $\nu=ka$ which corresponds to the dimensionless frequency Ω is obtained from (7.20), the latter being a dispersion relation, a term which is motivated in Appendix B. In the high frequency range $\Omega>2$, wave propagation is not possible, and we obtain a standing wave (7.22) which is exponentially attenuated as r increases. This result also justifies our choice of $\mu>0$, since a μ which is negative would result in an exponentially increasing standing wave amplitude with increasing r, a result which can be ruled out on physical grounds. The dimensionless frequency $\Omega=2$ which divides the resulting motion into traveling and standing waves is called the cut-off frequency of the lattice. Graphs of the dispersion relation (7.20) and the attenuation number μ defined by (7.21) are shown in Fig. 7.1. We note that for $\Omega>2$, the attenuation number increases with increasing frequency Ω.

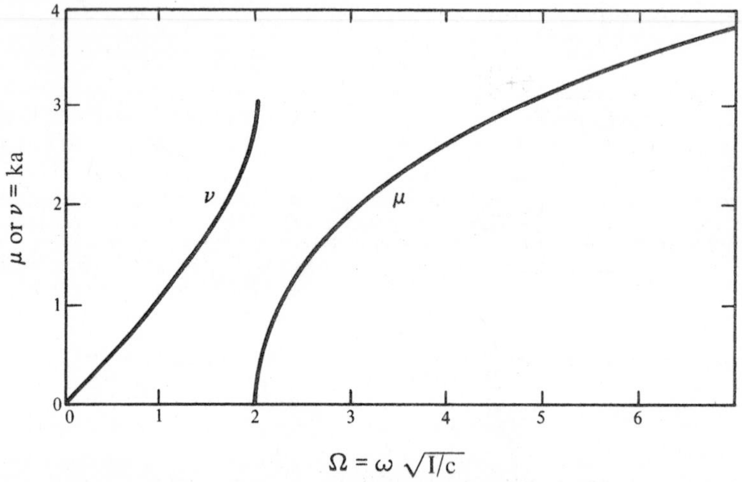

Figure 7.1

We shall now utilize (7.14) to calculate the velocity of energy transport in the iterated structure.

(a) The Low Frequency Band: $0<\Omega<2$.

In this case it is convenient to work entirely with real quantities, and therefore we take the real part of (7.19), i.e.,

$$\theta_r = A\cos(\omega t-\nu r) \tag{7.23}$$

By direct substitution of (7.23) into (7.6), we obtain

$$M_r = Ac\left[(\cos\nu-1)\cdot\cos(\omega t-\nu r) + \sin\nu\cdot\sin(\omega t- r)\right] \tag{7.24}$$

Further substitution of (7.23) and (7.24) into (7.10) results in

$$S_r=\omega A^2c\left[\sin\nu\cdot\sin^2(\omega t-\nu r)+(\cos\nu-1)\cdot\sin(\omega t-\nu r)\cdot\cos(\omega t-\nu r)\right] \tag{7.25}$$

Substitution of (7.23) into (7.13) results in

$$T_r=\tfrac{1}{2}cA^2\Omega^2\left[\cos^2\nu\cdot\sin^2(\omega t-\nu r)+\sin^2\nu\cdot\cos^2(\omega t-\nu r)\right.$$
$$\left.-2\sin\nu\cdot\cos\nu\cdot\sin(\omega t-\nu r)\cdot\cos(\omega t-\nu r)\right] \tag{7.26}$$

$$V_r = \frac{M_r^2}{2c} = \frac{A^2c}{2}\,[(\cos\nu-1)^2\cdot\cos^2(\omega t-\nu r)+\sin^2\nu\cdot\sin^2(\omega t-\nu r)$$

$$+\;2\sin\nu\cdot(\cos\nu-1)\cdot\sin(\omega t-\nu r)\cdot\cos(\omega t-\nu r)] \tag{7.27}$$

We now recall the following expressions for the time averages of presently encountered trigonometric functions:

$$\langle\sin^2(\omega t+\phi)\rangle = \frac{1}{\frac{n}{2}T}\int_t^{t+\frac{n}{2}T}\sin^2(\omega\tau+\phi)d\tau = \tfrac{1}{2} \tag{7.28a}$$

$$\langle\cos^2(\omega t+\phi)\rangle = \frac{1}{\frac{n}{2}T}\int_t^{t+\frac{n}{2}T}\cos^2(\omega\tau+\phi)d\tau = \tfrac{1}{2} \tag{7.28b}$$

$$\langle\sin(\omega t+\phi)\cdot\cos(\omega t+\phi)\rangle=\frac{1}{\frac{n}{2}T}\int_t^{t+\frac{n}{2}T}\sin(\omega\tau+\phi)\cdot\cos(\omega\tau+\phi)\cdot d\tau = 0 \tag{7.28c}$$

where $T=2\pi/\omega$, $n=1,2,\ldots$, and ϕ is an arbitrary phase angle. We now take time averages of (7.25), (7.26), and (7.27), and apply (7.28) as needed. The result is

$$\langle S_r\rangle = \tfrac{1}{2}\omega A^2 c\sin\nu \tag{7.29}$$

$$\langle T_r\rangle = \langle V_r\rangle = \tfrac{1}{4}cA^2\Omega^2 \tag{7.30}$$

and therefore

$$\langle a\varepsilon_r\rangle = \langle T_r\rangle + \langle V_r\rangle = \tfrac{1}{2}cA^2\Omega^2 \tag{7.31}$$

Upon substitution of (7.29) and (7.31) into (7.14) we obtain the desired result

$$\text{velocity of energy transport} = a\sqrt{c/I}\cos\frac{\nu}{2} \tag{7.32}$$

where we have used (7.20) to facilitate the reduction. In the present case, the group velocity v_G as defined in Appendix B is

$$v_G = \frac{d\omega}{dk} = a\sqrt{\frac{c}{I}}\frac{d\Omega}{d\nu} = a\sqrt{\frac{c}{I}}\cos\frac{\nu}{2} = a\sqrt{\frac{c}{I}}\sqrt{1-\frac{\Omega^2}{4}} \tag{7.33}$$

i.e., for the present case, the velocity of energy transport is equal to the group velocity. With reference to (7.23) and the discussion in Appendix B, the phase velocity is given by

$$v = \frac{\omega}{k} = a\sqrt{\frac{c}{I}}\left(\frac{\sin\frac{\nu}{2}}{\frac{\nu}{2}}\right) = a\sqrt{\frac{c}{I}}\;\frac{\Omega}{\cos^{-1}\left(1-\frac{\Omega^2}{2}\right)} \tag{7.34}$$

(b) The High Frequency Range: $2<\Omega$.

In this case we take the real part of (7.22):

$$\theta_r = A(-1)^r\cdot\exp(-\mu r)\cdot\cos\omega t \tag{7.35}$$

By direct substitution of (7.35) into (7.6), we obtain

$$M_r = -cA(-1)^r(1+e^{-\mu})e^{-\mu r}\cos\omega t \tag{7.36}$$

Further substitution of (7.35) and (7.36) into (7.10) results in

$$S_r = -\omega cA^2(1+e^{-\mu})e^{-2\mu r}\cdot\sin\omega t\cos\omega t \tag{7.37}$$

Substitution of (7.35) into (7.13) results in

$$T_r = \tfrac{1}{2}c\Omega^2A^2e^{-2\mu(r+1)}\cdot\sin^2\omega t \tag{7.38}$$

$$V_r = \tfrac{1}{2}cA^2(1+e^{-\mu})^2e^{-2\mu r}\cdot\cos^2\omega t \tag{7.39}$$

If we now take time averages of (7.37), (7.38), and (7.39) and utilize (7.28), we obtain

$$\langle S_r\rangle = 0 \tag{7.40}$$

$$\langle T_r\rangle = \tfrac{1}{4}c\Omega^2A^2e^{-2\mu(r+1)} \tag{7.41a}$$

$$\langle V_r\rangle = \tfrac{1}{4}cA^2(1+e^{-\mu})^2e^{-2\mu r} \tag{7.41b}$$

and therefore, after a certain amount of reduction,

$$\langle a\varepsilon_r\rangle = \langle T_r\rangle + \langle V_r\rangle = \tfrac{1}{4}cA^2(1+e^{-\mu})^3e^{-2\mu r} > 0 \tag{7.42}$$

Substitution of (7.40) and (7.42) into (7.14) reveals that the velocity of energy transport vanishes in the frequency range $2<\Omega$.

The results of the preceding analysis are embodied in Fig. 7.2, which shows dimensionless plots of (7.33) and (7.34). In the low frequency band $0<\Omega<2$, (or wave number band $0<\nu\leq\pi$), the lattice structure will allow a wave to propagate and to transport energy in the process. In the high frequency range $2<\Omega$, propagating waves are not possible, and only standing, exponentially attenuated waves can exist. Such a "wave" does not transport energy, and its phase and group velocities vanish. We also observe that in the low frequency range the phase velocity differs from the group velocity, and within the context of Appendix B, the lattice structure constitutes a dispersive medium. Since traveling waves are transmitted in the low frequency range only, and since these are ruled out in the high frequency range, the present lattice structure is also referred to as a mechanical low pass filter with a well defined cut-off frequency $\Omega=2$.

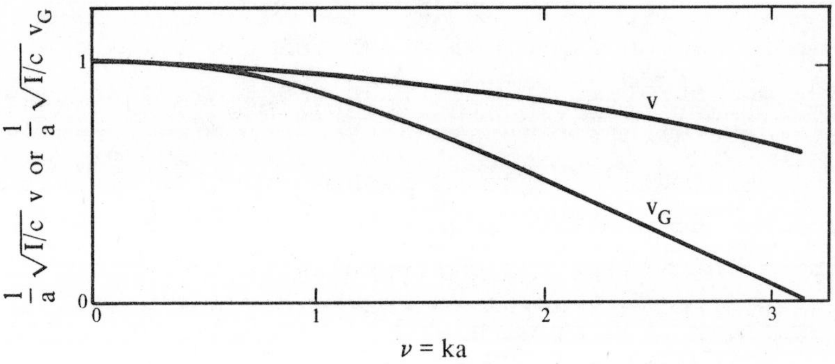

(a)

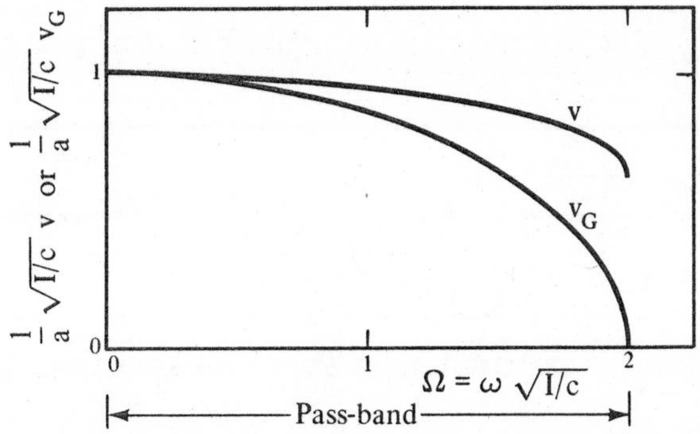

(b)

Figure 7.2

7.4 TRANSMISSION AND REFLECTION OF HARMONIC WAVES IN AN ITERATED STRUCTURE

Let us consider the iterated structure shown in Fig. 7.3. It will be convenient to designate as region 1 and region 2 those parts of the structure for which $z>0$ and $z\leq 0$, respectively. The discs in regions 1 and 2 are rigid.

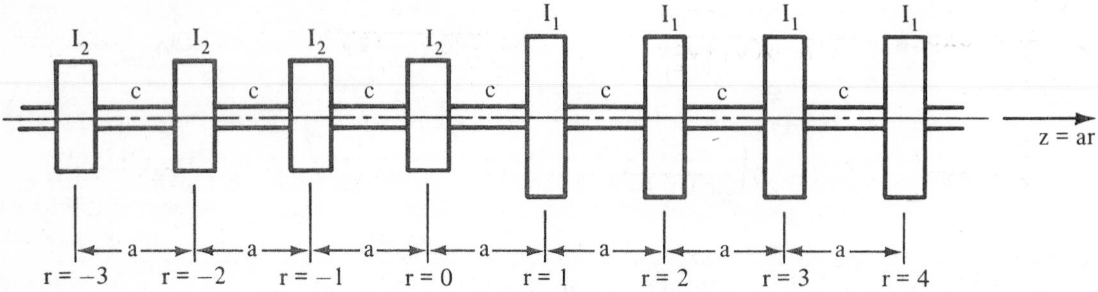

Figure 7.3

Each disc in region 1 has moment of inertia I_1, while each disc in region 2 has a moment of inertia I_2. The discs are connected by massless shaft segments with torsional stiffness c. It is now assumed that there is an energy source at $z\to\infty$, causing a time-harmonic torsional wave to travel in the negative z-direction. In the steady state, the "incoming" traveling wave $A \exp i(\omega t+r\nu_1)$ will occupy all of region 1. The wave is characterized by the frequency ω and (real) amplitude A, and both of these quantities are assumed to be specified. In the general case, the incoming wave gives rise to a "reflected" traveling wave $B \exp i(\omega t-r\nu_1)$, also in region 1. In addition, there will be a "transmitted" wave $C \exp i(\omega t+r\nu_2)$ in region 2. Consequently we have

$$\theta_r = A \exp i(\omega t+r\nu_1) + B \exp i(\omega t-r\nu_1), \quad r=1,2,.... \tag{7.43a}$$

$$\theta_r = C \exp i(\omega t+r\nu_2), \quad r=0, -1, -2,.... \tag{7.43b}$$

To determine the (complex) constants B and C, we write the equations of motion for the discs at r=0 and at r=1:

$$I_2\ddot{\theta}_0 = c(\theta_1-2\theta_0+\theta_{-1}) \tag{7.44a}$$

$$I_1\ddot{\theta}_1 = c(\theta_2-2\theta_1+\theta_0) \tag{7.44b}$$

Upon substitution of (7.43) into (7.44) and subsequent simplification utilizing the dispersion relation (7.20), we obtain

$$A + B = C$$

$$A \exp i\nu_1 + B \exp(-i\nu_1) = C \exp i\nu_2$$

Solving

$$\frac{B}{A} = \exp i\nu_1 \cdot \frac{\sin \frac{1}{2}(\nu_1-\nu_2)}{\sin \frac{1}{2}(\nu_1+\nu_2)} = \exp i\nu_1 \cdot \frac{v_G^{(2)}-v_G^{(1)}}{v_G^{(2)}+v_G^{(1)}} \tag{7.45a}$$

$$\frac{C}{A} = \exp\left[i\ \tfrac{1}{2}(\nu_1-\nu_2)\right] \cdot \frac{\sin \nu_1}{\sin \frac{1}{2}(\nu_1+\nu_2)} = 2\sqrt{\frac{I_1}{I_2}} \cdot \exp\left[i\tfrac{1}{2}(\nu_1-\nu_2)\right] \cdot \frac{v_G^{(1)}}{v_G^{(2)}+v_G^{(1)}} \tag{7.45b}$$

where $v_G^{(1)}$ and $v_G^{(2)}$ denote the group velocities in regions 1 and 2 of the structure, respectively (see (7.33)). Substitution of (7.45) into (7.43) results in

$$\frac{\theta_r}{A} = \exp i(\omega t+r\nu_1) + \frac{v_G^{(2)}-v_G^{(1)}}{v_G^{(2)}+v_G^{(1)}} \exp i(\omega t-r\nu_1+\nu_1),\ r=1,2,\ldots. \tag{7.46a}$$

$$\frac{\theta_r}{A} = 2\sqrt{\frac{I_1}{I_2}}\ \frac{v_G^{(1)}}{v_G^{(2)}+v_G^{(1)}} \exp i(\omega t+r\nu_2 + \frac{\nu_1}{2} - \frac{\nu_2}{2}) \tag{7.46b}$$

$r=0,\ -1,\ -2\ldots.$

The representation (7.46) is useful when $0<\omega<2\sqrt{c/I_1}$ and $0<\omega<2\sqrt{c/I_2}$, i.e., when $0<\omega^2<4c/\sqrt{I_1 I_2}$. In this case we shall have a traveling transmitted wave. When $2\sqrt{c/I_2}<\omega<2\sqrt{c/I_1}$, $I_2>I_1$ and we have an exponentially attenuated, standing wave. In this case the entire incoming energy will be reflected at $r=0$, and no energy will be transmitted to $z\to-\infty$. In this case we set $\nu_2=\pi-i\mu_2$, where $\cosh \mu_2=\frac{1}{2}\left(I_2/c\right)\omega^2-1$ according to (7.21). A certain amount of manipulation reveals that in this instance

$$\frac{\sin \frac{1}{2}(\nu_1-\nu_2)}{\sin \frac{1}{2}(\nu_1+\nu_2)} = -\exp(-i2\phi_1) \tag{7.47}$$

$$\frac{\sin \nu_1}{\sin \frac{1}{2}(\nu_1+\nu_2)} = \frac{\sin \nu_1 \exp(-i\phi_1)}{\sqrt{\cosh^2\frac{\mu_2}{2} - \sin^2\frac{\nu_1}{2}}} = \exp(-i\phi_1)\sqrt{\frac{(4-\Omega_1^2)}{\left(\frac{I_2}{I_1}-1\right)}} \tag{7.48}$$

where

$$\tan \phi_1 = \tan \frac{\nu_1}{2} \tanh \frac{\mu_2}{2} \tag{7.49}$$

Thus, when there is a standing wave in region 2, the appropriate solution is given by

$$\frac{\theta_r}{A} = \exp i(\omega t + r\nu_1) - \exp i(\omega t - r\nu_1 + \nu_1 - 2\phi_1) \qquad r=1,2,\ldots. \tag{7.50a}$$

$$\frac{\theta_r}{A} = \sqrt{\frac{(4-\Omega_1^2)}{\left(\frac{I_2}{I_1}-1\right)}} \cdot (-1)^r \cdot \exp \mu_2(r-\tfrac{1}{2}) \cdot \exp i\left(\omega t + \frac{\nu_1}{2} - \phi_1 - \frac{\pi}{2}\right) \qquad r=0,\ -1,\ -2,\ldots. \tag{7.50b}$$

The following limiting cases of the foregoing solution are noteworthy. When $I_1=I_2=I$ and $\Omega=\omega\sqrt{c/I} < 2$, we have $v_G^{(1)}=v_G^{(2)}$, $\nu_1=\nu_2$, and with reference to (7.46a) the amplitude of the reflected wave vanishes. With reference to (7.46b), we note that the incoming wave and the transmitted wave are now one and the same, and the situation is precisely the one treated in Section 3, except that the wave now travels in the direction of the negative z-axis.

When $I_2 \to \infty$, we have $\mu_2 \to \infty$ and because of (7.49), $2\phi_1=\nu_1$. In this case the solution (7.50) reduces to

$$\frac{\theta_r}{A} = \exp i(\omega t + r\nu_1) - \exp i(\omega t - r\nu_1) \qquad r=1,2,\ldots. \tag{7.51a}$$

$$\frac{\theta_r}{A} = 0, \quad r=0,\ -1,\ -2,\ \ldots. \tag{7.51b}$$

Equation (7.51a) satisfies the boundary condition $\theta_o(t)=0$ for all t. Therefore the limiting case characterized by (7.51a) corresponds to wave motion in the semi-infinite, iterated structure which is rigidly fixed at r=0, as shown in Fig. 7.4. Inspection of (7.51a) reveals that

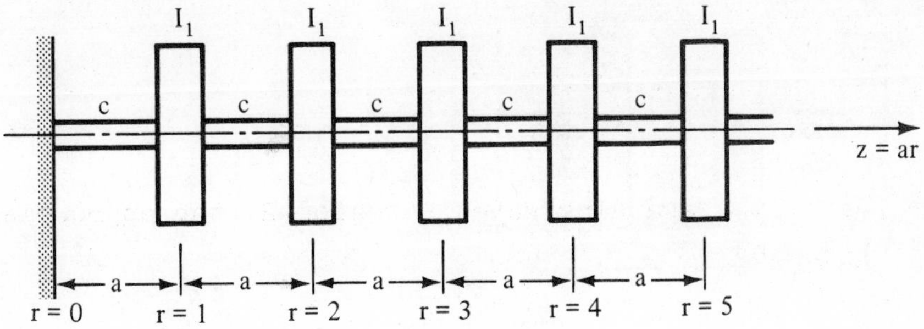

Figure 7.4

the amplitudes of the incoming and reflected waves are equal, and that these waves are exactly 180° out of phase. Taking the real part of (7.51a), we obtain

$$\frac{\theta_r}{A} = -2 \sin r\nu_1 \cdot \sin \omega t. \tag{7.52}$$

Thus, in this instance, the superposition of the incoming and reflected waves results in a standing wave.

Finally, we consider the limiting case $I_2=0$, resulting in $\nu_2=0$ and $v_G^{(2)} \to \infty$. In this case (7.46a) reduces to

$$\frac{\theta_r}{A} = \exp i(\omega t + r\nu_1) + \exp i(\omega t - r\nu_1 + \nu_1) \qquad r=1,2,\ldots. \tag{7.53}$$

Equation (7.53) satisfies the free-end boundary condition $M_0 = c(\theta_1 - \theta_0) = 0$ and therefore (7.53) characterizes the wave motion in the semi-infinite iterated structure which has a free disc at r=1. With reference to (7.53), we note that the amplitudes of the reflected and incoming waves are equal. Moreover, the reflected wave suffers a phase lag of magnitude ν_1 in the process of reflection.

7.5 TRANSIENT WAVE MOTION IN A SIMPLE PERIODIC STRUCTURE

Let us consider torsional wave motion in the semi-infinite, simple periodic structure shown in Fig. 7.5. We shall assume that the wave is generated by a disturbance and/or initial condition on the disc at r=0. For the remainder of the structure we assume quiescent initial conditions, i.e.,

$$\theta_r(0) = \dot{\theta}_r(0) = 0, \quad r=1,2,\ldots. \tag{7.54}$$

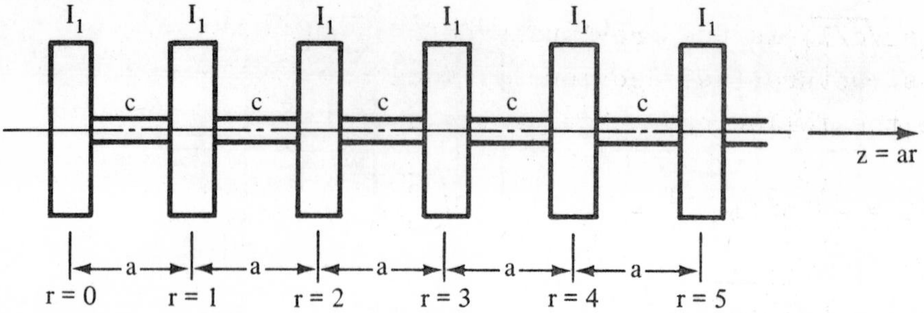

Figure 7.5

If T(t) denotes the time-dependent, external torque applied to the disc at r=0, the appropriate equations of motion are (see (7.5) and (7.6))

$$I_1\ddot{\theta}_o = c(\theta_1-\theta_o)+T(t) \tag{7.55a}$$

$$I_1\ddot{\theta}_r = c(\theta_{r+1}-2\theta_r+\theta_{r-1}) \tag{7.55b}$$

Taking Laplace transforms of (7.55) as explained in Chapter 1, we obtain

$$I_1p^2\overline{\theta}_o = c(\overline{\theta}_1-\overline{\theta}_o) +\overline{T}+I_1(\alpha p+\dot{\alpha}) \tag{7.56a}$$

$$I_1p^2\overline{\theta}_r = c(\overline{\theta}_{r+1}-2\overline{\theta}_r+\overline{\theta}_{r-1}) \tag{7.56b}$$

where

$$\alpha = \theta_o(0), \quad \dot{\alpha} = \dot{\theta}_o(0) \tag{7.57}$$

We now assume a solution of (7.56b) in the form

$$\overline{\theta}_r(p) = Ae^{-\Gamma r} \tag{7.58}$$

where $\Gamma=\Gamma(p)$, and upon substitution of (7.58) into (7.56b) we obtain

$$\sinh \tfrac{1}{2}\Gamma = \frac{p}{\omega_o} \tag{7.59}$$

where $\omega_0 = 2\sqrt{c/I_1}$ is the previously defined cut-off frequency of the lattice structure (see Section 1). Equation (7.59) can be used to generate the following, subsequently useful expressions:

$$e^{\Gamma}-1 = \frac{I_1}{c}\, p^2+1-e^{-\Gamma} = \frac{2}{\omega_0^2}\, p\left(p + \sqrt{\omega_0^2+p^2}\right) \tag{7.60a}$$

$$e^{-\Gamma r} = \left[\frac{\sqrt{\omega_0^2+p^2}\;-p}{\omega_0}\right]^{2r} \tag{7.60b}$$

Substitution of (7.58) into (7.56a) and utilization of (7.60a) results in

$$A = \frac{\overline{T}(p)+I_1\alpha p+I_1\dot{\alpha}}{\frac{2c}{\omega_0^2}\, p\left(p+\sqrt{\omega_0^2+p^2}\right)} \tag{7.61}$$

Case (a): The disc at r=0 is initially displaced an angle α from its equilibrium position and released from rest at t=0. In this case $\overline{T}(p)=0$, $\dot{\alpha}=0$, and

$$A = \frac{2\alpha}{p+\sqrt{\omega_0^2+p^2}} \tag{7.62}$$

Substituting (7.62) and (7.60b) into (7.58), we obtain

$$\overline{\theta}_r(p) = \frac{2(2r+1)\alpha}{\omega_0}\left[\frac{\left(\sqrt{\omega_0^2+p^2}-p\right)^{2r+1}}{(2r+1)\,\omega_0^{2r+1}}\right] \tag{7.63}$$

Inversion of (7.63) is readily accomplished with the aid of entry no. 17 in Table 1.1 of Chapter 1:

$$\theta_r(t) = 2(2r+1)\alpha\cdot\frac{J_{2r+1}(\omega_0 t)}{\omega_0 t} \tag{7.64}$$

where $J_{2r+1}(\omega_0 t)$ denotes the Bessel function of the first kind, of order 2r+1, with argument $\omega_0 t$. The power series expansion of the Bessel function is

$$J_{2r+1}(\omega_0 t) = \frac{(\tfrac{1}{2}\omega_0 t)^{2r+1}}{(2r+1)!}\left\{1 + 0\,(\tfrac{1}{2}\omega_0 t)^2\right\} \tag{7.65}$$

and substitution of (7.65) into (7.64) results in

$$\frac{\theta_r(t)}{\alpha} = \frac{1}{(2r)!} (\tfrac{1}{2}\omega_o t)^{2r} \quad \left\{1 + O\,(\tfrac{1}{2}\omega_o t)^2\right\} \tag{7.66}$$

Thus the motion of the disc at r=0 is given by

$$\frac{\theta_o(t)}{\alpha} = 1 + O[(\tfrac{1}{2}\omega_o t)^2]$$

and $\lim_{t\to 0} \theta_o(t)=\alpha$, which provides a check on the solution (7.64). When t is sufficiently small, the solution is approximated by

$$\theta_r(t) \simeq \frac{\alpha}{(2r)!} (\tfrac{1}{2}\omega_o t)^{2r}, \qquad r=1,2,\ldots. \tag{7.67}$$

When t is sufficiently large, we can use the asymptotic representation

$$J_{2r+1}(\omega_o t) \simeq \sqrt{2/\pi}\, \frac{1}{\sqrt{\omega_o t}} \cos\left[\omega_o t - \frac{\pi}{2}(2r+1) - \frac{\pi}{4}\right]$$

$$= \sqrt{2/\pi}\, \frac{(-1)^r}{\sqrt{\omega_o t}} \sin(\omega_o t - \frac{\pi}{4}) \tag{7.68}$$

Substitution of (7.68) into (7.64) results in the asymptotic form of our solution valid for t sufficiently large:

$$\theta_r(t) \simeq \frac{4(2r+1)(-1)^r\alpha}{\sqrt{2\pi}\,(\omega_o t)^{3/2}} \sin(\omega_o t - \frac{\pi}{4}) \tag{7.69}$$

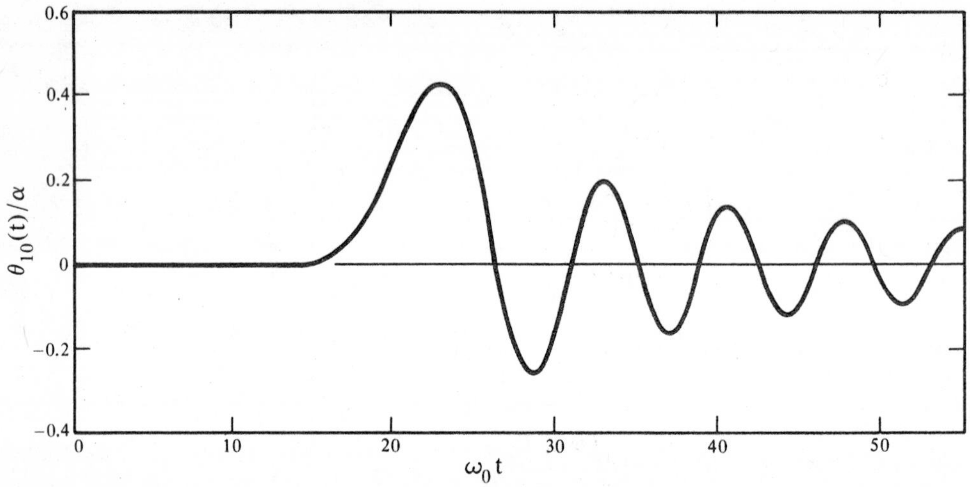

Figure 7.6

The response of the disc at $r=10$ is shown in Fig. 7.6. The general character of the motion can now be inferred with the aid of (7.64), (7.67), and (7.69). All discs begin to rotate immediately after $t=0$, and the discs at $r=1,2,\ldots,$ begin to rotate in the positive direction if $\alpha>0$. The greater the distance from $r=0$, the slower will be the beginning of the motion of a disc. The initially displaced disc at $r=0$ oscillates about the position $\theta_o=0$, and its maximum rotation angle monotonically decreases with each succeeding cycle of the motion. Its motion approaches simple harmonic motion, with frequency initially greater than ω_o, but approaching this frequency and remaining close to it after a few oscillations. Its amplitude gradually decreases and is approximately proportional to $(t)^{-3/2}$ after a few cycles. This decrease in amplitude comes about because the energy of vibratory motion is gradually transferred to its neighboring disc at $r=1$, which in turn, transfers energy to its neighbor at $r=2$, and so on. For sufficiently large t, neighboring discs oscillate in opposite phase. Thus the (strain) energy initially stored in the shaft segment from $z=0$ to $z=a$ gradually spreads over the entire structure in the form of kinetic and potential (strain) energy. Because the initial (strain) energy is finite and is ultimately distributed over a semi-infinite structure, its local effect at any given point $z\geq 0$ will ultimately become negligible.

Case (b): The disc at $r=0$ is given an initial angular velocity $\dot{\theta}_o(0)=\dot{\alpha}$. In this case $\overline{T}(p)=0$, $\alpha=0$, and (7.61) reduces to

$$A = \frac{2\dot{\alpha}}{p(p+\sqrt{\omega_o^2+p^2})} \tag{7.70}$$

Substitution of (7.70) and (7.60b) into (7.58) results in

$$p\overline{\theta}_r = \frac{2(2r+1)\dot{\alpha}}{\omega_o}\left[\frac{\left(\sqrt{\omega_o^2+p^2}-p\right)^{2r+1}}{(2r+1)\omega_o^{2r+1}}\right] \tag{7.71}$$

Inversion of (7.71) is readily accomplished with the aid of Table 1.1. We obtain

$$\dot{\theta}_r(t) = 2(2r+1)\dot{\alpha}\,\frac{J_{2r+1}(\omega_o t)}{\omega_o t} \tag{7.72}$$

and therefore

$$\theta(t) = 2(2r+1)\dot{\alpha} \int_0^t \frac{J_{2r+1}(\omega_0\tau)}{\omega_0\tau} d\tau \tag{7.73}$$

Case (c): The disc at r=0 is at rest and in its equilibrium position at t=0, i.e., $\alpha=\dot{\alpha}=0$. An impulsive torque $T(t)=T^*\delta(t)$ is applied to the disc at t=0, where $T^*=\dot{\alpha}I_1$ is the magnitude of the angular impulse. It can be readily verified that this case is identical to Case (b), and the resulting motion is given by

$$\theta_r(t) = 2(2r+1)\frac{T^*}{I_1}\int_0^t \frac{J_{2r+1}(\omega_0\tau)}{\omega_0\tau} d\tau \tag{7.74}$$

7.6 THE MECHANICAL BAND-PASS FILTER

In Section 1 of the present chapter, we showed how a simple, iterated structure acts as a low-pass filter, i.e., it transmits energy by means of time-harmonic waves as long as their frequency is below the cut-off frequency of the structure. Above the cut-off frequency, traveling waves are not possible, and energy will not be propagated. By a relatively simple modification of the iterated structure treated in Section 1, we can obtain the effect of a mechanical band-pass filter.

Let us assume that each disc of the iterated structure shown in Fig. 4.7 is restrained from rotating by means of an added torsional spring, such that the magnitude of the restraining torque is directly proportional to the rotation angle θ_r of the disc, and its sense is opposite to the sense of θ_r. Then the appropriate expressions for kinetic and potential energy are

$$T = \sum_j \tfrac{1}{2}I\dot{\theta}_j^2 \tag{7.75a}$$

$$V = \sum_j \tfrac{1}{2}c(\theta_j-\theta_{j-1})^2+\tfrac{1}{2}h\theta_j^2 \tag{7.75b}$$

where h is the torsional spring constant of the added elastic restraint. If there are no external, applied torques, $Q_r=0$, and substitution of (7.75) into Lagrange's equations (2.39) results in the differential-difference equations of motion

$$I\ddot{\theta}_r = c\left[\theta_{r+1}-2(1+\rho)\theta_r+\theta_{r-1}\right]$$
$$r = \ldots,-2,\ -1,\ 0,\ 1,\ 2,\ \ldots. \tag{7.76}$$

where $\rho = h/2c$.

We shall next investigate the possibility of traveling waves in our modified, iterated structure. Assuming a solution

$$\theta_r(t) = A \exp i(\omega t - \nu r) \tag{7.77}$$

and substituting (7.77) into (7.76), we obtain the dispersion relation

$$\cos \nu = 1 + \rho - \Omega^2 \tag{7.78a}$$

where $\Omega = \sqrt{I/2c}\,\omega$. In order that (7.77) characterize a traveling wave, the dimensionless wave number $\nu = ak$ must be a real quantity, and therefore we require that $-1 < \cos \nu < 1$, or in view of (7.78a), the wave number ν will be real provided

$$\sqrt{\rho} < \Omega < \sqrt{2+\rho} \tag{7.78b}$$

With reference to (7.78a), when $0 < \Omega < \sqrt{\rho}$ we have $1 < \cos \nu < 1+\rho$, and it is clear that ν can no longer be real. In this case we set $\nu = \eta - i\mu$, where $\mu > 0$ to insure attenuation. Then

$$1 < \cos \nu = \cos \eta \cdot \cosh \mu + i \sin \eta \cdot \sinh \mu < 1+\rho$$

and these inequalities will be satisfied if we set $\eta = 0$. In that case

$$1 < \cos \nu = \cosh \mu < 1+\rho$$

and (7.78a) becomes

$$\cosh \mu = 1+\rho-\Omega^2 \quad \text{for } 0<\Omega<\sqrt{\rho} \tag{7.79}$$

Again, in view of (7.78a), when $\sqrt{2+\rho} < \Omega$ we have $\cos \nu < -1$ and we require a complex wave number $\nu = \eta - i\mu$, where $\mu > 0$. In this case

$$\cos \nu = \cos \eta \cosh \mu + i \sin \eta \sinh \mu < -1$$

and to satifsy this inequality we set $\eta = \pi$. Thus

$$\cos \nu = -\cosh \mu < -1$$

and (7.78a) assumes the form

$$\cosh \mu = \Omega^2 - 1 - \rho \quad \text{for } \sqrt{2+\rho} < \Omega \tag{7.80}$$

With the aid of (7.78), (7.79), and (7.80) we have summarized our present analysis in Fig. 7.7. When $0 < \Omega < \sqrt{\rho}$ we have a standing, attenuated wave, and its attenuation number is characterized by (7.79). When $\sqrt{\rho} < \Omega < \sqrt{2+\rho}$, there will be a traveling wave with wave number ν which is characterized by the dispersion relation (7.78). For $\sqrt{2+\rho} < \Omega$ we again

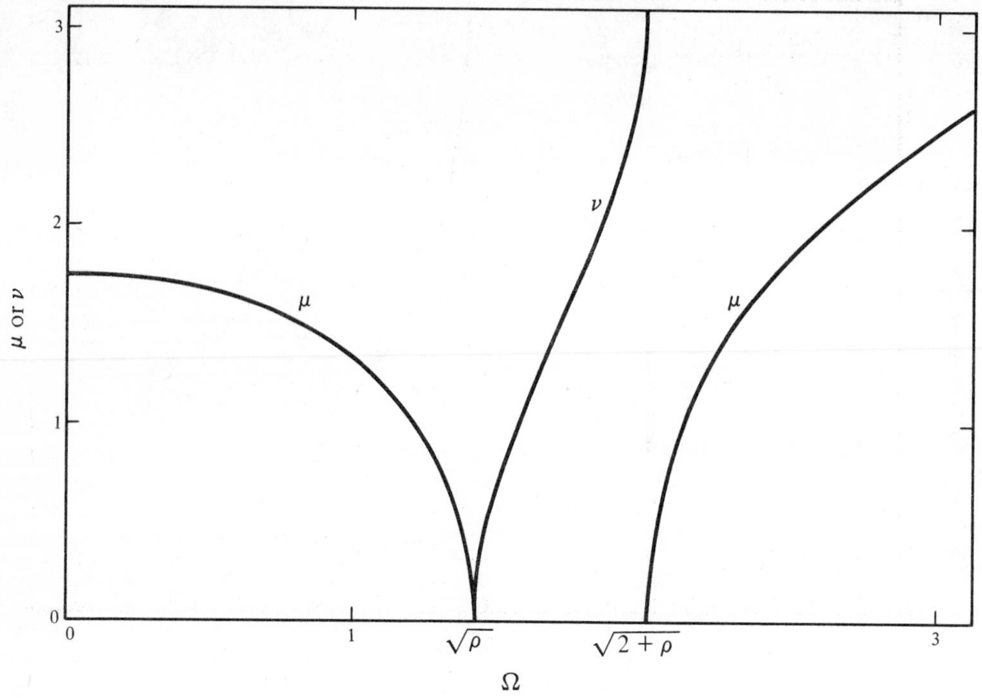

Figure 7.7

have a standing, attenuated wave and its attenuation number is given by (7.80). With reference to Fig. 7.7, it is now clear why our modified, iterated structure is called a mechanical band-pass filter: Mechanical energy can be transmitted by a time-harmonic torsional wave only as long as the (dimensionless) frequency is in the pass-band characterized by $\sqrt{\rho} < \Omega < \sqrt{2+\rho}$. For frequencies outside this pass-band no energy can be transmitted by time-harmonic waves.

We may utilize the dispersion relation (7.78) to compute phase and group velocities (see Appendix B):

$$\frac{1}{a}\sqrt{\frac{I}{2c}}\, V = \frac{\Omega}{\nu} = \frac{\sqrt{\rho+1-\cos\nu}}{\nu} = \frac{\Omega}{\sin^{-1}\left[\sqrt{\Omega^2-\rho)}\ \sqrt{(2+\rho)-\Omega^2}\right]} \tag{7.81}$$

$$\frac{1}{a}\sqrt{\frac{I}{2c}}\, V_G = \frac{d\Omega}{d\nu} = \frac{\sin\nu}{2\sqrt{\rho+1-\cos\nu}} = \frac{\sqrt{(\Omega^2-\rho)}\sqrt{(2+\rho)-\Omega^2}}{2\Omega} \tag{7.82}$$

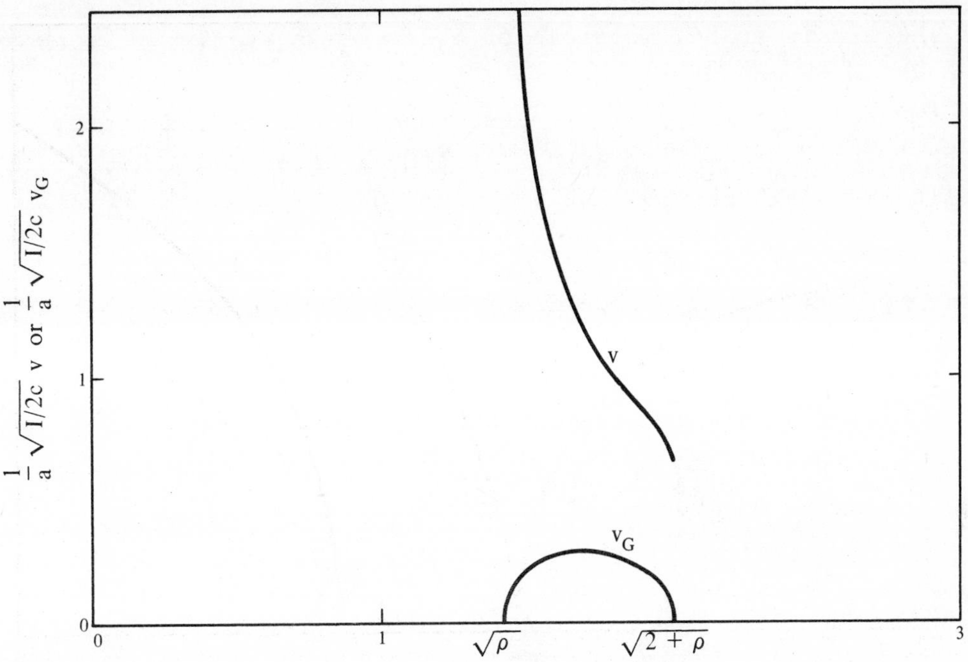

Figure 7.8

Equations (7.81) and (7.82) are valid (and meaningful) only for the frequency pass-band $\sqrt{\rho} < \Omega < \sqrt{2+\rho}$, and graphs of phase and group velocity as a function of frequency are shown in Fig. 7.8. It is obvious that our modified iterated structure can be classified as a dispersive medium with respect to torsional waves according to the definition given in Appendix B.

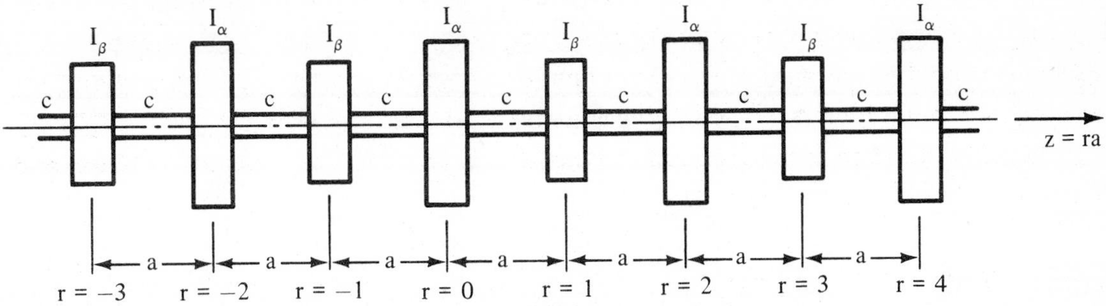

Figure 7.9

A somewhat more complex mechanical band-pass filter is shown in Fig. 7.9. In this case, we have rigid discs with moment of inertia I_α at points where z is an even multiple of a, z=...−2a,0,2a,4a,.... At points where z is an odd multiple of a, z=...-3a,-a,a,3a,..., there are rigid discs with moment of inertia I_β. The distance between any two adjacent discs is a and the mass-less shaft segments connecting the discs have torsional stiffness c. For convenience and without loss of generality we shall assume that $I_\alpha > I_\beta$ so that

$$\lambda = \sqrt{\frac{I_\alpha}{I_\beta}} > 1; \quad \frac{1}{\lambda} = \sqrt{\frac{I_\beta}{I_\alpha}} < 1.$$

The expressions for the kinetic and potential energy of the structure shown in Fig. 7.9 in the case of torsional motion are

$$2T = \sum_j I_\alpha \dot{\theta}_{2j}^2 + I_\beta \dot{\theta}_{2j+1}^2 \tag{7.83a}$$

$$V = \sum_j \tfrac{1}{2}c(\theta_{2j+1}-\theta_{2j})^2 + \tfrac{1}{2}c(\theta_{2j+2}-\theta_{2j+1})^2 \tag{7.83b}$$

We now write separate Lagrange equations of motion for discs with even indices 2r and odd indices 2r+1, r=...-2, -1, 0, 1, 2, ... In the absence of external forces we have

$$\frac{d}{dt}\left(\frac{\partial T}{\partial \dot{\theta}_{2r}}\right) + \frac{\partial V}{\partial \theta_{2r}} = 0 \tag{7.84a}$$

$$\frac{d}{dt}\left(\frac{\partial T}{\partial \dot{\theta}_{2r+1}}\right) + \frac{\partial V}{\partial \theta_{2r+1}} = 0 \tag{7.84b}$$

Upon substitution of (7.83) into (7.84), we obtain the system of differential-difference equations

$$I_\alpha \ddot{\theta}_{2r} - c(\theta_{2r+1}-\theta_{2r}) + c(\theta_{2r}-\theta_{2r-1}) = 0 \tag{7.85a}$$

$$I_\beta \ddot{\theta}_{2r+1} - c(\theta_{2r+2}-\theta_{2r+1}) + c(\theta_{2r+1}-\theta_{2r}) = 0 \tag{7.85b}$$

where r=...,-2,-1,0,1,2,... . To explore the possibility of traveling waves in the structure shown in Fig. 7.9, we assume a solution in the form

$$\theta_{2r} = A_\alpha \exp i(\omega t - 2\nu r) \tag{7.86a}$$

$$\theta_{2r+1} = A_\beta \exp i[\omega t - \nu(2r+1)] \tag{7.86b}$$

where $\nu=ka$ is the dimensionless wave number, ω the frequency, and A_α and A_β are constants. Upon substitution of (7.86) into (7.85) we obtain

$$A_\alpha\left(\frac{I_\alpha\omega^2}{2c}-1\right)+A_\beta\cos\nu=0 \tag{7.87a}$$

$$A_\alpha\cos\nu+A_\beta\left(\frac{I_\beta\omega^2}{2c}-1\right)=0 \tag{7.87b}$$

and a non-trivial solution of (7.87) for the constants A_α and A_β results in the dispersion relation

$$4\sin^2\nu=2(\lambda+\frac{1}{\lambda})\Omega^2-\Omega^4 \tag{7.88a}$$

where $\Omega^2=\frac{\sqrt{I_\alpha I_\beta}}{c}\omega^2$.

For traveling waves we require a real wave number, resulting in the restriction $0<4\sin^2\nu<4$, and in view of (7.88a) this implies that

$$0<2(\lambda+\frac{1}{\lambda})\Omega^2-\Omega^4<4 \tag{7.89}$$

The inequality on the left of (7.89) leads to

$$0<\Omega^2<2(\lambda+\frac{1}{\lambda}) \tag{7.90}$$

while the inequality on the right of (7.89) results in

$$2\lambda<\Omega^2 \text{ and } \Omega^2<\frac{2}{\lambda} \tag{7.91}$$

Taken together, (7.90) and (7.91) express the fact that the wave number ν as given by (7.88a) will be real provided

$$0<\Omega^2<\frac{2}{\lambda} \text{ and } 2\lambda<\Omega^2<2(\lambda+\frac{1}{\lambda}). \tag{7.88b}$$

With the aid of (7.87) and (7.88) we can obtain the amplitude ratio as a function of frequency:

$$\frac{A_\alpha}{A_\beta}=\frac{2\cos\nu}{2-\lambda\Omega^2}=\frac{1}{\lambda}\sqrt{\frac{2\lambda-\Omega^2}{\frac{2}{\lambda}-\Omega^2}}>0 \text{ for } 0<\Omega^2<\frac{2}{\lambda} \tag{7.92a}$$

$$\frac{A_\alpha}{A_\beta}=-\frac{1}{\lambda}\sqrt{\frac{\Omega^2-2\lambda}{\Omega^2-\frac{2}{\lambda}}}<0 \text{ for } 2\lambda<\Omega^2<2(\lambda+\frac{1}{\lambda}) \tag{7.92b}$$

With reference to (7.88a), $1<\sin^2\nu$ when $\sqrt{2/\lambda}<\Omega<\sqrt{2\lambda}$, and therefore we have a complex wave number $\nu=\eta-i\mu$, where $\mu>0$ to insure attenuation. In this case

$$1<\sin^2\nu = (\sin\eta\cos\mu - i\cos\eta\sinh\mu)^2$$

and this inequality is satisfied by setting $\eta=\pi/2$. Thus $1<\sin^2\nu=\cosh^2\mu$, and the dispersion relation (7.88a) assumes the form

$$4\cosh^2\mu = 2(\lambda+\tfrac{1}{\lambda})\Omega^2-\Omega^4 \quad \text{for} \sqrt{\tfrac{2}{\lambda}} < \Omega < \sqrt{2\lambda} \tag{7.93}$$

With reference to (7.88a), when $\sqrt{2(\lambda+1/\lambda)}<\Omega$, $\sin^2\nu<0$ and ν is again complex. We therefore let $\nu=\eta-i\mu$, where $\mu>0$, and obtain

$$\sin^2\nu = (\sin\eta\cosh\mu - i\cos\eta\sinh\mu)^2 < 0.$$

This inequality is satisfied by the choice $\eta=0$, and therefore $\sin^2\nu = -\sinh^2\mu<0$. In this case the dispersion relation (7.88a) is transformed to

$$4\sinh^2\mu = \Omega^4-2(\lambda+\tfrac{1}{\lambda})\Omega^2 \quad \text{for} \quad \sqrt{2(\lambda+\tfrac{1}{\lambda})} <\Omega. \tag{7.94}$$

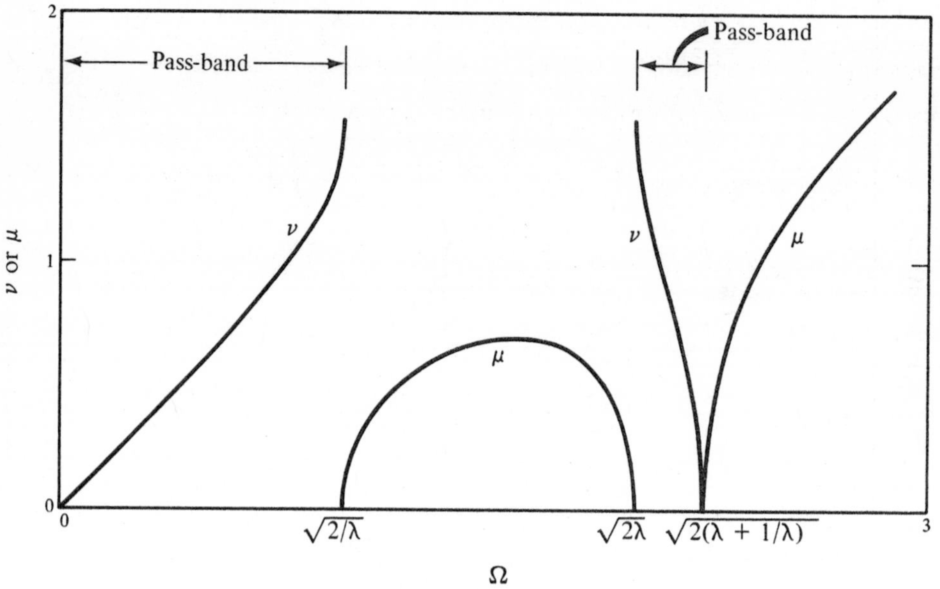

Figure 7.10

Graphs of the relations (7.88), (7.93), and (7.94) are shown in Fig. 7.10. With reference to that figure it is now clear that the structure shown in Fig. 7.9 acts as a band-pass filter with two separate and distinct pass-bands, i.e., mechanical energy can be transported by time-harmonic waves with wave number ν as long as $0<\Omega<\sqrt{2/\lambda}$ or $\sqrt{2\lambda}<\Omega<\sqrt{2(\lambda+1/\lambda)}$. When $\sqrt{2/\lambda}<\Omega<\sqrt{2\lambda}$ or $\sqrt{2\lambda+1/\lambda}<\Omega$, a propagating wave is not possible and we have a standing, attenuated wave with attenuation number μ.

To calculate the phase and group velocities in the pass-bands, we utilize (B.6) and (B.18) in conjunction with the dispersion relation (7.88):

$$\frac{(I_\alpha I_\beta)^{\frac{1}{4}}}{a\sqrt{c}}\, v = \frac{\Omega}{\nu} = \frac{\Omega}{\sin^{-1}\left[\frac{1}{2}\Omega\sqrt{2(\lambda+\frac{1}{\lambda})-\Omega^2}\right]} \tag{7.95}$$

$$\frac{(I_\alpha I_\beta)^{\frac{1}{4}}}{a\sqrt{c}}\, v_G = \frac{d\Omega}{d\nu} = \frac{\sqrt{\left[2(\lambda+\frac{1}{\lambda})-\Omega^2\right](2\lambda-\Omega^2)(\frac{2}{\lambda}-\Omega^2)}}{\left[2\ (\lambda+\frac{1}{\lambda})-\Omega^2\right]} \tag{7.96}$$

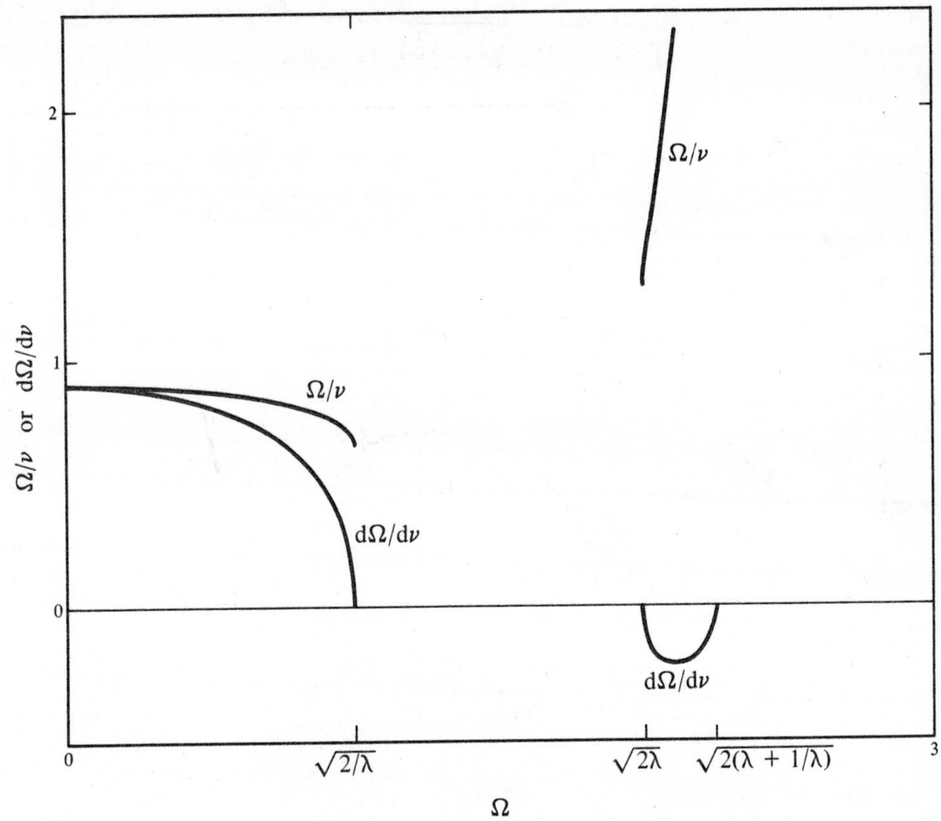

Figure 7.11

Dimensionless plots of phase and group velocity are shown in Fig. 7.11. According to Appendix B, the iterated structure under consideration is seen to be dispersive. We also observe that in the pass-band $\sqrt{2\lambda} < \Omega < \sqrt{2(\lambda+1/\lambda)}$, phase and group velocity have opposite signs. This implies that when the wave propagates in the direction of the positive (negative) z-axis, energy will be transported in the direction of the negative (positive) z-axis.

EXERCISES

7.1 Carry out the details necessary to prove the validity of (7.45).

7.2 Carry out the details necessary to prove the validity of (7.47) and (7.48).

7.3 Carry out the details necessary to prove the validity of (7.60).

7.4 Describe two iterated structures (different from the one shown in Fig. 4.7) which are characterized by equations (7.5), (7.6), and (7.7).

7.5 Describe two iterated structures (different from the one shown in Fig. 7.9) which are characterized by equations (7.85).

7.6 Consider the semi-infinite, iterated structure shown in Fig. 7.5. The rigid discs with moment of inertia I_1 are connected by massless shaft segments, each with torsional stiffness c. Assume that an external torque $T(t)=A \cos \omega t$ acts on the disc at r=0.

(a) Neglecting transients, give a full discussion of the steady-state torsional motion of the structure, with particular reference to its cut-off frequency.

(b) Calculate the energy flux in the structure when the excitation frequency is above and below the cut-off frequency.

(c) Show that the exciting torque T(t) and the angular velocity $\dot{\theta}_0$ are exactly 90° out of phase when the excitation frequency ω is above the cut-off frequency. Hence show that the average energy input to the structure vanishes when the excitation frequency is above the cut-off frequency.

7.7 Consider the unbounded, iterated structure shown in Fig. 4.7. The rigid discs with moment of inertia I are connected by mass-less shaft segments of torsional stiffness c. The distance between any two adjacent discs is a. Discuss the possibility of time-harmonic, torsional wave propagation through the structure if a viscous damping torque acts on each disc.

7.8 Consider the unbounded, iterated structure shown in the figure. The discs are connected by mass-less shaft segments, each with torsional stiffness c. Every disc has moment of inertia I, except the disc at r=0 which has moment of inertia $I_o \neq I$. Give a full discussion of the steady-state torsional motion of the structure if a torque $T(t)=A \cos \omega t$ is applied to the disc at r=0.

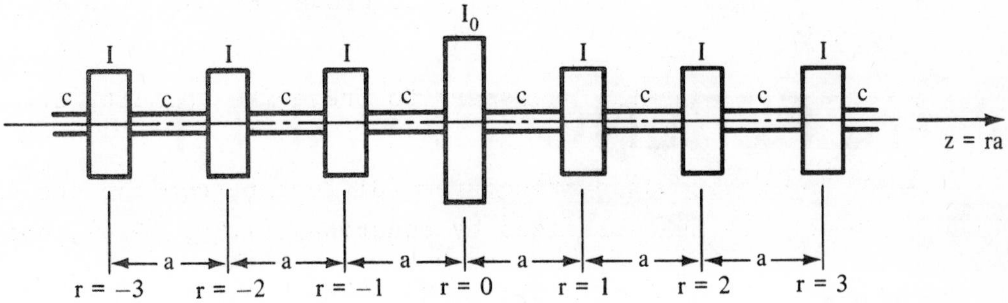

Figure: Exercise 7.8

7.9 Consider the structure in Exercise 7.8. If $\theta_r(0)=\dot{\theta}_r(0)=0$, r=1,2,....,r=-1,-2,...., and $\theta_o(0)=\alpha, \dot{\theta}_o(0)=0$, describe the transient wave motion for t>0 upon release of the initial constraints.

7.10 Consider the structure in Exercise 7.8. If $\theta_r(0)=\dot{\theta}_r(0)=0$, r=1,2,...., r=-1,-2,...., and $\theta_o(0)=0$, $\dot{\theta}_o(0)=\dot{\alpha}$, describe the ensuing transient wave motion for t>0.

7.11 Consider the unbounded, iterated structure in Exercise 7.8. An energy source at $z \to \infty$ gives rise to an incoming, traveling torsional wave $A \exp i(\omega t + r\nu)$, where A and ω are real constants. Give a quantitative description of the wave reflection and transmission process which takes place at r=0. Provide a complete discussion of the problem, including significant limiting cases.

REFERENCES: PART I

ANALYTICAL MECHANICS

S. Banach, Mechanics, Monografie Matematyczne, Warsaw, 1951.

H. Goldstein, Classical Mechanics, Addison-Wesley, Reading, Mass., 1950.

J.H. Jeans, An Elementary Treatise on Theoretical Mechanics, Ginn and Co., Boston, 1907.

L.D. Landau and E.M. Lifshitz, Mechanics, Second Edition, Pergamon Press, Oxford, 1969.

C. Lanczos, The Variational Principles of Mechanics, University of Toronto Press, Toronto, 1949.

J.B. Marion, Classical Dynamics, Academic Press, New York, 1965.

L.A. Pars, A Treatise on Analytical Dynamics, John Wiley and Sons, New York, 1965.

A. Sommerfeld, Mechanics, Academic Press, New York, 1952.

E.T. Whittaker, A Treatise on the Analytical Dynamics of Particles and Rigid Bodies, Fourth Edition, Cambridge University Press, London, 1937.

VIBRATION THEORY

R.E.D. Bishop and D.C. Johnson, The Mechanics of Vibration, Cambridge University Press, London, 1960.

Y. Chen, Vibrations: Theoretical Methods, Addison Wesley, Reading, Mass., 1966.

L. Meirovitch, Analytical Methods in Vibrations, The Macmillan Co., New York, 1967.

W.T. Thomson, Vibration Theory and Applications; Prentice Hall, Englewood Cliffs, New Jersey, 1965.

S. Timoshenko and D.H. Young, Vibration Problems in Engineering, Third Edition, D. Van Nostrand, New York, 1955.

ENGINEERING DYNAMICS

S.H. Crandall, D.C. Karnopp, E.F. Kurtz, Jr., and D.C. Pridmore-Brown, Dynamics of Mechanical and Electromechanical Systems, Mc-Graw Hill, New York, 1968.

D. Greenwood, Principles of Dynamics, Prentice Hall, Englewood Cliffs, New Jersey, 1965.

G.W. Housner and D.E. Hudson, Applied Mechanics-Dynamics, Second Edition, Van Nostrand Reinhold, New York, 1959.

Sir Charles Inglis, Applied Mechanics for Engineers, Cambridge University Press, London, 1951.

I.H. Shames, Engineering Mechanics-Dynamics, Second Edition, Prentice Hall, Englewood Cliffs, New Jersey, 1966.

S. Timoshenko and D.H. Young, Advanced Dynamics, Mc-Graw Hill, New York, 1948.

WAVE PROPAGATION

L. Brillouin, Wave Propagation in Periodic Structures, Second Edition, Dover Publications, New York, 1953.

T.H. Havelock, The Propagation of Disturbances in Dispersive Media, Cambridge University Press, London, 1914.

H.J. Pain, The Physics of Vibrations and Waves, John Wiley and Sons, London, 1968.

MATHEMATICAL METHODS

(a) General

R. Courant and D. Hilbert, Methods of Mathematical Physics, Vol. 1, Interscience Publishers, New York, 1953.

P. Frank and R.v. Mises, Die Differential und Integralgleichungen der Mechanik und Physik, Friedrich Vieweg and Sohn, Braunschweig, 1931.

E.A. Guillemin, The Mathematics of Circuit Analysis, John Wiley, New York, 1949.

T.v. Kármán and M.A. Biot, Mathematical Methods in Engineering, Mc-Graw Hill, New York, 1940.

(b) Variational Methods

C. Fox, An Introduction to the Calculus of Variations, Oxford University Press, London, 1950.

B.L. Moiseiwitsch, Variational Principles, Interscience Publishers, London, 1966.

R. Weinstock, Calculus of Variations, Mc-Graw Hill, New York, 1952.

(c) Matrix Methods

R.E.D. Bishop, G.M.L. Gladwell, and S. Michaelson, The Matrix Analysis of Vibration, Cambridge University Press, London, 1965.

R.A. Frazer, W.J. Duncan, and A.R. Collar, Elementary Matrices, Cambridge University Press, London, 1947.

Louis A. Pipes, Matrix Methods for Engineering, Prentice Hall, Englewood Cliffs, New Jersey, 1963.

(d) Laplace Transform Theory

H.S. Carslaw and J.C. Jaeger, Operational Methods in Applied Mathematics, Second Edition, Oxford University Press, London, 1947.

H. Jeffreys, Operational Methods in Mathematical Physics, Cambridge University Press, London, 1927.

E.J. Scott, Transform Calculus with an Introduction to Complex Variables, Harper and Brothers, New York, 1955.

M.R. Spiegel, Laplace Transforms, Schaum Publishing Co., New York, 1965

K.W. Wagner and A. Thoma, Operatorenrechnung und Laplacesche Transformation, Johann Ambrosius Barth, Leipzig, 1962.

(e) Tables of Laplace Transforms

A. Erdélyi, Ed., Tables of Integral Transforms, Vol. I, Mc-Graw Hill, New York, 1954.

G.E. Roberts and H. Kaufman, Table of Laplace Transforms, W.B. Saunders Philadelphia, 1966.

Part II

CONTINUOUS SYSTEMS

Chapter

8

CARTESIAN TENSORS

The general laws of nature are to be expressed by equations which hold good for all systems of co-ordinates, that is, are co-variant with respect to any substitutions whatever (generally co-variant).

Albert Einstein, Annalen der Physik, 49, 1916.

8.1 INTRODUCTION

One of the most convenient and efficient means of describing the deformations and stresses in continua is achieved with the aid of tensors. The present chapter provides an introductory treatment of Cartesian tensor algebra and analysis. The topics treated were selected primarily because of their utility in the subsequent analysis of continua. An understanding of the presentation requires a certain familiarity with vector algebra and analysis, but no prior knowledge of tensors is assumed. Those who would like to obtain a more rigorous or more extensive coverage of this topic are urged to consult the standard references.

Scalars are quantities which characterize magnitude only, while vectors are mathematical objects used to characterize both magnitude and direction. The use of scalars and vectors to represent physical entities is well established. For example, Newton's first and second laws of motion for a single particle are summarized by the equation $\vec{F} = m\vec{a}$. The mass of the particle, m, is a scalar, while the force, $\vec{F}$, and acceleration, $\vec{a}$, are vectors. Throughout the present and previous developments we use a letter with an arrow above it to signify a vector quantity.

The information required to specify the magnitude of a scalar or vector is provided by a single real number once a system of units has been selected. However, in order to specify the direction of a vector we need a set of reference directions which is readily provided by a coordinate system. There are many different types of coordinate systems which may be selected for this purpose, but for the present we shall consider only rectangular Cartesian coordinate systems, henceforth referred to by the abbreviation R.C.S. . A typical R.C.S. is shown in Fig. 8.1. In order to achieve some economy in our use of symbols we have labeled the coordinate axes X_1, X_2 and X_3 rather than the usual X, Y, Z. The unit vectors along the X_1, X_2 and X_3 axes are called base vectors and are represented by $\hat{e}_1$, $\hat{e}_2$ and $\hat{e}_3$ respectively. Their common point of intersection O is called the origin of the coordinate system. A letter with a caret above it such as $\hat{e}$ will be used to denote a unit vector (a vector whose magnitude is one).

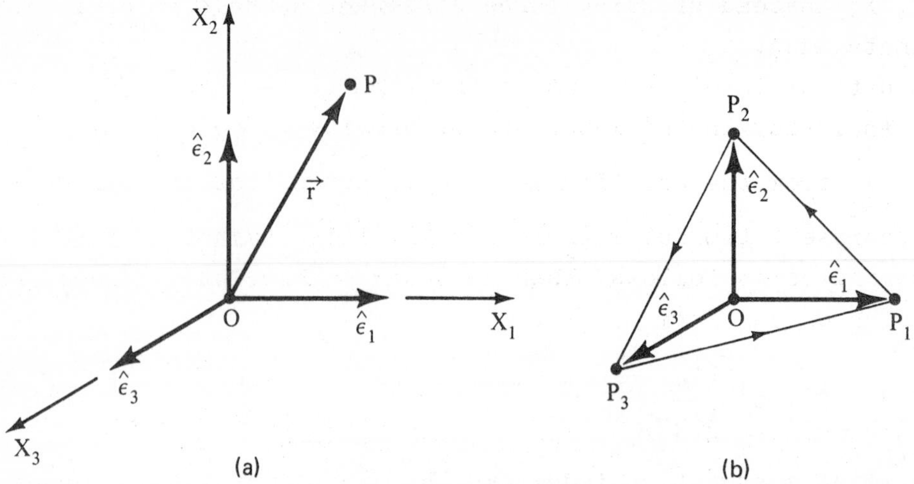

Figure 8.1

Coordinate systems may be classified as being right-handed (dextral) or left-handed (sinistral). We distinguish one from the other by employing the following device. Referring to Fig. 8.1b, draw a plane closed curve through the endpoints P_1, P_2 and P_3 of the base vectors $\hat{\varepsilon}_1$, $\hat{\varepsilon}_2$ and $\hat{\varepsilon}_3$ respectively. The sense of this curve is defined by traversing it from P_1 to P_2 to P_3. The ordered set $\{\hat{\varepsilon}_1, \hat{\varepsilon}_2, \hat{\varepsilon}_3\}$ is right-handed (dextral) if the sense of the curve is clockwise when viewed from the origin O. The set is left-handed (sinistral) if the sense is counterclockwise when viewed from O. The coordinate system is said to be right-handed or left-handed if its base vectors form a dextral or sinistral set respectively. Thus the coordinate system shown in Fig. 8.1 is right-handed.

8.2 INDEX NOTATION

Consider a generic point P whose coordinates in the R.C.S. of Fig. 8.1a are $\{x_1, x_2, x_3\}$. The vector $\vec{r}$, joining O to P is called the position vector of the point P and its component form is

$$\vec{r} = x_1\hat{\varepsilon}_1 + x_2\hat{\varepsilon}_2 + x_3\hat{\varepsilon}_3 = \sum_{i=1}^{3} x_i\hat{\varepsilon}_i. \tag{8.1}$$

The standard summation notation used in (8.1) allows us to express the component form of the vector very compactly. This is a direct consequence of our earlier agreement to append subscripts to a single symbol, X, instead of using three different symbols to designate the coordinate axes.

Equations involving sums of the form (8.1) occur frequently in tensor analysis. In the interests of brevity we delete the summation symbol $\sum_{i=1}^{3}$ from the equation and simply agree that the expression $x_i\hat{e}_i$ is to represent the sum $x_1\hat{e}_1+x_2\hat{e}_2+x_3\hat{e}_3$. This summation convention may be generalized as follows. When an index (subscript) is repeated in an expression of the form

$$A_iB_i,\ C_{jj} \text{ or } \frac{\partial f_k}{\partial x_k}$$

it is called a summation index and the expression itself represents a sum of three terms which may be obtained by summing the expression over the range {1,2,3} of the index. Thus,

$$\left.\begin{aligned} A_iB_i &\equiv A_1B_1+A_2B_2+A_3B_3 \\ C_{jj} &\equiv C_{11}+C_{22}+C_{33} \\ \frac{\partial f_k}{\partial x_k} &\equiv \frac{\partial f_1}{\partial x_1}+\frac{\partial f_2}{\partial x_2}+\frac{\partial f_3}{\partial x_3} \end{aligned}\right\} \tag{8.2}$$

The summation convention applies only to the three types of expressions listed above. It does not apply to an expression of the type (A_i+B_i). The same summation index may not appear more than twice in any expression. Thus the expression $A_iB_iC_i$ is meaningless by our convention.

Using this summation convention, we may now write the component form of the position vector $\vec{r}$ in the form

$$\vec{r} = x_i\hat{e}_i. \tag{8.3}$$

An additional advantage of the index or subscript notation we have adopted is illustrated by the following example. Suppose we are given two vectors, $\vec{A}$ and $\vec{B}$, whose components in the R.C.S. of Fig. 8.1 are (A_1, A_2, A_3) and (B_1, B_2, B_3), respectively. Then $\vec{A}=A_i\hat{e}_i$ and $\vec{B}=B_i\hat{e}_i$. We say that the two vectors are equal if and only if $A_1=B_1$, $A_2=B_2$ and $A_3=B_3$. These three equations may also be written as

$$A_i = B_i, \quad i = (1,2,3). \tag{8.4}$$

By letting i assume successively the values 1, 2 and 3 the three desired equations are obtained from the single indicial equation (8.4). The index i in (8.4) is called a free index since it is free to assume any value of its range (1,2,3). Unless stated to the contrary, the range of all free indices appearing in this text is (1,2,3). Thus it will not be necessary for us to rewrite the range of the free index after each indicial equation. The following rule must always be followed when writing an indicial equation. A free index must appear once and only once in each expression of the equation. For example, if $\vec{A}$ is the sum of $\vec{B}$ plus $\vec{C}$, then its components are $A_i=B_i+C_i$, or in unabridged form, $A_1=B_1+C_1$, $A_2=B_2+C_2$ and $A_3=B_3+C_3$. However, $A_i=B_i+C_j$ is not a valid or meaningful indicial equation.

Free indices may be used together with summation indices in the same equation as long as the preceding rules are not violated. The meaning of the indicial equation

$$y_i = b_i+a_{ij}x_j$$

is unambiguous and well defined. The symbol j in the last expression on the right-hand side represents a summation index since it is repeated. As a result of the summation convention, this indicial equation is equivalent to the following equation

$$y_i = b_i+a_{i1}x_1+a_{i2}x_2+a_{i3}x_3$$

By letting the free index i assume the values 1, 2 and 3 successively, we obtain the unabridged form

$$y_1 = b_1+a_{11}x_1+a_{12}x_2+a_{13}x_3$$
$$y_2 = b_2+a_{21}x_1+a_{22}x_2+a_{23}x_3$$
$$y_3 = b_3+a_{31}x_1+a_{32}x_2+a_{33}x_3.$$

8.3 VECTOR ALGEBRA

The reader is probably aware of the fact that there are several ways to define the product of two vectors. From the standpoint of mechanics, the two which are most convenient and physically meaningful are the scalar or dot product and the vector or cross product. Given two vectors, $\vec{A}$ and $\vec{B}$, their dot product is defined as follows:

$$\vec{A}\cdot\vec{B} = \vec{B}\cdot\vec{A} = |\vec{A}||\vec{B}| \cos \alpha \tag{8.5}$$

where $\alpha = \sphericalangle(\vec{A},\vec{B})$ and $|\vec{A}|$ represents the magnitude of the vector $\vec{A}$. The cross product of $\vec{A}$ into $\vec{B}$ is a vector $\vec{D}$ which is defined by the following.

$$\text{If} \quad \vec{D} = \vec{A} \times \vec{B} \qquad \text{then} \quad |\vec{D}| = |\vec{A}||\vec{B}| \sin\alpha \tag{8.6}$$

where $\alpha = \sphericalangle(\vec{A},\vec{B}) \leq \pi$. In addition, $\vec{D} \perp (\vec{A},\vec{B})$ and the ordered set $\{\vec{A},\vec{B},\vec{D}\}$ is dextral, thus $\vec{B}\times\vec{A} = -\vec{A}\times\vec{B}$. Geometrically, $|\vec{D}|$ is the area of the parallelogram having adjacent sides $\vec{A}$ and $\vec{B}$ with included angle α as shown in Fig. 8.2. If we form the dot product of $\vec{D}$ with a third vector $\vec{C}$ we obtain the scalar triple product defined by

$$\vec{D}\cdot\vec{C} = \vec{A}\times\vec{B}\cdot\vec{C} = |\vec{A}||\vec{B}||\vec{C}| \sin\alpha \cos\beta \tag{8.7}$$

where $\alpha = \sphericalangle(\vec{A},\vec{B}) \leq \pi$ and $\beta = \sphericalangle(\vec{D},\vec{C})$. Geometrically, $|\vec{D}\cdot\vec{C}|$ is the volume of the parallelopiped whose sides are $\vec{A}$, $\vec{B}$ and $\vec{C}$ as shown in Fig. 8.3.

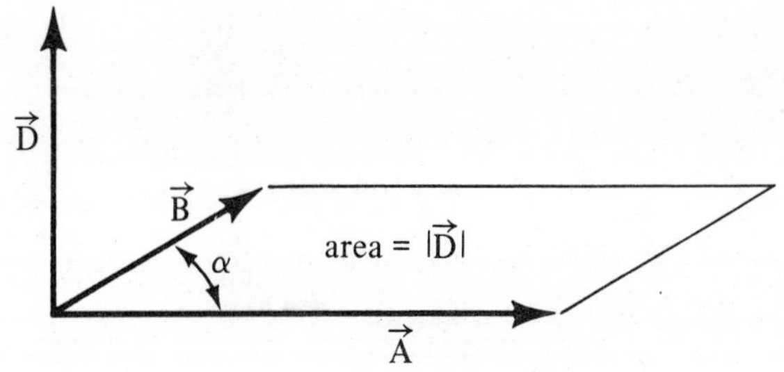

Figure 8.2

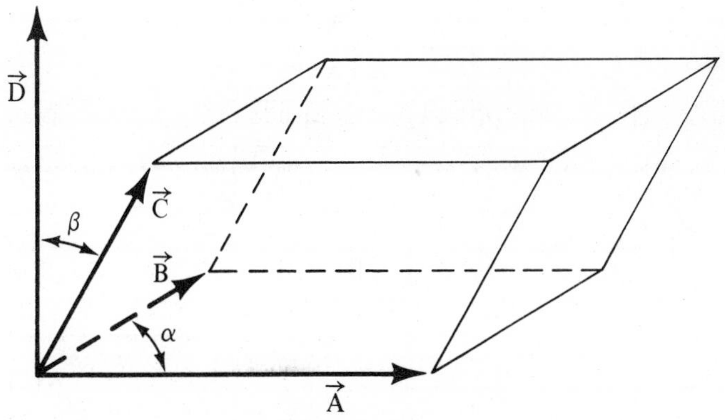

Figure 8.3

By definition, the volume V of this parallelopiped is a positive scalar quantity. Therefore,

$$V = \vec{A}\times\vec{B}\cdot\vec{C} \tag{8.8a}$$

if the ordered set $\{\vec{A},\vec{B},\vec{C}\}$ is dextral and

$$V = -\vec{A}\times\vec{B}\cdot\vec{C} \qquad (8.8b)$$

if it is sinistral.

Let us next examine the various products of the base vectors $\hat{\varepsilon}_1$, $\hat{\varepsilon}_2$ and $\hat{\varepsilon}_3$. They are mutually orthogonal unit vectors, thus according to (8.5) their dot products are

$$\hat{\varepsilon}_1\cdot\hat{\varepsilon}_1 = \hat{\varepsilon}_2\cdot\hat{\varepsilon}_2 = \hat{\varepsilon}_3\cdot\hat{\varepsilon}_3 = 1$$

$$\hat{\varepsilon}_1\cdot\hat{\varepsilon}_2 = \hat{\varepsilon}_2\cdot\hat{\varepsilon}_1 = \hat{\varepsilon}_2\cdot\hat{\varepsilon}_3 = \hat{\varepsilon}_3\cdot\hat{\varepsilon}_2 = \hat{\varepsilon}_3\cdot\hat{\varepsilon}_1 = \hat{\varepsilon}_1\cdot\hat{\varepsilon}_3 = 0.$$

These nine equations may be written in the indicial form

$$\hat{\varepsilon}_i\cdot\hat{\varepsilon}_j = \delta_{ij} \qquad (8.9)$$

where i and j are free indices with range (1,2,3) and

$$\delta_{ij} = \begin{Bmatrix} 1 \text{ if } i=j \\ 0 \text{ if } i\neq j \end{Bmatrix}. \qquad (8.10a)$$

We call δ_{ij} the Kronecker delta symbol. Equation (8.9) may be verified by substituting specific values for i and j, and then employing (8.10a) For example, if i=3 and j=2, $\hat{\varepsilon}_3\cdot\hat{\varepsilon}_2=\delta_{32}=0$. The nine components of the Kronecker delta may be visualized as the elements of a matrix. Thus $[\delta_{ij}]$ is the matrix whose elements are

$$[\delta_{ij}] = \begin{bmatrix} \delta_{11} & \delta_{12} & \delta_{13} \\ \delta_{21} & \delta_{22} & \delta_{23} \\ \delta_{31} & \delta_{32} & \delta_{33} \end{bmatrix} = \begin{bmatrix} 1 & 0 & 0 \\ 0 & 1 & 0 \\ 0 & 0 & 1 \end{bmatrix} \qquad (8.10b)$$

Note that the first index specifies the row number while the second index specifies the column number of each element of the matrix.

The Kronecker delta has the following useful property. Let

$$B_i = \delta_{ij}A_j = \delta_{i1}A_1+\delta_{i2}A_2+\delta_{i3}A_3.$$

By letting i assume the values 1, 2 and 3 successively we obtain, with the aid of (8.10)

$$B_1 = \delta_{11}A_1 + \delta_{12}A_2 + \delta_{13}A_3 = A_1$$

$$B_2 = \delta_{21}A_1 + \delta_{22}A_2 + \delta_{23}A_3 = A_2$$

$$B_3 = \delta_{31}A_1 + \delta_{32}A_2 + \delta_{33}A_3 = A_3.$$

Thus $B_i = A_i$ and we have shown that

$$A_i = \delta_{ij} A_j = \delta_{ji} A_j. \tag{8.11}$$

The second equality in (8.11) follows from the observation that $\delta_{ij} = \delta_{ji}$, e.g., $\delta_{12} = \delta_{21} = 0$, etc. . Equation (8.11) illustrates the substitution property of the Kronecker delta.

The ordered set $\{\hat{\varepsilon}_1, \hat{\varepsilon}_2, \hat{\varepsilon}_3\}$ is dextral, hence from (8.6) we obtain the following expressions for their cross products:

$$\hat{\varepsilon}_1 \times \hat{\varepsilon}_2 = -\hat{\varepsilon}_2 \times \hat{\varepsilon}_1 = \hat{\varepsilon}_3, \quad \hat{\varepsilon}_2 \times \hat{\varepsilon}_3 = -\hat{\varepsilon}_3 \times \hat{\varepsilon}_2 = \hat{\varepsilon}_1,$$

$$\hat{\varepsilon}_3 \times \hat{\varepsilon}_1 = -\hat{\varepsilon}_1 \times \hat{\varepsilon}_3 = \hat{\varepsilon}_2, \quad \hat{\varepsilon}_1 \times \hat{\varepsilon}_1 = \hat{\varepsilon}_2 \times \hat{\varepsilon}_2 = \hat{\varepsilon}_3 \times \hat{\varepsilon}_3 = 0.$$

These nine equations may be written in the indicial form

$$\hat{\varepsilon}_i \times \hat{\varepsilon}_j = e_{ijk} \hat{\varepsilon}_k \tag{8.12}$$

where

$$e_{ijk} = \left\{ \begin{array}{l} 1 \text{ if } (ijk) = (123), (312), (231) \\ -1 \text{ if } (ijk) = (321), (132), (213) \\ 0 \text{ if two or more subscripts are equal} \end{array} \right\}. \tag{8.13}$$

The symbol e_{ijk} is called the permutation tensor since it is nonzero only if (ijk) is a permutation of (123). Its correct application in (8.12) may be verified by direct substitution. For example, if $i=3$ and $j=2$,

$$\hat{\varepsilon}_3 \times \hat{\varepsilon}_2 = e_{32k} \hat{\varepsilon}_k = e_{321} \hat{\varepsilon}_1 + e_{322} \hat{\varepsilon}_2 + e_{323} \hat{\varepsilon}_3.$$

According to (8.13), $e_{321} = -1$, $e_{322} = e_{323} = 0$, thus $\hat{\varepsilon}_3 \times \hat{\varepsilon}_2 = -\hat{\varepsilon}_1$, which is the desired result.

The permutation tensor may also be expressed as the determinant of an array of Kronecker deltas in the following manner:

$$e_{ijk} = \begin{vmatrix} \delta_{i1} & \delta_{i2} & \delta_{i3} \\ \delta_{j1} & \delta_{j2} & \delta_{j3} \\ \delta_{k1} & \delta_{k2} & \delta_{k3} \end{vmatrix} = \begin{vmatrix} \delta_{i1} & \delta_{j1} & \delta_{k1} \\ \delta_{i2} & \delta_{j2} & \delta_{k2} \\ \delta_{i3} & \delta_{j3} & \delta_{k3} \end{vmatrix} \tag{8.14}$$

The equivalence of (8.13) and (8.14) can be readily verified. If, for example, (ijk)=(312), then

$$e_{312}=1=\begin{vmatrix}\delta_{31} & \delta_{32} & \delta_{33}\\ \delta_{11} & \delta_{12} & \delta_{13}\\ \delta_{21} & \delta_{22} & \delta_{23}\end{vmatrix}=\begin{vmatrix}0 & 0 & 1\\ 1 & 0 & 0\\ 0 & 1 & 0\end{vmatrix}=1.$$

A very useful identity, relating the permutation tensor to the Kronecker delta, may be obtained by applying (8.14) to the product $e_{ijk}e_{\alpha\beta\gamma}$.

$$e_{ijk}e_{\alpha\beta\gamma}=\begin{vmatrix}\delta_{i1} & \delta_{i2} & \delta_{i3}\\ \delta_{j1} & \delta_{j2} & \delta_{j3}\\ \delta_{k1} & \delta_{k2} & \delta_{k3}\end{vmatrix}\begin{vmatrix}\delta_{\alpha1} & \delta_{\beta1} & \delta_{\gamma1}\\ \delta_{\alpha2} & \delta_{\beta2} & \delta_{\gamma2}\\ \delta_{\alpha3} & \delta_{\beta3} & \delta_{\gamma3}\end{vmatrix}$$

By performing the usual row into column multiplication and noting that, as a result of (8.11), $\delta_{i1}\delta_{\alpha1}+\delta_{i2}\delta_{\alpha2}+\delta_{i3}\delta_{\alpha3}=\delta_{ip}\delta_{\alpha p}=\delta_{i\alpha}$, etc., we obtain

$$e_{ijk}e_{\alpha\beta\gamma}=\begin{vmatrix}\delta_{i\alpha} & \delta_{i\beta} & \delta_{i\gamma}\\ \delta_{j\alpha} & \delta_{j\beta} & \delta_{j\gamma}\\ \delta_{k\alpha} & \delta_{k\beta} & \delta_{k\gamma}\end{vmatrix}. \tag{8.15}$$

If any two rows or columns of a determinant are interchanged, its sign is reversed. Thus from (8.14) we conclude that

$$e_{ijk}=-e_{jik}=-e_{ikj}=-e_{kji}. \tag{8.16a}$$

Since a second interchange will again reverse the sign, we obtain with the aid of (8.16a)

$$e_{ijk}=e_{jki}=e_{kij}. \tag{8.16b}$$

In view of (8.16a) we say that e_{ijk} is skew-symmetric.

Consider three arbitrary vectors, $\vec{A}$, $\vec{B}$ and $\vec{C}$ whose components in the R.C.S. of Fig. 8.1 are A_i, B_i and C_i, respectively. Their component representations are

$$\vec{A}=A_i\hat{e}_i,\ \vec{B}=B_i\hat{e}_i \text{ and } \vec{C}=C_i\hat{e}_i.$$

With the aid of (8.9) and (8.11) the scalar product of $\vec{A}$ and $\vec{B}$ is given by

$$\vec{A}\cdot\vec{B} = (A_i\hat{\varepsilon}_i)\cdot(B_j\hat{\varepsilon}_j) = A_iB_j(\hat{\varepsilon}_i\cdot\hat{\varepsilon}_j)$$

$$\vec{A}\cdot\vec{B} = A_iB_j\delta_{ij} = A_iB_i = A_jB_j \ . \tag{8.17}$$

In unabridged form, (8.17) becomes

$$\vec{A}\cdot\vec{B} = A_1B_1 + A_2B_2 + A_3B_3.$$

The vector product of $\vec{A}$ into $\vec{B}$ in component form is

$$\vec{D} = \vec{A}\times\vec{B} = (A_i\hat{\varepsilon}_i)\times(B_j\hat{\varepsilon}_j) = A_iB_j(\hat{\varepsilon}_i\times\hat{\varepsilon}_j).$$

With the aid of (8.12), we obtain the following expression for $\vec{D}$.

$$\vec{D} = A_iB_je_{ijk}\hat{\varepsilon}_k$$

If we let $\vec{D} = D_k\hat{\varepsilon}_k$, then

$$D_k = e_{ijk}A_iB_j. \tag{8.18}$$

After performing the sums indicated in (8.18) and substituting the appropriate values for e_{ijk} from (8.13) we obtain the unabridged form

$$\left.\begin{aligned} D_1 &= A_2B_3-A_3B_2 \\ D_2 &= A_3B_1-A_1B_3 \\ D_3 &= A_1B_2-A_2B_1 \end{aligned}\right\} \tag{8.19}$$

or

$$D_k=e_{ijk}A_iB_j = \begin{vmatrix} A_1 & A_2 & A_3 \\ B_1 & B_2 & B_3 \\ \delta_{k1} & \delta_{k2} & \delta_{k3} \end{vmatrix} \ . \tag{8.20}$$

As an exercise the reader will be asked to show that (8.20) may be derived directly from (8.14).

With the aid of (8.17) and (8.18) we may express the scalar triple product of the three vectors $\vec{A}$, $\vec{B}$ and $\vec{C}$ as

$$\vec{A}\times\vec{B}\cdot\vec{C}=\vec{D}\cdot\vec{C}=D_kC_k=A_iB_jC_ke_{ijk} \ . \tag{8.21}$$

A second form of the scalar triple product which is convenient for computational purposes may be obtained from (8.20) as follows:

$$D_kC_k = \begin{vmatrix} A_1 & A_2 & A_3 \\ B_1 & B_2 & B_3 \\ \delta_{k1}C_k & \delta_{k2}C_k & \delta_{k3}C_k \end{vmatrix} = \begin{vmatrix} A_1 & A_2 & A_3 \\ B_1 & B_2 & B_3 \\ C_1 & C_2 & C_3 \end{vmatrix}$$

Thus,

$$\vec{A}\times\vec{B}\cdot\vec{C} = \begin{vmatrix} A_1 & A_2 & A_3 \\ B_1 & B_2 & B_3 \\ C_1 & C_2 & C_3 \end{vmatrix} \quad . \tag{8.22a}$$

Another triple product of some interest is the vector triple product

$$(\vec{A}\times\vec{B})\times\vec{C} = \vec{D}\times\vec{C} = e_{k\ell m}D_kC_\ell\hat{\varepsilon}_m$$
$$= e_{k\ell m}e_{ijk}A_iB_jC_\ell\hat{\varepsilon}_m.$$

With the aid of (8.15) one may readily verify that

$$e_{ijk}e_{k\ell m} = \delta_{i\ell}\delta_{jm}-\delta_{im}\delta_{j\ell}.$$

Therefore,

$$(\vec{A}\times\vec{B})\times\vec{C} = (\delta_{i\ell}\delta_{jm}-\delta_{im}\delta_{j\ell})A_iB_jC_\ell\hat{\varepsilon}_m$$
$$= (A_iC_i)B_m\hat{\varepsilon}_m-(B_jC_j)A_m\hat{\varepsilon}_m$$
$$(\vec{A}\times\vec{B})\times\vec{C} = (\vec{A}\cdot\vec{C})\vec{B}-(\vec{B}\cdot\vec{C})\vec{A}. \tag{8.22b}$$

As an exercise, the reader may show that

$$\vec{A}\times(\vec{B}\times\vec{C}) = (\vec{C}\cdot\vec{A})\vec{B} - (\vec{B}\cdot\vec{A})\vec{C} \ . \tag{8.22c}$$

Thus, $(\vec{A}\times\vec{B})\times\vec{C} \neq \vec{A}\times(\vec{B}\times\vec{C})$.

8.4 COORDINATE TRANSFORMATIONS

Indicial notation provides us with an efficient means to deal with equations and operations involving the components of vectors. However, whereas a vector is an entity independent of coordinate systems its components are not, In fact, the same vector will be represented by two different sets of components in two different coordinate systems. For example, consider the position vector $\vec{r}$ which joins the point O to the point P as shown in Fig. 8.4. This figure depicts two different

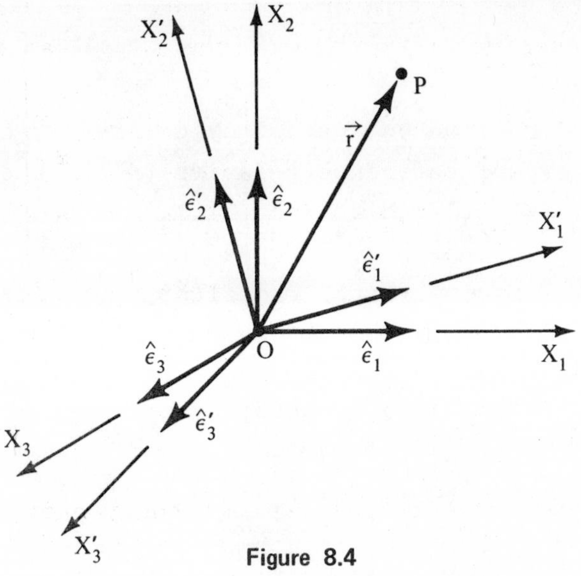

Figure 8.4

rectangular Cartesian coordinate systems sharing the common origin O. The coordinate axes and corresponding base vectors of the "unprimed" reference are labeled (X_1,X_2,X_3) and $(\hat{\varepsilon}_1,\hat{\varepsilon}_2,\hat{\varepsilon}_3)$, respectively. Similarly, the coordinate axes and corresponding base vectors of the "primed" reference are labeled (X_1',X_2',X_3') and $(\hat{\varepsilon}_1',\hat{\varepsilon}_2',\hat{\varepsilon}_3')$, respectively. If the coordinats of the point P are (x_1,x_2,x_3) in the unprimed reference and (x_1',x_2',x_3') in the primed reference, then

$$\vec{r} = x_i\hat{\varepsilon}_i \quad \text{in the unprimed reference} \tag{8.23a}$$

and

$$\vec{r} = x_i'\hat{\varepsilon}_i' \quad \text{in the primed reference.} \tag{8.23b}$$

Since the components (x_1,x_2,x_3) and the components (x_1',x_2',x_3') both describe the same vector they must be related. According to (8.23b), (8.9) and (8.11)

$$\hat{\varepsilon}_i'\cdot\vec{r} = \hat{\varepsilon}_i'\cdot(x_j'\hat{\varepsilon}_j') = x_j'(\hat{\varepsilon}_i'\cdot\hat{\varepsilon}_j') = x_j'\delta_{ij}' = x_i'.$$

Similarly, from (8.23a) and (8.5) we obtain

$$\hat{\varepsilon}_i'\cdot\vec{r} = \hat{\varepsilon}_i'\cdot(x_j\hat{\varepsilon}_j) = x_j(\hat{\varepsilon}_i'\cdot\hat{\varepsilon}_j) = a_{ij}x_j$$

where
$$\left.\begin{aligned} a_{ij} &\equiv \hat{\varepsilon}_i'\cdot\hat{\varepsilon}_j = \cos\alpha_{ij} \\ \text{and} \quad \alpha_{ij} &= \measuredangle\,(X_i',X_j). \end{aligned}\right\} \tag{8.24}$$

Thus,

$$x_i' = a_{ij}x_j. \tag{8.25a}$$

The a_{ij} are called direction cosines. They completely describe the orientation of the primed reference relative to the unprimed reference. For example, $a_{23}=\cos(X_2',X_3)$ and $a_{32}=\cos(X_3',X_2)$. Note that $a_{23}\neq a_{32}$ and in general $a_{ij}\neq a_{ji}$. Thus the order of the subscripts i and j in the direction cosines a_{ij} is important. The first subscript is always associated with a primed axis and the second subscript with an unprimed axis.

The inverse of (8.25a) may be obtained by a similar procedure. From (8.23a)

$$\hat{\varepsilon}_i\cdot\vec{r} = \hat{\varepsilon}_i\cdot(x_j\hat{\varepsilon}_j) = x_j(\hat{\varepsilon}_i\cdot\hat{\varepsilon}_j) = x_j\delta_{ij} = x_i$$

while according to (8.23b) and (8.24)

$$\hat{\varepsilon}_i\cdot\vec{r} = \hat{\varepsilon}_i\cdot(x_j'\hat{\varepsilon}_j') = x_j'(\hat{\varepsilon}_j'\cdot\hat{\varepsilon}_i) = a_{ji}x_j'.$$

Thus,

$$x_i = a_{ji}x_j'. \tag{8.25b}$$

We have shown that if two sets of components x_i and x_i' are to represent the same vector in the unprimed and primed R.C.S., respectively, then they must be related by equations (8.25). The converse of this statement is also true. If two sets of components, x_i and x_i', are related by (8.25), then they are the components of the same vector referred to the unprimed and primed coordinate systems, respectively. In fact, when dealing with the components of vectors, (8.25) is usually adopted as the definition of a vector.

The base vectors of the primed and unprimed references are related by

$$\hat{\varepsilon}_i' = a_{ij}\hat{\varepsilon}_j \tag{8.26a}$$

and

$$\hat{\varepsilon}_i = a_{ji}\hat{\varepsilon}_j' \tag{8.26b}$$

These relations are readily verified. For example, from (8.24) and (8.26a) we obtain

$$a_{ij} \equiv \hat{\varepsilon}_i'\cdot\hat{\varepsilon}_j = (a_{ik}\hat{\varepsilon}_k)\cdot\hat{\varepsilon}_j = a_{ik}\delta_{kj} = a_{ij}.$$

Equation (8.26b) may be verified in a similar manner. We may use (8.26) to derive some useful identities involving the a_{ij}. Since the $\hat{\varepsilon}_i'$ are mutually orthogonal unit vectors it follows from (8.10) that

$$\hat{\varepsilon}_i' \cdot \hat{\varepsilon}_j' = \delta_{ij}' \tag{8.27}$$

where $\delta_{ij}'=\delta_{ij}$ is the Kronecker delta symbol for the primed reference. With the aid of (8.9), (8.26) and (8.27) we obtain

$$\left.\begin{aligned} \delta_{ij}' &= \hat{\varepsilon}_i' \cdot \hat{\varepsilon}_j' = (a_{ik}\hat{\varepsilon}_k)\cdot(a_{j\ell}\hat{\varepsilon}_\ell) = a_{ik}a_{j\ell}\delta_{k\ell} \\ &\text{and} \\ \delta_{ij} &= \hat{\varepsilon}_i \cdot \hat{\varepsilon}_j = (a_{ki}\hat{\varepsilon}_k')\cdot(a_{\ell j}\hat{\varepsilon}_\ell') = a_{ki}a_{\ell j}\delta_{k\ell}'. \end{aligned}\right\} \tag{8.28}$$

Equations (8.28) relate the components of the Kronecker delta in the primed reference to its components in the unprimed reference, and vice versa. Equations (8.25) define a vector by means of a transformation law satisfied by its components, and, in a similar manner, we shall adopt (8.28) as the definition of a second order tensor, i.e., an entity which transforms in accordance with (8.28) will be called a second order tensor. Thus we can say that the Kronecker delta is a second order tensor. An arbitrary second order tensor would be described by a set of nine different components, T_{ij} in the unprimed reference or T_{ij}' in the primed reference, such that

$$\left.\begin{aligned} T_{ij}' &= a_{ik}a_{j\ell}T_{k\ell} \\ &\text{and} \\ T_{ij} &= a_{ki}a_{\ell j}T_{k\ell}'. \end{aligned}\right\} \tag{8.29}$$

If $T_{ji}=T_{ij}$ we say the tensor is symmetric and if $T_{ji}=-T_{ij}$ we say the tensor is skew-symmetric. The Kronecker delta is obviously a symmetric second order tensor since $\delta_{ij}=\delta_{ji}$ by definition. The Kronecker delta is a very special second order tensor because its components are the same in all rectangular Cartesian coordinate systems, i.e., $\delta_{ij}'=\delta_{ij}$. A tensor whose components are the same in all R.C.S. is called isotropic. Therefore, the Kronecker delta is an isotropic second order tensor. By employing this isotropy in (8.28) we obtain, with the aid of (8.11),

$$a_{ik}a_{jk} = a_{ki}a_{kj} = \delta_{ij}. \tag{8.30}$$

Equations (8.30) are the orthogonality relations satisfied by the direction cosines. They express the fact that the base vectors $\hat{\varepsilon}_1, \hat{\varepsilon}_2$ and $\hat{\varepsilon}_3$ are mutually orthogonal, and the base vectors $\hat{\varepsilon}_1'$, $\hat{\varepsilon}_2'$, and $\hat{\varepsilon}_3'$ are also mutually orthogonal.

We note that the primed as well as the unprimed references shown in Fig. 8.4 are right-handed (dextral), and the results of the present section obtained so far do not depend upon this property. However, in the following we shall assume that all coordinate systems are right-handed. Accordingly, we may show with the aid of (8.12), (8.9) and (8.11) that

$$\hat{\varepsilon}_i \times \hat{\varepsilon}_j \cdot \hat{\varepsilon}_k = (e_{ij\ell}\hat{\varepsilon}_\ell)\cdot\hat{\varepsilon}_k = e_{ij\ell}\delta_{\ell k} = e_{ijk} \tag{8.31}$$

and similarly that

$$\hat{\varepsilon}_i' \times \hat{\varepsilon}_j' \cdot \hat{\varepsilon}_k' = e_{ijk}' \tag{8.32}$$

where $e_{ijk}'=e_{ijk}$ is the permutation tensor for the primed reference. By thrice substituting (8.26a) into (8.32) we obtain, with the aid of (8.31)

$$\begin{aligned} e_{ijk}' &= (a_{i\ell}\hat{\varepsilon}_\ell)\times(a_{jm}\hat{\varepsilon}_m)\cdot(a_{kn}\hat{\varepsilon}_n) \\ &= a_{i\ell}a_{jm}a_{kn}(\hat{\varepsilon}_\ell \times \hat{\varepsilon}_m \cdot \hat{\varepsilon}_n) \end{aligned}$$

or

$$e_{ijk}' = a_{i\ell}a_{jm}a_{kn}e_{\ell mn}. \tag{8.33}$$

Similarly, if (8.26b) is thrice substituted into (8.31), we obtain, with the aid of (8.32)

$$\begin{aligned} e_{ijk} &= (a_{\ell i}\hat{\varepsilon}_\ell')\times(a_{mj}\hat{\varepsilon}_m')\cdot(a_{nk}\hat{\varepsilon}_n') \\ &= a_{\ell i}a_{mj}a_{nk}(\hat{\varepsilon}_\ell' \times \hat{\varepsilon}_m' \cdot \hat{\varepsilon}_n') \end{aligned}$$

or

$$e_{ijk} = a_{\ell i}a_{mj}a_{nk}e_{\ell mn}'. \tag{8.34}$$

We have in the past referred to e_{ijk} as the permutation "tensor" and we can now provide some justification for this term. We may define an entity called a third order tensor by a transformation law in a manner similar to that used in (8.25) and (8.29) for vectors and second order tensors, respectively. Thus an arbitrary third order tensor is described by a set of twenty-seven components T_{ijk} in the unprimed reference, or T'_{ijk} in the primed reference such that

$$\left.\begin{aligned} T'_{ijk} &= a_{i\ell}a_{jm}a_{kn}T_{\ell mn} \\ T_{ijk} &= a_{\ell i}a_{mj}a_{nk}T'_{\ell mn}. \end{aligned}\right\} \tag{8.35}$$

Comparing (8.33) and (8.34) with (8.35) reveals that the e_{ijk} are indeed the components of a third order tensor.

Since $e'_{ijk}=e_{ijk}$, the permutation tensor is said to be isotropic. Therefore (8.33) and (8.34) may be rewritten as

$$e_{ijk} = a_{ip}a_{jq}a_{kr}e_{pqr} = a_{pi}a_{qj}a_{rk}e_{pqr}. \tag{8.36}$$

As a result of (8.14)

$$e_{pqr} = \begin{vmatrix} \delta_{p1} & \delta_{p2} & \delta_{p3} \\ \delta_{q1} & \delta_{q2} & \delta_{q3} \\ \delta_{r1} & \delta_{r2} & \delta_{r3} \end{vmatrix}.$$

Therefore,

$$a_{ip}a_{jq}a_{kr}e_{pqr} = \begin{vmatrix} a_{ip}\delta_{p1} & a_{ip}\delta_{p2} & a_{ip}\delta_{p3} \\ a_{jq}\delta_{q1} & a_{jq}\delta_{q2} & a_{jq}\delta_{q3} \\ a_{kr}\delta_{r1} & a_{kr}\delta_{r2} & a_{kr}\delta_{r3} \end{vmatrix}$$

$$= \begin{vmatrix} a_{i1} & a_{i2} & a_{i3} \\ a_{j1} & a_{j2} & a_{j3} \\ a_{k1} & a_{k2} & a_{k3} \end{vmatrix}.$$

Thus (8.36) may also be written in the form

$$e_{ijk} = \begin{vmatrix} a_{i1} & a_{i2} & a_{i3} \\ a_{j1} & a_{j2} & a_{j3} \\ a_{k1} & a_{k2} & a_{k3} \end{vmatrix} = \begin{vmatrix} a_{1i} & a_{2i} & a_{3i} \\ a_{1j} & a_{2j} & a_{3j} \\ a_{1k} & a_{2k} & a_{3k} \end{vmatrix}. \tag{8.37}$$

Since $e_{123}=1$, (8.37) reveals the fact that

$$\begin{vmatrix} a_{11} & a_{12} & a_{13} \\ a_{21} & a_{22} & a_{23} \\ a_{31} & a_{32} & a_{33} \end{vmatrix} = 1. \tag{8.38}$$

Occasionally we shall write (8.38) in the abbreviated form,

$$\det(a_{ij}) = 1. \tag{8.39}$$

As a result of the orthogonality relations (8.30) and the rules governing the multiplication of determinants one can show that

$$\det(\delta_{ij}) = \det(a_{ik}a_{jk}) = [\det(a_{ij})]^2 .$$

Since $\det(\delta_{ij})=1$, we conclude that $\det(a_{ij})=\pm1$ for an orthogonal transformation. Orthogonal transformations for which $\det(a_{ij})=+1$ are called proper orthogonal transformations. With reference to (8.38) it is seen that any two right-handed, rectangular Cartesian coordinate systems may be related through a proper orthogonal transformation. Note that for a proper orthogonal transformation, the primed reference may be brought into coincidence with the unprimed reference by a rotation about the origin. If one were to consider only left-handed, rectangular Cartesian coordinate systems then one could show, by following the procedure used in (8.31) through (8.38), that again $\det(a_{ij})=+1$, i.e., any two left-handed, rectangular Cartesian coordinate systems may be related through a proper orthogonal transformation. However, if one admits both right and left-handed coordinates, then it could happen that the primed reference is left-handed while the unprimed is right-handed, or vice versa. In either case one would find that $\det(a_{ij})=-1$. This is called an improper orthogonal transformation and we shall not consider these in our development.

8.5 TENSOR ALGEBRA

In the previous section we defined second and third order tensors in terms of a transformation law for their components. A useful generalization of this procedure becomes evident in view of (8.25), (8.29) and (8.35). Thus we define a fourth order tensor in terms of a set of eighty-one components, T_{ijkl} in the unprimed reference or T'_{ijkl} in the primed reference, such that

$$\left.\begin{aligned} T'_{ijk\ell} &= a_{ip}a_{jq}a_{kr}a_{\ell s}T_{pqrs} \\ &\text{and} \\ T_{ijk\ell} &= a_{pi}a_{qj}a_{rk}a_{s\ell}T'_{pqrs} \end{aligned}\right\} \tag{8.40}$$

with similar expressions for higher order tensors. To complete this hierarchy of tensors we define a tensor of order one to be a vector and a tensor of order zero to be a scalar. The various tensor transformation laws are summarized in Table 8.1.

TABLE 8.1

TENSOR TRANSFORMATION LAWS

Tensor of Order	Number of Components	Law of Transformation	Sometimes Called a
0	1	$T'=T$	Scalar
1	3	$T'_i=a_{ip}T_p$	Vector
2	9	$T'_{ij}=a_{ip}a_{jq}T_{pq}$	Dyad, Bisor
3	27	$T'_{ijk}=a_{ip}a_{jq}a_{kr}T_{pqr}$	Triad, Trisor
4	81	$T'_{ijk\ell}=a_{ip}a_{jq}a_{kr}a_{\ell s}T_{pqrs}$	Tetrad, Tetror
5	243	etc.	

At this point we digress to consider the mathematical characterization of the laws of nature. For any given natural phenomenon, the laws of nature operate independently of the coordinate systems which are arbitrarily introduced for our convenience. Thus the form of an equation purporting to represent a law of nature must be the same in all coordinate systems, i.e., the form of the equation must be invariant with respect to any coordinate transformation (generally co-variant). We shall demonstrate that all Cartesian tensor equations possess this property. This form invariance of tensor equations is the basic reason for their importance in the study of mechanics. With the aid of Table 8.1 we observe that each tensor transformation is a linear homogeneous function of the tensor's components. Thus if all the components of a tensor vanish in one R.C.S. then they will all vanish in every other R.C.S. For example, if the components C_i of a vector are such that $C_1=C_2=C_3=0$, then in any other R.C.S. $C_i'=a_{ij}C_j=0$. Now consider the vector equation $A_i=B_i$. If we let $C_i=A_i-B_i$, then $C_i=0$ and by the above argument $C_i'=0$. However,

$$C_i' = a_{ij}C_j=a_{ij}A_j-a_{ij}B_j=A_i'-B_i'.$$

Therefore, $A_i'=B_i'$ and we have shown that vector equations are form invariant. This same procedure can be used to show that a tensor equation of any order is form invariant.

The basic algebraic properties of tensors are illustrated by the examples which follow. Let A_i be a vector and let B_{ij} and C_{ij} be two second order tensors. Strictly speaking, A_i are the components of a vector. However, these components together with the transformation law in Table 8.1 define a vector. Therefore, we shall at times speak of the vector A_i or the tensor B_{ij} when we actually mean the vector whose components in a particular reference are A_i, with similar stipulations for B_{ij}. The sum and difference of two tensors of the same order are also tensors of that order. For example, $F_{ij}=B_{ij}+C_{ij}$ and $G_{ij}=B_{ij}-C_{ij}$ are both second order tensors. The outer product of a tensor of order M with a tensor of order N is a tensor of order M+N. For example, $H_{ijk}=A_iB_{jk}$ is called the outer product of A_i and B_{jk}. It is a tensor of order three. If two indices of a tensor are set equal and the result summed according to our summation convention then we say the two indices have been contracted. A contraction performed on a tensor of order $N\geq 2$ results in a tensor of order N-2. For example, B_{ii} is the contraction of B_{ij} and it is a tensor of order zero, i.e., a scalar. The contraction of a second order tensor is also called the trace of the tensor. An inner product is an outer product followed by a contraction. For example, $H_{iji}=A_iB_{ji}$ is an inner product of A_i and B_{jk}.

The tensor character of the basic objects of mechanics such as force, displacement, velocity, acceleration, etc. is a result of the form invariance of the laws of nature. By applying the laws of nature to the hypotheses of continuum mechanics we obtain new entities such as stress and strain whose tensor character is of fundamental importance in subsequent developments. A considerable saving in effort will result if we can establish their tensor character directly from these laws without having to appeal to the transformations of Table 8.1 in every case. This will be made possible as a result of the quotient law of tensors which we will now consider. Suppose a set of quantities $\{T_1, T_2, T_3\}$ together with an arbitrary vector n_i satisfy the equation

$$T_i n_i = S \tag{8.41a}$$

where S is known to be a scalar. Under these conditions one may conclude that T_i is a vector. This can be shown as follows. If (8.41) represents a physical law, then it is form invariant. Thus in a primed reference the corresponding equation is

$$T_i' n_i' = S' ,$$

but according to Table 8.1 and (8.41a)

$S' = S = T_i n_i$ and $n_i' = a_{ij} n_j$. Therefore,

$$T_i' a_{ij} n_j = T_i n_i$$

or

$$(T_i - T_k' a_{ki}) n_i = 0.$$

Because n_i is arbitrary, the above equation must be satisfied for the three particular cases $n_i = \delta_{i1}$, $n_i = \delta_{i2}$ and $n_i = \delta_{i3}$. This leads us to conclude that $T_i - a_{ki} T_k' = 0$ or $T_i = a_{ki} T_k'$. Thus we conclude that T_i is a vector. We next consider the set of quantities T_{ij} which together with an arbitrary vector N_i satisfy the equations

$$T_{ij} N_i = T_j \tag{8.41b}$$

where T_j is known to be a vector. Under these conditions one may conclude that T_{ij} is a second order tensor. The proof is similar to the one used for vectors. In a primed reference,

$$T_{ij}' N_i' = T_j' = a_{jk} T_k = a_{jk} T_{mk} N_m .$$

However, $N_i' = a_{im} N_m$, and therefore

$$T_{ij}' a_{im} N_m = a_{jk} T_{mk} N_m \quad \text{or}$$

$$(T_{ij}' a_{im} - T_{mk} a_{jk}) N_m = 0.$$

By letting $N_m=\delta_{m1}$, δ_{m2}, and δ_{m3} successively in the above equations we conclude that

$$T'_{ij}a_{im}-T_{mk}a_{jk} = 0.$$

If this equation is multiplied by a_{lm} we obtain, with the aid of (8.30)

$$T'_{\ell j} = a_{\ell m}a_{jk}T_{mk},$$

thus proving that T_{ij} is a second order tensor. By combining (8.41a) with (8.41b) we may conclude that T_{ij} is a second order tensor provided

$$T_{ij}N_in_j = S \qquad (8.41c)$$

where N_i and n_j are independent and arbitrary vectors and S is a scalar. It can also be shown that T_{ij} is a symmetric second order tensor provided

$$T_{ij}n_in_j = S \qquad (8.41d)$$

where n_i is an arbitrary vector, S is a scalar and $T_{ij}=T_{ji}$. It should by now be evident that the quotient law can be extended to tensors of any order.

In the next chapter we will show that various physical quantities, most notably stress and strain, are characterized by second order tensors. Thus it behooves us to examine them further. For instance, a symmetric second order tensor can have at most six different components. This is obvious from the relations $T_{ji}=T_{ij}$. On the other hand, a skew-symmetric second order tensor can have at most three different components which also define a vector. In this case $T_{ji}=-T_{ij}$, and we at once conclude that $T_{11}=T_{22}=T_{33}=0$, $T_{12}=-T_{21}$, $T_{13}=-T_{31}$ and $T_{23}=-T_{32}$. If we let

$$T_{ij} = e_{ijk}T_k \qquad (8.42a)$$

then clearly $T_{ij}=-T_{ji}$ and by multiplying this equation by $e_{ij\ell}$ we obtain

$$e_{ij\ell}T_{ij} = e_{ij\ell}e_{ijk}T_k = 2\delta_{\ell k}T_k = 2T_\ell.$$

The identity $e_{ij1}e_{ijk}=2\delta_{1k}$ follows from (8.15). Thus the components of the so-called dual vector T_k are given by

$$T_k = \tfrac{1}{2}e_{ijk}T_{ij}. \qquad (8.42b)$$

If A_{ij} and B_{ij} are symmetric and skew-symmetric second order tensors respectively, then $A_{ij}B_{ij}=0$. This can be shown as follows. $A_{ij}B_{ij}=A_{ji}B_{ji}$, i.e., the value of the sum is independent of the choice of summation index. However, $A_{ji}=A_{ij}$ and $B_{ji}=-B_{ij}$ so that $A_{ij}B_{ij}=-A_{ij}B_{ij}$ and the desired result follows at once.

An arbitrary second order tensor may always be decomposed uniquely into the sum of a symmetric and a skew-symmetric tensor as follows.

$$T_{ij} = \varepsilon_{ij}+\omega_{ij} \tag{8.43}$$

where

$$\varepsilon_{ij} = \varepsilon_{ji} = \tfrac{1}{2}(T_{ij}+T_{ji})$$

$$\omega_{ij} = -\omega_{ji} = \tfrac{1}{2}(T_{ij}-T_{ji}) = e_{ijk}T_k$$

and

$$T_k = \tfrac{1}{2}e_{ijk}T_{ij}.$$

The components of a second order tensor may be displayed as the elements of a matrix. Thus the matrix corresponding to the tensor whose components are T_{ij} is

$$\left[T_{ij}\right] = \begin{bmatrix} T_{11} & T_{12} & T_{13} \\ T_{21} & T_{22} & T_{23} \\ T_{31} & T_{32} & T_{33} \end{bmatrix} . \tag{8.44}$$

This correspondence is very fruitful indeed since it allows one to apply the considerable body of knowledge concerning matrices to the study of second order tensors. Those interested in persuing this matter further should consult the references listed under Tensor Analysis and Matrix Theory.

In the previous section we pointed out that δ_{ij} and e_{ijk} are isotropic tensors. It can be shown that all isotropic second and third order tensors have the form $\lambda\delta_{ij}$ and μe_{ijk} respectively where λ and μ are arbitrary scalars. It is obvious from the transformation law for scalars $T'=T$ that all scalars are isotropic and, although it is less obvious, it can be shown that there are no nonzero isotropic vectors. Isotropic tensors of the fourth order play an important role in the linear theory of elasticity and therefore deserve our further consideration. If C_{ijkl} are the components of an isotropic fourth order tensor, then in the primed reference $C'_{ijkl}=C_{ijkl}$ and the transformation law (8.40) becomes

$$C_{ijk\ell} = a_{ip}a_{jq}a_{kr}a_{\ell s}C_{pqrs}. \tag{8.45}$$

It can be shown (see H. Jeffreys, "Cartesian Tensors," Cambridge University Press, 1963, pp. 66-70) that all fourth order isotropic tensors may be written in the form

$$C_{ijk\ell} = \lambda\delta_{ij}\delta_{k\ell} + \mu(\delta_{ik}\delta_{j\ell}+\delta_{i\ell}\delta_{jk})+\kappa(\delta_{ik}\delta_{j\ell}-\delta_{i\ell}\delta_{jk}). \tag{8.46}$$

The fact that (8.46) is isotropic may be verified by substituting it into the right-hand side of (8.45) which then reduces to an identity.

8.6 TENSOR CALCULUS

A tensor point function is a rule which uniquely assigns to every point in some region of space a tensor of the same order and type. A tensor point function thus defines a tensor field. For example, the temperature at each point in a body defines a scalar field, the velocity at each point defines a vector field, and the stress at each point defines a second order tensor field.

Let us examine some of the differential and integral properties of tensor fields. These properties are independent of the order of the tensors involved and therefore we shall use the symbol $\underset{\sim}{T}$ to characterize a tensor of any order. This is analogous to the notation employed earlier for vectors where, for instance, we denoted the position of a generic point by the symbol $\vec{r}$. The components of $\underset{\sim}{T}$ in a

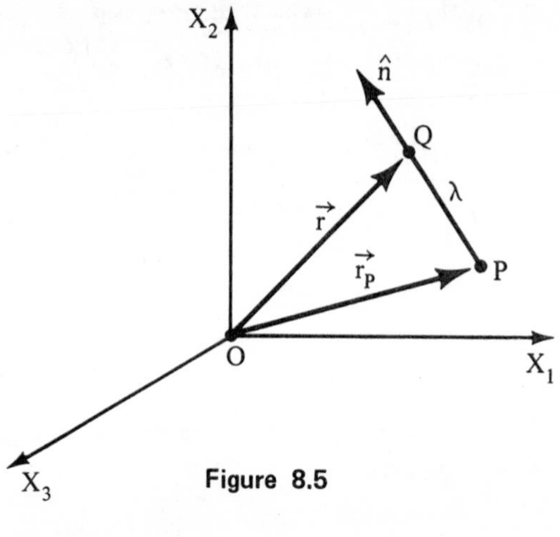

Figure 8.5

rectangular Cartesian coordinate system will be denoted by $T_{ij\ldots}$. A tensor point function will be represented by $\underset{\sim}{T}(\vec{r})$ where $\vec{r}=x_i\hat{\varepsilon}_i$ is the position vector of a generic point in the field. The component form of the tensor point function is $T_{ij\ldots}(x_1,x_2,x_3)$ or more briefly $T_{ij\ldots}(x)$ If each component of a tensor is a continuous function of the coordinates at the point P then we say the tensor is continuous at P. If, in addition, the three first partial derivatives of each component of the tensor exist and are continuous at P we say that the tensor is differentiable at P. We define the derivative of the tensor $\underset{\sim}{T}$ in the $\hat{n}$ direction at P as follows. Referring to Fig. 8.5, it is seen that the position vector of any point along the ray emanating from P in the $\hat{n}$ direction is given by

$$\vec{r} = \vec{r}_p + \lambda\hat{n} \tag{8.47}$$

where $\lambda = |\vec{r}-\vec{r}_p|$ is the distance between the two points. As one moves along this ray the only quantity which varies is λ. Thus, $\underset{\sim}{T}(\vec{r})$ is actually $\underset{\sim}{T}(\lambda)$ along this ray. The derivative of $\underset{\sim}{T}$ in the $\hat{n}$ direction at P is therefore defined as

$$\frac{d\underset{\sim}{T}}{d\lambda} \equiv \lim_{\lambda\to 0}\left[\frac{\underset{\sim}{T}(\vec{r}_P+\lambda\hat{n})-\underset{\sim}{T}(\vec{r}_P)}{\lambda}\right] . \tag{8.48}$$

If $\underset{\sim}{T}$ and its components are expressed as functions of x_1, x_2 and x_3, then

$$\frac{d\underset{\sim}{T}}{d\lambda} = \frac{\partial\underset{\sim}{T}}{\partial x_1}\frac{dx_1}{d\lambda} + \frac{\partial\underset{\sim}{T}}{\partial x_2}\frac{dx_2}{d\lambda} + \frac{\partial\underset{\sim}{T}}{\partial x_3}\frac{dx_3}{d\lambda} = \frac{\partial\underset{\sim}{T}}{\partial x_i}\frac{dx_i}{d\lambda} .$$

From (8.47), $x_i=x_i^P+\lambda n_i$, where x_i^P are the coordinates of the point P. Therefore $\frac{dx_i}{d\lambda}=n_i$ and we obtain

$$\frac{d\underset{\sim}{T}}{d\lambda} = n_i\frac{\partial\underset{\sim}{T}}{\partial x_i} \tag{8.49a}$$

or in component form

$$\frac{dT_{ij\ldots}}{d\lambda} = n_k\,\frac{\partial T_{ij\ldots}}{\partial x_k} . \tag{8.49b}$$

The derivatives $\partial T_{ij\ldots}/\partial x_k$ are the components of a tensor of order one greater than $\underset{\sim}{T}$ called the gradient of $\underset{\sim}{T}$. This is evident from an application of the chain rule to the derivative $\partial \underset{\sim}{T}/\partial x_i'$. With the aid of (8.25b),

$$\frac{\partial \underset{\sim}{T}}{\partial x_i'} = \frac{\partial \underset{\sim}{T}}{\partial x_j} \frac{\partial x_j}{\partial x_i'} = a_{ij} \frac{\partial \underset{\sim}{T}}{\partial x_j} .$$

If, in the expression for the gradient $\partial T_{ij\ldots}/\partial x_k$, we contract the index k with one of the indices i,j,... we obtain a tensor of order one less than $\underset{\sim}{T}$ called a divergence of $\underset{\sim}{T}$. For example, $\partial T_{ij\ldots}/\partial x_i$, $\partial T_{ij\ldots}/\partial x_j$, etc. are each called a divergence of $\underset{\sim}{T}$. There are obviously n different ways to define the divergence of a tensor of order n. Similarly, we may construct expressions of the type

$$e_{\ell ki} \frac{\partial T_{ij\ldots}}{\partial x_k} , \quad e_{\ell kj} \frac{\partial T_{ij\ldots}}{\partial x_k} , \text{ etc.}$$

These are the components of the curl or rotation of the tensor $\underset{\sim}{T}$ and again there are n different ways to define the curl of a tensor of order n. One can easily show that a curl of $\underset{\sim}{T}$ is a tensor of the same order as $\underset{\sim}{T}$. For a vector, n=1, and in this case each of these operations is uniquely defined. If $\vec{T}=T_i\hat{e}_i$, the components of the gradient, divergence and curl of $\vec{T}$ are $\partial T_i/\partial x_j$, $\partial T_i/\partial x_i$, and $e_{ijk}\ \partial T_k/\partial x_j$, respectively. We often write these expressions in vector form with the aid of the del operator defined by

$$\vec{\nabla} \equiv \hat{e}_1 \frac{\partial}{\partial x_1} + \hat{e}_2 \frac{\partial}{\partial x_2} + \hat{e}_3 \frac{\partial}{\partial x_3} = \hat{e}_i \frac{\partial}{\partial x_i} \tag{8.50a}$$

Thus the gradient, divergence, and curl of $\vec{T}$ may also be written as $\vec{\nabla}\vec{T}$, $\vec{\nabla}\cdot \vec{T}$, and $\vec{\nabla}\times\vec{T}$, respectively. The expression $\vec{\nabla}\vec{T}$ $(=\hat{e}_i\ \partial\vec{T}/\partial x_i = \hat{e}_i\hat{e}_j\ \partial T_j/\partial x_i)$ is called a dyadic. It is the invariant representation of a second order tensor, just as $\vec{r}(=\hat{e}_i x_i)$ is the invariant representation of a vector. The operator

$$(\vec{\nabla}\cdot\vec{\nabla}) = \frac{\partial^2}{\partial x_1^2} + \frac{\partial^2}{\partial x_2^2} + \frac{\partial^2}{\partial x_3^2} = \frac{\partial^2}{\partial x_i \partial x_i} \tag{8.50b}$$

is called the Laplacian operator and is usually written as ∇^2.

In the study of continuum mechanics we often encounter integrals of the type

$$\int_V \frac{\partial \tau_{ji}}{\partial x_j}\, dV$$

where V is a closed, bounded region in space. There is a very useful theorem which is associated, by various authors, with the names of Gauss and/or Ostrogradskii that relates such an integral to a surface integral over the boundary of the region under consideration. This theorem may be stated as follows: Let V be a closed, bounded region in space whose boundary S is a piecewise smooth, orientable surface. Let the tensor point function $\underset{\sim}{T}(\vec{r})$ be differentiable in a domain containing V. Then

$$\int_V \vec{\nabla}\underset{\sim}{T}dV = \oint_S \hat{n}\underset{\sim}{T}dS \qquad (8.51a)$$

or in component form

$$\int_V \frac{\partial T_{ij\ldots}}{\partial x_k} dV = \oint_S n_k T_{ij\ldots} dS \qquad (8.51b)$$

where $\hat{n}$ is the unit outward normal to S. As shown in Fig. 8.6, the unit outward normal to S is directed outward from the enclosed region V. The symbol $\oint_S$ represents the integral over the closed surface S. The proof of the preceding theorem is similar to the proof of the divergence theorem given in most texts on advanced calculus or vector analysis, see for example, L. Brand, "Vector Analysis," John Wiley and Sons, Inc. New York, 1957, pp. 137-143. As pointed out in this reference, the theorem may be extended to include regions bounded by several closed surfaces, for example, a cube containing a spherical cavity.

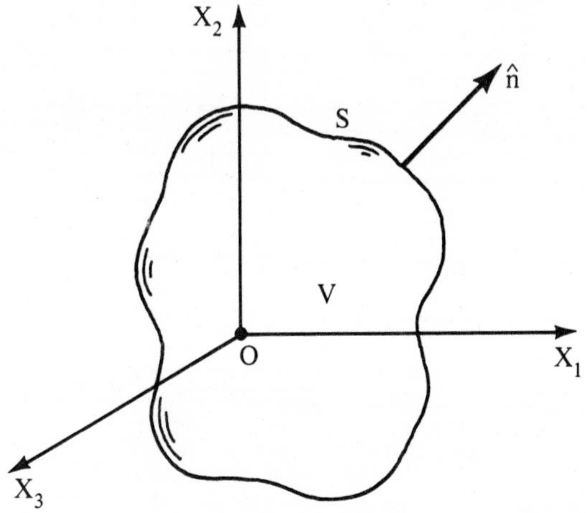

Figure 8.6

Two very useful corollaries of this theorem are:

$$\int_V \frac{\partial T_{ij\ldots}}{\partial x_i}\,dV = \oint_S n_i T_{ij\ldots}\,dS \tag{8.52}$$

and

$$\int_V e_{\ell ki} \frac{\partial T_{ij\ldots}}{\partial x_k}\,dV = \oint_S e_{\ell ki} n_k T_{ij\ldots}\,dS. \tag{8.53}$$

Equation (8.52), which is called the divergence theorem, is used extensively in continuum mechanics.

8.7 CURVILINEAR COORDINATES

We will at times find it convenient to use curvilinear coordinates, e.g., cylindrical, spherical, etc. rather than rectangular Cartesian coordinates to identify points in space. In a given region R, these curvilinear coordinates may be obtained from rectangular coordinates by means of the transformation

$$y_i = f_i(x_1,x_2,x_3),\ i=1,2,3. \tag{8.54}$$

Here $\{x_1,x_2,x_3\}$ are the rectangular Cartesian coordinates and $\{y_1,y_2,y_3\}$ the corresponding curvilinear coordinates of the points in R. We will assume that the partial derivatives $\partial f_i/\partial x_j$ all exist and are continuous on R, i.e., they exist and are continuous at every point of R. Furthermore, the Jacobian

$$J = \det\left(\frac{\partial f_i}{\partial x_j}\right) = \begin{vmatrix} \frac{\partial f_1}{\partial x_1} & \frac{\partial f_1}{\partial x_2} & \frac{\partial f_1}{\partial x_3} \\ \frac{\partial f_2}{\partial x_1} & \frac{\partial f_2}{\partial x_2} & \frac{\partial f_2}{\partial x_3} \\ \frac{\partial f_3}{\partial x_1} & \frac{\partial f_3}{\partial x_2} & \frac{\partial f_3}{\partial x_3} \end{vmatrix} \neq 0 \text{ on R.}$$

Since J is continuous and nonzero on R it is either positive on R or negative on R. Without loss of generality we can assume that $J>0$ on R. With these restrictions the functions f_1,f_2,f_3 are independent and the transformation (8.54) has an inverse on R given by

$$x_i = g_i(y_1,y_2,y_3),\ i=1,2,3. \tag{8.55}$$

Furthermore, each $\partial g_i/\partial y_j$ exists and is continuous on R.

According to the currently accepted convention the symbols y_i and x_i may be used to represent both the coordinates and the functions from which they are determined. Thus (8.54) and (8.55) may be written as

$$y_i = y_i(x_1,x_2,x_3) = y_i(x) \tag{8.56a}$$

and $$x_i = x_i(y_1,y_2,y_3) = x_i(y). \tag{8.56b}$$

The Jacobian of the transformation is

$$J = \det\left(\frac{\partial y_i}{\partial x_j}\right) > 0 \text{ on R.} \tag{8.57}$$

The Jacobian of the inverse transformation $J^{-1}=\det(\partial x_i/\partial y_j)$ is related to J by the identity

$$JJ^{-1}=\det\left(\frac{\partial y_i}{\partial x_j}\right)\det\left(\frac{\partial x_k}{\partial y_\ell}\right) = \det\left(\frac{\partial y_i}{\partial x_j}\frac{\partial x_j}{\partial y_k}\right)$$

$$JJ^{-1}=\det\left(\frac{\partial y_i}{\partial y_k}\right) = \det(\delta_{ik}) = 1$$

Thus,

$$J^{-1} = \frac{1}{J} > 0 \text{ on R.} \tag{8.58}$$

The position vector $\vec{r}$ of a point in R may now be expressed either in terms of the rectangular Cartesian coordinates or by means of (8.56) in terms of the curvilinear coordinates. In Cartesian coordinates

$$\vec{r} = \vec{r}(x_1,x_2,x_3) = x_i\hat{\varepsilon}_i$$

where $\hat{\varepsilon}_i$ are the base vectors of the Cartesian reference. Therefore $\hat{\varepsilon}_i = \partial\vec{r}/\partial x_i$. If $\vec{r}$ is expressed in terms of the curvilinear coordinates, $\vec{r}=\vec{r}(y_1,y_2,y_3)$ we define the base vectors of the curvilinear coordinate system by

$$\vec{g}_i = \frac{\partial\vec{r}}{\partial y_i} = \frac{\partial x_j}{\partial y_i}\hat{\varepsilon}_j. \tag{8.59}$$

Note that these base vectors are not necessarily unit vectors. In fact

$$|\vec{g}_i| = \sqrt{\vec{g}_{(i)}\cdot\vec{g}_{(i)}} = \sqrt{\frac{\partial x_j}{\partial y_{(i)}}\frac{\partial x_j}{\partial y_{(i)}}}\ . \tag{8.60}$$

When we wish to suspend the summation convention for a particular subscript, we will enclose it in parenthesis as in (8.60). Thus the equation $A_i=B_{(i)}C_{(i)}$ represents the three equations $A_1=B_1C_1$, $A_2=B_2C_2$ and $A_3=B_3C_3$. It is possible to construct a geometrical interpretation of (8.59). According to (8.56a), the curvilinear coordinates of the point $P_o=\{x_1^o,x_2^o,x_3^o\}$ are $y_i^o=y_i(x_1^o,x_2^o,x_3^o)$. This is the point of intersection of the three surfaces

$$y_1(x_1,x_2,x_3) = y_1^o, \qquad y_2(x_1,x_2,x_3) = y_2^o, \qquad y_3(x_1,x_2,x_3) = y_3^o$$

and these are called the coordinate surfaces. The pairwise intersections of these surfaces define the coordinate curves. For example, only y_1 varies along the curve defined by the intersection of the coordinate surfaces $y_2(x_1,x_2,x_3)=y_2^o$ and $y_3(x_1,x_2,x_3)=y_3^o$. If the position vector $\vec{r}$ of the point P_o is expressed in terms of the curvilinear coordinates, then

$$d\vec{r} = \frac{\partial \vec{r}}{\partial y_i}\, dy_i = \vec{g}_i dy_i .$$

Along the y_1 coordinate curve y_2 and y_3 are constant and therefore $d\vec{r}=\vec{g}_1dy_1$. Similarly, along the y_2 and y_3 coordinate curves, $d\vec{r}=\vec{g}_2dy_2$ and $d\vec{r}=\vec{g}_3dy_3$, respectively. Thus the base vectors $\vec{g}_i$ are tangent to the coordinate curves through the point P_o.

The base vectors $\hat{\varepsilon}_i$ may be expressed in terms of the $\vec{g}_i$ by

$$\hat{\varepsilon}_i = \frac{\partial \vec{r}}{\partial x_i} = \frac{\partial \vec{r}}{\partial y_j}\frac{\partial y_j}{\partial x_i} = \frac{\partial y_j}{\partial x_i}\vec{g}_j . \tag{8.61}$$

The various dot products of the base vectors $\vec{g}_i$ may be expressed as

$$\vec{g}_i\cdot\vec{g}_j = \frac{\partial x_k}{\partial y_i}\frac{\partial x_\ell}{\partial y_j}(\hat{\varepsilon}_k\cdot\hat{\varepsilon}_\ell) = \frac{\partial x_k}{\partial y_i}\frac{\partial x_k}{\partial y_j} = g_{ij} . \tag{8.62}$$

For reasons which will subsequently become apparent, g_{ij} is called the metric tensor. As a result of the rules governing the product of two determinants

$$g \equiv \det(g_{ij}) = \det\left(\frac{\partial x_k}{\partial y_i}\frac{\partial x_k}{\partial y_j}\right) = \det\left(\frac{\partial x_k}{\partial y_i}\right)\det\left(\frac{\partial x_\ell}{\partial y_j}\right)$$

$$g = \left[\det\left(\frac{\partial x_i}{\partial y_j}\right)\right]^2 = [J^{-1}]^2 = \frac{1}{J^2} .$$

Since J>0 we conclude that

$$J = \frac{1}{\sqrt{g}} \quad \text{or} \quad J^{-1} = \sqrt{g}\ . \tag{8.63}$$

By comparing (8.60) with (8.62) we conclude that

$$|\vec{g}_i| = \sqrt{g_{(i)(i)}}\ . \tag{8.64}$$

The scalar triple product of the base vectors is

$$\vec{g}_i \times \vec{g}_j \cdot \vec{g}_k = \frac{\partial x_\ell}{\partial y_i}\frac{\partial x_m}{\partial y_j}\frac{\partial x_n}{\partial y_k}(\hat{\varepsilon}_\ell \times \hat{\varepsilon}_m \cdot \hat{\varepsilon}_n)$$

$$= e_{\ell mn}\frac{\partial x_\ell}{\partial y_i}\frac{\partial x_m}{\partial y_j}\frac{\partial x_n}{\partial y_k}$$

$$= \begin{vmatrix} \frac{\partial x_1}{\partial y_i} & \frac{\partial x_2}{\partial y_i} & \frac{\partial x_3}{\partial y_i} \\ \frac{\partial x_1}{\partial y_j} & \frac{\partial x_2}{\partial y_j} & \frac{\partial x_3}{\partial y_j} \\ \frac{\partial x_1}{\partial y_k} & \frac{\partial x_2}{\partial y_k} & \frac{\partial x_3}{\partial y_k} \end{vmatrix} = e_{ijk}J^{-1}$$

Thus, $\vec{g}_i \times \vec{g}_j \cdot \vec{g}_k = \frac{e_{ijk}}{J}\ .$ (8.65)

Since J>0 we see that the ordered set $\{\vec{g}_1, \vec{g}_2, \vec{g}_3\}$ is dextral.

Consider the three vectors defined as

$$\vec{G}_i = \frac{\partial y_i}{\partial x_j}\hat{\varepsilon}_j\ . \tag{8.66}$$

These vectors have the following important relationship with the base vectors:

$$\vec{G}_i \cdot \vec{g}_j = \left(\frac{\partial y_i}{\partial x_k}\hat{\varepsilon}_k\right)\cdot\left(\frac{\partial x_\ell}{\partial y_j}\hat{\varepsilon}_\ell\right) = \frac{\partial y_i}{\partial x_k}\frac{\partial x_\ell}{\partial y_j}\delta_{k\ell}.$$

$$\vec{G}_i \cdot \vec{g}_j = \frac{\partial y_i}{\partial x_k}\frac{\partial x_k}{\partial y_j} = \frac{\partial y_i}{\partial y_j} = \delta_{ij} \tag{8.67}$$

As a result of this property the $\vec{G}_i$ are called the reciprocal base vectors. The reciprocal base vectors are normal to the coordinate surfaces at the point P_o. This is illustrated by the following argument.

$$\vec{g}_i \times \vec{g}_j = \left(\frac{\partial x_k}{\partial y_i}\hat{\varepsilon}_k\right)\times\left(\frac{\partial x_\ell}{\partial y_j}\hat{\varepsilon}_\ell\right) = e_{k\ell m}\frac{\partial x_k}{\partial y_i}\frac{\partial x_\ell}{\partial y_j}\hat{\varepsilon}_m$$

According to (8.66), we have

$$\frac{\partial x_m}{\partial y_n}\vec{G}_n = \frac{\partial x_m}{\partial y_n}\frac{\partial y_n}{\partial x_j}\hat{\varepsilon}_j = \frac{\partial x_m}{\partial x_j}\hat{\varepsilon}_j = \delta_{mj}\hat{\varepsilon}_j = \hat{\varepsilon}_m. \tag{8.68}$$

Therefore, with the aid of (8.65), we obtain

$$\vec{g}_i \times \vec{g}_j = e_{k\ell m}\frac{\partial x_k}{\partial y_i}\frac{\partial x_\ell}{\partial y_j}\frac{\partial x_m}{\partial y_n}\vec{G}_n = \frac{1}{J}e_{ijn}\vec{G}_n. \tag{8.69}$$

Consider the surface $y_3(x_1,x_2,x_3)=y_3^o$. The base vectors $\vec{g}_1$ and $\vec{g}_2$ are tangent to the coordinate curves defined by the pairwise intersections of the surfaces $(y_3=y_3^o,\ y_2=y_2^o)$ and $(y_3=y_3^o,\ y_1=y_1^o)$ respectively. Thus $\vec{g}_1$ and $\vec{g}_2$ are both tangent to the surface $y_3=y_3^o$ at P_o and their cross product is therefore normal to this surface. Thus, according to (8.69) $\vec{G}_3$ is normal to the surface $y_3(x_1,x_2,x_3)=y_3^o$ at P_o. In general each $\vec{G}_i$ is normal to the corresponding coordinate surface defined by $y_i(x_1,x_2,x_3)=y_i^o$ at P_o.

The various dot products of the reciprocal base vectors are given by

$$\vec{G}_i\cdot\vec{G}_j = \left(\frac{\partial y_i}{\partial x_k}\hat{\varepsilon}_k\right)\cdot\left(\frac{\partial y_j}{\partial x_\ell}\hat{\varepsilon}_\ell\right) = \frac{\partial y_i}{\partial x_k}\frac{\partial y_j}{\partial x_k} \equiv G_{ij}. \tag{8.70}$$

The determinant of the metric G_{ij} is

$$G = \det(G_{ij}) = \det\left(\frac{\partial y_i}{\partial x_k}\frac{\partial y_j}{\partial x_k}\right) = \left[\det\left(\frac{\partial y_i}{\partial x_j}\right)\right]^2$$

$$G = J^2 = \frac{1}{g}$$

Thus,

$$Gg = 1. \tag{8.71}$$

By following the procedure used to derive (8.65) one can show that the scalar triple product

$$\vec{G}_i \times \vec{G}_j \cdot \vec{G}_k = Je_{ijk}. \tag{8.72}$$

The various cross products of the reciprocal base vectors are given by

$$\vec{G}_i \times \vec{G}_j = \left(\frac{\partial y_i}{\partial x_k}\,\hat{\varepsilon}_k\right) \times \left(\frac{\partial y_j}{\partial x_\ell}\,\hat{\varepsilon}_\ell\right) = e_{k\ell m}\,\frac{\partial y_i}{\partial x_k}\,\frac{\partial y_j}{\partial x_\ell}\,\hat{\varepsilon}_m$$

According to (8.61), $\hat{\varepsilon}_m = (\partial y_n/\partial x_m)\vec{g}_n$, therefore

$$\vec{G}_i \times \vec{G}_j = e_{k\ell m}\,\frac{\partial y_i}{\partial x_k}\,\frac{\partial y_j}{\partial x_\ell}\,\frac{\partial y_n}{\partial x_m}\,\vec{g}_n$$

or

$$\vec{G}_i \times \vec{G}_j = Je_{ijn}\vec{g}_n. \tag{8.73}$$

The relationship between the metrics g_{ij} and G_{ij} is given by

$$g_{ik}G_{kj} = \left(\frac{\partial x_m}{\partial y_i}\,\frac{\partial x_m}{\partial y_k}\right)\left(\frac{\partial y_k}{\partial x_\ell}\,\frac{\partial y_j}{\partial x_\ell}\right)$$

$$= \frac{\partial x_m}{\partial y_i}\,\frac{\partial x_m}{\partial x_\ell}\,\frac{\partial y_j}{\partial x_\ell} = \frac{\partial x_m}{\partial y_i}\,\delta_{m\ell}\,\frac{\partial y_j}{\partial x_\ell}$$

$$g_{ik}G_{kj} = \frac{\partial x_m}{\partial y_i}\,\frac{\partial y_j}{\partial x_m} = \frac{\partial y_j}{\partial y_i} = \delta_{ij} \tag{8.74}$$

As a result of (8.74) we say that the matrix $[G_{ij}]$ is the inverse of the matrix $[g_{ij}]$ and vice versa.

Our discussion up to this point has been quite general. We now wish to specialize these results to the case of orthogonal curvilinear coordinates. For this case

$$g_{ij} = \vec{g}_i \cdot \vec{g}_j = 0 \text{ if } i \neq j.$$

Let,

$$H_i = |\vec{g}_i| = \sqrt{g_{(i)(i)}}\,. \tag{8.75}$$

Then at any point P_o, the unit vectors

$$\hat{\varepsilon}'_i = \frac{\vec{g}_{(i)}}{H_{(i)}} \tag{8.76}$$

define a right-handed R.C.S. . According to (8.59) and (8.76) this primed reference is related to the unprimed reference by means of the orthogonal transformation

$$\left.\begin{aligned} \hat{\varepsilon}'_i &= a_{ij}\hat{\varepsilon}_j \\ \text{where} \quad a_{ij} &= \frac{1}{H_{(i)}}\frac{\partial x_j}{\partial y_{(i)}} \,. \end{aligned}\right\} \qquad (8.77)$$

Similarly, from (8.61) we obtain

$$\left.\begin{aligned} \hat{\varepsilon}_i &= a_{ji}\hat{\varepsilon}'_j \\ \text{where} \quad a_{ji} &= H_{(j)}\frac{\partial y_{(j)}}{\partial x_i} \,. \end{aligned}\right\} \qquad (8.78)$$

From (8.75), $g_{11}=H_1^2$, $g_{22}=H_2^2$, $g_{33}=H_3^2$ and therefore

$$g = H_1^2H_2^2H_3^2 \text{ and } J^{-1} = H_1H_2H_3. \qquad (8.79)$$

With the aid of (8.76) and (8.79), (8.69) yields

$$\vec{G}_i = \frac{1}{H_{(i)}}\,\hat{\varepsilon}'_{(i)}. \qquad (8.80)$$

Thus, $G_{11} = \frac{1}{H_1^2}$, $G_{22} = \frac{1}{H_2^2}$, $G_{33} = \frac{1}{H_3^2}$ and $G = \frac{1}{H_1^2H_2^2H_3^2}$

We now wish to examine the differential operations in orthogonal curvilinear coordinates which correspond to the gradient, divergence and curl discussed in Section 6. Consider the scalar point function $T(y_1,y_2,y_3)$. In view of (8.66) and (8.80),

$$\vec{\nabla}T \equiv \hat{\varepsilon}_j\frac{\partial T}{\partial x_j} = \hat{\varepsilon}_j\frac{\partial T}{\partial y_i}\frac{\partial y_i}{\partial x_j} = \vec{G}_i\frac{\partial T}{\partial y_i} = \sum_{i=1}^{3}\frac{\hat{\varepsilon}'_{(i)}}{H_{(i)}}\frac{\partial T}{\partial y_i} \equiv \frac{\hat{\varepsilon}'_i}{H_{(i)}}\frac{\partial T}{\partial y_i} \qquad (8.81)$$

In order to circumvent the rule that the same summation index not appear more than twice in any expression we will enclose the index on the H-factor in parenthesis as in (8.81). Using this modified summation convention the del operator in curvilinear coordinates can be written as follows:

$$\vec{\nabla} = \vec{G}_i\frac{\partial}{\partial y_i} = \sum_{i=1}^{3}\frac{\hat{\varepsilon}'_{(i)}}{H_{(i)}}\frac{\partial}{\partial y_i} \equiv \frac{\hat{\varepsilon}'_i}{H_{(i)}}\frac{\partial}{\partial y_i} \,. \qquad (8.82)$$

We next consider the vector point function

$$\vec{u}(\vec{r}) = u_i(x_1,x_2,x_3)\hat{\varepsilon}_i = u_i'(y_1,y_2,y_3)\hat{\varepsilon}_i'. \tag{8.83}$$

In Cartesian coordinates the directional derivative is given by

$$\frac{d\vec{u}}{d\lambda} = (\hat{n}\cdot\vec{\nabla})\vec{u} = \hat{n}\cdot\hat{\varepsilon}_j \frac{\partial\vec{u}}{\partial x_j} = n_j \frac{\partial\vec{u}}{\partial x_j}$$

or in component form

$$\frac{d\vec{u}}{d\lambda} = n_j \frac{\partial u_i}{\partial x_j}\hat{\varepsilon}_i = n_j F_{ij}\hat{\varepsilon}_i \tag{8.84}$$

where $F_{ij} = \partial u_i/\partial x_j$ are the components of the gradient of $\vec{u}$. The divergence and curl of $\vec{u}$ are defined as

$$\vec{\nabla}\cdot\vec{u} = \frac{\partial u_i}{\partial x_i} = F_{ii} \tag{8.85a}$$

and

$$\vec{\nabla}\times\vec{u} = e_{kji}\frac{\partial u_i}{\partial x_j}\hat{\varepsilon}_k = e_{kji}F_{ij}\hat{\varepsilon}_k \tag{8.85b}$$

respectively. The corresponding quantities in curvilinear coordinates are

$$\left.\begin{aligned} &\frac{d\vec{u}}{d\lambda} = n_j'F_{ij}'\hat{\varepsilon}_i', \qquad \vec{\nabla}\cdot\vec{u} = F_{ii}' \\ &\text{and} \\ &\vec{\nabla}\times\vec{u} = e_{kji}F_{ij}'\hat{\varepsilon}_k'. \end{aligned}\right\} \tag{8.86}$$

The F_{ij}' may be determined as follows.

$$\frac{d\vec{u}}{d\lambda} = (\hat{n}\cdot\vec{\nabla})\vec{u} = \frac{n_j'}{H_{(j)}}\frac{\partial}{\partial y_j}(u_i'\hat{\varepsilon}_i')$$

where $n_j' = \hat{n}\cdot\hat{\varepsilon}_j'$. With the aid of (8.76) we obtain

$$\begin{aligned} \frac{d\vec{u}}{d\lambda} &= \frac{n_j'}{H_{(j)}}\frac{\partial}{\partial y_j}\left(\frac{u_i'}{H_{(i)}}\vec{g}_i\right) \\ &= \frac{n_j'}{H_{(j)}}\left[\frac{\partial}{\partial y_j}\left(\frac{u_i'}{H_{(i)}}\right)\vec{g}_i + \frac{u_k'}{H_{(k)}}\frac{\partial\vec{g}_k}{\partial y_j}\right] \end{aligned}$$

In view of (8.59), (8.78) and (8.77),

$$\frac{\partial \vec{g}_k}{\partial y_j} = \frac{\partial}{\partial y_j}\left(\frac{\partial x_\ell}{\partial y_k}\right)\hat{\varepsilon}_\ell = \frac{\partial^2 x_\ell}{\partial y_j \partial y_k} a_{i\ell}\hat{\varepsilon}'_i$$

$$= \frac{\partial^2 x_\ell}{\partial y_j \partial y_k}\frac{\partial x_\ell}{\partial y_i}\frac{1}{H_{(i)}}\hat{\varepsilon}'_i .$$

Thus, with the aid of (8.76) we obtain

$$\frac{d\vec{u}}{d\lambda} = \frac{n'_j}{H_{(j)}}\left[H_{(i)}\frac{\partial}{\partial y_j}\left(\frac{u'_i}{H_{(i)}}\right) + \frac{u'_k}{H_{(i)}H_{(k)}}\frac{\partial x_\ell}{\partial y_i}\frac{\partial^2 x_\ell}{\partial y_j \partial y_k}\right]\hat{\varepsilon}'_i .$$

By comparing the above equation to (8.86) we conclude that

$$F'_{ij} = \frac{H_{(i)}}{H_{(j)}}\frac{\partial}{\partial y_{(j)}}\left(\frac{u'_{(i)}}{H_{(i)}}\right) + \frac{u'_k}{H_{(i)}H_{(j)}H_{(k)}}\frac{\partial x_\ell}{\partial y_{(i)}}\frac{\partial^2 x_\ell}{\partial y_{(j)}\partial y_k} . \qquad (8.87)$$

A considerable simplification of the last term in the above equation is effected by noting that

$$\frac{\partial x_\ell}{\partial y_i}\frac{\partial^2 x_\ell}{\partial y_j \partial y_k} = \tfrac{1}{2}\left[\frac{\partial g_{ik}}{\partial y_j} + \frac{\partial g_{ij}}{\partial y_k} - \frac{\partial g_{jk}}{\partial y_i}\right] . \qquad (8.88)$$

This result may be verified by differentiating (8.62) as indicated. This group of derivatives is called the Christoffel symbol of the first kind and is usually denoted symbolically as

$$[kj,\, i] = [jk,\, i] = \frac{\partial x_\ell}{\partial y_i}\frac{\partial^2 x_\ell}{\partial y_j \partial y_k} . \qquad (8.89)$$

Thus, (8.87) may be written as

$$F'_{ij} = \frac{H_{(i)}}{H_{(j)}}\frac{\partial}{\partial y_{(j)}}\left(\frac{u'_{(i)}}{H_{(i)}}\right) + \frac{u'_k\,[k(j),(i)]}{H_{(i)}H_{(j)}H_{(k)}} \qquad (8.90)$$

Since $g_{ij}=\vec{g}_i\cdot\vec{g}_j=0$ when $i\neq j$ and $g_{(i)(i)}=H^2_{(i)}$, we conclude from (8.88) and (8.89) that, for $i\neq j\neq k$

$$[kj,i] = 0, \quad [(k)(k),i] = -H_{(k)}\frac{\partial H_{(k)}}{\partial y_i}$$

$$[(k)j,(k)] = H_{(k)}\frac{\partial H_{(k)}}{\partial y_j}, \quad [(k)(k),(k)] = H_{(k)}\frac{\partial H_{(k)}}{\partial y_{(k)}} .$$

As a result of these relations, one obtains, after some simplification, the following unabridged form of (8.90).

$$
\begin{aligned}
F'_{11} &= \frac{1}{H_1}\frac{\partial u'_1}{\partial y_1} + \frac{u'_2}{H_1H_2}\frac{\partial H_1}{\partial y_2} + \frac{u'_3}{H_1H_3}\frac{\partial H_1}{\partial y_3} \\
F'_{22} &= \frac{1}{H_2}\frac{\partial u'_2}{\partial y_2} + \frac{u'_1}{H_1H_2}\frac{\partial H_2}{\partial y_1} + \frac{u'_3}{H_2H_3}\frac{\partial H_2}{\partial y_3} \\
F'_{33} &= \frac{1}{H_3}\frac{\partial u'_3}{\partial y_3} + \frac{u'_1}{H_1H_3}\frac{\partial H_3}{\partial y_1} + \frac{u'_2}{H_2H_3}\frac{\partial H_3}{\partial y_2} \\
F'_{12} &= \frac{1}{H_2}\frac{\partial u'_1}{\partial y_2} - \frac{u'_2}{H_1H_2}\frac{\partial H_2}{\partial y_1} \\
F'_{21} &= \frac{1}{H_1}\frac{\partial u'_2}{\partial y_1} - \frac{u'_1}{H_1H_2}\frac{\partial H_1}{\partial y_2} \\
F'_{23} &= \frac{1}{H_3}\frac{\partial u'_2}{\partial y_3} - \frac{u'_3}{H_2H_3}\frac{\partial H_3}{\partial y_2} \\
F'_{32} &= \frac{1}{H_2}\frac{\partial u'_3}{\partial y_2} - \frac{u'_2}{H_2H_3}\frac{\partial H_2}{\partial y_3} \\
F'_{13} &= \frac{1}{H_3}\frac{\partial u'_1}{\partial y_3} - \frac{u'_3}{H_1H_3}\frac{\partial H_3}{\partial y_1} \\
F'_{31} &= \frac{1}{H_1}\frac{\partial u'_3}{\partial y_1} - \frac{u'_1}{H_1H_3}\frac{\partial H_1}{\partial y_3}
\end{aligned}
\tag{8.91}
$$

According to (8.86), the divergence of $\vec{u}$ is given by

$$\vec{\nabla}\cdot\vec{u} = F'_{ii} = F'_{11} + F'_{22} + F'_{33}. \tag{8.92a}$$

With the aid of (8.91) this may also be written as

$$\vec{\nabla}\cdot\vec{u} = \frac{1}{H_1H_2H_3}\frac{\partial}{\partial y_i}\left(\frac{H_1H_2H_3}{H_{(i)}}u'_i\right) . \tag{8.92b}$$

The components of the curl of $\vec{u}$ are given as follows. If $\vec{\Omega} = \vec{\nabla}\times\vec{u}$, then according to (8.86), $\Omega'_k=e_{kji}F'_{ij}$, or in unabridged form,

$$
\begin{aligned}
\Omega'_1 &= F'_{32}-F'_{23} = \frac{1}{H_2H_3}\left[\frac{\partial}{\partial y_2}(H_3u'_3)-\frac{\partial}{\partial y_3}(H_2u'_2)\right] \\
\Omega'_2 &= F'_{13}-F'_{31} = \frac{1}{H_1H_3}\left[\frac{\partial}{\partial y_3}(H_1u'_1)-\frac{\partial}{\partial y_1}(H_3u'_3)\right] \\
\Omega'_3 &= F'_{21}-F'_{12} = \frac{1}{H_1H_2}\left[\frac{\partial}{\partial y_1}(H_2u'_2)-\frac{\partial}{\partial y_2}(H_1u'_1)\right]
\end{aligned}
\tag{8.93}
$$

The curl of $\vec{u}$ may also be expressed in the determinant form

$$\vec{\nabla}\times\vec{u} = \frac{1}{H_1H_2H_3}\begin{vmatrix} H_1\hat{\varepsilon}'_1 & H_2\hat{\varepsilon}'_2 & H_3\hat{\varepsilon}'_3 \\ \frac{\partial}{\partial y_1} & \frac{\partial}{\partial y_2} & \frac{\partial}{\partial y_3} \\ H_1u'_1 & H_2u'_2 & H_3u'_3 \end{vmatrix} \tag{8.94}$$

If this determinant is expanded in terms of the minors of the first row we obtain (8.93). By letting $\vec{u}=\vec{\nabla}T$ in (8.92b) we obtain, with the aid of (8.81), an expression for the Laplacian of T, i.e.,

$$\nabla^2 T = \frac{1}{H_1H_2H_3}\frac{\partial}{\partial y_i}\left(\frac{H_1H_2H_3}{H_{(i)}^2}\frac{\partial T}{\partial y_i}\right) \tag{8.95a}$$

or in unabridged form

$$\nabla^2 T = \frac{1}{H_1H_2H_3}\left[\frac{\partial}{\partial y_1}\left(\frac{H_2H_3}{H_1}\frac{\partial T}{\partial y_1}\right) + \frac{\partial}{\partial y_2}\left(\frac{H_1H_3}{H_2}\frac{\partial T}{\partial y_2}\right) + \frac{\partial}{\partial y_3}\left(\frac{H_1H_2}{H_3}\frac{\partial T}{\partial y_3}\right)\right] \tag{8.95b}$$

According to (8.43), F'_{ij} may be represented by the sum of a symmetric and a skew-symmetric tensor as follows.

$$F'_{ij} = \varepsilon'_{ij} + \omega'_{ij} \tag{8.96}$$

where

$$\varepsilon'_{ij} = \tfrac{1}{2}(F'_{ij}+F'_{ji})$$

$$\omega'_{ij} = \tfrac{1}{2}(F'_{ij}-F'_{ji}) = e_{ijk}F'_k$$

and

$$F'_k = \tfrac{1}{2}\, e_{ijk}F'_{ij} = -\tfrac{1}{2}\, e_{kji}F'_{ij} = -\tfrac{1}{2}\, \Omega'_k .$$

Thus,

$$\omega'_{ij} = -\tfrac{1}{2}e_{ijk}\Omega'_k \tag{8.97}$$

where Ω'_k is given by (8.93). In the theory of small deformations of continua, ε'_{ij} is called the strain tensor and ω'_{ij} the rotation tensor when $\vec{u}$ represents a displacement field.

The preceding discussion was concerned with the representation of differential operations in curvilinear coordinates. We now wish to consider certain elements associated with integral operations. Consider the curve whose parametric representation is $x_i = x_i(t)$, $t_0 \le t \le t_1$. The length of this curve is given by the integral

$$L = \int_{t_0}^{t_1} \left[\frac{d\vec{r}}{dt}\bullet\frac{d\vec{r}}{dt}\right]^{\frac{1}{2}} dt = \int_{s_0}^{s_1} ds$$

where

$$ds^2 = \left[\frac{d\vec{r}}{dt}\bullet\frac{d\vec{r}}{dt}\right] dt^2 = d\vec{r}\bullet d\vec{r}.$$

In curvilinear coordinates $d\vec{r} = \vec{g}_i dy_i$, and therefore

$$\left.\begin{aligned} ds^2 &= (\vec{g}_i dy_i)\cdot(\vec{g}_j dy_j) = g_{ij} dy_i dy_j \\ &\text{or} \\ ds^2 &= H_1^2 dy_1^2 + H_2^2 dy_2^2 + H_3^2 dy_3^2 \ . \end{aligned}\right\} \tag{8.98}$$

The increments of arc along the y_1, y_2-and y_3 coordinate curves are $d\vec{r}_{(1)} = \vec{g}_1 dy_1$, $d\vec{r}_{(2)} = \vec{g}_2 dy_2$, and $d\vec{r}_{(3)} = \vec{g}_3 dy_3$, respectively. In view of (8.69) and (8.80) the area elements on the coordinate surfaces are given by

$$\left.\begin{aligned} d\vec{A}_{(1)} = d\vec{r}_{(2)} \times d\vec{r}_{(3)} &= \frac{\vec{G}_1}{J} dy_2 dy_3 = H_2 H_3 dy_2 dy_3 \hat{e}_1' \\ d\vec{A}_{(2)} = d\vec{r}_{(3)} \times d\vec{r}_{(1)} &= \frac{\vec{G}_2}{J} dy_3 dy_1 = H_1 H_3 dy_1 dy_3 \hat{e}_2' \\ d\vec{A}_{(3)} = d\vec{r}_{(1)} \times d\vec{r}_{(2)} &= \frac{\vec{G}_3}{J} dy_1 dy_2 = H_1 H_2 dy_1 dy_2 \hat{e}_3'. \end{aligned}\right\} \tag{8.99}$$

With the aid of (8.65), the volume element in curvilinear coordinates is given by

$$dV = d\vec{r}_{(1)} \times d\vec{r}_{(2)} \cdot d\vec{r}_{(3)} = \frac{dy_1 dy_2 dy_3}{J}$$

or

$$dV = H_1 H_2 H_3 dy_1 dy_2 dy_3. \tag{8.100}$$

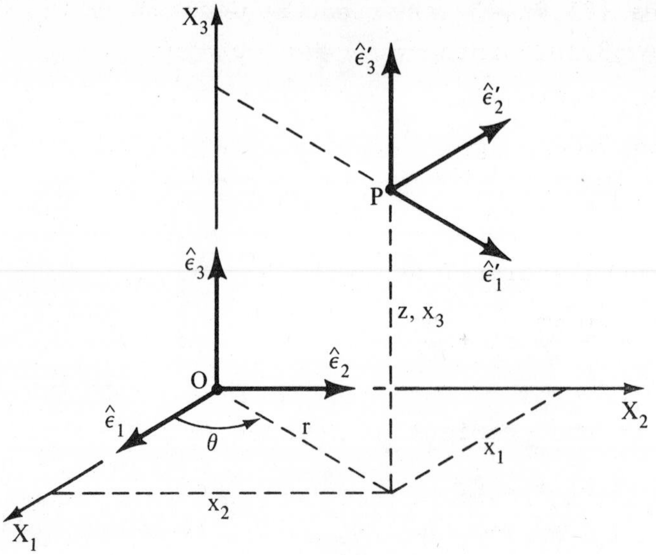

Figure 8.7

To illustrate the preceding results let us consider the cylindrical coordinates shown in Fig. 8.7. If we let $y_1 \equiv r$, $y_2 \equiv \theta$ and $y_3 \equiv z$, the inverse transformation is seen to be

$$x_1 = r \cos \theta, \; x_2 = r \sin \theta, \; x_3 = z$$

for

$$0 < r < \infty, \; -\pi < \theta < \pi \text{ and } -\infty < z < \infty. \tag{8.101}$$

With the aid of (8.59) we obtain from (8.101)

$$\begin{aligned} \vec{g}_1 &= \cos \theta \hat{\epsilon}_1 + \sin \theta \hat{\epsilon}_2 \\ \vec{g}_2 &= -r \sin \theta \hat{\epsilon}_1 + r \cos \theta \hat{\epsilon}_2 \\ \vec{g}_3 &= \hat{\epsilon}_3. \end{aligned} \tag{8.102}$$

Thus,

$$\begin{aligned} H_1 &= |\vec{g}_1| = 1 \\ H_2 &= |\vec{g}_2| = r \\ H_3 &= |\vec{g}_3| = 1. \end{aligned} \tag{8.103}$$

The unit vectors $\{\hat{e}'_1, \hat{e}'_2, \hat{e}'_3\}$ are usually denoted by $\{\hat{e}_r, \hat{e}_\theta, \hat{e}_z\}$. Upon substitution of (8.102) into (8.76) we obtain

$$\begin{aligned} \hat{e}_r &= \cos\theta\hat{e}_1 + \sin\theta\hat{e}_2 \\ \hat{e}_\theta &= -\sin\theta\hat{e}_1 + \cos\theta\hat{e}_2 \\ \hat{e}_z &= \hat{e}_3. \end{aligned} \tag{8.104}$$

The gradient of the scalar T becomes

$$\vec{\nabla}T = \frac{\partial T}{\partial r}\hat{e}_r + \frac{1}{r}\frac{\partial T}{\partial \theta}\hat{e}_\theta + \frac{\partial T}{\partial z}\hat{e}_z. \tag{8.105}$$

The components of the gradient of the vector

$$\vec{u} = u_r\hat{e}_r + u_\theta\hat{e}_\theta + u_z\hat{e}_z$$

are

$$\left.\begin{aligned} &F_{rr} = \frac{\partial u_r}{\partial r}, \quad F_{\theta\theta} = \frac{1}{r}\frac{\partial u_\theta}{\partial \theta} + \frac{u_r}{r}, \quad F_{zz} = \frac{\partial u_z}{\partial z} \\ &F_{r\theta} = \frac{1}{r}\frac{\partial u_r}{\partial \theta} - \frac{u_\theta}{r}, \quad F_{\theta r} = \frac{\partial u_\theta}{\partial r} \\ &F_{\theta z} = \frac{\partial u_\theta}{\partial z}, \quad F_{z\theta} = \frac{1}{r}\frac{\partial u_z}{\partial \theta} \\ &F_{rz} = \frac{\partial u_r}{\partial z}, \quad F_{zr} = \frac{\partial u_z}{\partial r}. \end{aligned}\right\} \tag{8.106}$$

The divergence and curl of $\vec{u}$ are

$$\vec{\nabla}\cdot\vec{u} = \frac{\partial u_r}{\partial r} + \frac{u_r}{r} + \frac{1}{r}\frac{\partial u_\theta}{\partial \theta} + \frac{\partial u_z}{\partial z} \tag{8.107}$$

and

$$\begin{aligned} \Omega_r &= \frac{1}{r}\frac{\partial u_z}{\partial \theta} - \frac{\partial u_\theta}{\partial z} \\ \Omega_\theta &= \frac{\partial u_r}{\partial z} - \frac{\partial u_z}{\partial r} \\ \Omega_z &= \frac{\partial u_\theta}{\partial r} + \frac{u_\theta}{r} - \frac{1}{r}\frac{\partial u_r}{\partial \theta} \end{aligned} \tag{8.108}$$

where $\vec{\nabla}\times\vec{u} = \vec{\Omega}$.

The Laplacian of T becomes

$$\nabla^2 T = \frac{1}{r}\frac{\partial}{\partial r}\left(r\,\frac{\partial T}{\partial r}\right) + \frac{1}{r^2}\frac{\partial^2 T}{\partial \theta^2} + \frac{\partial^2 T}{\partial z^2}\,. \tag{8.109}$$

The symmetric part of the gradient of $\vec{u}$ is

$$\varepsilon_{rr} = \frac{\partial u_r}{\partial r},\quad \varepsilon_{\theta\theta} = \frac{1}{r}\frac{\partial u_\theta}{\partial \theta} + \frac{u_r}{r},\quad \varepsilon_{zz} = \frac{\partial u_z}{\partial z}$$

$$\varepsilon_{r\theta} = \tfrac{1}{2}\left[\frac{1}{r}\frac{\partial u_r}{\partial \theta} - \frac{u_\theta}{r} + \frac{\partial u_\theta}{\partial r}\right]$$

$$\varepsilon_{\theta z} = \tfrac{1}{2}\left[\frac{\partial u_\theta}{\partial z} + \frac{1}{r}\frac{\partial u_z}{\partial \theta}\right] \tag{8.110}$$

$$\varepsilon_{zr} = \tfrac{1}{2}\left[\frac{\partial u_z}{\partial r} + \frac{\partial u_r}{\partial z}\right]$$

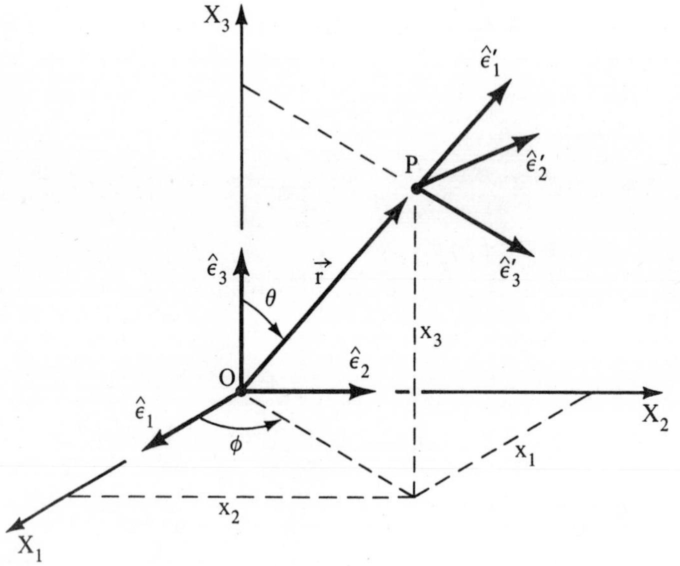

Figure 8.8

It may be readily verified that for the spherical coordinates shown in Fig. 8.8

$$x_1 = r\sin\theta\cos\phi,\quad x_2 = r\sin\theta\sin\phi,\quad x_3 = r\cos\theta$$

where

$$0<r<\infty,\qquad 0<\theta<\pi,\qquad -\pi<\phi<\pi. \tag{8.111}$$

If we replace $\{y_1,y_2,y_3\}$ by $\{r,\theta,\phi\}$ and $\{\hat{\varepsilon}'_1,\hat{\varepsilon}'_2,\hat{\varepsilon}'_3\}$ by $\{\hat{\varepsilon}_r,\hat{\varepsilon}_\theta,\hat{\varepsilon}_\phi\}$ respectively, then

$$H_1=1,\ H_2=r,\ H_3=r\sin\theta. \tag{8.112}$$

The gradient of T becomes

$$\vec{\nabla}T = \frac{\partial T}{\partial r}\hat{\varepsilon}_r + \frac{1}{r}\frac{\partial T}{\partial\theta}\hat{\varepsilon}_\theta + \frac{1}{r\sin\theta}\frac{\partial T}{\partial\phi}\hat{\varepsilon}_\phi. \tag{8.113}$$

The divergence and curl of $\vec{u}$ are

$$\vec{\nabla}\cdot\vec{u} = \frac{1}{r^2}\frac{\partial}{\partial r}(r^2 u_r) + \frac{1}{r\sin\theta}\frac{\partial}{\partial\theta}(u_\theta\sin\theta) + \frac{1}{r\sin\theta}\frac{\partial u_\phi}{\partial\phi} \tag{8.114}$$

and

$$\Omega_r = \frac{1}{r\sin\theta}\left[\frac{\partial}{\partial\theta}(u_\phi\sin\theta) - \frac{\partial u_\theta}{\partial\phi}\right]$$

$$\Omega_\theta = \frac{1}{r\sin\theta}\frac{\partial u_r}{\partial\phi} - \frac{1}{r}\frac{\partial}{\partial r}(ru_\phi) \tag{8.115}$$

$$\Omega_\phi = \frac{1}{r}\left[\frac{\partial}{\partial r}(ru_\theta) - \frac{\partial u_r}{\partial\theta}\right]$$

respectively, where $\vec{\Omega}=\vec{\nabla}\times\vec{u}$. The Laplacian of T becomes

$$\nabla^2 T = \frac{1}{r^2}\frac{\partial}{\partial r}\left(r^2\frac{\partial T}{\partial r}\right) + \frac{1}{r^2\sin\theta}\frac{\partial}{\partial\theta}\left(\sin\theta\frac{\partial T}{\partial\theta}\right) + \frac{1}{r^2\sin^2\theta}\frac{\partial^2 T}{\partial\phi^2} \tag{8.116}$$

The symmetric part of the gradient $\vec{u}$ is

$$\varepsilon_{rr} = \frac{\partial u_r}{\partial r}$$

$$\varepsilon_{\theta\theta} = \frac{1}{r}\frac{\partial u_\theta}{\partial\theta} + \frac{u_r}{r}$$

$$\varepsilon_{\phi\phi} = \frac{1}{r\sin\theta}\frac{\partial u_\phi}{\partial\phi} + \frac{u_r}{r} + u_\theta\frac{\cot\theta}{r} \tag{8.117}$$

$$\varepsilon_{r\theta} = \tfrac{1}{2}\left[\frac{1}{r}\frac{\partial u_r}{\partial\theta} - \frac{u_\theta}{r} + \frac{\partial u_\theta}{\partial r}\right]$$

$$\varepsilon_{r\phi} = \tfrac{1}{2}\left[\frac{1}{r\sin\theta}\frac{\partial u_r}{\partial\phi} - \frac{u_\phi}{r} + \frac{\partial u_\phi}{\partial r}\right]$$

$$\varepsilon_{\theta\phi} = \tfrac{1}{2}\left[\frac{1}{r}\frac{\partial u_\phi}{\partial\theta} - \frac{u_\phi\cot\theta}{r} + \frac{1}{r\sin\theta}\frac{\partial u_\theta}{\partial\phi}\right]$$

EXERCISES

8.1 The three vectors $\vec{A}$, $\vec{B}$ and $\vec{C}$ intersect at O. $\vec{A}$ and $\vec{C}$ are parallel to the plane of the paper and $\vec{B}$ is directed into the plane of the paper.

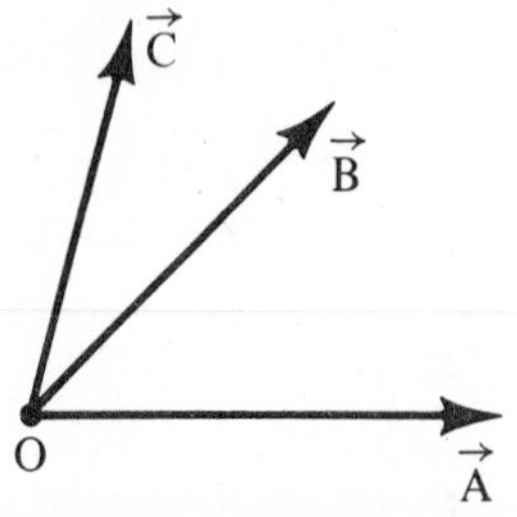

Figure: Exercise 8.1

(a) Which of the following ordered sets are dextral and which are sinistral.

(i) $\{\vec{A},\vec{B},\vec{C}\}$ (iv) $\{\vec{C},\vec{B},\vec{A}\}$

(ii) $\{\vec{C},\vec{A},\vec{B}\}$ (v) $\{\vec{A},\vec{C},\vec{B}\}$

(iii) $\{\vec{B},\vec{C},\vec{A}\}$ (vi) $\{\vec{B},\vec{A},\vec{C}\}$

(b) If the direction of $\vec{C}$ is reversed how are the results of (a) affected.

8.2 What is the corresponding unabridged form of each expression listed below? If the indicial notation is used incorrectly in any expression explain why.

(a) $A_i=B_i$

(b) $F_i=G_i+H_{ji}A_j$

(c) $U_i=V_j$

(d) $A_i=B_i+C_iD_i$

(e) $F_i=A_i+B_{ij}C_jD_j$

(f) $\phi = \dfrac{\partial F_i}{\partial x_i}$

(g) $d = \sqrt{x_i x_i}$

8.3 If $\{\hat{\varepsilon}_1',\hat{\varepsilon}_2',\hat{\varepsilon}_3'\}$ is a mutually orthogonal, sinistral set of unit vectors as shown in the figure verify that $\hat{\varepsilon}_i'\cdot\hat{\varepsilon}_j'=\delta_{ij}$ and $\hat{\varepsilon}_i'\times\hat{\varepsilon}_j'=-e_{ijk}\hat{\varepsilon}_k'$.

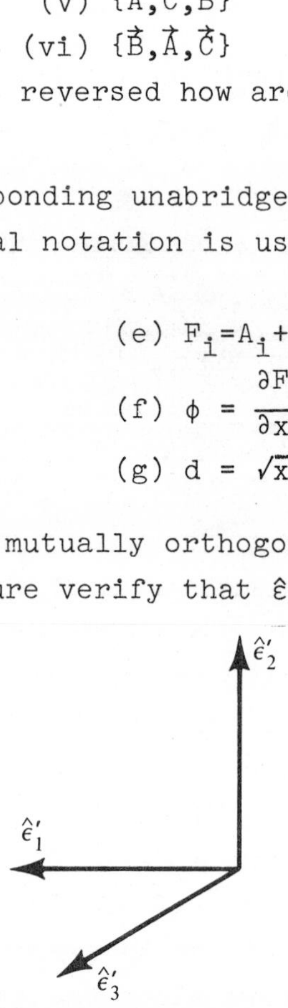

Figure: Exercise 8.3

8.4 With the aid of (8.15) show that

(a) $e_{ijk}e_{\alpha\beta k} = \delta_{i\alpha}\delta_{j\beta}-\delta_{i\beta}\delta_{j\alpha}$

(b) $e_{ijk}e_{\alpha jk} = 2\delta_{i\alpha}$

(c) $e_{ijk}e_{ijk} = 6$

8.5 Prove the following identities.

(a) $\delta_{ii} = 3$

(b) $\delta_{ij}C_{ij} = C_{ii}$

(c) $A_{ij}e_{ijk} = 0$ if $A_{ij} = A_{ji}$

(d) $A_{ij}B_{k\ell}\delta_{ik} = A_{ij}B_{i\ell}$

8.6 Show that (8.20) follows from (8.14).

8.7 Two coordinate systems are connected by the direction cosines a_{ij} where

$$[a_{ij}] = \begin{vmatrix} \frac{1}{\sqrt{3}} & \frac{1}{\sqrt{3}} & \frac{1}{\sqrt{3}} \\ \frac{1}{\sqrt{2}} & 0 & -\frac{1}{\sqrt{2}} \\ -\frac{1}{\sqrt{6}} & \sqrt{\frac{2}{3}} & -\frac{1}{\sqrt{6}} \end{vmatrix}$$

The unprimed reference is rectangular, Cartesian and right-handed.

(a) Are the primed coordinate axes mutually orthogonal?

(b) Is the primed reference right-handed or left-handed?

(c) The origin of these coordinates is a point within a large vessel containing a fluid. By placing thermometers at one inch intervals along the X_1, X_2 and X_3 axes respectively we find that the temperature of the fluid increases at a rate of $\sqrt{6}^{\circ}$F/in. in each of the positive coordinate directions. Similarly by placing thermometers at one inch intervals along the X_1', X_2' and X_3' axis respectively we find that the temperature increases at a rate of $3\sqrt{2}^{\circ}$ F/in. along the X_1' axis but remains constant along the X_2' and X_3' axes respectively. Do the three rates of change of temperature along the coordinate axes define a vector?

(d) Estimate the rate of change of temperature in the $\hat{n}$ direction at the origin if

$$n = \frac{1}{\sqrt{6}}\,\hat{\varepsilon}_1 + \sqrt{\frac{2}{3}}\,\hat{\varepsilon}_2 + \frac{1}{\sqrt{6}}\,\hat{\varepsilon}_3.$$

8.8 The mass M shown in the accompanying figure is connected by a system of springs and levers to the point A.

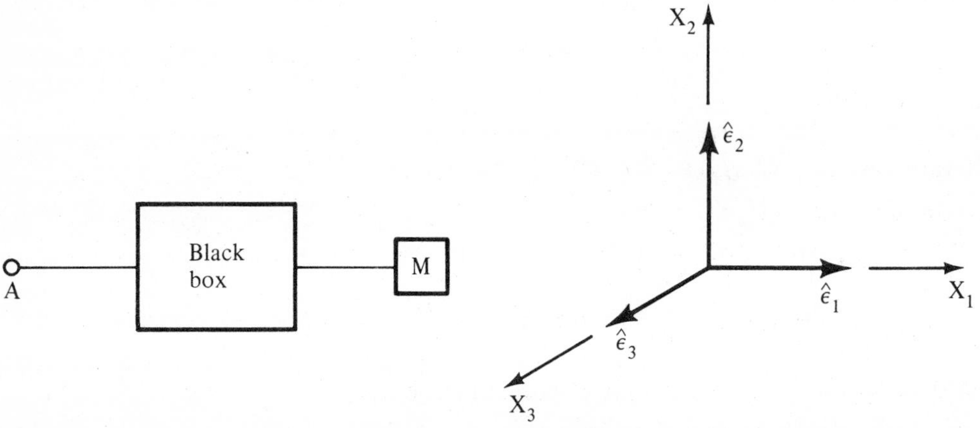

Figure: Exercise 8.8

If a unit force is applied to A in the $\hat{\varepsilon}_1$ direction the displacement of M is found to be

$$\vec{u}_M^{(1)} = k_{11}\hat{\varepsilon}_1 + k_{21}\hat{\varepsilon}_2 + k_{31}\hat{\varepsilon}_3$$

Similarly a unit force in the $\hat{\varepsilon}_2$ and $\hat{\varepsilon}_3$ directions produces displacements

$$\vec{u}_M^{(2)} = k_{12}\hat{\varepsilon}_1 + k_{22}\hat{\varepsilon}_2 + k_{32}\hat{\varepsilon}_3 \quad \text{and}$$

$$\vec{u}_M^{(3)} = k_{13}\hat{\varepsilon}_1 + k_{23}\hat{\varepsilon}_2 + k_{33}\hat{\varepsilon}_3 \quad \text{respectively.}$$

(a) Assuming that the device is linear write down an indicial equation relating the components of an arbitrary force F_i to its corresponding displacement u_i.

(b) Are the influence coefficients k_{ij} the components of a tensor? Why?

(c) If $[k_{ij}] = [a_{ij}]$ (lb/in) from problem (8.7) find u_i if $F_1 = F_2 = F_3 = \sqrt{6}$ (lb).

8.9 Given a vector A_i and two second order tensors B_{ij} and C_{ij}, prove

(a) $F_{ij}=B_{ij}+C_{ij}$ is a second order tensor
(b) $H_{ijk}=A_iB_{jk}$ is a third order tensor
(c) B_{ii} is a scalar
(d) $H_{iji}=A_iB_{ji}$ is a vector

8.10 Given a matrix $[T_{ij}]$, two arbitrary vectors N_i and n_i and a scalar S, prove:
(a) If $T_{ij}N_in_j=S$ then T_{ij} is a second order tensor
(b) If $T_{ij}n_in_j=S$ and $T_{ij}=T_{ji}$ then T_{ij} is a symmetric second order tensor.

8.11 Given: $F_{ij}=\frac{\partial u_i}{\partial x_j}$
(a) Compute ε_{ij}, ω_{ij} and F_k in (8.43).
(b) Show that $\vec{F} = F_k\hat{e}_k = -\frac{1}{2}\vec{\nabla}\times\vec{u}$.
(c) If τ_{ij} is a symmetric second order tensor show that $\tau_{ij}F_{ij}=\tau_{ij}\varepsilon_{ij}$.

8.12 Prove that C_{ijkl} given by (8.46) is isotropic.

8.13 Given a scalar point function $\phi(x_1,x_2,x_3)$
(a) Show that $\vec{\nabla}\times(\vec{\nabla}\phi) = 0$
(b) Show that $\nabla^2\phi = \vec{\nabla}\cdot(\vec{\nabla}\phi)$

8.14 Given a vector field $\vec{u}(x_1,x_2,x_3)$, prove:
(a) $\vec{\nabla}\cdot(\vec{\nabla}\times\vec{u}) = 0$
(b) $\nabla^2\vec{u} = \vec{\nabla}\cdot(\vec{\nabla}\vec{u})$
(c) $\vec{\nabla}\times(\vec{\nabla}\times\vec{u}) = \vec{\nabla}(\vec{\nabla}\cdot\vec{u})-\nabla^2\vec{u}$

8.15 Verify (8.72)

8.16) Verify equations (8.111) through (8.117) for spherical coordinates.

8.17 Given $\vec{\omega}=\vec{\nabla}\times\vec{\Omega}$
(a) In orthogonal curvilinear coordinates how are the ω_i' related to the Ω_i'.
(b) If $\vec{\Omega}=\vec{\nabla}\times\vec{u}$, how are the ω_i' related to the u_i'.
(c) Using the results of (a) and (b) together with the result of problem (8.14c) compute the components of the vector $(\nabla^2\vec{u})$ in orthogonal curvilinear coordinates.

8.18 If $\vec{D}=\nabla^2\vec{u}$ show that

(a) In cylindrical coordinates

$$D_r = \nabla^2 u_r - \frac{u_r}{r^2} - \frac{2}{r^2}\frac{\partial u_\theta}{\partial \theta}$$

$$D_\theta = \nabla^2 u_\theta + \frac{2}{r^2}\frac{\partial u_r}{\partial \theta} - \frac{u_\theta}{r^2}$$

$$D_z = \nabla^2 u_z$$

and

(b) in spherical coordinates

$$D_r = \nabla^2 u_r - \frac{2u_r}{r^2} - \frac{2}{r^2}\frac{\partial u_\theta}{\partial \theta} - \frac{2u_\theta \cot\theta}{r^2} - \frac{2}{r^2 \sin\theta}\frac{\partial u_\phi}{\partial \phi}$$

$$D_\theta = \nabla^2 u_\theta + \frac{2}{r^2}\frac{\partial u_r}{\partial \theta} - \frac{u_\theta}{r^2 \sin^2\theta} - \frac{2\cos\theta}{r^2 \sin^2\theta}\frac{\partial u_\phi}{\partial \phi}$$

$$D_\phi = \nabla^2 u_\phi - \frac{u_\phi}{r^2 \sin^2\theta} + \frac{2}{r^2 \sin\theta}\frac{\partial u_r}{\partial \phi} + \frac{2\cos\theta}{r^2 \sin^2\theta}\frac{\partial u_\theta}{\partial \phi}$$

Chapter 9

MECHANICS OF CONTINUA

These are the Phenomena of Springs and springy bodies, which as they have not hitherto been by any that I know reduced to Rules, so have all the attempts for the explications of the reason of their power, and of springiness in general, been very insufficient.

Robert Hooke, De Potentia Restitutiva, 1678.

9.1 INTRODUCTION

To analyze the mechanical behavior of a physical system we must first construct a mathematical model of it. When this model consists of a finite or a countably infinite set of discrete inertial elements it is called a discrete system. On the other hand, if we assume that the inertia or mass is continuously distributed throughout the system then the model is called a continuum. The material properties of rigidity, elasticity, etc. are characterized, in discrete systems, by massless connections between the inertial elements. In continuous systems the material properties are intrinsic to the inertial elements and cannot be separated from them.

In some cases the continuous model of a system may be viewed as the limit of a sequence of discrete models. This approach is illustrated by the example of Section 4.4. Each discrete model in this sequence is characterized by a system of ordinary differential equations whereas its limit, the continuous model, is characterized by a single partial differential equation. As a result, the motion of each model consists of a countable set $\{\theta_1(t), \theta_2(t),\ldots,\theta_n(t)\}$ of displacements while the motion of the continuous model is characterized by a time varying field $\{\theta(z,t),\ 0\leq z\leq L\}$. This example illustrates the basic difference between discrete and continuous models of nature. The discrete model deals with countable sets of forces, displacements, etc. whereas the continuous model deals with fields of force, displacement, etc. .

The study of continuous models of nature is called continuum mechanics and this chapter contains a brief account of some basic concepts from continuum mechanics. As in Part I we will use Newtonian mechanics to study the motion of bodies in a Euclidean space, usually of three dimensions. However, we now assume that the matter comprising the body continuously fills the region of space defined by its instantaneous configuration. Since we are primarily interested in the dynamics of elastic solids this chapter contains no general discussion on constitutive relations or thermodynamics. We will carefully examine the kinematics and dynamics of continua. The last section contains some useful energy theorems together with a discussion on the constitution of elastic solids. The chapter concludes by demonstrating the applicability of Hamilton's principle to elastic solids.

9.2 KINEMATICS OF CONTINUA

In this section we will examine various techniques for analytically describing the motion of a continuous body. Let us assume that at some fixed instant of time t_o we can stop all motion and observe the configuration of the body. This configuration will describe a region B_o in space. We assume space to be Euclidean and the region B_o to be closed, bounded and connected. At this instant, each material particle of the body has a definite place in space. Since two different material particles cannot simultaneously occupy the same place, there is a one-to-one correspondence between the material particles of the body and the geometrical points of the region B_o. If we assume that the space is spanned by a rectangular Cartesian coordinate system, as illustrated by Fig. 9.1, then each point of B_o is associated with an ordered triple of real numbers called its Cartesian coordinates. The Cartesian coordinates of a generic point P_o in B_o will be denoted by $\{a_1,a_2,a_3\}$. Thus the position vector of the point P_o has the following component representation.

$$\vec{r}_o = a_i\hat{\varepsilon}_i \tag{9.1a}$$

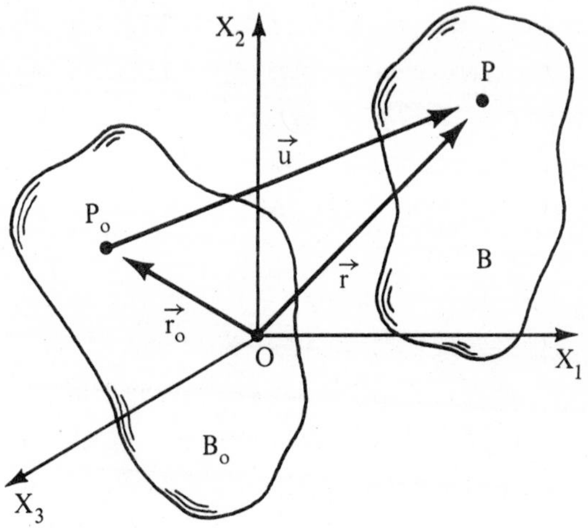

Figure 9.1

As a result of the preceding observations, the configuration of the body at the instant t_o may be described geometrically by the set of points P_o contained in B_o or, analytically, by the set of position vectors $\vec{r}_o$ of these points. Before permitting the motion to resume we will record the coordinates of each material particle in the body and refer to these as the Lagrangian coordinates of the body. Each material particle of the body is now uniquely identifiable for all time by its Lagrangian coordinates.

After completing these observations we permit the motion of the body to resume. At the instant t we again stop all motion in order to observe the current configuration of the body. The current configuration describes the region B in space, as illustrated in Fig. 9.1. The material particle whose Lagrangian coordinates are $\{a_1,a_2,a_3\}$ is now located at the geometrical point P in B whose cartesian coordinates we denote by $\{x_1,x_2,x_3\}$. The position vector of the point P is thus

$$\vec{r} = x_i\hat{e}_i. \tag{9.1b}$$

We now have established a correspondence between the points P_o and P, i.e., each one identifies a place occupied by the same material particle. If the body remains continuous during the motion, i.e., it does not crack or tear or fold over upon itself, then we see that each point P in B will correspond to one and only one point P_o in B_o and, conversely, each point P_o in B_o will correspond to one and only point P in B. Such a correspondence is called a one-to-one mapping of the region B_o onto the region B. As a result of this one-to-one mapping, curves, surfaces and neighborhoods in B_o are mapped onto curves, surfaces and neighborhoods respectively in B. This mapping or correspondence may be expressed analytically by saying that the coordinates of the point P are a function of the coordinates of the point P_o, i.e.,

$$x_i = x_i(a_1,a_2,a_3,t) \equiv x_i(a,t) \tag{9.2a}$$

or in vector form

$$\vec{r} = \vec{r}(\vec{r}_o,t). \tag{9.2b}$$

As a result of the body's motion, the form of these functions will vary with time. This variation is taken into account by including the time t as an independent variable in (9.2). The continuity of the body is insured by requiring that the x_i appearing in (9.2a) be continuous functions of the Lagrangian coordinates $\{a_1,a_2,a_3\}$ for all P_o contained in B_o. The continuity of the motion will be insured by requiring the x_i in (9.2a) to be continuous functions of the time t. The functions

$x_i(a_1,a_2,a_3,t)$ or $\vec{r}(\vec{r}_o,t)$ provide us with a complete, analytical characterization of the body's motion. This is called the Lagrangian description of the motion since it is based upon an (assumed) knowledge of the Lagrangian reference configuration B_o. At this time, it becomes apparent that the Lagrangian reference configuration need not be one actually assumed by the body during its motion. All that we require for (9.2) is that B_o be a possible configuration of the body and that it be fixed in time.

The one-to-one correspondence between B_o and B may also be expressed analytically by saying that the coordinates of the point P_o are a function of the coordinates of the point P, i.e.,

$$a_i = a_i(x_1,x_2,x_3,t) \equiv a_i(x,t) \tag{9.3a}$$

or in vector form

$$\vec{r}_o = \vec{r}_o(\vec{r},t). \tag{9.3b}$$

This is called the Eulerian description of the motion and it is seen to be based upon an assumed knowledge of the current configuration of the body. By assuming that the a_i in (9.3a) are continuous functions of the Eulerian coordinates $\{x_1,x_2,x_3\}$ for all P contained in B and that they are also continuous functions of the time t we assure the required continuity of the body and its motion.

The vector shown in Fig. 9.1 joining P_o to P is called a displacement vector and it admits the following representation

$$\vec{u} = u_i\hat{\varepsilon}_i = \vec{r}-\vec{r}_o = (x_i-a_i)\hat{\varepsilon}_i. \tag{9.4}$$

If the Lagrangian reference configuration is known then the motion is characterized by the displacement field

$$u_i(a_1,a_2,a_3,t) = x_i(a_1,a_2,a_3,t)-a_i \tag{9.5a}$$

or in vector form

$$\vec{u}(\vec{r}_o,t) = \vec{r}(\vec{r}_o,t)-\vec{r}_o. \tag{9.5b}$$

Similarly, if the current configuration of the body is known then its motion may be characterized by the displacement field

$$u_i(x_1,x_2,x_3,t) = x_i-a_i(x_1,x_2,x_3,t) \tag{9.6a}$$

or in vector form

$$\vec{u}(\vec{r},t) = \vec{r}-\vec{r}_o(\vec{r},t). \tag{9.6b}$$

The velocity of the material particle that is currently located at the point P is, by definition, the time rate of change of its position, i.e.,

$$\vec{v} \equiv \left(\frac{\partial \vec{r}}{\partial t}\right)_a = \left(\frac{\partial x_i}{\partial t}\right)_a \hat{\varepsilon}_i. \tag{9.7a}$$

Here the notation $(\partial\vec{r}/\partial t)_a$ is used to emphasize the fact that the derivative is to be evaluated for a particular material particle, i.e., the Lagrangian coordinates $\{a_1,a_2,a_3\}$ are fixed. This is called the material derivative. Since $\vec{r}=\vec{u}+\vec{r}_o$ and $(\partial\vec{r}_o/\partial t)_a=0$, the velocity may also be written as

$$\vec{v} = \left(\frac{\partial \vec{u}}{\partial t}\right)_a = \left(\frac{\partial u_i}{\partial t}\right)_a \hat{\varepsilon}_i. \tag{9.7b}$$

The acceleration of the particle is given by

$$\vec{\alpha} = \left(\frac{\partial \vec{v}}{\partial t}\right)_a.$$

Note that if $\vec{v}=\vec{v}(x,t)$, then

$$\left(\frac{\partial \vec{v}}{\partial t}\right)_a = \left(\frac{\partial \vec{v}}{\partial t}\right)_x + \left(\frac{\partial \vec{v}}{\partial x_i}\right)_t \left(\frac{\partial x_i}{\partial t}\right)_a = \frac{\partial \vec{v}}{\partial t} + v_i \frac{\partial \vec{v}}{\partial x_i} .$$

If attention is focused on a particular material particle, then the material derivative of any property pertaining to this particle is, in fact, an ordinary derivative with respect to time, and for this reason we shall usually denote it by the symbol d/dt, e.g.,

$$\vec{v} = \frac{d\vec{r}}{dt} \equiv \left(\frac{\partial \vec{r}}{\partial t}\right)_a.$$

The simplest motion we will consider is a rigid body motion as illustrated in Fig. 9.2. This motion may be described as follows.

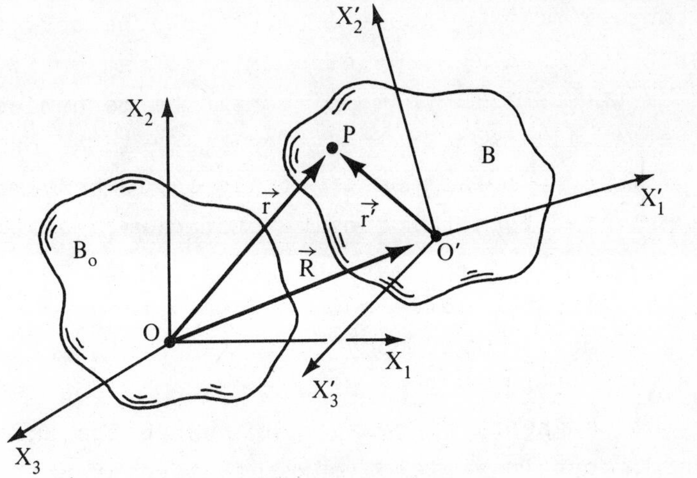

Figure 9.2

Imagine a primed reference to be rigidly attached to the body such that at the instant t_o, $x_i'=a_i$, i.e., it coincides with the Lagrangian reference at t_o. Since this primed reference moves with the body, the coordinates of the point P are fixed with respect to the $O'X_1'X_2'X_3'$ reference. With the aid of (8.26a) we obtain

$$\vec{r}' = a_i\hat{\varepsilon}_i' = a_i a_{ij}\hat{\varepsilon}_j$$

where a_{ij} are the direction cosines relating the primed base vectors to the unprimed base vectors at the instant t. We also note that $a_{ij}(t_o)=\delta_{ij}$. With reference to Fig. 9.2,

$$\vec{r} = \vec{R}+\vec{r}'$$

where $\vec{R}=C_i(t)\hat{\varepsilon}_i$ is the position vector to the origin of the primed reference. Thus,

$$x_i(a,t) = C_i(t) + a_{ji}(t)a_j \tag{9.8a}$$

or

$$a_i(x,t) = a_{ij}(t)\left[x_j-C_j(t)\right] \tag{9.8b}$$

where $a_{ij}a_{ik}=a_{ji}a_{ki}=\delta_{jk}$. The displacement field is

$$\left.\begin{aligned} u_i(a,t) &= C_i(t) + \left[a_{ji}(t)-\delta_{ij}\right]a_j \\ &\text{or} \\ u_i(x,t) &= \left[\delta_{ij}-a_{ij}(t)\right]x_j + a_{ij}(t)C_j(t) \end{aligned}\right\} \tag{9.9}$$

A more complex motion ensues if we allow the body to deform as it travels through space. The study of such deformations together with their causes is the central theme of continuum mechanics. To facilitate this objective we will seek an analytical technique for describing the deformation of a body independently of its rigid body motion. One such technique is provided by the following procedure. Let us fix our attention on two neighboring material particles in the body. As the body deforms the distance between these two particles may vary, however during a purely rigid body motion this distance remains fixed. The change in this distance is thus a measure of the deformation occurring in the neighborhood containing the points under consideration. In order to obtain a field representation of the deformation we will assume that the neighborhood containing the two particles of

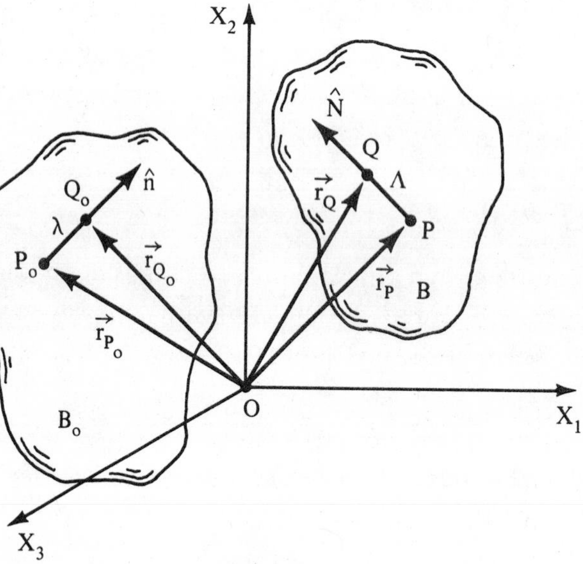

Figure 9.3

interest is arbitrarily small. As shown in Fig. 9.3, the particles are located at the points P_0 and Q_0 at the instant t_0. The coordinates of the points P_0 and Q_0 are

$$P_0 = \{a_1, a_2, a_3\}$$

and

$$Q_0 = \{a_1+\lambda n_1,\ a_2+\lambda n_2,\ a_3+\lambda n_3\} \tag{9.10}$$

respectively. Therefore, the position vectors of the points P_o and Q_o are

$$\vec{r}_{P_o} = a_i \hat{\varepsilon}_i \tag{9.11a}$$

and

$$\vec{r}_{Q_o} = \vec{r}_{P_o} + \lambda\hat{n} \tag{9.11b}$$

where $\hat{n}$ is a unit vector in the direction of the segment joining P_o to Q_o and $\lambda=|\vec{r}_{Q_o}-\vec{r}_{P_o}|$ is the distance of Q_o from P_o. At the instant t, the same two material particles are located at the points P and Q with coordinates

$$P = \{x_1, x_2, x_3\}$$

and (9.12)

$$Q = \{x_1+\Lambda N_1, x_2+\Lambda N_2, x_3+\Lambda N_3\}$$

respectively. The position vectors of P and Q are

$$\vec{r}_P = x_i \hat{\varepsilon}_i \tag{9.13a}$$

and

$$\vec{r}_Q = \vec{r}_P + \Lambda\hat{N} \tag{9.13b}$$

respectively, where $\hat{N}$ is a unit vector in the direction of the segment joining P to Q and $\Lambda=|\vec{r}_Q-\vec{r}_P|$ is the distance of Q from P. To obtain a measure of the deformation, we shall concentrate on the difference $(\Lambda-\lambda)$. This difference may be expressed in terms of either the Lagrangian or the Eulerian coordinates. Let us first examine the Lagrangian representation of the deformation. Since Q_o is in an arbitrarily small neighborhood of P_o we select as our measure of the deformation the extension $E(\hat{n})$ defined as follows.

$$E(\hat{n}) \equiv \lim_{\lambda\to 0} \left(\frac{\Lambda-\lambda}{\lambda}\right) \tag{9.14}$$

The extension is a scalar point function of the Lagrangian coordinates and therefore provides us with a field representation of the deformation. The functional notation $E(\hat{n})$ is used to emphasize the fact that the extension depends on the direction $\hat{n}$ as well as the point P_o under consideration. We note that $\Lambda=|\vec{r}_Q-\vec{r}_P|$ or

$$\Lambda^2 = (\vec{r}_Q-\vec{r}_P)\cdot(\vec{r}_Q-\vec{r}_P). \tag{9.15}$$

The position vector of a generic point in B is $\vec{r}=x_i\hat{\varepsilon}_i$. As a result of this representation and in view of (9.2) we conclude that

$$\left.\begin{aligned} \vec{r}_P &= x_i(a_1,a_2,a_3,t)\hat{\varepsilon}_i = \vec{r}(\vec{r}_{P_o},t) \\ \text{and} \\ \vec{r}_Q &= x_i(a_1+\lambda n_1,a_2+\lambda n_2,a_3+\lambda n_3,t)\hat{\varepsilon}_i = \vec{r}(\vec{r}_{P_o}+\lambda\hat{n},t). \end{aligned}\right\} \tag{9.16}$$

With the aid of (9.16) and (8.48) we obtain

$$\lim_{\lambda\to 0}\left(\frac{\vec{r}_Q-\vec{r}_P}{\lambda}\right) = \lim_{\lambda\to 0}\left[\frac{\vec{r}\left(\vec{r}_{P_o}+\lambda\hat{n},t\right)-\vec{r}\left(\vec{r}_{P_o},t\right)}{\lambda}\right] = \frac{\partial\vec{r}}{\partial\lambda} \tag{9.17}$$

The symbol for partial differentiation is used in (9.17) because $\vec{r}$ is a function of t as well as λ. In view of (9.15) and (9.17) together with (8.49) we obtain

$$\begin{aligned} \lim_{\lambda\to 0}\frac{\Lambda^2}{\lambda^2} &= \frac{\partial\vec{r}}{\partial\lambda}\cdot\frac{\partial\vec{r}}{\partial\lambda} = \left(n_i\,\frac{\partial\vec{r}}{\partial a_i}\right)\cdot\left(n_j\,\frac{\partial\vec{r}}{\partial a_j}\right) \\ &= n_i n_j\,\frac{\partial x_k}{\partial a_i}\,\frac{\partial x_k}{\partial a_j}\,. \end{aligned} \tag{9.18}$$

Since, $n_i n_j \delta_{ij}=n_i n_i=1=\lambda^2/\lambda^2$, (9.18) leads to the following result.

$$\lim_{\lambda\to 0}\left(\frac{\Lambda^2-\lambda^2}{\lambda^2}\right) = n_i n_j\left(\frac{\partial x_k}{\partial a_i}\,\frac{\partial x_k}{\partial a_j} - \delta_{ij}\right) \tag{9.19}$$

The right-hand side of (9.19) is a homogeneous, quadratic function of the direction numbers n_i. We will represent this quadratic form as follows. Let

$$E(\hat{n},\hat{n}) = E_{ij}n_i n_j \tag{9.20a}$$

where

$$E_{ij} = \tfrac{1}{2}\left(\frac{\partial x_k}{\partial a_i}\,\frac{\partial x_k}{\partial a_j} - \delta_{ij}\right). \tag{9.20b}$$

Then, as a result of (9.19) and (9.20)

$$\lim_{\lambda\to 0}\left(\frac{\Lambda^2-\lambda^2}{\lambda^2}\right) = 2E(\hat{n},\hat{n}). \tag{9.21}$$

It is clear from (9.21) that $E(\hat{n},\hat{n})$ is a scalar and from (9.20b) that $E_{ij}=E_{ji}$. Since (9.20a) is valid for arbitrary choice of n_i such that $n_i n_i=1$, we conclude from the quotient law (8.41d) that E_{ij} are the components of a symmetric second order tensor called the Lagrangian or Green's strain tensor. The quadratic form $E(\hat{n},\hat{n})$ is called the normal strain in the $\hat{n}$ direction at the point P_o. Comparison of (9.21) with (9.14) reveals that the normal strain measures the change of the square of the length of an infinitesimal line segment whereas the extension measures the change of the length itself. We may relate the two as follows. As a result of (9.21),

$$\lim_{\lambda\to 0} \frac{\Lambda}{\lambda} = [1+2E(\hat{n},\hat{n})]^{\frac{1}{2}} \tag{9.22}$$

The positive square root is used since Λ and λ are both positive distances. Substituting the above result into (9.14) yields

$$E(\hat{n}) = [1+2E(\hat{n},\hat{n})]^{\frac{1}{2}}-1. \tag{9.23}$$

In view of (9.20a) and (9.23) we see that the deformation of the neighborhood of a point P_o is completely characterized by the Lagrangian strain tensor at that point, or more generally, the deformation of the body is completely characterized by the strain field. Thus, the strain tensor is of fundamental importance in the study of continuum mechanics. With the aid of (9.4) we may express the Lagrangian strain tensor in terms of the displacement field. According to (9.4),

$$\frac{\partial x_i}{\partial a_j} = \frac{\partial u_i}{\partial a_j} + \delta_{ij}.$$

Substituting this result into (9.20b) yields

$$E_{ij} = \frac{1}{2}\left[\frac{\partial u_i}{\partial a_j} + \frac{\partial u_j}{\partial a_i} + \frac{\partial u_k}{\partial a_i}\frac{\partial u_k}{\partial a_j}\right]. \tag{9.24}$$

As an exercise the reader may demonstrate that $E_{ij}=0$ for an arbitrary rigid body motion.

A deformation produces a change in the relative positions of the points in the body. Thus, in addition to a change in the distance between points, angular relationships within the body may also be altered. We now wish to examine this aspect of the deformation. To accomplish this we will consider the change in relative position of the three neighboring points P_o, Q_o and R_o shown in Fig. 9.4. The symbols

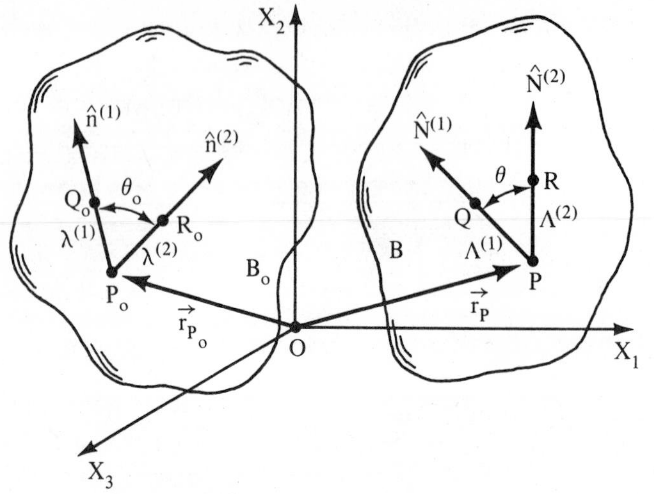

Figure 9.4

$\hat{n}^{(1)}$ and $\hat{n}^{(2)}$ denote unit vectors in the directions of the segments joining P_o to Q_o and P_o to R_o, respectively. The distances from P_o to Q_o and from P_o to R_o are

$$\lambda^{(1)} = |\vec{r}_{Q_o} - \vec{r}_{P_o}| \text{ and } \lambda^{(2)} = |\vec{r}_{R_o} - \vec{r}_{P_o}|$$

respectively. The angle θ_o is given by $\cos\theta_o = \hat{n}^{(1)}\cdot\hat{n}^{(2)}$. The material particles which were located at P_o, Q_o and R_o at the instant t_o are located at P, Q and R, respectively, at the instant t. The symbols $\hat{N}^{(1)}$ and $\hat{N}^{(2)}$ denote unit vectors in the directions of the segments joining P to Q and P to R, respectively. The distances from P to Q and P to R are $\Lambda^{(1)}=|\vec{r}_Q-\vec{r}_P|$ and $\Lambda^{(2)}=|\vec{r}_R-\vec{r}_P|$ respectively. The angle θ is given by $\cos\theta=\hat{N}^{(1)}\cdot\hat{N}^{(2)}$. We now wish to compare the angle θ with the angle θ_o. We note that

$$(\Lambda^{(1)}\hat{N}^{(1)})\cdot(\Lambda^{(2)}\hat{N}^{(2)}) = \Lambda^{(1)}\Lambda^{(2)}\cos\theta \ .$$

Divide both sides of this equation by $(\lambda^{(1)}\lambda^{(2)})$ to obtain

$$\left(\frac{\Lambda^{(1)}\hat{N}^{(1)}}{\lambda^{(1)}}\right)\cdot\left(\frac{\Lambda^{(2)}\hat{N}^{(2)}}{\lambda^{(2)}}\right) = \left(\frac{\Lambda^{(1)}}{\lambda^{(1)}}\right)\left(\frac{\Lambda^{(2)}}{\lambda^{(2)}}\right)\cos\theta. \qquad (9.25)$$

According to (9.14)

$$\lim_{\lambda^{(1)}\to 0} \frac{\Lambda^{(1)}}{\lambda^{(1)}} = 1+E(\hat{n}^{(1)})$$

and

$$\lim_{\lambda^{(2)}\to 0} \frac{\Lambda^{(2)}}{\lambda^{(2)}} = 1+E(\hat{n}^{(2)})$$

We note that

$$\Lambda^{(1)}\hat{N}^{(1)} = \vec{r}_Q-\vec{r}_P = \vec{r}(\vec{r}_{P_o}+\lambda^{(1)}\hat{n}^{(1)},t)-\vec{r}(\vec{r}_{P_o},t)$$

and thus, as a result of (9.17)

$$\lim_{\lambda^{(1)}\to 0} \frac{\Lambda^{(1)}\hat{N}^{(1)}}{\lambda^{(1)}} = \frac{\partial\vec{r}}{\partial\lambda^{(1)}} = n_i^{(1)} \frac{\partial\vec{r}}{\partial a_i}$$

and similarly

$$\lim_{\lambda^{(2)}\to 0} \frac{\Lambda^{(2)}\hat{N}^{(2)}}{\lambda^{(2)}} = \frac{\partial\vec{r}}{\partial\lambda^{(2)}} = n_j^{(2)} \frac{\partial\vec{r}}{\partial a_j}$$

Hence, the limit of (9.25) as $\lambda^{(1)}\to 0$ and $\lambda^{(2)}\to 0$ is given by

$$n_i^{(1)}n_j^{(2)} \frac{\partial\vec{r}}{\partial a_i}\cdot\frac{\partial\vec{r}}{\partial a_j} = \left[1+E(\hat{n}^{(1)})\right]\left[1+E(\hat{n}^{(2)})\right]\cos\theta \tag{9.26}$$

As a result of (9.20b)

$$2E_{ij} = \frac{\partial x_k}{\partial a_i}\frac{\partial x_k}{\partial a_j} - \delta_{ij} = \frac{\partial\vec{r}}{\partial a_i}\cdot\frac{\partial\vec{r}}{\partial a_j} - \delta_{ij}$$

or

$$\frac{\partial\vec{r}}{\partial a_i}\cdot\frac{\partial\vec{r}}{\partial a_j} n_i^{(1)}n_j^{(2)} = (\delta_{ij}+2E_{ij})n_i^{(1)}n_j^{(2)}$$

$$= \cos\theta_o + 2E(\hat{n}^{(1)},\hat{n}^{(2)})$$

where

$$\cos\theta_o = n_i^{(1)}n_i^{(2)} = \delta_{ij}n_i^{(1)}n_j^{(2)}$$

and

$$E(\hat{n}^{(1)},\hat{n}^{(2)}) = E(\hat{n}^{(2)},\hat{n}^{(1)}) = E_{ij}n_i^{(1)}n_j^{(2)}. \tag{9.27}$$

As a result of these observations, (9.26) may be written as

$$\cos\theta = \frac{\cos\theta_0 + 2E\left(\hat{n}^{(1)},\hat{n}^{(2)}\right)}{\left[1+E(\hat{n}^{(1)})\right]\left[1+E(\hat{n}^{(2)})\right]} \tag{9.28}$$

If we take $\theta_0=\pi/2$ then $\cos\theta_0=0$. For this case we define a shear angle $\Gamma(\hat{n}^{(1)},\hat{n}^{(2)})$ as follows. $\Gamma=\theta_0-\theta=\pi/2-\theta$, hence $\cos\theta=\sin\Gamma$ and (9.28) becomes

$$\sin\Gamma = \frac{2E(\hat{n}^{(1)},\hat{n}^{(2)})}{\left[1+E(\hat{n}^{(1)})\right]\left[1+E(\hat{n}^{(2)})\right]}\,. \tag{9.29}$$

The shear angle $\Gamma(\hat{n}^{(1)},\hat{n}^{(2)})$ measures the decrease in the angle included between the originally orthogonal segments P_0Q_0 and P_0R_0 as the deformation proceeds.

We conclude from (9.29) that the change in angular relationships within the body is also completely characterized by the Lagrangian strain tensor. Changes in distance are related through (9.23) to the normal strain $E(\hat{n},\hat{n})$, and changes in angular relationships are related through (9.29) to the associated bilinear form $E(\hat{n}^{(1)},\hat{n}^{(2)})$ which is called the shear strain.

In many cases of practical importance we find that the strains are very small, i.e.,

$$|E(\hat{n}^{(1)},\hat{n}^{(2)})| \ll 1 \tag{9.30}$$

for all $\hat{n}^{(1)}$ and $\hat{n}^{(2)}$ such that $\hat{n}^{(1)}\cdot\hat{n}^{(1)}=\hat{n}^{(2)}\cdot\hat{n}^{(2)}=1$. For this case (9.23) and (9.29) are simplified considerably by noting that

$$[1+x]^{\frac{1}{2}} = 1 + \frac{x}{2} - \frac{x^2}{8} \pm \ldots \tag{9.31a}$$

and

$$\sin^{-1}x = x + \frac{x^3}{6} + \ldots \tag{9.31b}$$

for $x^2<1$. Hence, as a first approximation

$$E(\hat{n}) \simeq E(\hat{n},\hat{n}) = E_{ij}n_i n_j \tag{9.32a}$$

and

$$\Gamma(\hat{n}^{(1)},\hat{n}^{(2)}) \simeq 2E(\hat{n}^{(1)},\hat{n}^{(2)}) = 2E_{ij}n_i^{(1)}n_j^{(2)}. \tag{9.32b}$$

By letting $\hat{n}=\hat{\varepsilon}_1,\hat{\varepsilon}_2$ and $\hat{\varepsilon}_3$ successively in (9.32a) we obtain $E(\hat{\varepsilon}_1)\simeq E_{11}$, $E(\hat{\varepsilon}_2)\simeq E_{22}$ and $E(\hat{\varepsilon}_3)\simeq E_{33}$ respectively. Thus, the normal strains E_{11}, E_{22} and E_{33} are approximately equal to the extensions in the $\hat{\varepsilon}_1$, $\hat{\varepsilon}_2$ and $\hat{\varepsilon}_3$ directions respectively. Similarly, from (9.32b) we obtain $\Gamma(\hat{\varepsilon}_1,\hat{\varepsilon}_2)\simeq 2E_{12}$, $\Gamma(\hat{\varepsilon}_2,\hat{\varepsilon}_3)\simeq 2E_{23}$ and $\Gamma(\hat{\varepsilon}_3,\hat{\varepsilon}_1)\simeq 2E_{31}$. Therefore, the shear strains E_{12}, E_{23}, and E_{31} are approximately equal to half of the corresponding shear angles.

The Eulerian description of the deformation may be obtained by treating the current configuration, B, of the body as the reference configuration. In this case we define the extension as follows. With reference to Fig. 9.3,

$$e(\hat{N}) \equiv \lim_{\Lambda\to 0} \left(\frac{\Lambda-\lambda}{\Lambda}\right) . \tag{9.33}$$

By definition, $\lambda=|\vec{r}_{Q_o}-\vec{r}_{P_o}|$ and therefore

$$\lambda^2 = (\vec{r}_{Q_o}-\vec{r}_{P_o})\cdot(\vec{r}_{Q_o}-\vec{r}_{P_o}). \tag{9.34}$$

The position vector of a generic point of B_o is $\vec{r}_o=a_i\hat{\varepsilon}_i$, so that

$$\vec{r}_{P_o} = a_i(x_1,x_2,x_3,t)\hat{\varepsilon}_i = \vec{r}_o(\vec{r}_P,t) \tag{9.35a}$$

and

$$\vec{r}_{Q_o} = a_i(x_1+\Lambda N_1,x_2+\Lambda N_2,x_3+\Lambda N_3,t)\hat{\varepsilon}_i = \vec{r}_o(\vec{r}_P+\Lambda\hat{N},t) \tag{9.35b}$$

As a result of (9.35) and (8.48),

$$\lim_{\Lambda\to 0}\left(\frac{\vec{r}_{Q_o}-\vec{r}_{P_o}}{\Lambda}\right) = \lim_{\Lambda\to 0}\left[\frac{\vec{r}_o(\vec{r}_P+\Lambda\hat{N},t)-\vec{r}_o(\vec{r}_P,t)}{\Lambda}\right] = \frac{\partial\vec{r}_o}{\partial\Lambda} . \tag{9.36}$$

By applying (9.36) to (9.34) we obtain with the aid of (8.49)

$$\lim_{\Lambda\to 0}\frac{\lambda^2}{\Lambda^2} = \frac{\partial\vec{r}_o}{\partial\Lambda}\cdot\frac{\partial\vec{r}_o}{\partial\Lambda} = \left(N_i\,\frac{\partial\vec{r}_o}{\partial x_i}\right)\cdot\left(N_j\,\frac{\partial\vec{r}_o}{\partial x_j}\right)$$

$$= N_iN_j\,\frac{\partial a_k}{\partial x_i}\,\frac{\partial a_k}{\partial x_j} . \tag{9.37}$$

By noting that $N_i N_j \delta_{ij}=1$, (9.37) may also be expressed in the form

$$\lim_{\Lambda\to 0}\left(\frac{\Lambda^2-\lambda^2}{\Lambda^2}\right) = N_i N_j \left(\delta_{ij} - \frac{\partial a_k}{\partial x_i}\frac{\partial a_k}{\partial x_j}\right) . \tag{9.38}$$

As a result of (9.38), we define the normal strain $e(\hat{N},\hat{N})$ as

$$e(\hat{N},\hat{N}) = e_{ij} N_i N_j \tag{9.39a}$$

where

$$e_{ij} = \tfrac{1}{2}\left(\delta_{ij} - \frac{\partial a_k}{\partial x_i}\frac{\partial a_k}{\partial x_j}\right) . \tag{9.39b}$$

Comparing (9.38) with (9.39) reveals that

$$\lim_{\Lambda\to 0}\left(\frac{\Lambda^2-\lambda^2}{\Lambda^2}\right) = 2e(\hat{N},\hat{N}). \tag{9.40}$$

We conclude from (9.40) that $e(\hat{N},\hat{N})$ is a scalar and from (9.39b) that $e_{ij}=e_{ji}$. Since (9.39a) is valid for arbitrary N_i satisfying $N_i N_i=1$, we conclude from the quotient law (8.41d) that e_{ij} is a symmetric second order tensor which we call the Eulerian or Almansi's strain tensor. The quadratic form $e(\hat{N},\hat{N})$ is called the normal strain in the $\hat{N}$ direction at the point P. From (9.40),

$$\lim_{\Lambda\to 0} \frac{\lambda}{\Lambda} = \left[1-2e(\hat{N},\hat{N})\right]^{\frac{1}{2}} \tag{9.41}$$

and therefore

$$e(\hat{N}) = 1-\left[1-2e(\hat{N},\hat{N})\right]^{\frac{1}{2}}. \tag{9.42}$$

In view of (9.42) and (9.39) we conclude that the Eulerian strain field provides an alternative to the Lagrangian strain field for the description of the deformation. According to (9.4),

$$\frac{\partial a_i}{\partial x_j} = \delta_{ij} - \frac{\partial u_i}{\partial x_j} .$$

Substituting this result into (9.39b) reveals that

$$e_{ij} = \tfrac{1}{2}\left[\frac{\partial u_i}{\partial x_j} + \frac{\partial u_j}{\partial x_i} - \frac{\partial u_k}{\partial x_i}\frac{\partial u_k}{\partial x_j}\right]. \tag{9.43}$$

We will now show that the Eulerian strain tensor also characterizes the change in the angular relationships of the body. With reference to Fig. 9.4 we note that

$$(\lambda^{(1)}\hat{n}^{(1)})\cdot(\lambda^{(2)}\hat{n}^{(2)}) = \lambda^{(1)}\lambda^{(2)}\cos\theta_o. \tag{9.44}$$

In view of (9.33)

$$\lim_{\Lambda^{(1)}\to 0}\left(\frac{\lambda^{(1)}}{\Lambda^{(1)}}\right) = 1-e(\hat{N}^{(1)})$$

and

$$\lim_{\Lambda^{(2)}\to 0}\left(\frac{\lambda^{(2)}}{\Lambda^{(2)}}\right) = 1-e(\hat{N}^{(2)}).$$

With reference to Fig. 9.4 we observe that $\lambda^{(1)}\hat{n}^{(1)}=(\vec{r}_{Q_o}-\vec{r}_{P_o})$ and $\lambda^{(2)}\hat{n}^{(2)}=(\vec{r}_{R_o}-\vec{r}_{P_o})$, and with the aid of (9.36) and (8.49) we obtain

$$\lim_{\Lambda^{(1)}\to 0}\left(\frac{\lambda^{(1)}\hat{n}^{(1)}}{\Lambda^{(1)}}\right) = \frac{\partial\vec{r}_o}{\partial\Lambda^{(1)}} = N_i^{(1)}\frac{\partial\vec{r}_o}{\partial x_i}$$

and

$$\lim_{\Lambda^{(2)}\to 0}\left(\frac{\lambda^{(2)}\hat{n}^{(2)}}{\Lambda^{(2)}}\right) = \frac{\partial\vec{r}_o}{\partial\Lambda^{(2)}} = N_j^{(2)}\frac{\partial\vec{r}_o}{\partial x_j}$$

If we now divide equation (9.44) by $\Lambda^{(1)}\Lambda^{(2)}$ and then take the limit as $\Lambda^{(1)}\to 0$ and $\Lambda^{(2)}\to 0$ we obtain with the aid of the above results,

$$N_i^{(1)}N_j^{(2)}\frac{\partial\vec{r}_o}{\partial x_i}\cdot\frac{\partial\vec{r}_o}{\partial x_j} = \left[1-e(\hat{N}^{(1)})\right]\left[1-e(\hat{N}^{(2)})\right]\cos\theta_o. \tag{9.45}$$

With the aid of (9.39), the above equation may also be written as

$$\cos\theta_o = \frac{\cos\theta-2e(\hat{N}^{(1)},\hat{N}^{(2)})}{\left[1-e(\hat{N}^{(1)})\right]\left[1-e(\hat{N}^{(2)})\right]} \tag{9.46}$$

where

$$e(\hat{N}^{(1)},\hat{N}^{(2)}) = e_{ij}N_i^{(1)}N_j^{(2)}. \tag{9.47}$$

If we set $\theta=\pi/2$ then $\cos\theta=0$. In this case we define a shear angle $\gamma(\hat{N}^{(1)},\hat{N}^{(2)})$ by $\gamma=\theta_0-\theta=\theta_0-\pi/2$. Consequently $\cos\theta_0=-\sin\gamma$, and (9.46) becomes

$$\sin\gamma = \frac{2e(\hat{N}^{(1)},\hat{N}^{(2)})}{\left[1-e(\hat{N}^{(1)})\right]\left[1-e(\hat{N}^{(2)})\right]} \tag{9.48}$$

In order to restore the currently orthogonal segments PQ and PR to their original relative orientation, we must increase the currently right angle by an amount $\gamma(\hat{N}^{(1)},\hat{N}^{(2)})$. The bilinear form $e(\hat{N}^{(1)},\hat{N}^{(2)})$ is called the shear strain. For sufficiently small strains we have

$$|e(\hat{N}^{(1)},\hat{N}^{(2)})|<<1 \tag{9.49}$$

for all $\hat{N}^{(1)}$ and $\hat{N}^{(2)}$ such that $\hat{N}^{(1)}\cdot\hat{N}^{(1)}=\hat{N}^{(2)}\cdot\hat{N}^{(2)}=1$, and we can use (9.31) to obtain the following approximations for (9.42) and (9.48).

$$e(\hat{N}) \simeq e(\hat{N},\hat{N}) = e_{ij}N_iN_j \tag{9.50a}$$

$$\gamma(\hat{N}^{(1)},\hat{N}^{(2)}) \simeq 2e(\hat{N}^{(1)},\hat{N}^{(2)}) = 2e_{ij}N_i^{(1)}N_j^{(2)} \tag{9.50b}$$

By letting $\hat{N}=\hat{e}_1,\hat{e}_2$ and $\hat{e}_3$ successively in (9.50a) we obtain $e(\hat{e}_1)\simeq e_{11}$, $e(\hat{e}_2)\simeq e_{22}$ and $e(\hat{e}_3)\simeq e_{33}$, respectively. Thus the normal strains e_{11}, e_{22} and e_{33} are approximately equal to the extensions in the $\hat{e}_1$, $\hat{e}_2$ and $\hat{e}_3$ directions respectively. Similarly, from (9.50b) we conclude that $\gamma(\hat{e}_1,\hat{e}_2)\simeq 2e_{12}$, $\gamma(\hat{e}_2,\hat{e}_3)\simeq 2e_{23}$ and $\gamma(\hat{e}_3,\hat{e}_1)\simeq 2e_{31}$. Therefore, the shear strains e_{12}, e_{23} and e_{31} are approximately equal to half of the corresponding shear angles.

We shall next examine the effect which the deformation has on the differential properties of curves, surfaces and neighborhoods within the body by replacing the distances λ and Λ by infinitesimals of the first order. If we let $\lambda=ds_0$ and $\Lambda=ds$, then

$$\left.\begin{aligned} &\lambda\hat{n}=\hat{n}ds_0=\vec{r}_{Q_0}-\vec{r}_{P_0}=d\vec{r}_0=da_i\hat{e}_i \\ &\text{and} \\ &\Lambda\hat{N} = \hat{N}ds = \vec{r}_Q-\vec{r}_P = d\vec{r} = dx_i\hat{e}_i \end{aligned}\right\} \tag{9.51}$$

where $n_i ds_o = da_i$ and $N_i ds = dx_i$. The Lagrangian description of the deformation is obtained as follows. With the aid of (9.2) and (9.51) we obtain, after neglecting infinitesimals of order greater than one,

$$d\vec{r} = \frac{\partial\vec{r}}{\partial a_i}\, da_i = \frac{\partial\vec{r}}{\partial a_i}\, n_i ds_o = \frac{\partial x_j}{\partial a_i}\, n_i ds_o \hat{e}_j. \tag{9.52}$$

In view of (9.52) and (9.51),

$$ds^2 = d\vec{r}\cdot d\vec{r} = \frac{\partial\vec{r}}{\partial a_i}\cdot\frac{\partial\vec{r}}{\partial a_j}\, n_i n_j ds_o^2$$

or

$$ds^2 = g_{ij} n_i n_j ds_o^2 = g_{ij}\, da_i da_j \tag{9.53}$$

where

$$g_{ij} = \frac{\partial\vec{r}}{\partial a_i}\cdot\frac{\partial\vec{r}}{\partial a_j} = \frac{\partial x_k}{\partial a_i}\frac{\partial x_k}{\partial a_j} = 2E_{ij}+\delta_{ij} \tag{9.54}$$

is called the Lagrangian metric tensor or Green's deformation tensor. Note that the quadratic form $g_{ij} n_i n_j$ is positive definite. With the aid of (9.53) and (9.54) we see that the normal strain and extension are given by

$$ds^2 = [1+2E(\hat{n},\hat{n})]\, ds_o^2$$

and (9.55)

$$ds = [1+E(\hat{n})]\, ds_o$$

respectively.

A comparison of (9.52) with (9.51) reveals that the three vectors $d\vec{r}^{(1)} = (\partial\vec{r}/\partial a_1)da_1$, $d\vec{r}^{(2)} = (\partial\vec{r}/\partial a_2)da_2$ and $d\vec{r}^{(3)} = (\partial\vec{r}/\partial a_3)da_3$ in B correspond to the three vectors $d\vec{r}_o^{(1)} = da_1\hat{e}_1$, $d\vec{r}_o^{(2)} = da_2\hat{e}_2$ and $d\vec{r}_o^{(3)} = da_3\hat{e}_3$ respectively in B_o. Thus the area elements

$$\left.\begin{aligned} d\vec{A}_o^{(1)} &= d\vec{r}_o^{(2)}\times d\vec{r}_o^{(3)} = da_2 da_3\hat{e}_1 \\ d\vec{A}_o^{(2)} &= d\vec{r}_o^{(3)}\times d\vec{r}_o^{(1)} = da_1 da_3\hat{e}_2 \\ d\vec{A}_o^{(3)} &= d\vec{r}_o^{(1)}\times d\vec{r}_o^{(2)} = da_1 da_2\hat{e}_3 \end{aligned}\right\} \tag{9.56}$$

in B_o are deformed into the area elements

$$\left.\begin{aligned} d\vec{A}^{(1)} &= d\vec{r}^{(2)} \times d\vec{r}^{(3)} = \frac{\partial \vec{r}}{\partial a_2} \times \frac{\partial \vec{r}}{\partial a_3}\, da_2 da_3 \\ d\vec{A}^{(2)} &= d\vec{r}^{(3)} \times d\vec{r}^{(1)} = \frac{\partial \vec{r}}{\partial a_3} \times \frac{\partial \vec{r}}{\partial a_1}\, da_1 da_3 \\ d\vec{A}^{(3)} &= d\vec{r}^{(1)} \times d\vec{r}^{(2)} = \frac{\partial \vec{r}}{\partial a_1} \times \frac{\partial \vec{r}}{\partial a_2}\, da_1 da_2 \end{aligned}\right\} \tag{9.57}$$

respectively in B. The reader may verify that, as a result of (8.18) and Exercise (8.4a),

$$(\vec{A}\times\vec{B})\cdot(\vec{C}\times\vec{D}) = \begin{vmatrix} \vec{A}\cdot\vec{C} & \vec{A}\cdot\vec{D} \\ \vec{B}\cdot\vec{C} & \vec{B}\cdot\vec{D} \end{vmatrix}$$

for any four vectors $\vec{A}$, $\vec{B}$, $\vec{C}$ and $\vec{D}$. With the aid of this identity and equation (9.54) we obtain from (9.57)

$$\left.\begin{aligned} |d\vec{A}^{(1)}| &= \sqrt{g_{22}g_{33}-g_{23}^2}\; |d\vec{A}_o^{(1)}| \\ |d\vec{A}^{(2)}| &= \sqrt{g_{33}g_{11}-g_{31}^2}\; |d\vec{A}_o^{(2)}| \\ |d\vec{A}^{(3)}| &= \sqrt{g_{11}g_{22}-g_{12}^2}\; |d\vec{A}_o^{(3)}| \end{aligned}\right\} \tag{9.58}$$

The volume element, $dV_o = d\vec{r}_o^{(i)} \cdot d\vec{A}_o^{(i)} = da_1 da_2 da_3$ (no sum on i) in B_o is deformed into the volume element

$$dV = d\vec{r}^{(i)} \cdot d\vec{A}^{(i)} = \left(\frac{\partial \vec{r}}{\partial a_1} \times \frac{\partial \vec{r}}{\partial a_2} \cdot \frac{\partial \vec{r}}{\partial a_3}\right) da_1 da_2 da_3 \text{ in B.}$$

If we define the Jacobian determinant J as

$$J \equiv \begin{vmatrix} \frac{\partial x_1}{\partial a_1} & \frac{\partial x_2}{\partial a_1} & \frac{\partial x_3}{\partial a_1} \\ \frac{\partial x_1}{\partial a_2} & \frac{\partial x_2}{\partial a_2} & \frac{\partial x_3}{\partial a_2} \\ \frac{\partial x_1}{\partial a_3} & \frac{\partial x_2}{\partial a_3} & \frac{\partial x_3}{\partial a_3} \end{vmatrix} \tag{9.59a}$$

then as a result of (8.14) we obtain

$$J = e_{ijk} \frac{\partial x_i}{\partial a_1} \frac{\partial x_j}{\partial a_2} \frac{\partial x_k}{\partial a_3} = e_{ijk} \frac{\partial x_1}{\partial a_i} \frac{\partial x_2}{\partial a_j} \frac{\partial x_3}{\partial a_k} \tag{9.59b}$$

Thus the volume element dV may be expressed as

$$dV = J da_1 da_2 da_3 = J dV_0. \tag{9.60}$$

If we let $g = \det(g_{ij})$ then $g = J^2$ and $dV = \sqrt{g}\, dV_0$.

We may obtain a measure of the rate at which the deformation proceeds by evaluating the material derivative of the first of equations (9.55). Since ds_0 is a constant magnitude,

$$\frac{d}{dt}(ds^2) = 2ds_0^2 \frac{dE(\hat{n},\hat{n})}{dt}. \tag{9.61}$$

With the aid of (9.20) and (9.7a) we obtain

$$\frac{dE_{ij}}{dt} \equiv \left(\frac{\partial E_{ij}}{\partial t}\right)_a = \tfrac{1}{2}\left(\frac{\partial v_k}{\partial a_i}\frac{\partial x_k}{\partial a_j} + \frac{\partial x_k}{\partial a_i}\frac{\partial v_k}{\partial a_j}\right)$$

If v_k is expressed in Eulerian coordinates, then

$$\frac{\partial v_k}{\partial a_i} = \frac{\partial v_k}{\partial x_\ell}\frac{\partial x_\ell}{\partial a_i}$$

therefore

$$\frac{dE_{ij}}{dt} = \tfrac{1}{2}\left(\frac{\partial v_k}{\partial x_\ell}\frac{\partial x_\ell}{\partial a_i}\frac{\partial x_k}{\partial a_j} + \frac{\partial x_k}{\partial a_i}\frac{\partial x_\ell}{\partial a_j}\frac{\partial v_k}{\partial x_\ell}\right)$$

or

$$\frac{dE_{ij}}{dt} = \tfrac{1}{2}\left(\frac{\partial v_k}{\partial x_\ell} + \frac{\partial v_\ell}{\partial x_k}\right)\frac{\partial x_k}{\partial a_i}\frac{\partial x_\ell}{\partial a_j}$$

$$= \dot{e}_{k\ell}\frac{\partial x_k}{\partial a_i}\frac{\partial x_\ell}{\partial a_j} \tag{9.62a}$$

where

$$\dot{e}_{k\ell} = \tfrac{1}{2}\left(\frac{\partial v_k}{\partial x_\ell} + \frac{\partial v_\ell}{\partial x_k}\right) \tag{9.62b}$$

is called the rate of deformation tensor. Substitute (9.62) into (9.61) to obtain

$$\frac{d}{dt}(ds^2) = 2\dot{e}_{k\ell}\frac{\partial x_k}{\partial a_i}\frac{\partial x_\ell}{\partial a_j} n_i n_j ds_0^2$$

Comparison of (9.52) with (9.51) reveals that

$$N_k ds = dx_k = \frac{\partial x_k}{\partial a_i} n_i ds_0.$$

Thus,

$$\frac{d}{dt}(ds^2) = 2\dot{e}_{k\ell}N_kN_\ell ds^2 = 2\dot{e}_{k\ell}dx_kdx_\ell. \qquad (9.63)$$

It should be emphasized that the rate of deformation tensor is not the material derivative of the Eulerian strain tensor, i.e., $\dot{e}_{k\ell} \neq de_{k\ell}/dt$.

9.3 DYNAMICS OF CONTINUA

In mechanics, matter is characterized by its inertia, i.e., matter has the ability to resist attempts to change its state of motion. As pointed out by Newton, the inertia of a body is proportional to its quantity of matter. The measure of this quantity of matter is called the mass of the body. The mass of a body is, by definition, a positive quantity. Let B represent the region in space described by the configuration of the body at the instant of time t. In its role as the mathematical model of the physical body at this instant, the region B has two measures, its geometrical measure or volume which we denote by V(B) and its mechanical measure or mass which we denote by M(B). As a result of our continuum hypothesis, M(B) must vanish together with V(B), i.e., point masses do not exist in our continuous model of nature. Let us suppose that the body and therefore the region B are divided, in any manner imaginable, into several separate parts $B_1, B_2, \ldots, B_n$. Let $V(B_i)$ and $M(B_i)$ represent the volume and mass respectively of the i'th part. The whole body is the same as the union of all its separate parts, and therefore the volume and mass of the body are just the sums of the volumes and masses, respectively, of its separate parts. By induction, this is seen to be true even if the number of parts becomes countably infinite. Thus we have

$$V(B) = \sum_{i=1}^{n} V(B_i)$$

and

$$M(B) = \sum_{i=1}^{n} M(B_i) = \sum_{i=1}^{n} \rho(B_i)V(B_i) \qquad (9.64)$$

where

$$\rho(B_i) \equiv \frac{M(B_i)}{V(B_i)}$$

is called the average density of the body B_i. If we let $n \to \infty$ in such a manner that the largest of the $V(B_i)$ tends to zero, then (9.64) becomes

$$M(B) = \int_B \rho(P)dV \tag{9.65}$$

and the scalar point function $\rho(P)$ is called the density of the body at the point P. The density describes the manner in which the mass is distributed throughout the region B.

One of the fundamental axioms of Newtonian mechanics is that matter can neither be created nor destroyed. We will satisfy this condition by assuming that the mass of the body remains constant during its motion. If B_o and B represent the Lagrangian and the current configurations, respectively, of the same body, then

$$M(B_o) = M(B) \tag{9.66}$$

where

$$M(B_o) = \int_{B_o} \rho(a_1,a_2,a_3,t_o)dV_o$$

and

$$M(B) = \int_B \rho(x_1,x_2,x_3,t)dV.$$

As a result of (9.60) and (9.2)

$$M(B) = \int_{B_o} \rho(a,t)J(a,t)dV_o$$

where

$$\rho(a,t) = \rho[x(a,t),t].$$

Therefore (9.66) may be expressed as

$$\int_{B_o} [\rho(a,t_o)-\rho(a,t)J(a,t)]\, dV_o=0.$$

In order for this to be true for all bodies, the integrand must vanish at every point of B_o, thus

$$\rho(a,t_o) = \rho(a,t)J(a,t). \tag{9.67}$$

The material derivative of (9.67) yields

$$\frac{d}{dt}(\rho J) = \frac{\partial}{\partial t}\Big[\rho(a,t_o)\Big]_a = 0 \tag{9.68}$$

and (9.68) is a continuity equation. If ρ and J are expressed in terms of the Eulerian coordinates, then

$$J\,\frac{d\rho}{dt} + \rho\,\frac{dJ}{dt} = 0.$$

As a result of (9.59b),

$$\frac{dJ}{dt} = \frac{\partial}{\partial t}\left[e_{ijk}\,\frac{\partial x_1}{\partial a_i}\,\frac{\partial x_2}{\partial a_j}\,\frac{\partial x_3}{\partial a_k}\right]_a$$

$$= e_{ijk}\left[\frac{\partial v_1}{\partial a_i}\,\frac{\partial x_2}{\partial a_j}\,\frac{\partial x_3}{\partial a_k} + \frac{\partial x_1}{\partial a_i}\,\frac{\partial v_2}{\partial a_j}\,\frac{\partial x_3}{\partial a_k} + \frac{\partial x_1}{\partial a_i}\,\frac{\partial x_2}{\partial a_j}\,\frac{\partial v_3}{\partial a_k}\right].$$

If v_i is expressed in Eulerian coordinates, then $\dfrac{\partial v_i}{\partial a_j} = \dfrac{\partial v_i}{\partial x_m}\,\dfrac{\partial x_m}{\partial a_j}$ and

$$\frac{dJ}{dt} = e_{ijk}\left[\frac{\partial v_1}{\partial x_m}\,\frac{\partial x_m}{\partial a_i}\,\frac{\partial x_2}{\partial a_j}\,\frac{\partial x_3}{\partial a_k} + \frac{\partial v_2}{\partial x_m}\,\frac{\partial x_1}{\partial a_i}\,\frac{\partial x_m}{\partial a_j}\,\frac{\partial x_3}{\partial a_k} + \frac{\partial v_3}{\partial x_m}\,\frac{\partial x_1}{\partial a_i}\,\frac{\partial x_2}{\partial a_j}\,\frac{\partial x_m}{\partial a_k}\right]$$

Because of the skew-symmetry of e_{ijk} the above result simplifies to

$$\frac{dJ}{dt} = e_{ijk}\,\frac{\partial x_1}{\partial a_i}\,\frac{\partial x_2}{\partial a_j}\,\frac{\partial x_3}{\partial a_k}\left(\frac{\partial v_1}{\partial x_1} + \frac{\partial v_2}{\partial x_2} + \frac{\partial v_3}{\partial x_3}\right)$$

or

$$\frac{dJ}{dt} = J\,\frac{\partial v_i}{\partial x_i} \tag{9.69}$$

Hence the continuity equation becomes

$$J\left[\frac{d\rho}{dt} + \rho\,\frac{\partial v_i}{\partial x_i}\right] = 0$$

and since $J \neq 0$, we have

$$\frac{d\rho}{dt} + \rho\,\frac{\partial v_i}{\partial x_i} = 0. \tag{9.70a}$$

Noting that $d\rho/dt = \partial\rho/\partial t + v_i\,(\partial\rho/\partial x_i)$ we can rewrite (9.70a) to read

$$\frac{\partial\rho}{\partial t} + \frac{\partial(\rho v_i)}{\partial x_i} = 0. \tag{9.70b}$$

In the forthcoming analysis we will frequently encounter material derivatives of the type dF/dt where

$$F \equiv \int_B \rho f dV. \tag{9.71}$$

With the aid of (9.60) we obtain

$$\frac{dF}{dt} = \frac{d}{dt}\left[\int_{B_o} \rho f J dV_o\right]$$

and since the region B_o is fixed in space for all time the order of integration and differentiation may be interchanged. Thus,

$$\frac{dF}{dt} = \int_{B_o} \frac{d}{dt}(\rho J f) dV_o.$$

In view of the continuity equation (9.68) we obtain for the above integral

$$\frac{dF}{dt} = \int_{B_o} \frac{df}{dt} \rho J dV_o = \int_B \frac{df}{dt} \rho dV \tag{9.72a}$$

and if f is expressed in Eulerian coordinates we have

$$\frac{dF}{dt} = \int_B \left(\frac{\partial f}{\partial t} + v_i \frac{\partial f}{\partial x_i}\right) \rho dV. \tag{9.72b}$$

We will assume, as a fundamental axiom of mechanics, that the motion of bodies is governed by Newton's laws. The resulting equations of motion for a continuous body are called Euler's equations and may be expressed as follows. If $\vec{F}$ and $\vec{M}$ represent the resultants of all the forces and torques, respectively, acting on the body then

$$\vec{F} = \frac{d\vec{P}}{dt} \text{ and } \vec{M} = \frac{d\vec{H}}{dt} \tag{9.73a}$$

where

$$\vec{P} = \int_B \rho \vec{v} dV \tag{9.73b}$$

is the total linear momentum of the body and

$$\vec{H} = \int_B (\vec{r} \times \vec{v}) \rho dV \tag{9.73c}$$

is the total moment of momentum or angular momentum of the body. The position vector $\vec{r}$ is measured from the same point about which the torque is calculated, usually the origin of our coordinates. As a result of (9.72), (9.73) may also be written in the form

$$\vec{F} = \int_B \rho \frac{d\vec{v}}{dt} dV \tag{9.74a}$$

and

$$\vec{M} = \int_B \frac{d}{dt}(\vec{r}\times\vec{v})\rho dV = \int_B (\vec{r} \times\frac{d\vec{v}}{dt})\rho dV. \tag{9.74b}$$

The second equality in (9.74b) follows from (9.7a) and from the definition of the cross product.

As pointed out earlier, in continuum mechanics we are concerned with fields of force. The force fields we shall consider are of two distinct types. The first, called a body force field, exerts a direct influence on each element of matter in the body. For example, inertia, gravity and electromagnetic forces are all classified as body forces. We will characterize these body forces mathematically by a volume density of force, $\vec{B}(\vec{r},t)$ (force per unit volume), defined on the region B at the instant t. The second type, called a surface traction or contact force exerts its influence on the configuration of the body, i.e., it acts on the boundary of the body. A surface traction results from the direct physical contact of two different bodies. We will characterize such surface tractions mathematically by a surface density of force, $\vec{T}(\vec{r},t)$ (force per unit area), defined on the boundary S of the region B at the instant t. As a result of these observations, we may represent the total force acting on the body by

$$\vec{F} = \int_B \vec{B}dV + \int_S \vec{T}dS \tag{9.75a}$$

If we assume that the torque on the body arises solely from the action of these applied forces, then

$$\vec{M} = \int_B (\vec{r}\times\vec{B})dV + \int_S(\vec{r}\times\vec{T})dS. \tag{9.75b}$$

Combining (9.75) with (9.74) yields

$$\int_B \left[\vec{B}-\rho \frac{d\vec{v}}{dt}\right]dV + \int_S \vec{T}ds = 0 \tag{9.76a}$$

and

$$\int_B \vec{r}\times\left[\vec{B}-\rho \frac{d\vec{v}}{dt}\right]dV + \int_S (\vec{r}\times\vec{T})dS = 0. \tag{9.76b}$$

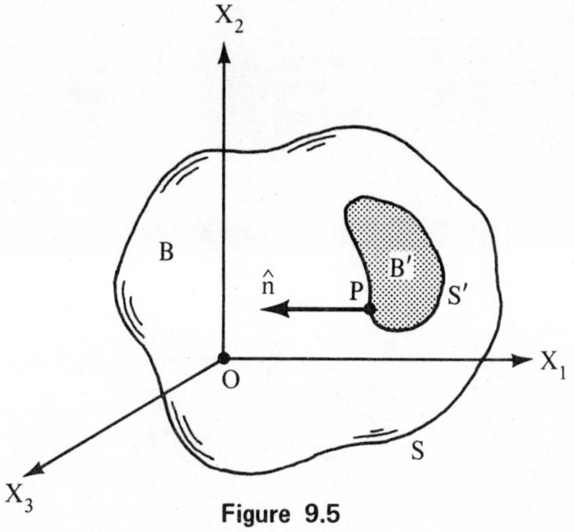

Figure 9.5

We will now examine the manner in which force is transmitted through a body. At the instant t, imagine a closed surface S' within the region B. As shown in Fig. 9.5, S' encloses a region B' which may itself be treated as a body. If this is done, then the interaction of the surrounding material with the material contained in B' may be characterized by a surface traction defined on S'. Let $\hat{n}$ be the unit vector normal to S' at P and directed outward from B'. We call $\hat{n}$ the positive normal or unit outer normal to S' at P. We shall assume that at any point on S', the traction vector $\vec{T}$ depends only on the direction of the positive normal at that point, i.e., at P, $\vec{T}=\vec{T}(\hat{n})$. This is called the stress principle of Euler and Cauchy and $\vec{T}(\hat{n})$ is called the stress vector on the surface whose positive normal at P is $\hat{n}$. If P is an interior point of B, then the normal to our imagined surface may have any direction whatever. Hence, there are an infinity of stress vectors at P, one corresponding to each direction $\hat{n}$. We will be particularly interested in those three which act on the surfaces whose positive normals at P are parallel to the base vectors of our coordinates. The nine components of these three stress vectors are called the stresses at P and are represented as follows.

$$\vec{T}(\hat{\varepsilon}_1) = \tau_{11}\hat{\varepsilon}_1+\tau_{12}\hat{\varepsilon}_2+\tau_{13}\hat{\varepsilon}_3 = \tau_{1j}\hat{\varepsilon}_j \tag{9.77a}$$

$$\vec{T}(\hat{\varepsilon}_2) = \tau_{21}\hat{\varepsilon}_1+\tau_{22}\hat{\varepsilon}_2+\tau_{23}\hat{\varepsilon}_3 = \tau_{2j}\hat{\varepsilon}_j \tag{9.77b}$$

$$\vec{T}(\hat{\varepsilon}_3) = \tau_{31}\hat{\varepsilon}_1+\tau_{32}\hat{\varepsilon}_2+\tau_{33}\hat{\varepsilon}_3 = \tau_{3j}\hat{\varepsilon}_j \tag{9.77c}$$

or

$$\vec{T}(\hat{\varepsilon}_i) = \tau_{ij}\hat{\varepsilon}_j \tag{9.77d}$$

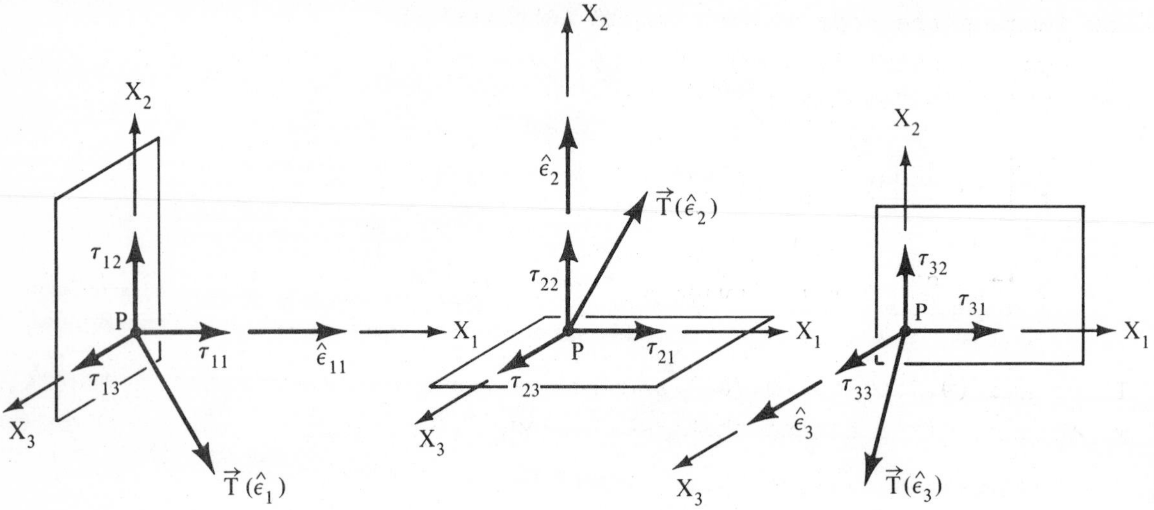

Figure 9.6

These stresses are illustrated graphically in Fig. 9.6. The component of the stress vector normal to the surface being considered is called the normal stress on that surface, e.g., τ_{11}, τ_{22} and τ_{33} are normal stresses. A component of the stress vector parallel to the surface is called a shear stress, e.g., τ_{12}, τ_{21}, τ_{13}, τ_{31}, τ_{23} and τ_{32} are shear stresses.

We may employ (9.76a) to determine the form of the function $\vec{T}(\hat{n})$. Let $\hat{\nu}$ represent the unit outer normal to the boundary S of the body shown in Fig. 9.7. According to (9.76a),

$$\int_B \left[\vec{B}-\rho\,\frac{d\vec{v}}{dt}\right]dV + \int_S \vec{T}(\hat{\nu})\,dS = 0. \tag{9.78a}$$

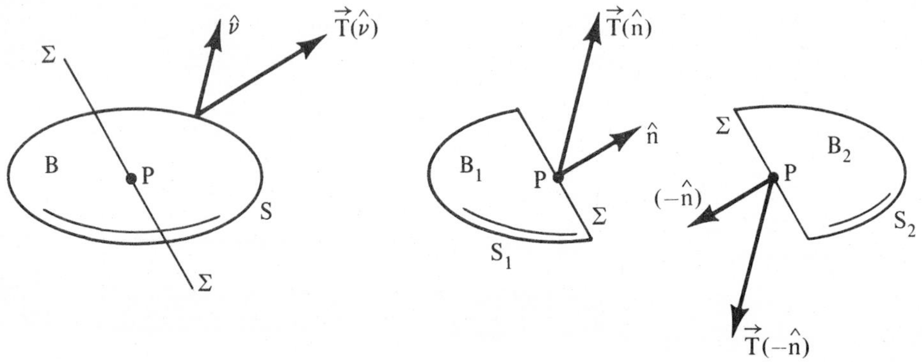

Figure 9.7

Now imagine the body to be divided into two parts by passing a plane Σ through the point P. Since each of these parts may also be considered as a body, (9.76a) must also apply to each part individually. Thus,

$$\int_{B_1}\left[\vec{B}-\rho\,\frac{d\vec{v}}{dt}\right]dV + \int_{S_1}\vec{T}(\hat{\nu})dS + \int_{\Sigma}\vec{T}(\hat{n})dS = 0 \qquad (9.78b)$$

$$\int_{B_2}\left[\vec{B}-\rho\,\frac{d\vec{v}}{dt}\right]dV + \int_{S_2}\vec{T}(\hat{\nu})dS + \int_{\Sigma}\vec{T}(-\hat{n})dS = 0. \qquad (9.78c)$$

If we add (9.78b) to (9.78c) and then subtract (9.78a) from the result, we obtain, after noting that $B=B_1+B_2$ and $S=S_1+S_2$,

$$\int_{\Sigma}\left[\vec{T}(\hat{n}) + \vec{T}(-\hat{n})\right]dS = 0.$$

Since this result must be true for all bodies and for all planes passing through P, we conclude that

$$\vec{T}(\hat{n}) + \vec{T}(-\hat{n}) = 0$$

or

$$\vec{T}(-\hat{n}) = -\vec{T}(\hat{n}). \qquad (9.79)$$

Thus $\vec{T}$ is an odd function of $\hat{n}$. This result is called Cauchy's lemma. It will be used to prove Cauchy's theorem on stress which relates the stress vector to the surface normal at any point. Before proceeding further we should point out that our proof of Cauchy's lemma assumes that the velocity varies continuously across the surface Σ. If this is not the case, then (9.79) does not hold true but in its place a jump condition is obtained relating the discontinuity in velocity to a discontinuity in stress.

Cauchy's theorem on stress may be stated as follows: The stress vector on any surface passing through the point P in B is a linear homogeneous function of the components of the positive normal to the surface at P. The coefficients in this linear function are the stress vectors which act on the planes passing through P parallel to the corresponding coordinate planes. Using the notation adopted in (9.77),

$$\vec{T}(\hat{n}) = n_i\vec{T}(\hat{\varepsilon}_i) = n_i\tau_{ij}\hat{\varepsilon}_j \qquad (9.80a)$$

or, in component form,

$$T_j(n_1,n_2,n_3) = n_i\tau_{ij}. \qquad (9.80b)$$

Since the direction of the unit vector $\hat{n}$ is arbitrary, we conclude after comparing (9.80) with (8.41b), the quotient law for second order tensors, that τ_{ij} are the components of a second order tensor which will be called the stress tensor. As a result of (9.80b), we see that the quadratic form

$$\tau(\hat{n},\hat{n}) = \tau_{ij}n_in_j \tag{9.81a}$$

is just the normal stress on the surface being considered and the corresponding bilinear form

$$\tau(\hat{n},\hat{s}) = \tau_{ij}n_is_j \tag{9.81b}$$

where $n_is_i=0$ is a shear stress on this surface. A proof of Cauchy's theorem may be constructed by applying (9.76a) to the tetrahedron shown in Fig. 9.8. In Fig. 9.8, h is the perpendicular distance from P

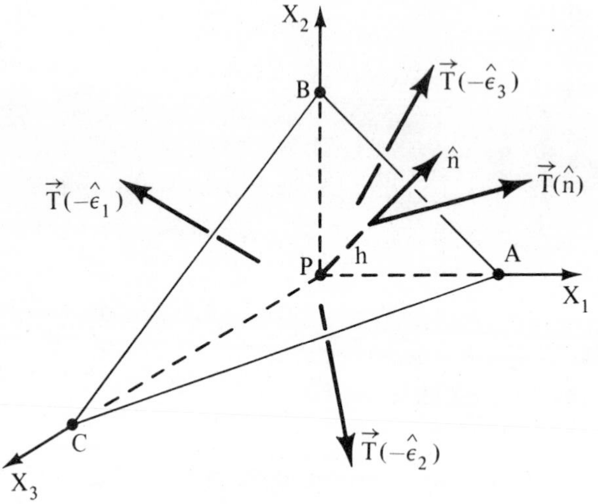

Figure 9.8

to the surface ABC whose area we denote by S. The volume of this tetrahedron is V = 1/3 hS. The areas of the remaining faces of the tetrahedron are related to S as follows:

$$S_1 = PBC = n_1S$$

$$S_2 = PAC = n_2S$$

$$S_3 = PAB = n_3S$$

where $\hat{n}$ is the unit outer normal to S. The average density of force acting on the body is defined as follows:

$$\langle\vec{F}\rangle \equiv \frac{1}{V}\int_B \left[\vec{B}-\rho\,\frac{d\vec{v}}{dt}\right]dV$$

$$\langle\vec{T}(\hat{n})\rangle \equiv \frac{1}{S}\int_S \vec{T}(\hat{n})dS$$

With the aid of these definitions, (9.76a) may be written for the tetrahedron in the form $V\langle\vec{F}\rangle+S\langle\vec{T}(\hat{n})\rangle+S_1\langle\vec{T}(-\hat{e}_1)\rangle+S_2\langle\vec{T}(-\hat{e}_2)\rangle+S_3\langle\vec{T}(-\hat{e}_3)\rangle=0$ Divide this equation by S to obtain

$$\langle\vec{T}(\hat{n})\rangle+n_1\langle\vec{T}(-\hat{e}_1)\rangle+n_2\langle\vec{T}(-\hat{e}_2)\rangle+n_3\langle\vec{T}(-\hat{e}_3)\rangle = -\frac{h}{3}\langle\vec{F}\rangle.$$

We will now take the limit of this equation as $h\to 0$, noting that $\lim_{h\to 0}\langle\vec{T}(\hat{n})\rangle=\vec{T}(\hat{n})$ at P and $\lim_{h\to 0}\langle\vec{F}\rangle=\vec{F}$ at P to obtain

$$\vec{T}(\hat{n}) + n_1\vec{T}(-\hat{e}_1) + n_2\vec{T}(-\hat{e}_2) + n_3\vec{T}(-\hat{e}_3) = 0$$

Applying Cauchy's lemma to the above equation yields

$$\vec{T}(\hat{n}) = n_1\vec{T}(\hat{e}_1) + n_2\vec{T}(\hat{e}_2) + n_3\vec{T}(\hat{e}_3)$$

or

$$\vec{T}(\hat{n}) = n_i\vec{T}(\hat{e}_i).$$

Thus, Cauchy's theorem is proved.

Cauchy's theorem may be used together with Euler's equations to derive the differential equations of motion for a continuum as follows: The indicial form of (9.76) as applied to the body shown in Fig. 9.9 is

$$\int_B \left[B_i-\rho\,\frac{dv_i}{dt}\right] dV + \int_S T_i dS = 0$$

$$\int_B e_{ijk}x_j \left[B_k-\rho\,\frac{dv_k}{dt}\right]dV + \int_S e_{ijk}x_j T_k dS = 0.$$

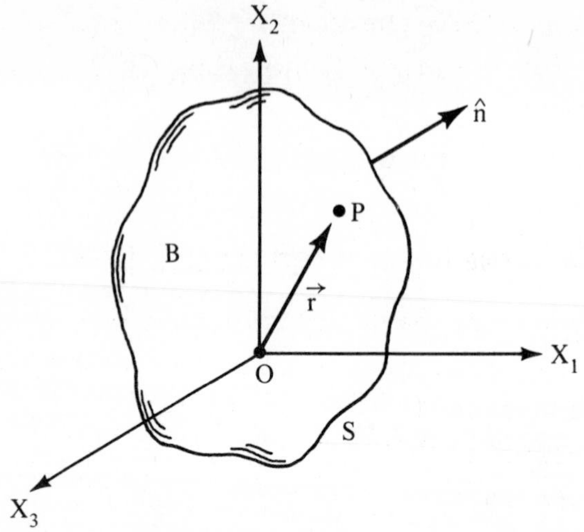

Figure 9.9

With the aid of Cauchy's theorem (9.80) and the divergence theorem (8.52) we may transform the surface integrals in the above equation into volume integrals as follows:

$$\int_S T_i dS = \int_S \tau_{ji} n_j dS = \int_B \frac{\partial \tau_{ji}}{\partial x_j} dV$$

$$\int_S e_{ijk} x_j T_k dS = \int_S e_{ijk} x_j \tau_{\ell k} n_\ell dS =$$

$$\int_B \frac{\partial}{\partial x_\ell} (e_{ijk} x_j \tau_{\ell k}) dV = \int_B e_{ijk}\left(\tau_{jk} + x_j \frac{\partial \tau_{\ell k}}{\partial x_\ell}\right) dV$$

Substituting these results into Euler's equations yields

$$\int_B \left[B_i + \frac{\partial \tau_{ji}}{\partial x_j} - \rho \frac{dv_i}{dt}\right] dV = 0$$

and

$$\int_B e_{ijk} \left\{x_j \left[B_k + \frac{\partial \tau_{\ell k}}{\partial x_\ell} - \rho \frac{dv_k}{dt}\right] + \tau_{jk}\right\} dV = 0$$

Since these results apply to all bodies, i.e., the region B is arbitrary, we conclude that

$$\frac{\partial \tau_{ji}}{\partial x_j} + B_i = \rho \frac{dv_i}{dt} \tag{9.82a}$$

and

$$e_{ijk} \tau_{jk} = 0. \tag{9.82b}$$

Equations (9.82a) are called the stress equations of motion or Cauchy's equations of motion for continua. Equations (9.82b) require that the stress tensor be symmetric, i.e.,

$$\tau_{ij} = \tau_{ji}. \tag{9.82c}$$

9.4 THEOREMS ON WORK AND ENERGY IN ELASTIC SOLIDS

Our discussions so far have dealt with those characteristics which are common to all continuous media. We would now like to limit the scope of our discussion to a very special type of continuum, the elastic solid. The elastic solid is a mathematical abstraction whose usefulness lies in the fact that it enables one to accurately predict the mechanical behavior of a large class of real bodies under commonly occurring circumstances. By definition, the elastic solid is a non-dissipative or conservative system which possesses a natural state. In the absence of any external forces, the natural state is a stable equilibrium state of the body and is free of stress and strain. In the developments which follow, we shall select the Lagrangian reference configuration in such a manner that it coincides with the natural state of the body.

If we consider purely mechanical processes, then as a result of the conservative nature of elastic solids, no thermodynamical arguments are required. The balance of work and energy may be expressed in the form

$$\frac{dE^*}{dt} = P(t) \tag{9.83}$$

where E^* is the total energy of the body and $P(t)$ is the power of the external forces, i.e., the rate at which the external forces do work on the body. Let B represent the region in space described by the current configuration of the body and imagine B to be divided into several separate parts $B_1, B_2, \ldots, B_n$. We now define a new measure for the region B called its energy content $E^*(B)$. Since the energy of a body is just the sum of the energies of its separate parts, we conclude that

$$E^*(B) = \sum_{i=1}^{n} E^*(B_i) = \sum_{i=1}^{n} e(B_i)M(B_i)$$

where $E^*(B_i)$ is the energy content of the region B_i and

$$e(B_i) \equiv \frac{E^*(B_i)}{M(B_i)}$$

is called the average specific energy of B_i. According to (9.64), $M(B_i)=\rho(B_i)V(B_i)$, thus

$$E^*(B) = \sum_{i=1}^{n} e(B_i)\rho(B_i)V(B_i). \tag{9.84}$$

If we let $n\to\infty$ in such a manner that the largest of the $V(B_i)$ tends to zero, then (9.84) becomes

$$E^*(B) = \int_B e\rho dV \tag{9.85}$$

where the scalar point function $e(P)$ is called the specific energy of the body. The total energy E^* consists of a kinetic energy T and a potential energy U, i.e.,

$$E^* = T+U. \tag{9.86}$$

The kinetic energy per unit mass of the body is, by definition $\frac{1}{2}|\vec{v}|^2$ where $\vec{v}$ represents the velocity field in the body. The potential energy per unit volume of the Lagrangian reference configuration B_o will be called the strain energy density and will be characterized by the symbol W^*. Thus, the potential energy per unit mass is represented by $1/\rho_o\ W^*$, where ρ_o is the density of the body in the natural state, i.e., in the Lagrangian reference configuration. As a result of the above definitions, the specific energy is given by

$$e = \tfrac{1}{2}v_i v_i + \frac{1}{\rho_o} W^* \tag{9.87}$$

Substituting (9.87) into (9.85) and comparing the result with (9.86) reveals that

$$T = \tfrac{1}{2}\int_B \rho v_i v_i dV \tag{9.88}$$

and

$$U = \int_B \frac{\rho}{\rho_o} W^* dV = \int_{B_o} W^* dV_o \tag{9.89}$$

where B_o represents the Lagrangian reference configuration of the body. The potential energy U is also called the strain energy of the body.

With the aid of (9.83) we may derive an energy continuity equation as follows. The power of the external forces may be expressed as

$$P(t) = \int_B B_i v_i dV + \int_S T_i v_i dS. \tag{9.90}$$

The first integral in (9.90) represents the power of the body forces and the second integral represents the power of the surface tractions acting on the boundary S of the region B. Differentiating (9.85) with respect to time yields, with the aid of (9.72a),

$$\frac{dE^*}{dt} = \int_B \frac{de}{dt} \rho dV. \tag{9.91}$$

If we substitute (9.91) and (9.90) into (9.83) we obtain, with the aid of Cauchy's theorem (9.80b),

$$\int_B \frac{de}{dt} \rho dV = \int_B B_i v_i dV - \int_S s_j n_j dS \tag{9.92}$$

where

$$s_j = -\tau_{ji} v_i \tag{9.93}$$

is called the energy flux vector. Equation (9.92) may be stated as follows: The rate at which the total energy of the body increases equals the rate at which energy is produced internally plus the rate at which energy flows into the body through its boundary. Thus, the quantity $s_j n_j$ represents the efflux of energy per unit area per unit time across the boundary S. We see that the flow of energy through the body is characterized by the energy flux vector s_j. By applying the divergence theorem to (9.92) we obtain the energy continuity equation

$$\rho \frac{de}{dt} = B_i v_i - \frac{\partial s_j}{\partial x_j} . \tag{9.94}$$

The stress equations of motion (9.82a) may also be interpreted as momentum continuity equations. If we define a momentum flux tensor J^*_{ij}

$$J^*_{ij} = -\tau_{ij} \tag{9.95}$$

then (9.82a) becomes

$$\rho \frac{dv_i}{dt} = B_i - \frac{\partial J^*_{ji}}{\partial x_j} . \tag{9.96}$$

Integrating (9.96) over the region B yields, with the aid of the divergence theorem (8.52) and (9.72a),

$$\frac{d}{dt}\int_B \rho v_i dV = \int_B B_i dV - \int_S J^*_{ji} n_j dS. \qquad (9.97)$$

Stated in words: The rate of increase of momentum in the body equals the rate at which momentum is produced internally plus the rate at which momentum flows into the body through its boundary. Thus, the quantity $J^*_{ji}n_j$ represents the efflux of momentum per unit area per unit time across the boundary S. Hence, the flow of momentum through the body is characterized by the momentum flux tensor $J^*_{ij}=-\tau_{ij}$, i.e., $-\tau_{ij}$ is the momentum per unit area per unit time flowing across a surface whose positive normal is $\hat{e}_i$.

As a result of (9.93) and (9.95), we see that the energy flux vector is related to the momentum flux tensor as follows:

$$s_j = J^*_{ji} v_i \qquad (9.93b)$$

The term flux, as used in the preceding discussion, is not to be construed to imply a flow of matter. The quantities which are "flowing" are mathematical abstractions, i.e., energy and momentum.

With the aid of our definitions of elastic solid and strain energy density, the constitutive relations for the elastic solid may be obtained as follows. By applying Cauchy's theorem and then the divergence theorem to (9.90) we obtain the following expression for the power of the external forces

$$P(t) = \int_B \left(\frac{\partial \tau_{ji}}{\partial x_j} + B_i\right) v_i dV + \int_B \tau_{ji} \frac{\partial v_i}{\partial x_j} dV.$$

The symmetry of the stress tensor together with the definition (9.62b) of the rate of deformation tensor indicate that

$$\tau_{ji} \frac{\partial v_i}{\partial x_j} = \tau_{ji} \dot{e}_{ij}.$$

This result together with the stress equations of motion (9.82a) permit us to express the power P(t) as follows:

$$P(t) = \int_B \rho \frac{dv_i}{dt} v_i dV + \int_B \tau_{ji} \dot{e}_{ij} dV$$

The time rate of change of kinetic energy is, according to (9.88) and (9.72a),

$$\frac{dT}{dt} = \int_B \rho \frac{dv_i}{dt} v_i dV \quad , \tag{9.98}$$

thus

$$P(t) = \frac{dT}{dt} + \int_B \tau_{ji}\dot{e}_{ij} dV \quad . \tag{9.99}$$

Substituting (9.86) into (9.83) yields

$$P(t) = \frac{dT}{dt} + \frac{dU}{dt} \; . \tag{9.100}$$

Comparing (9.99) with (9.100) reveals, with the aid of (9.89), that

$$\frac{dU}{dt} = \int_{B_o} \frac{dW^*}{dt} dV_o = \int_B \tau_{ji}\dot{e}_{ij} dV \tag{9.101}$$

In order to obtain the constitutive relations from (9.101) we must transform the second integral into Lagrangian coordinates. According to (9.62a),

$$\frac{dE_{k\ell}}{dt} = \frac{\partial x_\alpha}{\partial a_k}\frac{\partial x_\beta}{\partial a_\ell}\dot{e}_{\alpha\beta}$$

Forming the inner products of the above expression first with $\partial a_\ell/\partial x_j$ and then with $\partial a_k/\partial x_i$, we readily obtain

$$\dot{e}_{ij} = \frac{\partial a_k}{\partial x_i}\frac{\partial a_\ell}{\partial x_j}\frac{dE_{k\ell}}{dt}$$

Substituting this result into (9.101) yields, with the aid of (9.60),

$$\int_{B_o}\left[\frac{dW^*}{dt} - J\frac{\partial a_k}{\partial x_i}\frac{\partial a_\ell}{\partial x_j}\tau_{ji}\frac{dE_{k\ell}}{dt}\right] dV_o = 0.$$

Since this result is true for all elastic solids, the region B_o is arbitrary and, we conclude that

$$\frac{dW^*}{dt} = J\frac{\partial a_k}{\partial x_i}\frac{\partial a_\ell}{\partial x_j}\tau_{ji}\frac{dE_{k\ell}}{dt} \; . \tag{9.102}$$

In view of (9.102) we conclude that

$$W^* = W^*(a_1, a_2, a_3, E_{11}, E_{12}, E_{13}, \ldots, E_{33}). \tag{9.103}$$

Thus,

$$\frac{dW^*}{dt} = \frac{\partial W^*}{\partial E_{k\ell}} \frac{dE_{k\ell}}{dt}$$

and (9.102) may also be written as

$$\left(\frac{\partial W^*}{\partial E_{k\ell}} - J \frac{\partial a_k}{\partial x_i} \frac{\partial a_\ell}{\partial x_j} \tau_{ji} \right) \frac{dE_{k\ell}}{dt} = 0$$

Since this result is true for all processes, i.e., $dE_{k\ell}/dt$ is arbitrary, we conclude that

$$\frac{\partial W^*}{\partial E_{k\ell}} = J \frac{\partial a_k}{\partial x_i} \frac{\partial a_\ell}{\partial x_j} \tau_{ji}. \tag{9.104}$$

The expression on the right-hand side of (9.104), i.e., $J \frac{\partial a_k}{\partial x_i} \frac{\partial a_\ell}{\partial x_j} \tau_{ji}$, is called the Piola-Kirchhoff stress tensor. If we replace the summation indices i and j in (9.104) by α and β, respectively and then form the inner products of the result with $\partial x_j/\partial a_\ell$ and $\partial x_i/\partial a_k$ we will obtain

$$\tau_{ji} = \frac{1}{J} \frac{\partial x_i}{\partial a_k} \frac{\partial x_j}{\partial a_\ell} \frac{\partial W^*}{\partial E_{k\ell}} \tag{9.105}$$

These are the constitutive relations for the elastic solid. It is apparent that the elasticity of the material is characterized by the strain energy density function.

In order to obtain (9.105) we made use of the fact that an elastic solid is a conservative system. The existence of a natural state for the body will be insured by requiring the strain energy density function of the Lagrangian reference configuration to be a minimum with respect to all neighboring states of strain. The necessary condition for the existence of such a minimum is

$$\left(\frac{\partial W^*}{\partial E_{k\ell}} \right)_o = 0 \tag{9.106}$$

where the subscript $(\ )_o$ refers to the natural state. Comparing (9.105) with (9.106) reveals that

$$(\tau_{ji})_o = 0, \tag{9.107}$$

i.e., the stresses all vanish in the natural state. Since the Lagrangian reference configuration coincides with the natural state, the strains also vanish in the natural state. If the body is placed in its natural state with zero kinetic energy, then in the absence of any external forces according to (9.107) and (9.82a) it will remain there indefinitely. Thus, the natural state is an equilibrium state of the body. The stability of this equilibrium state for sufficiently small deformations results from the fact that the strain energy density function is a minimum in the natural state. The sufficient conditions for the existence of this minimum are given by relations similar to (3.3) and will be examined in greater detail in the next chapter for linear elastic solids.

Two solids of particular interest to us are the homogeneous and isotropic elastic solids. If a body is homogeneous in its natural state, then

$$W^* = W^*(E_{11},E_{12},E_{13},\ldots,E_{33}). \tag{9.108}$$

If the mechanical properties are the same in all directions at a point in the natural state of a body, then we say the body is isotropic. For an isotropic elastic solid,

$$W^* = W^*(a_1,a_2,a_3,K_1,K_2,K_3) \tag{9.109}$$

where

$$K_1 = E_{ii},\; K_2 = \frac{1}{2!} E_{ij}E_{ji}$$

and

$$K_3 = \frac{1}{3!} E_{ij}E_{jk}E_{ki}$$

are the three invariants of the strain tensor. If the body is both homogeneous and isotropic, then

$$W^* = W^*(K_1,K_2,K_3). \tag{9.110}$$

The stress equations of motion for an arbitrary continuum were derived earlier from Euler's equations. For an elastic solid, these equations of motion may also be obtained through the application of Hamilton's principle. The reader will recall that in Chapter 2, Hamilton's principle was established for discrete systems. We will now demonstrate that the statement and form of the principle remains unchanged when applied to elastic solids if the Lagrangian function L and the virtual work δW are properly defined.

Consider the motion of an elastic solid during some arbitrarily chosen time interval (t_1,t_2). The body force and surface traction influencing the motion are B_i and T_i respectively. The Eulerian description of the motion is provided by the displacement field $u_i(x,t)$ and the velocity field $v_i(x,t)$ which, together with the associated stress and strain fields, satisfy the stress equations of motion (9.82a), the strain displacement relation (9.43) and the constitutive relations (9.105). At any instant of time, the kinetic and potential energies associated with this motion are given by (9.88) and (9.89) respectively.

Let us assume at this point that during the interval (t_1,t_2) the body has a motion $u_i'(x,t)$ and $v_i'(x,t)$ which differs slightly from the actual motion described above but which is consistent with the constraints on the body, i.e., wherever the displacement is specified in an a priori manner, $u_i'(x,t)=u_i(x,t)$. Furthermore, this assumed displacement field will be chosen so as to coincide with the actual displacement field at the instants t_1 and t_2. If we define

$$\delta u_i(x,t) \equiv u_i'(x,t)-u_i(x,t) \tag{9.111a}$$

and

$$\delta v_i(x,t) \equiv v_i'(x,t)-v_i(x,t) \tag{9.111b}$$

then

$$\delta u_i(x,t_1) = \delta u_i(x,t_2) = 0 \tag{9.112}$$

and

$$\delta u_i(x,t) = 0 \text{ for } t_1<t<t_2$$

wherever displacement boundary conditions are specified. By a slight difference in motion we will mean that

(a) δu_i is a continuous function of x_1,x_2,x_3 and t,

$$\text{(b) } |\delta u_i|<<|u_i|, \quad \text{(c) } |\delta v_i|<<|v_i|, \text{ and (d) } \left|\frac{\partial \delta u_i}{\partial x_j}\right|<<1. \tag{9.113}$$

The last condition, i.e., (9.113d) states that at any given instant of time, the configuration of the body associated with the assumed motion may be obtained by a small deformation of the configuration associated with the actual motion.

The kinetic energy associated with the assumed motion is given by

$$T' = \tfrac{1}{2} \int_B \rho' v_i' v_i' dV'.$$

As a result of the conservation of mass $\rho' dV' = \rho dV$, and therefore

$$T' = \tfrac{1}{2} \int_B \rho v_i' v_i' dV.$$

In view of (9.111b) and (9.113c),

$$\begin{aligned} v_i' v_i' &= (v_i + \delta v_i)(v_i + \delta v_i) \\ &= v_i v_i + 2(v_i + \tfrac{1}{2}\delta v_i)\delta v_i \\ &\simeq v_i v_i + 2 v_i \delta v_i \end{aligned}$$

Thus, the difference in kinetic energy between the actual and assumed motions is approximately given by

$$\delta T = T' - T = \int_B \rho v_i \delta v_i dV. \qquad (9.114)$$

We also note that

$$\frac{d}{dt}(\delta u_i) = \frac{d}{dt}(u_i' - u_i) = v_i' - v_i = \delta v_i.$$

Thus (9.114) may be written as follows:

$$\begin{aligned} \delta T &= \int_B \rho v_i \frac{d}{dt}(\delta u_i) dV \\ &= \int_B \rho \frac{d}{dt}(v_i \delta u_i) dV - \int_B \rho \frac{dv_i}{dt} \delta u_i dV. \end{aligned}$$

In view of (9.72a) we conclude that

$$\delta T = \frac{d}{dt} \int_B \rho v_i \delta u_i dV - \int_B \rho \frac{dv_i}{dt} \delta u_i dV.$$

If the above equation is integrated over the interval (t_1, t_2) we obtain

$$\int_{t_1}^{t_2} \delta T dt = \left[\int_B \rho v_i \delta u_i dV \right]_{t_1}^{t_2} - \int_{t_1}^{t_2} \int_B \rho \frac{dv_i}{dt} \delta u_i dV dt$$

and in view of (9.112) the first expression will vanish, therefore

$$\int_{t_1}^{t_2} \delta T dt = -\int_{t_1}^{t_2} \int_B \rho \frac{dv_i}{dt} \delta u_i dV dt. \qquad (9.115)$$

The strain energy associated with the assumed motion is given by

$$U' = \int_{B_o} W^{*\prime} dV_o$$

where

$$W^{*\prime} = W^*(a_1, a_2, a_3, E'_{11}, E'_{12}, E'_{13}, \ldots, E'_{33}).$$

If we expand the strain energy density function $W^{*\prime}$ in a Taylor series about the actual strain energy density W^*, we obtain

$$W^{*\prime} = W^* + \frac{\partial W^*}{\partial E_{ij}} \delta E_{ij} + O(\delta E_{ij} \delta E_{k\ell})$$

where

$$\delta E_{ij} = E'_{ij} - E_{ij}.$$

Thus, the difference in strain energy between the actual and the assumed motion is approximately given by

$$\delta U = U' - U = \int_{B_o} \frac{\partial W^*}{\partial E_{ij}} \delta E_{ij} dV_o. \qquad (9.116)$$

In terms of the current configuration of the body,

$$\delta U = \int_B \frac{1}{J} \frac{\partial W^*}{\partial E_{ij}} \delta E_{ij} dV.$$

In view of (9.104), $\frac{1}{J} \frac{\partial W^*}{\partial E_{ij}} = \tau_{\ell k} \frac{\partial a_i}{\partial x_k} \frac{\partial a_j}{\partial x_\ell}$ and therefore

$$\delta U = \int_B \tau_{\ell k} \frac{\partial a_i}{\partial x_k} \frac{\partial a_j}{\partial x_\ell} \delta E_{ij} dV. \qquad (9.117)$$

As a result of (9.20b),

$$2\delta E_{ij} = 2(E'_{ij}-E_{ij}) = \frac{\partial x'_\alpha}{\partial a_i}\frac{\partial x'_\alpha}{\partial a_j} - \frac{\partial x_\alpha}{\partial a_i}\frac{\partial x_\alpha}{\partial a_j}$$

where,

$$x'_\alpha = a_\alpha + u'_\alpha = a_\alpha + u_\alpha + \delta u_\alpha = x_\alpha + \delta u_\alpha .$$

Thus,

$$2\delta E_{ij} = \frac{\partial x_\alpha}{\partial a_i}\frac{\partial \delta u_\alpha}{\partial a_j} + \frac{\partial x_\alpha}{\partial a_j}\frac{\partial \delta u_\alpha}{\partial a_i} + \frac{\partial \delta u_\alpha}{\partial a_i}\frac{\partial \delta u_\alpha}{\partial a_j}$$

The above result, together with the chain rule for partial differentiation yields

$$\frac{\partial a_i}{\partial x_k}\frac{\partial a_j}{\partial x_\ell}\delta E_{ij} = \tfrac{1}{2}\left[\frac{\partial \delta u_k}{\partial x_\ell} + \frac{\partial \delta u_\ell}{\partial x_k} + \frac{\partial \delta u_\alpha}{\partial x_k}\frac{\partial \delta u_\alpha}{\partial x_\ell}\right]$$

where

$$\frac{\partial \delta u_k}{\partial x_\ell} = \frac{\partial}{\partial x_\ell}(u'_k - u_k) = \frac{\partial u'_k}{\partial x_\ell} - \frac{\partial u_k}{\partial x_\ell} = \delta\left(\frac{\partial u_k}{\partial x_\ell}\right)$$

In view of (9.113d), we conclude that

$$\left|\frac{\partial \delta u_\alpha}{\partial x_k}\frac{\partial \delta u_\alpha}{\partial x_\ell}\right| << \left|\frac{\partial \delta u_k}{\partial x_\ell}\right|$$

and this nonlinear term may be neglected to obtain

$$\tau_{\ell k}\frac{\partial a_i}{\partial x_k}\frac{\partial a_j}{\partial x_\ell}\delta E_{ij} = \tfrac{1}{2}\left[\frac{\partial \delta u_k}{\partial x_\ell} + \frac{\partial \delta u_\ell}{\partial x_k}\right]\tau_{\ell k} = \frac{\partial \delta u_k}{\partial x_\ell}\tau_{\ell k}.$$

Substituting the above result into (9.117) yields the following result:

$$\delta U = \int_B \frac{\partial \delta u_k}{\partial x_\ell}\tau_{\ell k} dV \tag{9.118a}$$

or

$$\delta U = \int_B \frac{\partial}{\partial x_\ell}(\delta u_k \tau_{\ell k})dV - \int_B \frac{\partial \tau_{\ell k}}{\partial x_\ell}\delta u_k dV$$

With the aid of the divergence theorem, Cauchy's theorem and the stress equations of motion, the above result may be written as

$$\delta U = \int_S T_k \delta u_k dS + \int_B B_k \delta u_k dV - \int_B \rho \frac{dv_k}{dt}\delta u_k dV \tag{9.118b}$$

or

$$\delta U = \delta W - \int_B \rho \frac{dv_k}{dt} \delta u_k dV \tag{9.118c}$$

where

$$\delta W = \int_S T_k \delta u_k dS + \int_B B_k \delta u_k dV \tag{9.119}$$

is called the virtual work of the applied forces. If we define the Lagrangian function of the body as

$$L = T - U \tag{9.120}$$

then $\delta L = \delta T - \delta U$ and as a result of (9.115) and (9.118) we conclude that

$$\int_{t_1}^{t_2} \delta L dt = - \int_{t_1}^{t_2} \delta W dt. \tag{9.121}$$

This is one form of Hamilton's principle for elastic solids. If we let

$$I = \int_{t_1}^{t_2} L dt \tag{9.122a}$$

then

$$\delta I = \int_{t_1}^{t_2} L' dt - \int_{t_1}^{t_2} L dt = \int_{t_1}^{t_2} \delta L dt$$

and therefore

$$\delta I = - \int_{t_1}^{t_2} \delta W dt. \tag{9.122b}$$

The value of I associated with the actual motion of the body is given by (9.122a). Equation (9.122b) states that the variation of I produced by a sufficiently small variation in the displacement field consistent with the constraints on the body and vanishing at the instants t_1 and t_2 is equal to the negative of the integral of the virtual work performed by the external forces in passing from the actual to the varied configuration of the body.

Equation (9.121) with δW given by (9.119) is valid for arbitrary body forces and surface tractions. However, if, as is often the case, the external forces are independent of $\vec{u}$ and $\vec{v}$, then

$$\int_S T_k \delta u_k dS = \delta \int_S T_k u_k dS$$

and

$$\int_B B_k \delta u_k dS = \delta \int_B B_k u_k dV.$$

In this case we may define a potential energy of the applied loads as follows:

$$W = \int_S T_k u_k dS + \int_B B_k u_k dV. \tag{9.123}$$

If we define the action integral A by

$$A = \int_{t_1}^{t_2} (T-U+W)dt \tag{9.124}$$

then Hamilton's principle becomes

$$\delta A = 0, \tag{9.125}$$

i.e., the value of the action integral corresponding to the actual motion is stationary with respect to admissible variations of the displacement field. The admissible variations are those which satisfy (9.112), i.e., they vanish throughout the body at the instants t_1 and t_2 and they vanish over the entire interval (t_1,t_2), wherever displacement boundary conditions are specified.

In the derivation of Hamilton's principle we made use of the fact that the material was elastic in order to identify the symbol δU as the variation of the strain energy. However, if we adopt (9.118a) as the definition of δU and attach no further significance to the term then the results which follow are independent of any constitutive relation and therefore apply to all continua. If δU is not identified as a variation of strain energy then (9.121) can no longer be interpreted as a variational principle for work and energy. Instead, it becomes a variational equation of motion for continuous bodies. If we let ε_{ij} represent the symmetric part of the gradient of u_i, then

$$\delta\varepsilon_{ij} \equiv \tfrac{1}{2}\left[\frac{\partial\delta u_i}{\partial x_j} + \frac{\partial\delta u_j}{\partial x_i}\right], \tag{9.126}$$

and (9.118a) may be written as

$$\delta U = \int_B \tau_{ji}\delta\varepsilon_{ij}dV. \tag{9.127}$$

If we substitute (9.115), (9.119) and (9.127) into (9.121) we obtain, after making use of the fact that the interval (t_1,t_2) is arbitrary,

$$\int_B \tau_{ji}\delta\varepsilon_{ij}dV = \int_B\left[B_i-\rho\,\frac{dv_i}{dt}\right]\delta u_i dV + \int_S T_i\delta u_i dS. \tag{9.128}$$

This is called the variational equation of motion and it is valid for all continuous bodies. Since (9.128) is a scalar equation, it is valid in all coordinate systems. This fact can be put to immediate use in the derivation of the stress equations of motion and associated boundary conditions in various curvilinear coordinate systems.

To illustrate the applications of the variational equation of motion we will examine two examples. As our first example, we will derive the stress equations of motion and associated boundary conditions in Cartesian coordinates. From (9.127),

$$\begin{aligned}\delta U &= \int_B \tau_{ji}\delta\varepsilon_{ij}dV = \int_B \tau_{ji}\,\frac{\partial\delta u_i}{\partial x_j}\,dV\\ &= \int_B \frac{\partial}{\partial x_j}\,(\tau_{ji}\delta u_i)dV - \int_B \frac{\partial\tau_{ji}}{\partial x_j}\,\delta u_i dV\\ &= \int_S n_j\tau_{ji}\delta u_i dS - \int_B \frac{\partial\tau_{ji}}{\partial x_j}\,\delta u_i dV.\end{aligned}$$

Substituting this result into (9.128) yields, after some rearrangement of the terms,

$$\int_B\left[\frac{\partial\tau_{ji}}{\partial x_j} + B_i-\rho\,\frac{dv_i}{dt}\right]\delta u_i dV + \int_S [T_i-\tau_{ji}n_j]\,\delta u_i dS = 0.$$

Since the region B and its boundary S are arbitrary, the integrand of each of the above integrals must vanish. The variations δu_i are arbitrary in B, so that

$$\frac{\partial \tau_{ji}}{\partial x_j} + B_i = \rho \frac{dv_i}{dt} \qquad \text{in B.} \tag{9.129}$$

The surface S consists of the union of the sets S_u and S_T where S_u is the set of points on which the displacements are specified and S_T is the set of points on which the surface tractions are specified. Since the admissible variations of u_i vanish on S_u and are arbitrary on S_T we conclude that

$$\tau_{ji}n_j = T_i \qquad \text{on } S_T$$

and (9.130)

$$\delta u_i = 0 \qquad \text{on } S_u.$$

We note that in addition to yielding the stress equations of motion, the variational equation also yields the proper natural boundary conditions for those equations. This will be of great utility in subsequent chapters where we derive the equations of motion for some special approximate theories. In those cases the proper natural boundary conditions are not always as obvious as they are in this example. For our second example, we will derive the equations of motion and associated boundary conditions in cylindrical coordinates. The cylindrical coordinates r, θ and z are defined by (8.101) and illustrated in Fig. 8.7. To obtain the desired results, we will apply (9.128) to the region defined by $R_1<r<R_2$, $0<\theta<\theta_1$ and $0<z<z_1$. The cylindrical element occupying this region is illustrated in Fig. 9.10.

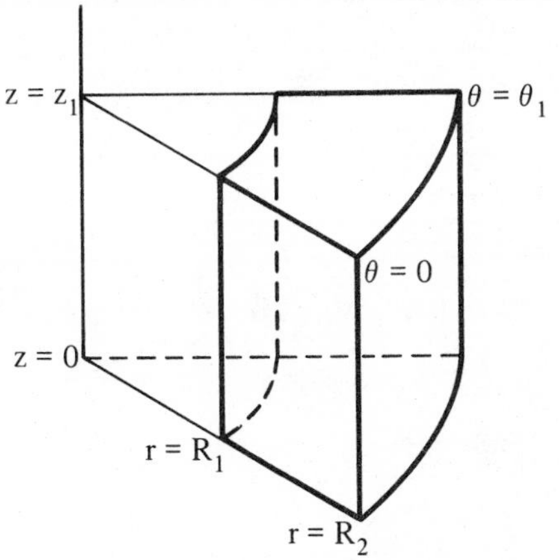

Fig. 9.10

The component form of the displacement vector in cylindrical coordinates is

$$\vec{u} = u_r\hat{\varepsilon}_r + u_\theta\hat{\varepsilon}_\theta + u_z\hat{\varepsilon}_z \tag{9.131}$$

where $\hat{\varepsilon}_r$, $\hat{\varepsilon}_\theta$ and $\hat{\varepsilon}_z$ are given by (8.104). The position vector to a generic point in the field is given by

$$\vec{r} = r\hat{\varepsilon}_r+z\hat{\varepsilon}_z. \tag{9.132}$$

The velocity vector is defined as

$$\vec{v} \equiv \frac{d\vec{r}}{dt} = \frac{dr}{dt}\,\hat{\varepsilon}_r + \frac{dz}{dt}\,\hat{\varepsilon}_z + r\,\frac{d\hat{\varepsilon}_r}{dt} + z\,\frac{d\hat{\varepsilon}_z}{dt}\;.$$

Differentiating (8.104) with respect to time yields

$$\frac{d\hat{\varepsilon}_r}{dt} = \frac{d\theta}{dt}\,\hat{\varepsilon}_\theta,\; \frac{d\hat{\varepsilon}_\theta}{dt} = -\,\frac{d\theta}{dt}\,\hat{\varepsilon}_r,\; \frac{d\hat{\varepsilon}_z}{dt} = 0. \tag{9.133}$$

Thus,

$$\vec{v} = \frac{dr}{dt}\,\hat{\varepsilon}_r + r\frac{d\theta}{dt}\,\hat{\varepsilon}_\theta + \frac{dz}{dt}\,\hat{\varepsilon}_z$$

or in component form

$$v_r = \frac{dr}{dt}\;,\; v_\theta=r\,\frac{d\theta}{dt},\; v_z= \frac{dz}{dt}. \tag{9.134}$$

The acceleration vector is defined as

$$\vec{\alpha} \equiv \frac{d\vec{v}}{dt} = \frac{d}{dt}\,(v_r\hat{\varepsilon}_r+v_\theta\hat{\varepsilon}_\theta+v_z\hat{\varepsilon}_z).$$

After differentiating the above expression, we obtain, with the aid of (9.133),

$$\vec{\alpha} = \left(\frac{dv_r}{dt} - \frac{v_\theta^2}{r}\right)\hat{\varepsilon}_r + \left(\frac{dv_\theta}{dt} + \frac{v_r v_\theta}{r}\right)\hat{\varepsilon}_\theta + \frac{dv_z}{dt}\,\hat{\varepsilon}_z$$

or in component form,

$$\alpha_r = \frac{dv_r}{dt} - \frac{v_\theta^2}{r},\; \alpha_\theta = \frac{dv_\theta}{dt} + \frac{v_r v_\theta}{r},\; \alpha_z = \frac{dv_z}{dt}\;. \tag{9.135}$$

The expressions appearing in (9.128) may be transformed to cylindrical coordinates as follows:

$$T_i\delta u_i = \vec{T}\cdot\delta\vec{u} = T_r\delta u_r + T_\theta\delta u_\theta + T_z\delta u_z \tag{9.136a}$$

$$B_i\delta u_i = \vec{B}\cdot\delta\vec{u} = B_r\delta u_r + B_\theta\delta u_\theta + B_z\delta u_z \tag{9.136b}$$

$$\frac{dv_i}{dt}\,\delta u_i = \vec{\alpha}\cdot\delta\vec{u} = \alpha_r\delta u_r + \alpha_\theta\delta u_\theta + \alpha_z\delta u_z \tag{9.136c}$$

$$\tau_{ji}\delta\varepsilon_{ij} = \tau_{rr}\delta\varepsilon_{rr} + \tau_{\theta\theta}\delta\varepsilon_{\theta\theta} + \tau_{zz}\delta\varepsilon_{zz} + 2\tau_{r\theta}\delta\varepsilon_{r\theta} + 2\tau_{\theta z}\delta\varepsilon_{\theta z} + 2\tau_{zr}\delta\varepsilon_{zr}$$

According to (8.110),

$$\delta\varepsilon_{rr} = \frac{\partial\delta u_r}{\partial r}\,,\quad \delta\varepsilon_{\theta\theta} = \frac{1}{r}\frac{\partial\delta u_\theta}{\partial\theta} + \frac{\delta u_r}{r}\,,\quad \delta\varepsilon_{zz} = \frac{\partial\delta u_z}{\partial z}\,,$$

$$2\delta\varepsilon_{r\theta} = \frac{1}{r}\frac{\partial\delta u_r}{\partial\theta} - \frac{\delta u_\theta}{r} + \frac{\partial\delta u_\theta}{\partial r}\,,\quad 2\delta\varepsilon_{\theta z} = \frac{\partial\delta u_\theta}{\partial z} + \frac{1}{r}\frac{\partial\delta u_z}{\partial\theta}\,,$$

$$2\delta\varepsilon_{zr} = \frac{\partial\delta u_z}{\partial r} + \frac{\partial\delta u_r}{\partial z}\,.$$

In view of the above results,

$$\int_B \tau_{ji}\delta\varepsilon_{ij}dV = \int_0^{z_1}\int_0^{\theta_1}\int_{R_1}^{R_2}\Bigg[\tau_{rr}\frac{\partial\delta u_r}{\partial r} + \tau_{\theta\theta}\left(\frac{1}{r}\frac{\partial\delta u_\theta}{\partial\theta} + \frac{\delta u_r}{r}\right)$$

$$+\tau_{zz}\frac{\partial\delta u_z}{\partial z} + \tau_{r\theta}\left(\frac{1}{r}\frac{\partial\delta u_r}{\partial\theta} - \frac{\delta u_\theta}{r} + \frac{\partial\delta u_\theta}{\partial r}\right) + \tau_{\theta z}\left(\frac{\partial\delta u_\theta}{\partial z} + \frac{1}{r}\frac{\partial\delta u_z}{\partial\theta}\right)$$

$$+\tau_{zr}\left(\frac{\partial\delta u_z}{\partial r} + \frac{\partial\delta u_r}{\partial z}\right)\Bigg]\,r\,dr\,d\theta\,dz.$$

We will now integrate by parts each term of the preceding equation that involves a derivative of the variation of a displacement component, e.g.,

$$\int_{R_1}^{R_2} r\tau_{rr}\frac{\partial\delta u_r}{\partial r}\,dr = \left[r\tau_{rr}\delta u_r\right]_{R_1}^{R_2} - \int_{R_1}^{R_2}\frac{\partial}{\partial r}(r\tau_{rr})\delta u_r dr.$$

After performing these integrations and collecting terms with common factors we obtain

$$\delta U = -\int_0^{z_1}\int_0^{\theta_1}\int_{R_1}^{R_2} [F_r\delta u_r + F_\theta\delta u_\theta + F_z\delta u_z]rdrd\theta dz$$

$$+ \int_0^{z_1}\int_0^{\theta_1}[r(\tau_{rr}\delta u_r + \tau_{r\theta}\delta u_\theta + \tau_{zr}\delta u_z)]_{R_1}^{R_2} d\theta dz$$

$$+ \int_0^{z_1}\int_{R_1}^{R_2}[\tau_{r\theta}\delta u_r + \tau_{\theta\theta}\delta u_\theta + \tau_{\theta z}\delta u_z]_0^{\theta_1} drdz$$

$$+ \int_0^{\theta_1}\int_{R_1}^{R_2}[\tau_{zr}\delta u_r + \tau_{\theta z}\delta u_\theta + \tau_{zz}\delta u_z]_0^{z_1} rdrd\theta \qquad (9.137)$$

where

$$\left.\begin{aligned} F_r &= \frac{\partial\tau_{rr}}{\partial r} + \frac{1}{r}\frac{\partial\tau_{r\theta}}{\partial\theta} + \frac{\partial\tau_{zr}}{\partial z} + \frac{\tau_{rr}-\tau_{\theta\theta}}{r} \\ F_\theta &= \frac{\partial\tau_{r\theta}}{\partial r} + \frac{1}{r}\frac{\partial\tau_{\theta\theta}}{\partial\theta} + \frac{\partial\tau_{\theta z}}{\partial z} + \frac{2\tau_{r\theta}}{r} \\ F_z &= \frac{\partial\tau_{zr}}{\partial r} + \frac{1}{r}\frac{\partial\tau_{\theta z}}{\partial\theta} + \frac{\partial\tau_{zz}}{\partial z} + \frac{\tau_{zr}}{r} \end{aligned}\right\} \qquad (9.138)$$

If we let

$$\hat{n} = n_r\hat{e}_r + n_\theta\hat{e}_\theta + n_z\hat{e}_z \qquad (9.139)$$

be the unit outer normal to S, then (9.137) may be written as follows:

$$\delta U = -\int_B[F_r\delta u_r + F_\theta\delta u_\theta + F_z\delta u_z]dV$$

$$+ \int_S[(\tau_{rr}n_r + \tau_{r\theta}n_\theta + \tau_{zr}n_z)\delta u_r + (\tau_{r\theta}n_r + \tau_{\theta\theta}n_\theta + \tau_{\theta z}n_z)\delta u_\theta$$

$$+ (\tau_{zr}n_r + \tau_{\theta z}n_\theta + \tau_{zz}n_z)\delta u_z]dS \qquad (9.140)$$

Substituting (9.136) and (9.140) into (9.128) yields the following result:

$$0 = \int_B \Big\{[F_r+B_r-\rho\alpha_r]\delta u_r+[F_\theta+B_\theta-\rho\alpha_\theta]\delta u_\theta+[F_z+B_z-\rho\alpha_z]\delta u_z\Big\}dV$$

$$+ \int_S \Big\{[T_r-(\tau_{rr}n_r+\tau_{r\theta}n_\theta+\tau_{zr}n_z)]\,\delta u_r$$

$$+ [T_\theta-(\tau_{r\theta}n_r+\tau_{\theta\theta}n_\theta+\tau_{\theta z}n_z)]\,\delta u_\theta$$
$$+ [T_z-(\tau_{zr}n_r+\tau_{\theta z}n_\theta+\tau_{zz}n_z)]\,\delta u_z\Big\}dS \qquad (9.141)$$

Since the above result applies to all bodies, the region B and its boundary S may now be considered to be arbitrary, thus the integrand of each of the above integrals must vanish. The variations δu_r, δu_θ and δu_z are independent and arbitrary in B, thus

$$\left.\begin{aligned} F_r + B_r &= \rho\alpha_r \\ F_\theta + B_\theta &= \rho\alpha_\theta \\ F_z + B_z &= \rho\alpha_z \end{aligned}\right\} \text{ in B.} \qquad (9.142)$$

Since δu_r, δu_θ and δu_z vanish on S_u and are independent and arbitrary on S_T we conclude that

$$\left.\begin{aligned} \tau_{rr}n_r + \tau_{\theta r}n_\theta + \tau_{zr}n_z &= T_r \\ \tau_{r\theta}n_r + \tau_{\theta\theta}n_\theta + \tau_{z\theta}n_z &= T_\theta \\ \tau_{rz}n_r + \tau_{\theta z}n_\theta + \tau_{zz}n_z &= T_z \end{aligned}\right\} \text{ on } S_T \qquad (9.142a)$$

and

$$\delta u_r = \delta u_\theta = \delta u_z = 0 \quad \text{on } S_u. \qquad (9.142b)$$

We note that mixed boundary conditions are also possible. For example, on some part of the surface, say S_M, we could specify

$$\left.\begin{aligned} &\tau_{rr}n_r + \tau_{\theta r}n_\theta + \tau_{zr}n_z = T_r \\ &\delta u_\theta = \delta u_z = 0 \end{aligned}\right\} \quad \text{on } S_M$$

or any other combination which makes each term of the surface integral in (9.141) vanish.

EXERCISES

9.1 The body shown in the figure below translates as a rigid body along the X_1 axis with a constant speed v(in/sec) and simultaneously rotates about the edge OA in the counterclockwise direction with a constant angular velocity ω(rad/sec).

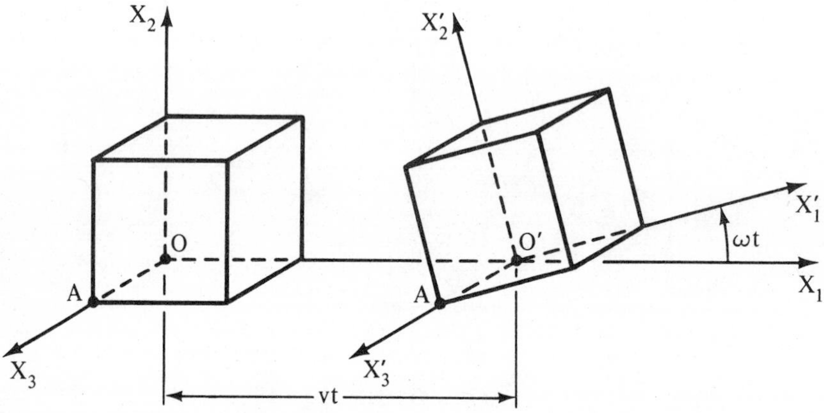

Figure: Exercise 9.1

(a) Find the direction cosines $a_{ij}(t)$.

(b) Find the displacement field in terms of both the Lagrangian and the Eulerian coordinates of the body.

(c) Find the velocity and acceleration fields in the body. Express your result in both Lagrangian and Eulerian coordinates.

9.2 Prove that $E_{ij}=e_{ij}=0$ for a rigid body motion as characterized by (9.8).

9.3 The sphere shown below is given a uniform dilation defined by

$$E(\hat{n}) = \frac{R-R_o}{R_o} = \lambda \text{ and } \Gamma(\hat{n}^{(1)}, \hat{n}^{(2)}) = 0$$

or

$$e(\hat{N}) = \frac{R-R_o}{R} \text{ and } \gamma(\hat{N}^{(1)}, \hat{N}^{(2)}) = 0.$$

Show that

$$E_{ij} = \lambda(1 + \frac{\lambda}{2})\delta_{ij}, \quad e_{ij} = \frac{E_{ij}}{(1+\lambda)^2} \text{ and } J = (1+\lambda)^3.$$

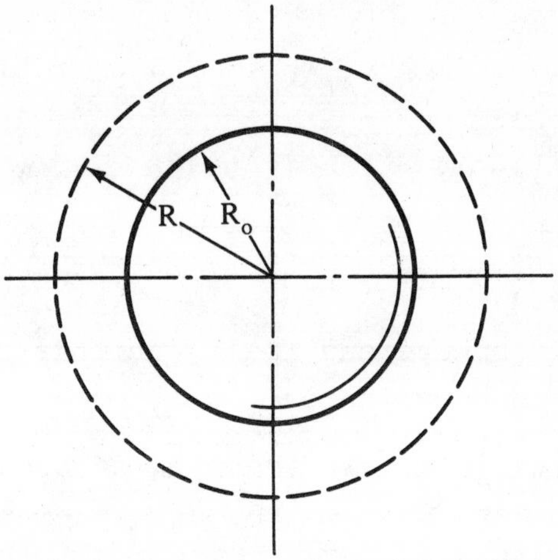

Figure: Exercise 9.3

9.4 Measurements of the tensile test specimen shown below indicate that

$$E(\hat{e}_1) = \frac{L-L_o}{L_o} = \lambda$$

$$E(\hat{e}_2) = E(\hat{e}_3) = \frac{D-D_o}{D_o} = -\nu\lambda$$

$$\Gamma(\hat{e}_1,\hat{e}_2) = \Gamma(\hat{e}_2,\hat{e}_3) = \Gamma(\hat{e}_3,\hat{e}_1) = 0.$$

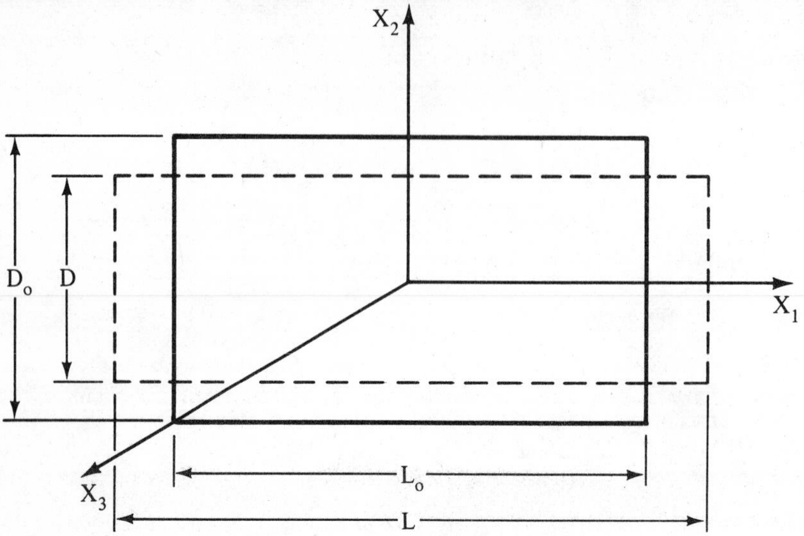

Figure: Exercise 9.4

(a) Show that

$$E_{11} = (1+\lambda)^2 e_{11} = \lambda(1+ \frac{\lambda}{2})$$
$$E_{22} = E_{33} = (1-\nu\lambda)^2 e_{22} = (1-\nu\lambda)^2 e_{33} = -\nu\lambda(1- \frac{\nu\lambda}{2})$$
$$E_{12} = E_{23} = E_{31} = e_{12} = e_{23} = e_{31} = 0$$
$$J = (1+\lambda)(1-\nu\lambda)^2.$$

(b) Express the displacement field for this problem in both Lagrangian and Eulerian coordinates.

9.5 A cube is given a simple shearing deformation as indicated below by the dashed lines. Express the displacement and strain fields which describe this deformation in both Lagrangian and Eulerian coordinates. Also compute the Lagrangian and Eulerian extensions and shear angles corresponding to the coordinate directions.

Figure: Exercise 9.5

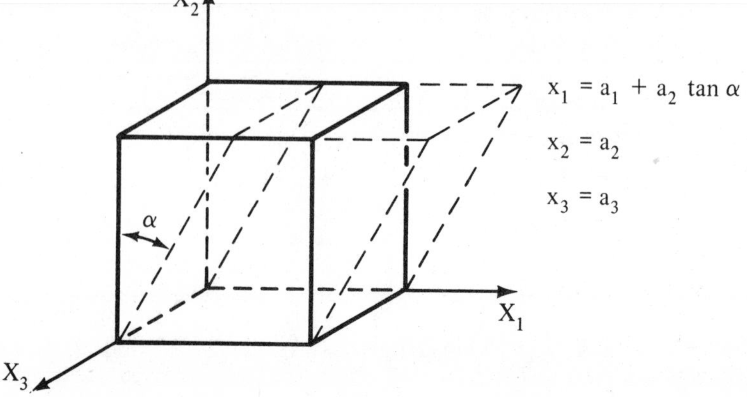

9.6 Given the displacement field $u_i(a_1,a_2,a_3,t)=c_i(t)+d_{ij}(t)a_j$ where $c_i(t)$ and $d_{ij}(t)$ are continuous functions of time, compute E_{ij}. This is called a homogeneous or affine deformation. Show that plane sections of the body remain plane, and straight lines in the body remain straight during an affine deformation. (Hint: A plane section of the current configuration of the body may be represented by $b_i x_i + b_4 = 0$.)

9.7 (a) Show that

$$\frac{d}{dt}\left(\frac{\partial u_i}{\partial x_j}\right) = \frac{\partial v_i}{\partial x_j} - \frac{\partial v_k}{\partial x_j}\frac{\partial u_i}{\partial x_k} .$$

(b) Show that

$$\frac{de_{ij}}{dt} = \dot{e}_{ij} - \left(e_{ki}\frac{\partial v_k}{\partial x_j} + e_{kj}\frac{\partial v_k}{\partial x_i}\right),$$ i.e., the rate of deformation tensor is not equal to the material derivative of the Eulerian strain tensor.

9.8 Given $F = \int_B \rho f dV$, derive the Reynold's transport theorem:

$$\frac{dF}{dt} = \int_B \frac{\partial}{\partial t}(\rho f)dV + \int_S \rho f v_i n_i dS$$

where S is the boundary of B and $\hat{n}$ is the unit outer normal to S. Hint: Begin with equation (9.72b).

9.9 Referring to Fig. 9.7, assume Σ is a surface of discontinuity in B with $\vec{v}=\vec{v}^+$ and $\rho=\rho^+$ to the left of Σ and $\vec{v}=\vec{v}^-$, $\rho=\rho^-$ to the right of Σ. The conservation of mass and linear momentum for B state that $dM/dt=0$ and $d\vec{P}/dt = \int_B \vec{B}dV + \int_S \vec{T}(\hat{\nu})dS$ where

$$M = \int_B \rho dV \text{ and } \vec{P} = \int_B \rho\vec{v}dV.$$

Apply the Reynold's transport theorem to these equations and then show by a procedure similar to that used to derive (9.79) that

$$\vec{T}(\hat{n}) + \vec{T}(-\hat{n}) = q(\vec{v}^+ - \vec{v}^-)$$

where $q=\rho^+ v_i^+ n_i = \rho^- v_i^- n_i$ is the mass flow per unit area, per unit time across Σ.

9.10 A conservative body force is one for which $B_i = -(\partial\phi/\partial x_i)$ where $\phi(x_1,x_2,x_3)$ is a scalar point function.

(a) Show that the continuity equation together with the stress equations of motion may be written as follows when the body force is conservative.

$$\frac{\partial \rho}{\partial t} + \frac{\partial}{\partial x_j}(\rho v_j) = 0$$

$$\frac{\partial(\rho v_i)}{\partial t} + \frac{\partial}{\partial x_j}(\phi\delta_{ij} + \rho v_i v_j - \tau_{ji}) = 0$$

(b) Show that the above equations are identically satisfied if

$$\tau_{11} = \phi + \rho v_1^2 + \frac{\partial^2\psi_2}{\partial x_3^2} + \frac{\partial^2\psi_3}{\partial x_2^2} + \frac{\partial^2\psi_4}{\partial t^2}$$

$$\tau_{22} = \phi + \rho v_2^2 + \frac{\partial^2\psi_3}{\partial x_1^2} + \frac{\partial^2\psi_1}{\partial x_3^2} + \frac{\partial^2\psi_5}{\partial t^2}$$

$$\tau_{33} = \phi + \rho v_3^2 + \frac{\partial^2\psi_1}{\partial x_2^2} + \frac{\partial^2\psi_2}{\partial x_1^2} + \frac{\partial^2\psi_6}{\partial t^2}$$

$$\tau_{12} = \rho v_1 v_2 - \frac{\partial^2\psi_3}{\partial x_1 \partial x_2}$$

$$\tau_{13} = \rho v_1 v_3 - \frac{\partial^2\psi_2}{\partial x_1 \partial x_3}$$

$$\tau_{23} = \rho v_2 v_3 - \frac{\partial^2\psi_1}{\partial x_2 \partial x_3}$$

$$\rho = -\frac{\partial^2\psi_4}{\partial x_1^2} - \frac{\partial^2\psi_5}{\partial x_2^2} - \frac{\partial^2\psi_6}{\partial x_3^2}$$

$$\rho v_1 = \frac{\partial^2\psi_4}{\partial x_1 \partial t}, \quad \rho v_2 = \frac{\partial^2\psi_5}{\partial x_2 \partial t}, \quad \rho v_3 = \frac{\partial^2\psi_6}{\partial x_3 \partial t}$$

where the "stress functions" $\psi_1, \psi_2, \ldots, \psi_6$ together with the partial derivatives required for the above demonstration are assumed to be continuous but otherwise arbitrary. This solution is due to C. Truesdell ("Invariant and Complete Stress Functions for General

Continua," Archive for Rational Mechanics and Analysis, vol. 4, 1959/60, p. 12) and is a special case of Finzi's general solution which is also given in the reference cited above. The above solution reduces to the well-known Maxwell's stress functions in the case of static equilibrium.

9.11 Use the variational equation of motion to show that the stress equations of motion and associated boundary conditions assume the following form in spherical coordinates.

$$F_r+B_r=\rho\alpha_r,\quad F_\theta+B_\theta=\rho\alpha_\theta,\quad F_\phi+B_\phi=\rho\alpha_\phi$$

where

$$F_r = \frac{\partial\tau_{rr}}{\partial r} + \frac{1}{r}\frac{\partial\tau_{r\theta}}{\partial\theta} + \frac{1}{r\sin\theta}\frac{\partial\tau_{r\phi}}{\partial\phi} + \frac{2\tau_{rr}-\tau_{\theta\theta}-\tau_{\phi\phi}+\tau_{r\theta}\cot\theta}{r}$$

$$F_\theta = \frac{\partial\tau_{r\theta}}{\partial r} + \frac{1}{r}\frac{\partial\tau_{\theta\theta}}{\partial\theta} + \frac{1}{r\sin\theta}\frac{\partial\tau_{\phi\theta}}{\partial\phi} + \frac{3\tau_{r\theta}+(\tau_{\theta\theta}-\tau_{\phi\phi})\cot\theta}{r}$$

$$F_\phi = \frac{\partial\tau_{r\phi}}{\partial r} + \frac{1}{r}\frac{\partial\tau_{\theta\phi}}{\partial\theta} + \frac{1}{r\sin\theta}\frac{\partial\tau_{\phi\phi}}{\partial\phi} + \frac{3\tau_{r\phi}+2\tau_{\phi\theta}\cot\theta}{r}$$

$$\left.\begin{aligned} T_r &= \tau_{rr}n_r + \tau_{\theta r}n_\theta + \tau_{\phi r}n_\phi \\ T_\theta &= \tau_{r\theta}n_r + \tau_{\theta\theta}n_\theta + \tau_{\phi\theta}n_\phi \\ T_\phi &= \tau_{r\phi}n_r + \tau_{\theta\phi}n_\theta + \tau_{\phi\phi}n_\phi \end{aligned}\right\} \text{ on } S_T$$

$$\delta u_r = \delta u_\theta = \delta u_\phi = 0 \qquad \text{on } S_u$$

Chapter

10

DYNAMICS OF ELASTIC BODIES (LINEAR THEORY)

The major part of man's direct contact with his environment is successfully described by the classical theories of linear elasticity and linear viscosity, or by the theory of perfect fluids, which is a special case of the latter, or by the theory of linear heat conduction, or by the theory of the linear dielectric. The phenomena of sound, wind, tides, flying, heat, light and much of the effects of small deformation of most kinds of solids are described well enough by one or another of the classical linear theories, which continue to be studied intensively and yield new results of major importance. They have won an indisputable permanence in man's interpretation of his environment.

C. Truesdell, The Principles of Continuum Mechanics, 1960.

10.1 LINEARIZATION OF THE BASIC EQUATIONS

The laws which govern the motion of elastic bodies are characterized mathematically by the strain-displacement equations (9.24) or (9.43), the continuity equation (9.67) or (9.70), Cauchy's equations (9.82) and the constitutive relation (9.105). Because of their non-linear structure, the solution of these equations for technically significant problems is a formidable task. However, if we assume that the deformation of the body is sufficiently small and that its elasticity is linear then a major simplification results: the equations are linearized. Fortunately, these linearizing assumptions are valid and meaningful in many problems of practical importance. Thus, in many instances the linear theory provides an excellent approximation to the body's true motion and state of stress.

The small deformation assumption may be stated as follows: at each point in the body, we assume that the magnitude of each of the eighteen partial derivatives $\frac{\partial u_i(a,t)}{\partial a_j}$ and $\frac{\partial u_i(x,t)}{\partial x_j}$ is negligibly small compared to unity. If we let

$$\delta = \underset{i,j}{\text{Max}}\left(\left|\frac{\partial u_i}{\partial x_j}\right| , \left|\frac{\partial u_i}{\partial a_j}\right|\right) \tag{10.1}$$

then the small deformation assumption states that $\delta \ll 1$ at each point of the body. The notation $A_{ij} \simeq c\delta^n$ where c and n are constants will be used in the forthcoming analysis to represent the inequality $\underset{i,j}{\text{Max}} \left|A_{ij}\right| < c\delta^n$. A consistent linear theory of elastic bodies will now result if we neglect terms of order δ^n and smaller compared to terms of order δ^{n-1} in each of the basic equations. For example, after neglecting terms of order δ^2 compared to terms of order δ in the definition of the Lagrangian strain tensor (9.24) and the Eulerian strain tensor (9.43) we obtain

$$E_{ij} = \tfrac{1}{2}\left(\frac{\partial u_i}{\partial a_j} + \frac{\partial u_j}{\partial a_i}\right) \simeq \delta \tag{10.2a}$$

and

$$e_{ij} = \tfrac{1}{2}\left(\frac{\partial u_i}{\partial x_j} + \frac{\partial u_j}{\partial x_i}\right) \simeq \delta. \tag{10.2b}$$

Differentiation of (9.5) and (9.6) reveals that

$$\frac{\partial x_i}{\partial a_j} = \delta_{ij} + \frac{\partial u_i}{\partial a_j}$$

and

$$\frac{\partial a_i}{\partial x_j} = \delta_{ij} - \frac{\partial u_i}{\partial x_j} \; .$$

Since $\delta_{ij} \simeq 1$ while $\frac{\partial u_i}{\partial a_j} \simeq \delta$ and $\frac{\partial u_i}{\partial x_j} \simeq \delta$ we conclude that, for a sufficiently small deformation of the body,

$$\frac{\partial x_i}{\partial a_j} = \frac{\partial a_i}{\partial x_j} = \delta_{ij} \; . \tag{10.3}$$

Consider a differentiable tensor field $\underset{\sim}{T}(x,t)$ defined in the current configuration of the body. In view of (10.3),

$$\frac{\partial \underset{\sim}{T}(x,t)}{\partial x_i} = \frac{\partial \underset{\sim}{\tau}(a,t)}{\partial a_j}\frac{\partial a_j}{\partial x_i} = \frac{\partial \underset{\sim}{\tau}(a,t)}{\partial a_i}$$

where,

$$\begin{aligned} \underset{\sim}{\tau}(a,t) &\equiv \underset{\sim}{T}\,[x_i(a,t),t] = \underset{\sim}{T}\,[a_i + u_i(a,t),t] \\ &= \underset{\sim}{T}(a,t) + u_j \left(\frac{\partial \underset{\sim}{T}}{\partial x_j}\right)_{x_j = a_j + \lambda u_j} \end{aligned}$$

for some $0<\lambda<1$. This last result follows from (9.5a) and the mean value theorem. We shall henceforth include in our small deformation assumption the additional restriction that the displacement components be sufficiently small so as to render the term

$u_j \left(\frac{\partial \underset{\sim}{T}}{\partial x_j}\right)_{x_j = a_j + \lambda u_j}$ negligible in the preceding equation. With this additional restriction to small displacements we conclude that

$$\underset{\sim}{\tau}(a,t) = \underset{\sim}{T}(a,t) \tag{10.4}$$

and

$$\frac{\partial \underset{\sim}{T}(x,t)}{\partial x_i} = \frac{\partial \underset{\sim}{T}(a,t)}{\partial a_i} \equiv \underset{\sim}{T}_{,i} \tag{10.5}$$

for all differentiable tensor fields defined in the body. In view of (10.2) and (10.5) it is now evident that, for small deformations,

$$E_{ij} = e_{ij} = \tfrac{1}{2}(u_{i,j} + u_{j,i}) \ .$$

In the future we will use the symbol ε_{ij} to represent the small deformation strain tensor, i.e.,

$$\varepsilon_{ij} \equiv \tfrac{1}{2}(u_{i,j} + u_{j,i}) \ . \tag{10.6}$$

The continuity equation assumes a particularly simple form for small deformations. By substituting (10.3) into (9.59b) we obtain

$$J = e_{ijk}\delta_{i1}\delta_{j2}\delta_{k3} = e_{123} = 1 \ . \tag{10.7}$$

Thus, (9.67) becomes

$$\rho(a,t) = \rho(a,t_o), \tag{10.8}$$

i.e., the density of the body does not vary with time. Since the density is usually known at some initial instant of time we shall henceforth assume that $\rho=\rho(a)$ is a given function of position in the body.

An important consequence of the small deformation assumption is that it is no longer necessary to distinguish between the deformed and natural configurations of the body when performing integrations. For volume integrals, we conclude from (9.60), (10.4), and (10.7) that

$$\int_B \tilde{T} dV = \int_{B_o} \tilde{T} J dV_o = \int_{B_o} \tilde{T} dV_o \tag{10.9a}$$

where $\tilde{T}$ is a tensor field defined in B. Similarly, for surface integrals,

$$\int_S \tilde{T} N_\gamma dS = \int_{S_o} \tilde{T} J \frac{\partial a_k}{\partial x_\gamma} n_k dS_o = \int_{S_o} \tilde{T} n_\gamma dS_o \tag{10.9b}$$

where $\tilde{T}$ is a tensor field defined on S, N_γ is the unit outer normal to S and n_γ is the unit outer normal to S_o. The second equality in the preceding equation follows from (10.3) and (10.7). The relation

$$N_\gamma dS = J \frac{\partial a_k}{\partial x_\gamma} n_k dS_o \tag{10.10}$$

may be obtained as follows. A surface element in the undeformed configuration of the body will be represented by

$$d\vec{S}_o = \hat{n}dS_o = d\vec{r} \times \delta\vec{r}$$

where $d\vec{r} = da_i\hat{\varepsilon}_i$ and $\delta\vec{r} = \delta a_j\hat{\varepsilon}_j$ are tangent vectors to the surface under consideration. Hence,

$$\hat{n}dS_o = e_{ijk}da_i\delta a_j\hat{\varepsilon}_k$$

or

$$n_k dS_o = e_{ijk}da_i\delta a_j \ .$$

As a result of the body's motion, the vectors $d\vec{r}$ and $\delta\vec{r}$ are deformed into the vectors $d\vec{R}$ and $\delta\vec{R}$, respectively, where

$$d\vec{R} = dx_\alpha\hat{\varepsilon}_\alpha = \frac{\partial x_\alpha}{\partial a_i} da_i\hat{\varepsilon}_\alpha$$

and

$$\delta\vec{R} = \delta x_\beta\hat{\varepsilon}_\beta = \frac{\partial x_\beta}{\partial a_j} \delta a_j\hat{\varepsilon}_\beta \ .$$

Thus, the surface element $d\vec{S}_o$ is deformed into the surface element $d\vec{S}$, where $d\vec{S} = \hat{N}dS = d\vec{R} \times \delta\vec{R} = e_{\alpha\beta\gamma}\dfrac{\partial x_\alpha}{\partial a_i}\dfrac{\partial x_\beta}{\partial a_j} da_i\delta a_j\hat{\varepsilon}_\gamma$. It can be verified, with the aid of (9.59b), that

$$Je_{ijk} = e_{\alpha\beta\gamma}\frac{\partial x_\alpha}{\partial a_i}\frac{\partial x_\beta}{\partial a_j}\frac{\partial x_\gamma}{\partial a_k}$$

from which it follows that

$$\frac{\partial a_k}{\partial x_\gamma} Je_{ijk} = e_{\alpha\beta\gamma}\frac{\partial x_\alpha}{\partial a_i}\frac{\partial x_\beta}{\partial a_j} \ .$$

The deformed surface element may now be represented as follows:

$$\hat{N}dS = \frac{\partial a_k}{\partial x_\gamma} Je_{ijk}da_i\delta a_j\hat{\varepsilon}_\gamma$$

or

$$N_\gamma dS = \frac{\partial a_k}{\partial x_\gamma} Jn_k dS_o \ .$$

For small deformations, J=1 and $\frac{\partial a_k}{\partial x_\gamma} = \delta_{k\gamma}$. Thus, $N_\gamma dS = n_\gamma dS_0$ or, in other words, the difference between $\hat{N}$ and $\hat{n}$ as well as the difference between dS and dS_0 is negligibly small. These observations lead to a considerable simplification of the stress equations of motion and associated boundary conditions. With the aid of (10.9), the variational equation of motion (9.128) may be written in Lagrangian coordinates as follows:

$$\int_{B_0} \tau_{ij}\delta\varepsilon_{ij}dV_0 = \int_{B_0}\left[B_i - \rho\,\frac{dv_i}{dt}\right]\delta u_i dV_0 + \int_{S_0} T_i\delta u_i dS_0$$

However, in Lagragian coordinates $\rho=\rho(a)$ and

$$\frac{dv_i}{dt} = \frac{\partial v_i(a,t)}{\partial t} \equiv \dot{v}_i$$

where

$$v_i = \frac{\partial u_i(a,t)}{\partial t} \equiv \dot{u}_i \ .$$

Furthermore,

$$\int_{B_0} \tau_{ij}\delta\varepsilon_{ij}dV_0 = \int_{B_0} \tau_{ij}\delta u_{i,j}dV_0 =$$

$$\int_{B_0} (\tau_{ij}\delta u_i)_{,j}dV_0 - \int_{B_0} \tau_{ij,j}\delta u_i dV_0 =$$

$$\int_{S_0} \tau_{ij}\delta u_i n_j dS_0 - \int_{B_0} \tau_{ij,j}\delta u_i dV_0 \ .$$

Thus, the variational equation of motion becomes

$$\int_{B_0}\left[\tau_{ij,j} + B_i - \rho\ddot{u}_i\right]\delta u_i dV_0 + \int_{S_0}\left[T_i - \tau_{ij}n_j\right]\delta u_i dS_0 = 0 \ .$$

Since the region B_0 is arbitrary, the integrand of each of the preceding integrals must vanish. Thus,

$$[\tau_{ij,j} + B_i - \rho\ddot{u}_i]\ \delta u_i = 0 \qquad \text{in } B_0 \tag{10.11a}$$

and

$$[T_i - \tau_{ij}n_j]\ \delta u_i = 0 \qquad \text{on } S_0 \ . \tag{10.11b}$$

However, δu_i is arbitrary in B_o and therefore we conclude that

$$\tau_{ij,j} + B_i = \rho\ddot{u}_i \qquad \text{in } B_o \ . \tag{10.12}$$

We note from (10.11) and (10.12) that the use of Lagrangian coordinates for small deformations results in three major simplifications: the density $\rho(a)$ is a specified function in B_o, the acceleration vector $\vec{\alpha}$ is simply related to the velocity and displacement vectors, i.e., $\vec{\alpha} = \frac{\partial \vec{v}}{\partial t} = \frac{\partial^2 \vec{u}}{\partial t^2}$, and the boundary conditions are applied for all time to the undeformed surface S_o whose geometry we assume is known.

For sufficiently small deformations of an elastic body, the constitutive relation (9.105) becomes

$$\tau_{ij} = \frac{\partial W^*}{\partial \varepsilon_{ij}} \tag{10.13}$$

where

$$W^* = W^*(a_1,a_2,a_3,\varepsilon_{11},\varepsilon_{12},\varepsilon_{13},\ldots,\varepsilon_{33}) \ .$$

Since the strains are small compared to one, i.e., $\varepsilon_{ij} \simeq \delta \ll 1$, we will assume that the strain energy density function W^* may be represented by the following Taylor series expansion:

$$W^* = W^*_o + \left(\frac{\partial W^*}{\partial \varepsilon_{ij}}\right)_o \varepsilon_{ij} + \frac{1}{2!}\left(\frac{\partial^2 W^*}{\partial \varepsilon_{ij}\partial \varepsilon_{kl}}\right)_o \varepsilon_{ij}\varepsilon_{kl}$$

$$+ \frac{1}{3!}\left(\frac{\partial^3 W^*}{\partial \varepsilon_{ij}\partial \varepsilon_{kl}\partial \varepsilon_{mn}}\right)_o \varepsilon_{ij}\varepsilon_{kl}\varepsilon_{mn} + \ldots$$

We shall select the natural state of the body as our datum for strain energy, thus

$$W^*_o \equiv W^*(a_1,a_2,a_3,0,0,0,\ldots,0) = 0 \ .$$

Since the stresses all vanish in the natural state we also conclude that

$$\left(\frac{\partial W^*}{\partial \varepsilon_{ij}}\right)_o = (\tau_{ij})_o = 0 \ .$$

In view of these results, the Taylor series expansion of W* may now be written as follows:

$$\left.\begin{aligned} &W^* = \tfrac{1}{2}\, C_{ijkl}(a)\varepsilon_{ij}\varepsilon_{kl} + \frac{1}{6}\, D_{ijklmn}(a)\varepsilon_{ij}\varepsilon_{kl}\varepsilon_{mn} + \dots \\ &\text{where} \\ &C_{ijkl}(a) \equiv \left(\frac{\partial^2 W^*}{\partial\varepsilon_{ij}\partial\varepsilon_{kl}}\right)_o \\ &\text{and} \\ &D_{ijklmn}(a) \equiv \left(\frac{\partial^3 W^*}{\partial\varepsilon_{ij}\partial\varepsilon_{kl}\partial\varepsilon_{mn}}\right)_o \;. \end{aligned}\right\} \qquad (10.14)$$

If, at each point in the body, the stresses are directly proportional to the strains then we say the elasticity of the body is linear. From (10.13) and (10.14) we see that this will be the case if W* is a quadratic function of the strains. We shall therefore assume that the strain energy density function is of the form

$$W^* = \tfrac{1}{2}\, C_{ijkl}(a)\varepsilon_{ij}\varepsilon_{kl} \;. \qquad (10.15)$$

It is evident from (10.14) and (10.15) that

$$C_{ijkl} = C_{jikl} = C_{ijlk} = C_{klij} \;. \qquad (10.16)$$

Thus, there are twenty-one material properties which characterize linear elasticity at each point in the body. By applying the quotient law for tensors to (10.15) we conclude, with the aid of (10.16), that C_{ijkl} are the components of a fourth order tensor. Substituting (10.15) into (10.13) yields, with the aid of (10.16) and (10.6),

$$\tau_{ij} = C_{ijkl}\varepsilon_{kl} = C_{ijkl}u_{k,l} \qquad (10.17a)$$

and therefore

$$W^* = \tfrac{1}{2}\,\tau_{ij}\varepsilon_{ij} \;. \qquad (10.17b)$$

If, as is often the case, the body is assumed to be homogeneous in its natural state then the C_{ijkl} will be constants. If the body is assumed to be isotropic in its natural state then the C_{ijkl} must be the components of an isotropic fourth order tensor of the form (8.46). Thus, for an isotropic body,

$$\tau_{ij} = \lambda\varepsilon\delta_{ij} + 2\mu\varepsilon_{ij} \qquad (10.18a)$$

and

$$W^* = \tfrac{1}{2}\,(\lambda\varepsilon^2 + 2\mu\varepsilon_{ij}\varepsilon_{ij}) \tag{10.18b}$$

where $\varepsilon \equiv \varepsilon_{kk} = u_{k,k}$ is called the dilatation. We note that an isotropic body is characterized by two material properties, λ and μ, which are called Lamé's elastic constants. Strictly speaking, they are constants only if the body is also homogeneous. Other parameters commonly used to describe the elasticity of a body are the shear modulus G, Young's modulus E and Poisson's ratio ν. Lamé's elastic constants are related to these as follows:

$$\mu = G,\; 2\mu = \frac{E}{1+\nu} \;\text{ and }\; \lambda = \frac{\nu E}{(1+\nu)(1-2\nu)} \;. \tag{10.19}$$

The reader will recall from Chapter 9 that the strain energy density function attains its minimum value in the natural state. In our Taylor series expansion of W^* we chose this minimum value to be zero. Therefore, the strain energy density is a positive definite quadratic function of the strains, i.e., $W^* \geq 0$ with $W^*=0$ if and only if $\varepsilon_{ij} = 0$. This condition places certain restrictions on the admissible values of the C_{ijkl}. For example, in the isotropic case we must have

$$\mu > 0 \text{ and } 3\lambda + 2\mu > 0 \;. \tag{10.20a}$$

These inequalities are obtained as follows. Let $q_1 = \varepsilon_{11}, q_2 = \varepsilon_{22}, q_3 = \varepsilon_{33}$, $q_4 = \varepsilon_{12}$, $q_5 = \varepsilon_{23}$ and $q_6 = \varepsilon_{31}$. Now rewrite (10.18b) in the form

$$W^* = \tfrac{1}{2}\,[k_{11}q_1^2 + k_{22}q_2^2 + k_{33}q_3^2 + k_{44}q_4^2 + k_{55}q_5^2 + k_{66}q_6^2$$

$$+ 2\,(k_{12}q_1q_2 + k_{23}q_2q_3 + k_{31}q_3q_1)]$$

where

$$k_{11} = k_{22} = k_{33} = \lambda + 2\mu$$

$$k_{12} = k_{23} = k_{31} = \lambda$$

and

$$k_{44} = k_{55} = k_{66} = 4\mu \;.$$

The necessary and sufficient conditions for W* to be positive definite are given by (3.3), i.e.,

$$\Delta_1 = k_{11} = \lambda + 2\mu > 0$$

$$\Delta_2 = \begin{vmatrix} k_{11} & k_{12} \\ k_{12} & k_{22} \end{vmatrix} = 4\mu(\lambda+\mu) > 0$$

$$\Delta_3 = \begin{vmatrix} k_{11} & k_{12} & k_{13} \\ k_{12} & k_{22} & k_{23} \\ k_{13} & k_{23} & k_{33} \end{vmatrix} = 4\mu^2(3\lambda+2\mu) > 0$$

and

$$\Delta_6 = 4\mu\Delta_5 = 16\mu^2\Delta_4 = 64\mu^3\Delta_3 > 0 \ .$$

Thus, the inequalities (10.20a) are both necessary and sufficient for W* to be positive definite. In terms of Young's modulus and Poisson's ratio, these inequalities become

$$E > 0 \text{ and } -1 < \nu < \tfrac{1}{2} \ . \qquad (10.20b)$$

In view of the preceding results the analysis of small deformations of linearly elastic bodies will be carried out by referring the body to a Lagrangian reference system. We shall henceforth represent the region in space described by the undeformed configuration of the body as the union of the open domain V and its boundary S. Furthermore, the symbols (a_1, a_2, a_3), (x_1, x_2, x_3) or (x, y, z) will be used interchangeably to represent the Lagrangian coordinates of the body.

The basic equations of the linear theory of elasticity are summarized below.

Strain-Displacement Relations: (10.21)

$$\varepsilon_{ij} = \tfrac{1}{2}(u_{i,j} + u_{j,i})$$

Stress-Strain Relations (Hooke's Law): (10.22)

$$\tau_{ij} = C_{ijkl}\varepsilon_{kl} \qquad \text{(anisotropic)}$$

$$\tau_{ij} = \lambda\varepsilon\delta_{ij} + 2\mu\varepsilon_{ij} \qquad \text{(isotropic)}$$

Equations of Motion: (10.23)

$$\tau_{ij,j} + B_i = \rho\ddot{u}_i \qquad \text{in } V$$

where

$$\tau_{ij,j} = (C_{ijkl}u_{k,l})_{,j} \qquad \text{(anisotropic)}$$

or

$$\tau_{ij,j} = (\lambda u_{j,j})_{,i} + [\mu(u_{i,j} + u_{j,i})]_{,j} \qquad \text{(isotropic)}$$

Boundary Conditions: (10.24)

$$u_i = u_i^*(x,t) \qquad \text{on } S_u$$

$$\tau_{ji}n_j = T_i^*(x,t) \qquad \text{on } S_T$$

$$u_i n_i = u_n^*, \tau_{ji}n_j t_i = T_t^*, \tau_{ji}n_j s_i = T_s^* \qquad \text{on } S_{u_n}$$

$$\tau_{ji}n_j n_i = T_n^*, u_i t_i = u_t^*, u_i s_i = u_s^* \qquad \text{on } S_{T_n}$$

where

$$S = S_u + S_T + S_{u_n} + S_{T_n} \ .$$

The unit vectors t_i and s_i are chosen so as to form an orthogonal triad with the unit outer normal vector n_i. An asterisk is used to denote a function which is specified on some part of the surface. Thus, S_u consists of all those parts of the surface S on which the displacement vector is specified, S_T consists of all those parts of S on which the traction vector is specified, S_{u_n} consists of all those parts of S on which the normal displacement and shear stress are specified and S_{T_n} consists of all those parts of S on which the normal stress and tangential displacement are specified. We should like to point out that the set S_u or S_T or S_{u_n} or S_{T_n} need not be connected and in any particular example one or more of these sets may be empty. Thus, $S = S_u$ or $S = S_T$ etc. are special cases of (10.24). By noting that $\delta u_I = 0$ wherever u_I is specified (I represents 1,2,3 or n,s,t) it can be shown that the boundary conditions (10.24) satisfy the variational condition (10.11b).

Another useful boundary condition which can be treated within the framework of a linear theory is the so called linear elastic foundation or restraint. The traction boundary condition on S_T will be generalized to include this possibility by assuming

$$T_i = T_i^* - \alpha_{ij}u_j \qquad \text{on } S_T \; . \qquad (10.25)$$

As before, $T_i^*(x,t)$ is specified on S_T. The linear elastic restoring force is provided by the traction $(-\alpha_{ij}u_j)$. The foundation moduli $\alpha_{ij}(x) = \alpha_{ji}(x)$ are specified so as to render the quadratic form $\alpha_{ij}u_iu_j$ positive definite on S_T. This restriction insures us that the potential energy U_α associated with the elastic restraint is a minimum when the body is in its natural state, i.e.,

$$U_\alpha = \tfrac{1}{2}\int_{S_T} \alpha_{ij}u_iu_j dS \geq 0 \; . \qquad (10.26)$$

In view of (10.25), the boundary condition on S_T may now be stated as follows:

$$T_i = \tau_{ji}n_j = T_i^* - \alpha_{ij}u_j \qquad \text{on } S_T$$

or

$$\tau_{ji}n_j + \alpha_{ij}u_j = T_i^*(x,t) \qquad \text{on } S_T \; . \qquad (10.27)$$

The essential feature of equations (10.21) through (10.24) and (10.27) is their linearity with respect to the displacement field and its partial derivatives. This property will be used in the forthcoming sections to show that the solution of these equations is unique, and it will also be used to show how that solution can be generated for a bounded medium.

10.2 F. NEUMANN'S UNIQUENESS THEOREM

Let us assume that we have a solution (displacement field) which satisfies the equations of motion (10.23) in V and the boundary conditions (10.24) and (10.27) on S. The question which naturally arises is the following: Is this solution unique? This question is answered by the Neumann Uniqueness Theorem which states that the solution of equations (10.23) subject to the boundary conditions (10.24) and (10.27) is unique provided that the initial displacement and velocity fields are specified throughout V at t=0, i.e.,

$$u_i(x,0) = u_i^{(o)}(x) \text{ and } \dot{u}_i(x,0) = \dot{u}_i^{(o)}(x) \quad \text{in V.} \qquad (10.28)$$

The theorem will be proved by showing that the assumption of non-uniqueness leads to a contradiction. Thus we assume there exists two different solutions $u_i^{(1)}(x,t)$ and $u_i^{(2)}(x,t)$ both of which satisfy (10.23) in V, (10.24) and (10.27) on S and (10.28) in V at t=0. As a result of the linearity of these equations and the associated subsidiary conditions, the difference of the two solutions is also a solution. Let $u_i \equiv u_i^{(1)} - u_i^{(2)}$, $\varepsilon_{ij} = \varepsilon_{ij}^{(1)} - \varepsilon_{ij}^{(2)}$ and $\tau_{ij} \equiv \tau_{ij}^{(1)} - \tau_{ij}^{(2)}$. Since both of the assumed solutions satisfy (10.23), (10.24) and (10.27), the difference solution will satisfy

$$\tau_{ij,j} = \rho\ddot{u}_i \qquad \text{in } V \qquad (10.23')$$

$$u_i = 0 \qquad \text{on } S_u$$

$$\tau_{ij}n_j + \alpha_{ij}u_j = 0 \qquad \text{on } S_T$$

$$u_i n_i = 0,\ \tau_{ji}n_j t_i = 0,\ \tau_{ji}n_j s_i = 0 \qquad \text{on } S_{u_n} \qquad (10.24')$$

$$\tau_{ji}n_j n_i = 0,\ u_i t_i = 0,\ u_i s_i = 0 \qquad \text{on } S_{T_n}$$

As the following argument shows, the total energy associated with the difference solution is a constant. The total energy of the system is

$$E^* = T + U + U_\alpha \qquad (10.29)$$

where T is the kinetic energy of the body, U is the strain energy of the body and U_α is the potential energy of the elastic foundation. According to (9.101),

$$\begin{aligned}\frac{dU}{dt} &= \int_V \tau_{ij}\dot{\varepsilon}_{ij}\,dV = \int_V \tau_{ij}\dot{u}_{i,j}\,dV \\ &= \int_V (\tau_{ij}\dot{u}_i)_{,j}\,dV - \int_V \tau_{ij,j}\dot{u}_i\,dV \\ &= \int_S \tau_{ij}n_j\dot{u}_i\,dS - \int_V \tau_{ij,j}\dot{u}_i\,dV\ .\end{aligned}$$

Substituting (10.23') and (10.24') into the preceding result yields

$$\frac{dU}{dt} = -\int_{S_T} \alpha_{ij}\dot{u}_i u_j\,dS - \int_V \rho\ddot{u}_i\dot{u}_i\,dV\ .$$

However, in view of (9.98) and (10.26),

$$\frac{dT}{dt} = \int_V \rho \ddot{u}_i \dot{u}_i dV$$

and

$$\frac{dU_\alpha}{dt} = \int_{S_T} \alpha_{ij} \dot{u}_i u_j dS \ .$$

Thus,

$$\frac{dU}{dt} = - \frac{dU_\alpha}{dt} - \frac{dT}{dt}$$

or

$$\frac{dE^*}{dt} = \frac{dT}{dt} + \frac{dU}{dt} + \frac{dU_\alpha}{dt} = 0 \ . \qquad (10.30)$$

If we integrate (10.30) from 0 to t we obtain

$$T(t) + U(t) + U_\alpha(t) = T(0) + U(0) + U_\alpha(0) \ .$$

However, both of the assumed solutions satisfy the same initial conditions, therefore the difference solution will satisfy the homogeneous initial conditions

$$u_i(x,0) = \dot{u}_i(x,0) = 0 \qquad \text{in } V \ . \qquad (10.28')$$

Thus, $T(0) = U(0) = U_\alpha(0) = 0$ and therefore $T(t) + U(t) + U_\alpha(t) = 0$. Since T, U and U_α are all positive quantities, each must vanish independently of the others, i.e.,

$$T(t) = 0, \ U(t) = 0 \text{ and } U_\alpha(t) = 0 \qquad \text{for } t > 0.$$

Thus,

$$\dot{u}_i(x,t) = \varepsilon_{ij}(x,t) = 0 \qquad \text{in } V \qquad (10.31a)$$

and

$$u_i(x,t) = 0 \qquad \text{on } S_T. \qquad (10.31b)$$

Equations (10.31a) may be integrated to obtain

$$u_i(x,t) = a_i + e_{ijk} b_j x_k \qquad \text{in } V$$

where a_i and b_j are constant vectors with $b_j \approx \delta$ (small deformation assumption). This corresponds to a rigid body displacement. However, $u_i(x,0) = 0$ in V and we must therefore choose $a_i = b_i = 0$. Thus, $u_i(x,t) = 0$ in V and on S_T or $u_i^{(1)}(x,t) = u_i^{(2)}(x,t)$. Since this result contradicts the initial hypothesis that the two solutions are different, we must conclude that the hypothesis is incorrect and that the solution is indeed unique.

10.3 RECIPROCAL RELATIONS

Some very useful reciprocal relations exist between the various states of stress and strain which a given linearly elastic body may experience. Let $\varepsilon_{ij}^{(1)}$ and $\varepsilon_{ij}^{(2)}$ represent two different states of strain which can exist in a given body and let $\tau_{ij}^{(1)}$ and $\tau_{ij}^{(2)}$ represent the corresponding stress fields in that body. In view of (10.22) and (10.16),

$$\tau_{ij}^{(1)}\varepsilon_{ij}^{(2)} = C_{ijkl}\varepsilon_{ij}^{(2)}\varepsilon_{kl}^{(1)} = C_{klij}\varepsilon_{kl}^{(2)}\varepsilon_{ij}^{(1)}$$

$$= C_{ijkl}\varepsilon_{ij}^{(1)}\varepsilon_{kl}^{(2)} = \tau_{ij}^{(2)}\varepsilon_{ij}^{(1)} \ .$$

Integrating this result over the region V yields

$$\int_V \tau_{ij}^{(1)}\varepsilon_{ij}^{(2)}dV = \int_V \tau_{ij}^{(2)}\varepsilon_{ij}^{(1)}dV \ . \tag{10.32a}$$

With the aid of (10.21) and the symmetry of the stress tensor we find that

$$\int_V \tau_{ij}^{(1)}\varepsilon_{ij}^{(2)}dV = \int_V \tau_{ij}^{(1)}u_{i,j}^{(2)}dV$$

$$= \int_V \left(\tau_{ij}^{(1)}u_i^{(2)}\right)_{,j}dV - \int_V \tau_{ij,j}^{(1)}u_i^{(2)}dV$$

$$= \int_S T_i^{(1)}u_i^{(2)}dS - \int_V \tau_{ij,j}^{(1)}u_i^{(2)}dV \tag{10.32b}$$

where

$$T_i^{(1)} = \tau_{ij}^{(1)}n_j \text{ and } T_i^{(2)} = \tau_{ij}^{(2)}n_j \ .$$

This result, together with a similar one for the integral $\int_V \tau_{ij}^{(2)}\varepsilon_{ij}^{(1)}dV$, substituted into (10.32a) yields

$$\int_V \left(\tau_{ij,j}^{(2)}u_i^{(1)} - \tau_{ij,j}^{(1)}u_i^{(2)}\right)dV = \int_S \left(T_i^{(2)}u_i^{(1)} - T_i^{(1)}u_i^{(2)}\right)ds \quad . \qquad (10.33)$$

This is called the Betti-Rayleigh reciprocal relation. It will play a prominent role in our study of the free harmonic vibrations of elastic solids to be presented in the next section. The Betti-Rayleigh relation is very useful in the study of static elastic problems in which case $\tau_{ij,j} = -B_i$ and (10.33) becomes

$$\int_V B_i^{(1)}u_i^{(2)}dV + \int T_i^{(1)}u_i^{(2)}dS = \int_V B_i^{(2)}u_i^{(1)}dV$$

$$+ \int T_i^{(2)}u_i^{(1)}dS \quad . \qquad (10.34)$$

Another reciprocal relation which is particularly well-suited to the study of dynamic elastic problems may be obtained by noting that (10.33) remains valid if all the quantities in that equation are replaced by their Laplace transforms, i.e.,

$$\int_V \left(\bar{\tau}_{ij,j}^{(2)}\bar{u}_i^{(1)} - \bar{\tau}_{ij,j}^{(1)}\bar{u}_i^{(2)}\right)dV = \int_S \left(\bar{T}_i^{(2)}\bar{u}_i^{(1)} - \bar{T}_i^{(1)}\bar{u}_i^{(2)}\right)dS \qquad (10.35)$$

where

$$\bar{u}_i \equiv \int_0^\infty u_i e^{-pt}dt$$

$$\bar{\varepsilon}_{ij} = \tfrac{1}{2}\left(\bar{u}_{i,j} + \bar{u}_{j,i}\right)$$

$$\bar{\tau}_{ij} = C_{ijkl}\bar{\varepsilon}_{ij} \text{ and } \bar{T}_i = \bar{\tau}_{ij}n_j \quad .$$

Taking the Laplace transform of the equations of motion (10.23) yields

$$\bar{\tau}_{ij,j} = -\bar{B}_i + \rho\left[p^2\bar{u}_i - pu_i(x,0) - \dot{u}_i(x,0)\right] \quad .$$

This result is now substituted into (10.35) to obtain:

$$\int_V \left(\bar{B}_i^{(1)}\bar{u}_i^{(2)} - \bar{B}_i^{(2)}\bar{u}_i^{(1)}\right)dV + \int_V \rho\left[u_i^{(1)}(x,0)p\bar{u}_i^{(2)} - u_i^{(2)}(x,0)p\bar{u}_i^{(1)}\right]dV$$

$$+\int_V \rho\left[\dot{u}_i^{(1)}(x,0)\bar{u}_i^{(2)} - \dot{u}_i^{(2)}(x,0)\bar{u}_i^{(1)}\right]dV = \int_S \left(\bar{T}_i^{(2)}\bar{u}_i^{(1)} - \bar{T}_i^{(1)}\bar{u}_i^{(2)}\right)dS$$

(10.36)

The inverse Laplace transform of (10.36) may be expressed in terms of convolution integrals as follows:

$$\int_0^t \int_V B_i^{(1)}(x,\tau)u_i^{(2)}(x,t-\tau)dVd\tau + \int_0^t \int_S T_i^{(1)}(x,\tau)u_i^{(2)}(x,t-\tau)dSd\tau$$

$$+\int_V \rho\left[u_i^{(1)}(x,0)\dot{u}_i^{(2)}(x,t)+\dot{u}_i^{(1)}(x,0)u_i^{(2)}(x,t)\right]dV=$$

$$\int_0^t \int_V B_i^{(2)}(x,t-\tau)u_i^{(1)}(x,\tau)dVd\tau + \int_0^t \int_S T_i^{(2)}(x,t-\tau)u_i^{(1)}(x,\tau)dSd\tau$$

$$+\int_V \rho\left[u_i^{(2)}(x,0)\dot{u}_i^{(1)}(x,t)+\dot{u}_i^{(2)}(x,0)u_i^{(1)}(x,t)\right]dV \quad . \tag{10.37}$$

This is the dynamic reciprocal relation for linearly elastic bodies. This derivation of the dynamic reciprocal theorem is due to D. Graffi, "Sul teorema di reciprocità nella dinamica dei corpi elastici," Memoria della Accademia delle Scienze, Bologna, series 10, Vol. 4, 1946/47, pp. 103-109. In the case of zero initial conditions, the form of (10.37) is very similar to that of (10.34).

With the aid of (10.37) we may define a Green's function for the various boundary value problems of linear elasticity. Consider the body force distribution characterized by $B_1^{(2)}(x,t) = \delta(x)\delta(t)$, $B_2^{(2)} = B_3^{(2)} = 0$, where $\delta(x) = 0$ if $x_i \neq 0$ and $\int_V f(x)\delta(x)dV = f(0)$ if $x_i = 0$ is in V.

In this case

$$\int_0^t \int_V B_i^{(2)}(x-x',t-\tau)u_i^{(1)}(x,\tau)dVd\tau = u_1^{(1)}(x',t) \quad .$$

Let us assume that the corresponding displacement field $u_i^{(2)}(x,t)$ satisfies the homogeneous initial conditions $u_i^{(2)}(x,0)=\dot{u}_i^{(2)}(x,0) = 0$ and the homogeneous boundary conditions

$$u_i^{(2)}(x,t) = 0 \text{ on } S_u \text{ and } T_i^{(2)}(x,t) = 0 \text{ on } S_T \ .$$

Substituting these results into (10.37) yields

$$u_i^{(1)}(x',t) = \int_0^t \int_V B_i^{(1)}(x,\tau)u_i^{(2)}(x-x',t-\tau)dVd\tau$$
$$+ \int_0^t \int_{S_T} T_i^*(x,\tau)u_i^{(2)}(x-x',t-\tau)dSd\tau$$
$$- \int_0^t \int_{S_u} u_i^*(x,\tau)T_i^{(2)}(x-x',t-\tau)dSd\tau$$
$$+ \int_V \rho\left[u_i^{(o)}(x)\dot{u}_i^{(2)}(x-x',t)+\dot{u}_i^{(o)}(x)u_i^{(2)}(x-x',t)\right] dV \qquad (10.38)$$

where

$$u_i^{(1)}(x,0)=u_i^{(o)}(x),\ \dot{u}_i^{(1)}(x,0)=\dot{u}_i^{(o)}(x) \qquad \text{in } V$$

$$u_i^{(1)}(x,t)=u_i^*(x,t) \qquad \text{on } S_u$$

and

$$T_i^{(1)}(x,t)=T_i^*(x,t) \qquad \text{on } S_T\ .$$

Similar expressions for $u_2^{(1)}$ and $u_3^{(1)}$ can be obtained by considering the body forces $\vec{B}^{(2)}= \delta(x)\delta(t)\hat{\varepsilon}_2$ and $\vec{B}^{(2)}= \delta(x)\delta(t)\hat{\varepsilon}_3$, respectively. It is apparent from (10.38) that a knowledge of the Green's function $u_i^{(2)}(x,t)$ enables one to express the displacement field $u_i^{(1)}(x,t)$ explicitly in terms of the body force, boundary conditions and initial conditions which cause it. Although this result is of considerable theoretical interest, its practical value is limited by the difficulty encountered in finding the Green's function for most bodies. Thus we shall not pursue this method of solution any further. Rather, in the sections which follow we shall consider another method which is better suited for the actual computation of displacement fields in bounded elastic solids.

10.4 FREE VIBRATIONS OF BOUNDED ELASTIC BODIES

In this section we wish to examine the possibility of harmonic oscillations of the form

$$u_i(x,t) = u_i(x) \exp (i\omega t) \qquad (10.39)$$

in bounded elastic bodies that are free from external forces and subject to the homogeneous boundary conditions

$$\begin{aligned} &u_i = 0 && \text{on } S_u \\ &\tau_{ji}n_j + \alpha_{ij}u_j = 0 && \text{on } S_T \\ &u_i n_i = 0,\ \tau_{ji}n_j t_i = 0,\ \tau_{ji}n_j s_i = 0 && \text{on } S_{u_n} \\ &\tau_{ji}n_i n_j = 0,\ u_i t_i = 0,\ u_i s_i = 0 && \text{on } S_{T_n} \end{aligned} \qquad (10.40)$$

Whenever i appears as a subscript it represents one of the positive integers (1,2,3), otherwise it represents $\sqrt{-1}$. In view of (10.21), (10.22) and (10.39) we conclude that, in the case of free vibrations,

$$\varepsilon_{ij}(x,t) = \varepsilon_{ij}(x) \exp(i\omega t)$$

and

$$\tau_{ij}(x,t) = \tau_{ij}(x) \exp(i\omega t) \qquad (10.41)$$

After substituting (10.39) and (10.41) into the equations of motion (10.23) we obtain

$$-\frac{1}{\rho(x)} \tau_{ij,j}(x) = \omega^2 u_i(x)$$

where

$$\tau_{ij,j} = (\lambda u_{k,k})_{,i} + [\mu(u_{i,j} + u_{j,i})]_{,j} \qquad (10.42)$$

We note that (10.42) and (10.40) are identically satisfied by the displacement field $u_i(x) \equiv 0$. This is called the trivial solution and it is not of interest to us. We wish to examine the properties of the nontrivial solutions of (10.42) and (10.40). We shall assume that such nontrivial solutions exist only for certain values of ω^2 called eigenvalues and that there are a denumerable infinity of these eigenvalues which, after excluding rigid body displacement fields from consideration, may be ordered as follows:

$$0 < \omega_1^2 \le \omega_2^2 \le \omega_3^2 \ldots \quad \text{and} \quad \lim_{n\to\infty} \omega_n^2 = \infty \qquad (10.43)$$

The general validity of these assumptions has been demonstrated for many of the boundary conditions contained in (10.40) by K. Friedrichs, D.M. Eidus and S.G. Mikhlin. These demonstrations are beyond the scope of this book but may be found in Mikhlin's book, "The Problem of The Minimum of a Quadratic Functional," Holden-Day, Inc., San Francisco, 1965, pp. 117-146. In many particular cases, including those we shall consider later, the validity of these assumptions may be established by actually exhibiting the eigenvalues. The solution $u_i^{(n)}(x)$ corresponding to the eigenvalue ω_n^2 is called the n'th eigenfunction or mode shape function of the body.

It is evident from (10.43) that the eigenvalues of a bounded elastic body are positive real numbers. We will now show that this is a consequence of the reciprocal relation of Betti and Rayleigh and the positive definite character of the total potential energy. In view of the homogeneous boundary conditions being imposed on the free vibrations of the body, the reciprocal relation (10.33) becomes

$$\int_V \left(\tau_{ij,j}^{(2)} u_i^{(1)} - \tau_{ij,j}^{(1)} u_i^{(2)}\right) dV = 0 \qquad (10.44)$$

and (10.32b) becomes

$$-\int_V \tau_{ij,j}^{(1)} u_i^{(2)} dV = \int_V \tau_{ij}^{(1)} \varepsilon_{ij}^{(2)} dV + \int_{S_T} \alpha_{ij} u_i^{(2)} u_j^{(1)} dS \quad . \qquad (10.45)$$

If we let $U_T(u_i) = U(u_i) + U_\alpha(u_i)$ represent the total potential energy associated with the displacement field $u_i(x)$, and we set state (1) = state (2), equation (10.45) will reveal that

$$-\int_V \tau_{ij,j} u_i dV = 2U_T(u_i) \geq 0 \ . \qquad (10.46a)$$

Since we are excluding rigid body displacement fields from our considerations, $U_T(u_i) = 0$ if and only if $u_i(x) \equiv 0$. Let us assume that the eigenvalue ω_n^2 and the associated eigenfunction $u_i^{(n)}$ are complex, i.e.,

$$u_i^{(n)} = \phi_i^{(n)} + i\psi_i^{(n)}$$

where $\phi_i^{(n)}$ and $\psi_i^{(n)}$ are real valued functions of x_i. In view of (10.21) and (10.22) the corresponding modal stress will assume the form

$$\tau_{ij}^{(n)} = \Phi_{ij}^{(n)} + i\Psi_{ij}^{(n)}$$

where

$$\Phi_{ij}^{(n)} = \lambda\phi_{k,k}^{(n)}\delta_{ij} + \mu(\phi_{i,j}^{(n)} + \phi_{j,i}^{(n)})$$

and

$$\Psi_{ij}^{(n)} = \lambda\psi_{k,k}^{(n)}\delta_{ij} + \mu(\psi_{i,j}^{(n)} + \psi_{j,i}^{(n)}) \quad .$$

An asterisk will be placed above a quantity to denote its complex conjugate, e.g.,

$$\overset{*}{u}{}_i^{(n)} = \phi_i^{(n)} - i\psi_i^{(n)} \quad .$$

With the aid of (10.42) we find that

$$-\tau_{ij,j}^{(n)}\overset{*}{u}{}_i^{(n)} = \rho\omega_n^2 u_i^{(n)}\overset{*}{u}{}_i^{(n)}$$

and therefore

$$-\int_V \tau_{ij,j}^{(n)}\overset{*}{u}{}_i^{(n)}dV = \omega_n^2 N^2(u_i^{(n)}) \tag{10.46b}$$

where

$$N(u_i^{(n)}) \equiv \left[\int_V \rho u_i^{(n)}\overset{*}{u}{}_i^{(n)}dV\right]^{\frac{1}{2}} \tag{10.47}$$

is called the norm of the function $u_i^{(n)}$ with respect to the weighting function $\rho(x)$. Since $\rho(x)>0$ in V we conclude that $N^2(u_i^{(n)})\geq 0$ with $N^2(u_i^{(n)})=0$ if and only if $u_i^{(n)}(x)=0$ almost everywhere in V, i.e., everywhere except, perhaps, on a set of zero volume. According to our hypothesis we may also write

$$-\tau_{ij,j}^{(n)}\overset{*}{u}{}_i^{(n)} = -(\Phi_{ij,j}^{(n)} + i\Psi_{ij,j}^{(n)})(\phi_i^{(n)} - i\psi_i^{(n)})$$

and therefore

$$-\int_V \tau_{ij,j}^{(n)}\overset{*}{u}{}_i^{(n)}dV = -\int_V(\Phi_{ij,j}^{(n)}\phi_i^{(n)} + \Psi_{ij,j}^{(n)}\psi_i^{(n)})\,dV$$

$$- i\int_V(\Psi_{ij,j}^{(n)}\phi_i^{(n)} - \Phi_{ij,j}^{(n)}\psi_i^{(n)})\,dV \quad .$$

However, as a result of (10.44) and (10.46a)

$$\int_V(\Psi_{ij,j}^{(n)}\phi_i^{(n)} - \Phi_{ij,j}^{(n)}\psi_i^{(n)})dV=0$$

$$-\int_V \Phi_{ij,j}^{(n)}\phi_i^{(n)}dV = 2U_T(\phi_i^{(n)})\geq 0$$

and

$$-\int_V \Psi_{ij,j}^{(n)}\psi_i^{(n)}dV = 2U_T(\psi_i^{(n)})\geq 0$$

Therefore,

$$-\int_V \tau_{ij,j}^{(n)}\overset{*}{u}_i^{(n)}dV = 2U_T(\phi_i^{(n)})+ 2U_T(\psi_i^{(n)}) \quad . \tag{10.48}$$

Equating (10.46b) to (10.48) yields the result

$$\omega_n^2= \frac{2}{N^2(u_i^{(n)})}\left[U_T(\phi_i^{(n)})+U_T(\psi_i^{(n)})\right] \quad . \tag{10.49}$$

Since $u_i^{(n)}(x)\neq 0$ in V, $N^2(u_i^{(n)})>0$. Furthermore, since we do not accept rigid body displacement fields as eigenfunctions, $U_T(\phi_i^{(n)})+U_T(\psi_i^{(n)})>0$. Thus, the right-hand side of (10.49) is real and positive, and therefore the eigenvalue ω_n^2 is real and positive. The positive square root of an eigenvalue is called a natural frequency of the system. In view of the preceding discussion the natural frequencies (ω_n,n=1,2,3,...) are all real. If the eigenfunction $u_i^{(n)}(x)$ is real then (10.49) can be written as

$$\omega_n^2 = \frac{2U_T(u_i^{(n)})}{N^2(u_i^{(n)})} > 0 \quad . \tag{10.50}$$

The right-hand side of (10.50) is called the Rayleigh Quotient. It can be shown, via the calculus of variations, that the eigenvalues correspond to minima of the Rayleigh Quotient. This minimum property of the eigenvalues forms the basis for several methods for their approximate determination. Since we are primarily interested in direct analytical techniques we shall not pursue this matter further, however, an excellent account of these methods can be found in the book by S.G. Mikhlin, "Variational Methods in Mathematical Physics," Pergamon Press, 1964.

The preceding arguments do not preclude the possibility of complex eigenfunctions, however, we will now show that the linearity of our theory together with the reality of the eigenvalues permits us to assume, without any loss in generality, that all the eigenfunctions are real. Suppose there exists two real eigenfunctions $\phi_i^{(n)}$ and $\psi_i^{(n)}$, both corresponding to the same eigenvalue ω_n^2. Then,

$$-\frac{1}{\rho}\,\Phi_{ij,j}^{(n)}= \omega_n^2\phi_i^{(n)} \quad\text{and}\quad -\frac{1}{\rho}\,\Psi_{ij,j}^{(n)}= \omega_n^2\psi_i^{(n)}$$

where

$$\Phi_{ij}^{(n)} = \lambda\phi_{k,k}^{(n)}\delta_{ij} + \mu(\phi_{i,j}^{(n)} + \phi_{j,i}^{(n)})$$

and

$$\Psi_{ij}^{(n)} = \lambda\psi_{k,k}^{(n)}\delta_{ij} + \mu(\psi_{i,j}^{(n)} + \psi_{j,i}^{(n)}) .$$

This situation is called a degeneracy and we say that the eigenvalue ω_n^2 is two-fold degenerate. Let $u_i^{(n)} = A_n\phi_i^{(n)} + B_n\psi_i^{(n)}$ where A_n and B_n are arbitrary constants which may be real or complex. The corresponding modal stress is

$$\tau_{ij}^{(n)} = \lambda u_{k,k}^{(n)}\delta_{ij} + \mu(u_{i,j}^{(n)} + u_{j,i}^{(n)}) = A_n\Phi_{ij}^{(n)} + B_n\Psi_{ij}^{(n)} .$$

Thus,

$$-\frac{1}{\rho}\tau_{ij,j}^{(n)} = -\frac{1}{\rho}(A_n\Phi_{ij,j}^{(n)} + B_n\Psi_{ij,j}^{(n)}) = \omega_n^2(A_n\phi_i^{(n)} + B_n\psi_i^{(n)}) = \omega_n^2 u_i^{(n)} .$$

The above result shows that any linear combination of the eigenfunctions $\phi_i^{(n)}$ and $\psi_i^{(n)}$ is also an eigenfunction corresponding to the eigenvalue ω_n^2, in particular,

$$u_i^{(n)} = \phi_i^{(n)} + i\psi_i^{(n)} \text{ and } u_i^{(n)} = \phi_i^{(n)} - i\psi_i^{(n)}$$

are eigenfunctions. Conversely, if $u_i^{(n)} = \phi_i^{(n)} \pm i\psi_i^{(n)}$ is a complex eigenfunction, then $\phi_i^{(n)}$ and $\psi_i^{(n)}$ are each real eigenfunctions, for,

$$-\frac{1}{\rho}\tau_{ij,j}^{(n)} = -\frac{1}{\rho}\left[\Phi_{ij,j}^{(n)} \pm i\Psi_{ij,j}^{(n)}\right] = \omega_n^2(\phi_i^{(n)} \pm i\psi_i^{(n)})$$

and since ω_n^2 is real

$$-\frac{1}{\rho}\Phi_{ij,j}^{(n)} = \omega_n^2\phi_i^{(n)}$$

and

$$-\frac{1}{\rho}\Psi_{ij,j}^{(n)} = \omega_n^2\psi_i^{(n)} .$$

Thus, a complex eigenfunction is completely equivalent to, and can always be replaced by, two real eigenfunctions.

Another important property of the eigenfunctions may be deduced from the reciprocal relation (10.44). If we let states (1) and (2) in (10.44) be the eigenstates (n) and (m) respectively we obtain

$$\int_V (\tau_{ij,j}^{(m)} u_i^{(n)} - \tau_{ij,j}^{(n)} u_i^{(m)})\, dv = 0.$$

However, in view of (10.42)

$$\tau_{ij,j}^{(m)} = -\rho\omega_m^2 u_i^{(m)} \text{ and } -\tau_{ij,j}^{(n)} = \rho\omega_n^2 u_i^{(n)} \quad . \quad \text{Thus,}$$

$$(\omega_n^2 - \omega_m^2) \int_V \rho u_i^{(n)} u_i^{(m)} dV = 0$$

and if $\omega_n^2 \neq \omega_m^2$ we conclude that

$$\int_V \rho u_i^{(n)} u_i^{(m)} dV = 0 \quad . \tag{10.51}$$

Equation (10.51) is called an orthogonality relation. It states that the eigenfunctions corresponding to different eigenvalues are orthogonal with respect to the weighting function $\rho(x)$. This result can easily be extended to include all eigenfunctions regardless of whether or not they correspond to different eigenvalues. For example, if two eigenfunctions $\phi_i^{(n)}$ and $\psi_i^{(n)}$ both correspond to the same eigenvalue ω_n^2, i.e., if ω_n^2 is two-fold degenerate, then, as shown earlier, any linear combination of $\phi_i^{(n)}$ and $\psi_i^{(n)}$ is also an eigenfunction corresponding to ω_n^2. Let

$$u_i^{(n,1)} = \phi_i^{(n)} \text{ and } u_i^{(n,2)} = \psi_i^{(n)} - C_n u_i^{(n,1)}$$

where C_n is a constant to be determined so as to render $u_i^{(n,1)}$ and $u_i^{(n,2)}$ orthogonal. Clearly, $u_i^{(n,1)}$ and $u_i^{(n,2)}$ are linear combinations of $\phi_i^{(n)}$ and $\psi_i^{(n)}$ and therefore are eigenfunctions. Since we wish them to be orthogonal we set

$$\int_V \rho u_i^{(n,1)} u_i^{(n,2)} dV = \int_V \rho u_i^{(n,1)} \psi_i^{(n)} dV - C_n N^2(u_i^{(n,1)}) = 0.$$

Thus,

$$C_n = \frac{1}{N^2(u_i^{(n,1)})} \int_V \rho u_i^{(n,1)} \psi_i^{(n)} dV \quad .$$

Assuming that the preceding Gram-Schmidt orthogonalization process has been applied wherever necessary (it is easily generalized to accommodate K-fold degeneracies) we can conclude that all the eigenfunctions of an elastic body are orthogonal, regardless of whether or not they correspond to different eigenvalues.

It is evident from (10.40) and (10.42) that if $u_i^{(n)}$ is an eigenfunction then $A_n u_i^{(n)}$, where A_n is an arbitrary nonzero constant, is also an eigenfunction. Thus, each eigenfunction is determined by these equations only to within an arbitrary, nonzero, multiplicative constant. For the sake of convenience in the subsequent analysis we will fix the value of this constant by adjoining the normalization condition

$$N(u_i^{(n)}) = \left[\int_V \rho u_i^{(n)} u_i^{(n)} dV\right]^{\frac{1}{2}} = 1, \quad n = 1,2,3,\ldots \tag{10.52}$$

As a result of this normalization condition, (10.50) becomes $\omega_n^2 = 2U_T(u_i^{(n)})$, i.e., the eigenvalue ω_n^2 is equal to twice the total potential energy of the system in its n'th normalized mode of vibration. We wish to point out that, as a result of this normalization condition, the dimension of the normalized eigenfunctions will be the reciprocal of the square root of mass. Equations (10.51) and (10.52) may be combined to obtain the orthonormality relation

$$\int_V \rho u_i^{(m)} u_i^{(n)} dV = \delta_{nm} = \begin{Bmatrix} 1 \text{ if } n=m \\ 0 \text{ if } n \neq m \end{Bmatrix} . \tag{10.53}$$

10.5 FORCED MOTION OF BOUNDED ELASTIC BODIES

We will now consider a method which can be used to resolve the forced motion problem characterized by the equations of motion (10.23), the boundary conditions (10.24) and/or (10.27) and the initial conditions (10.28). Let $u_i^{(s)}(x,t)$ represent the solution of the quasi-static problem which results when the inertia term is deleted from (10.23). If we let

$$\tau_{ij}^{(s)} = \lambda\delta_{ij}u_{k,k}^{(s)} + \mu(u_{i,j}^{(s)} + u_{j,i}^{(s)})$$

then,

$$\tau_{ij,j}^{(s)} = - B_i(x,t) \qquad \text{in } V \tag{10.54}$$

$$u_i^{(s)} = u_i^*(x,t) \qquad \text{on } S_u$$

$$\tau_{ji}^{(s)}n_j + \alpha_{ij}u_j^{(s)} = T_i^*(x,t) \qquad \text{on } S_T$$

$$u_i^{(s)}n_i = u_n^*,\ \tau_{ji}^{(s)}n_jt_i = T_t^*,\ \tau_{ji}^{(s)}n_js_i = T_s^* \qquad \text{on } S_{u_n} \tag{10.55}$$

$$\tau_{ji}^{(s)}n_in_j = T_n^*, u_i^{(s)}t_i = u_t^*,\ u_i^{(s)}s_i = u_s^* \qquad \text{on } S_{T_n}$$

In this quasi-static problem the time appears as a parameter rather than as a variable. Thus, it is considerably simpler than the original forced motion problem. In fact, in many cases of practical interest the solution $u_i^{(s)}(x,t)$ is expressible as a finite combination of elementary functions. The solution of the forced motion problem may be constructed with the aid of this quasi-static solution as follows. Let $u_i^{(D)}(x,t)$ represent the difference between the forced motion solution $u_i(x,t)$ and the quasi-static solution $u_i^{(s)}(x,t)$, i.e.,

$$u_i^{(D)}(x,t) = u_i(x,t) - u_i^{(s)}(x,t) \quad . \tag{10.56}$$

Substituting (10.56) into (10.22) through (10.28) yields, with the aid of (10.54) and (10.55),

$$\tau_{ij,j}^{(D)} - \rho\ddot{u}_i^{(D)} = \rho\ddot{u}_i^{(s)} \qquad \text{in } V \tag{10.57}$$

$$\begin{aligned} &u_i^{(D)} = 0 && \text{on } S_u \\ &\tau_{ij}^{(D)} n_j + \alpha_{ij} u_j^{(D)} = 0 && \text{on } S_T \\ &u_i^{(D)} n_i = 0, \ \tau_{ji}^{(D)} n_j t_i = 0, \ \tau_{ji}^{(D)} n_j s_i = 0 && \text{on } S_{u_n} \\ &\tau_{ji}^{(D)} n_j n_i = 0, \ u_i^{(D)} t_i = 0, \ u_i^{(D)} s_i = 0 && \text{on } S_{T_n} \end{aligned} \tag{10.58}$$

$$\left.\begin{aligned} u_i^{(D)}(x,0) &= u_i^{(o)}(x) - u_i^{(s)}(x,0) \\ \dot{u}_i^{(D)}(x,0) &= \dot{u}_i^{(o)}(x) - \dot{u}_i^{(s)}(x,0) \end{aligned}\right\} \quad \text{in } V \tag{10.59}$$

where

$$\tau_{ij}^{(D)} = \tau_{ij} - \tau_{ij}^{(s)} = \lambda\delta_{ij}u_{k,k}^{(D)} + \mu\left(u_{i,j}^{(D)} + u_{j,i}^{(D)}\right) \quad .$$

We note that the difference solution $u_i^{(D)}(x,t)$ satisfies the same homogeneous boundary conditions on S as do the eigenfunctions $u_i^{(n)}(x)$ (compare (10.40) with (10.58)). This observation leads us to the assumption that the difference solution may be represented as an eigenfunction expansion the coefficients of which are time dependent, i.e.,

$$u_i^{(D)}(x,t) = \sum_{n=1}^{\infty} u_i^{(n)}(x) q_n(t) \quad . \tag{10.60}$$

The validity of this assumption will be established if we can determine the coefficients $q_n(t)$ so as to satisfy (10.57) and (10.59). This is accomplished in the following manner. Substitute (10.60) into (10.57) and (10.59) to obtain, with the aid of (10.42),

$$\sum_{n=1}^{\infty} (\ddot{q}_n + \omega_n^2 q_n)\rho u_i^{(n)} = -\rho\ddot{u}_i^{(s)} \tag{10.61a}$$

$$\sum_{n=1}^{\infty} q_n(0)u_i^{(n)}(x) = u_i^{(o)}(x) - u_i^{(s)}(x,0) \tag{10.61b}$$

$$\sum_{n=1}^{\infty} \dot{q}_n(0)u_i^{(n)}(x) = \dot{u}_i^{(o)}(x) - \dot{u}_i^{(s)}(x,0) \quad . \tag{10.61c}$$

Multiply (10.61a) by $u_i^{(m)}(x)$, (10.61b) by $\rho u_i^{(m)}(x)$ and (10.61c) by $\rho u_i^{(m)}(x)$. Now integrate the resulting equations over the region V, making use of the orthonormality relations (10.53), to obtain

$$\ddot{q}_m + \omega_m^2 q_m = \ddot{Q}_m(t) \tag{10.62a}$$

$$q_m(0) = Q_m(0) + \int_V \rho u_i^{(m)}(x)u_i^{(o)}(x)dV \tag{10.62b}$$

$$\dot{q}_m(0) = \dot{Q}_m(0) + \int_V \rho u_i^{(m)}(x)\dot{u}_i^{(o)}(x)dV \tag{10.62c}$$

where

$$Q_m(t) = -\int_V \rho u_i^{(s)}(x,t)u_i^{(m)}(x)dV \quad . \tag{10.62d}$$

The solution of (10.62) for $t \geq 0$ is

$$q_n(t) = q_n(0)\cos\omega_n t + \frac{\dot{q}_n(0)}{\omega_n}\sin\omega_n t$$
$$+ \frac{1}{\omega_n}\int_o^t \ddot{Q}_n(\tau)\sin\omega_n(t-\tau)d\tau \quad . \tag{10.63}$$

If the integral on the right-hand side of (10.63) is integrated by parts twice the following alternate expression is obtained for $q_n(t)$:

$$q_n(t) = A_n\cos\omega_n t + \frac{B_n}{\omega_n}\sin\omega_n t$$
$$+ Q_n(t) - \omega_n\int_o^t Q_n(\tau)\sin\omega_n(t-\tau)d\tau \tag{10.64a}$$

where

$$A_n = q_n(0) - Q_n(0) = \int_V \rho u_i^{(n)}u_i^{(o)}dV \tag{10.64b}$$

and

$$B_n = \dot{q}_n(0) - \dot{Q}_n(0) = \int_V \rho u_i^{(n)}\dot{u}_i^{(o)}dV \quad . \tag{10.64c}$$

The forcing function $Q_n(t)$ may be expressed in terms of the body force and the surface tractions by means of the following transformations. According to (10.62d) and (10.42),

$$\omega_n^2 Q_n(t) = \int_V \left[-\rho\omega_n^2 u_i^{(n)}\right] u_i^{(s)} dV = \int_V \tau_{ij,j}^{(n)} u_i^{(s)} dV$$
$$= \int_V (\tau_{ij}^{(n)} u_i^{(s)})_{,j} \, dV - \int_V \tau_{ij}^{(n)} u_{i,j}^{(s)} dV \quad .$$

In view of (10.32a),

$$\int_V \tau_{ij}^{(n)} u_{i,j}^{(s)} dV = \int_V \tau_{ij}^{(s)} u_{i,j}^{(n)} dV$$
$$= \int_V (\tau_{ij}^{(s)} u_i^{(n)})_{,j} \, dV - \int_V \tau_{ij,j}^{(s)} u_i^{(n)} dV \quad .$$

Thus,

$$\omega_n^2 Q_n(t) = \int_V (\tau_{ij}^{(n)} u_i^{(s)} - \tau_{ij}^{(s)} u_i^{(n)})_{,j} \, dV + \int_V \tau_{ij,j}^{(s)} u_i^{(n)} dV \quad .$$

If Gauss' theorem is applied to the preceding result we obtain, with the aid of (10.54),

$$\omega_n^2 Q_n(t) = \int_S (\tau_{ij}^{(n)} u_i^{(s)} - \tau_{ij}^{(s)} u_i^{(n)}) n_j dS - \int_V B_i u_i^{(n)} dV \quad .$$

The boundary conditions (10.55) for the quasi-static solution together with the corresponding homogeneous boundary conditions (10.40) for the modal solution are now substituted into the preceding result to obtain

$$\omega_n^2 Q_n(t) = - \int_V B_i u_i^{(n)} dV + \int_{S_u} T_i^{(n)} u_i^* dS - \int_{S_T} T_i^* u_i^{(n)} dS$$
$$+ \int_{S_{u_n}} (T_n^{(n)} u_n^* - T_t^* u_t^{(n)} - T_s^* u_s^{(n)}) dS$$
$$+ \int_{S_{T_n}} (-T_n^* u_n^{(n)} + T_t^{(n)} u_t^* + T_s^{(n)} u_s^*) dS \qquad (10.65)$$

Substitution of (10.60) into (10.56) yields the complete solution to the forced motion problem:

$$u_i(x,t) = u_i^{(s)}(x,t) + \sum_{n=1}^{\infty} u_i^{(n)}(x) q_n(t) \quad . \qquad (10.66)$$

The three principal parts of this solution are the quasi-static solution $u_i^{(s)}$, the free vibration solution $(u_i^{(n)}, \omega_n^2)$ and the expansion coefficients q_n. Once the eigenfunctions are known, the expansion coefficients are obtained from (10.63) or (10.64) by straightforward integrations of the body force, initial conditions and boundary conditions. Thus, the major effort required for this method of solution is in the resolution of the free vibration problem and the quasi-static problem. This method for the solution of boundary value problems in three-dimensional elastokinetics was first presented by H. Reismann in his paper, "On The Forced Motion of Elastic Solids," Applied Scientific Research, Vol. 18, 1967, pp. 156-165.

The representation (10.66) of the forced motion solution may be viewed as a generalization to continuous systems of the eigenfunction expansion (3.33) for discrete systems with an infinite number of degrees of freedom. In this analogy the mode shape functions $u_i^{(n)}(x)$ take the place of the modal coefficients a_{ir} and the time functions $q_n(t)$ act as the normal coordinates of the system corresponding to $\xi_r(t)$ in (3.33). Equations (10.62a) are, in effect, Lagrange's equations for the continuous system corresponding to (3.45) for the discrete system. These equations may be obtained directly from Hamilton's principle as follows. According to (9.120) and (9.121), Hamilton's principle for continuous systems takes the form

$$\int_{t_1}^{t_2}(\delta T - \delta U_T + \delta W)dt = 0 \quad . \qquad (10.67)$$

If we assume a displacement field of the form (10.66), then $T = T(\dot{q}_1, \dot{q}_2, \dot{q}_3 \ldots)$ and

$$\delta T = \sum_{n=1}^{\infty} \frac{\partial T}{\partial \dot{q}_n} \delta \dot{q}_n \quad .$$

Since $\delta q_n = 0$ at $t = t_1$ and $t = t_2$, we have

$$\int_{t_1}^{t_2} \delta T dt = \sum_{n=1}^{\infty} \int_{t_1}^{t_2} \frac{\partial T}{\partial \dot{q}_n} \delta \dot{q}_n dt$$

$$= \sum_{n=1}^{\infty} \left[\left(\frac{\partial T}{\partial \dot{q}_n} \delta q_n \right)_{t_1}^{t_2} - \int_{t_1}^{t_2} \frac{d}{dt}\left(\frac{\partial T}{\partial \dot{q}_n} \right) \delta q_n dt \right]$$

$$= - \int_{t_1}^{t_2} \sum_{n=1}^{\infty} \frac{d}{dt} \left(\frac{\partial T}{\partial \dot{q}_n} \right) \delta q_n dt \quad .$$

Similarly, the total potential energy U_T will be a function of the normal coordinates $q_n(t)$ and therefore,

$$\delta U_T = \sum_{n=1}^{\infty} \frac{\partial U_T}{\partial q_n} \delta q_n \quad .$$

The work done by the external forces, excluding the elastic restoring force $(-\alpha_{ij}u_j)$, is

$$\delta W = \int_V B_i \delta u_i dV + \int_{S_T} T_i^* \delta u_i dS + \int_{S_{T_n}} T_n^* \delta u_n dS + \int_{S_{u_n}} (T_t^* \delta u_t + T_s^* \delta u_s) dS \quad .$$

According to (10.66)

$$\delta u_i = \sum_{n=1}^{\infty} u_i^{(n)} \delta q_n$$

and therefore

$$\delta W = \sum_{n=1}^{\infty} \Theta_n \delta q_n$$

where

$$\Theta_n = \int_V B_i u_i^{(n)} dV + \int_{S_T} T_i^* u_i^{(n)} dS + \int_{S_{T_n}} T_n^* u_n^{(n)} dS + \int_{S_{u_n}} (T_t^* u_t^{(n)} + T_s^* u_s^{(n)}) dS \quad .$$

In the preceding equation, $u_n^{(n)} = u_i^{(n)} n_i$, $u_t^{(n)} = u_i^{(n)} t_i$ and $u_s^{(n)} = u_i^{(n)} s_i$. Substituting the above results into (10.67) yields the following form of Hamilton's principle.

$$- \int_{t_1}^{t_2} \left\{ \sum_{n=1}^{\infty} \left[\frac{d}{dt}\left(\frac{\partial T}{\partial \dot{q}_n}\right) + \frac{\partial U_T}{\partial q_n} - \Theta_n \right] \delta q_n \right\} dt = 0$$

Since the interval (t_1, t_2) is arbitrary and the variations δq_n are independent, we obtain Lagrange's equations for the system

$$\frac{d}{dt}\left(\frac{\partial T}{\partial \dot{q}_n}\right) + \frac{\partial U_T}{\partial q_n} = \Theta_n \quad . \qquad (10.68)$$

With the aid of (10.66), the kinetic energy may be expressed as follows:

$$T = \tfrac{1}{2} \int_V \rho \dot{u}_i \dot{u}_i dV = \tfrac{1}{2} \int_V \rho \Big[\dot{u}_i^{(s)} \dot{u}_i^{(s)} + 2 \sum_{n=1}^{\infty} \dot{q}_n \dot{u}_i^{(s)} u_i^{(n)} + \sum_{n=1}^{\infty} \sum_{m=1}^{\infty} \dot{q}_n \dot{q}_m u_i^{(n)} u_i^{(m)} \Big] dV \; .$$

In view of the orthonormality relations (10.53),

$$T = T^{(s)} + \sum_{n=1}^{\infty} \left[\tfrac{1}{2} \dot{q}_n^2 - \dot{Q}_n \dot{q}_n \right]$$

where

$$T^{(s)} = \tfrac{1}{2} \int_V \rho \dot{u}_i^{(s)} \dot{u}_i^{(s)} dV$$

and $Q_n(t)$ is given by (10.62d). Hence,

$$\frac{d}{dt}\left(\frac{\partial T}{\partial \dot{q}_n}\right) = \frac{d}{dt}(\dot{q}_n - \dot{Q}_n) = \ddot{q}_n - \ddot{Q}_n \quad . \tag{10.69}$$

Similarly, with the aid of (10.66), the total potential energy of the body may be expressed as

$$U_T = \tfrac{1}{2}\int_V \tau_{ij}u_{i,j}dV + \tfrac{1}{2}\int_{S_T}\alpha_{ij}u_iu_j dS$$

$$= \tfrac{1}{2}\int_V\left(\tau_{ij}^{(s)} + \sum_{n=1}^{\infty}\tau_{ij}^{(n)}q_n\right)\left(u_{i,j}^{(s)} + \sum_{n=1}^{\infty}u_{i,j}^{(n)}q_n\right)dV$$

$$+ \tfrac{1}{2}\int_{S_T}\alpha_{ij}\left(u_i^{(s)} + \sum_{n=1}^{\infty}u_i^{(n)}q_n\right)\left(u_j^{(s)} + \sum_{n=1}^{\infty}u_j^{(n)}q_n\right)dS$$

$$U_T = U_T^{(s)} + \sum_{n=1}^{\infty}q_n\left[\tfrac{1}{2}\int_V\left(\tau_{ij}^{(s)}u_{i,j}^{(n)} + \tau_{ij}^{(n)}u_{i,j}^{(s)}\right)dV + \int_{S_T}\alpha_{ij}u_i^{(n)}u_j^{(s)}dS\right]$$

$$+ \tfrac{1}{2}\sum_{n=1}^{\infty}\sum_{m=1}^{\infty}q_nq_m\left[\int_V\tau_{ij}^{(n)}u_{i,j}^{(m)}dV + \int_{S_T}\alpha_{ij}u_i^{(n)}u_j^{(m)}dS\right]$$

In view of (10.32a), (10.40), (10.54) and (10.55),

$$\tfrac{1}{2}\int_V(\tau_{ij}^{(s)}u_{i,j}^{(n)} + \tau_{ij}^{(n)}u_{i,j}^{(s)})dV + \int_{S_T}\alpha_{ij}u_i^{(n)}u_j^{(s)}dS$$

$$= \int_V\tau_{ij}^{(s)}u_{i,j}^{(n)}dV + \int_{S_T}\alpha_{ij}u_i^{(n)}u_j^{(s)}dS$$

$$= \int_V(\tau_{ij}^{(s)}u_i^{(n)})_{,j}dV - \int_V\tau_{ij,j}^{(s)}u_i^{(n)}dV + \int_{S_T}\alpha_{ij}u_i^{(n)}u_j^{(s)}dS$$

$$= \int_S\tau_{ij}^{(s)}n_ju_i^{(n)}dS + \int_{S_T}\alpha_{ij}u_i^{(n)}u_j^{(s)}dS + \int_V B_iu_i^{(n)}dV$$

$$= \int_{S_T}T_i^*u_i^{(n)}dS + \int_{S_{T_n}}T_n^*u_n^{(n)}dS + \int_{S_{u_n}}(T_t^*u_t^{(n)} + T_s^*u_s^{(n)})dS + \int_V B_iu_i^{(n)}dV = \Theta_n \ .$$

Furthermore, with the aid of (10.40), (10.42) and (10.53), the two integrals appearing in the double sum may be transformed as follows:

$$\int_V \tau_{ij}^{(n)} u_{i,j}^{(m)} dV + \int_{S_T} \alpha_{ij} u_i^{(n)} u_j^{(m)} dS$$

$$= \int_V (\tau_{ij}^{(n)} u_i^{(m)})_{,j}\, dV + \int_{S_T} \alpha_{ij} u_i^{(n)} u_j^{(m)} dS - \int_V \tau_{ij,j}^{(n)} u_i^{(m)} dV$$

$$= \int_S \tau_{ij}^{(n)} n_j u_i^{(m)} dS + \int_{S_T} \alpha_{ij} u_i^{(m)} u_j^{(n)} dS + \int_V \rho\omega_n^2 u_i^{(n)} u_i^{(m)} dV$$

$$= \int_{S-S_T} \tau_{ij}^{(n)} n_j u_i^{(m)} dS + \int_{S_T} (\tau_{ij}^{(n)} n_j + \alpha_{ij} u_j^{(n)}) u_i^{(m)} dS + \omega_n^2 \delta_{nm} = \omega_n^2 \delta_{nm}$$

The total potential energy of the body is thus given by

$$U_T = U_T^{(s)} + \sum_{n=1}^{\infty} \left[\tfrac{1}{2}\omega_n^2 q_n^2 + q_n \Theta_n \right]$$

where

$$U_T^{(s)} = \tfrac{1}{2} \int_V \tau_{ij}^{(s)} u_{i,j}^{(s)} dV + \tfrac{1}{2} \int_{S_T} \alpha_{ij} u_i^{(s)} u_j^{(s)} dS \quad .$$

Therefore,

$$\frac{\partial U_T}{\partial q_n} = \omega_n^2 q_n + \Theta_n \quad . \tag{10.70}$$

Substituting (10.69) and (10.70) into (10.68) yields the following form of Lagrange's equations

$$\ddot{q}_n + \omega_n^2 q_n = \ddot{Q}_n$$

which is precisely what we obtained earlier as (10.62a) by an entirely different route.

The eigenfunctions and eigenvalues of the free vibration problem may also be used to represent the solution of static elastic problems. For example, consider the case

$$\tau_{ij,j} = - B_i(x) \qquad \text{in } V \tag{10.71}$$

$$u_i = u_i^*(x) \qquad \text{on } S_u$$

$$\tau_{ij} n_j = T_i^*(x) \qquad \text{on } S_T \tag{10.72}$$

where

$$S = S_u + S_T \quad .$$

Let us assume that

$$u_i(x) = G_i(x) + \sum_{n=1}^{\infty} q_n u_i^{(n)}(x) \tag{10.73}$$

where $G_i(x)$ is any twice differentiable vector field defined in V which satisfies the boundary conditions

$$G_i(x) = u_i^*(x) \qquad \text{on } S_u$$

and (10.74)

$$G_{ij}n_j = T_i^*(x) \qquad \text{on } S_T$$

with

$$G_{ij} = \lambda G_{k,k}\delta_{ij} + \mu(G_{i,j} + G_{j,i}) \quad .$$

The eigenfunctions $u_i^{(n)}(x)$ satisfy

$$\tau_{ij,j}^{(n)} = -\rho\omega_n^2 u_i^{(n)} \qquad \text{in V} \tag{10.75}$$

$$u_i^{(n)} = 0 \quad \text{on } S_u \text{ and } \tau_{ij}^{(n)} n_j = 0 \qquad \text{on } S_T \quad .$$

Substituting (10.73) into (10.71) yields, with the aid of (10.75),

$$\sum_{n=1}^{\infty} \rho\omega_n^2 u_i^{(n)} q_n = B_i + G_{ij,j} \quad .$$

Multiply both sides of this equation by $u_i^{(m)}(x)$ and then integrate the result over the region V, making use of the orthonormality relations (10.53) to obtain

$$\omega_n^2 q_n = \int_V u_i^{(n)}(B_i + G_{ij,j})dV \quad . \tag{10.76}$$

This completes the solution of the static elastic example.

10.6 FORCED MOTION OF A SPHERICAL SHELL

To illustrate the method presented in the preceding section we will consider the point symmetric forced motion of a homogeneous, isotropic spherical shell. In view of the assumed point symmetry, the displacement field will have the form

$$u_r = u_r(r,t), \; u_\Theta = u_\phi = 0 \quad . \tag{10.77}$$

By substituting (10.77) into (8.117) we obtain

$$\varepsilon_{rr} = \frac{\partial u_r}{\partial r} , \quad \varepsilon_{\Theta\Theta} = \varepsilon_{\phi\phi} = \frac{u_r}{r}$$

$$\varepsilon_{r\Theta} = \varepsilon_{\Theta\phi} = \varepsilon_{\phi r} = 0 \quad .$$

The dilatation ε is given by

$$\varepsilon = \varepsilon_{rr} + \varepsilon_{\Theta\Theta} + \varepsilon_{\phi\phi} = \frac{\partial u_r}{\partial r} + \frac{2u_r}{r}$$

and with the aid of Hooke's law (10.22) we obtain the following stress-displacement relations:

$$\left.\begin{aligned} \tau_{rr} &= \lambda\varepsilon + 2\mu\varepsilon_{rr} = (\lambda+2\mu)\frac{\partial u_r}{\partial r} + 2\lambda\frac{u_r}{r} \\ \tau_{\Theta\Theta} &= \tau_{\phi\phi} = \lambda\varepsilon + 2\mu\varepsilon_{\Theta\Theta} = \lambda\frac{\partial u_r}{\partial r} + 2(\lambda+\mu)\frac{u_r}{r} \\ \tau_{r\Theta} &= \tau_{\Theta\phi} = \tau_{\phi r} = 0 \end{aligned}\right\} \tag{10.79}$$

The stress equations of motion (see Exercise 11, Chapter 9) in this case reduce to

$$\frac{\partial\tau_{rr}}{\partial r} + \frac{2\tau_{rr} - \tau_{\Theta\Theta} - \tau_{\phi\phi}}{r} + B_r = \rho\frac{\partial^2 u_r}{\partial t^2} \quad . \tag{10.80}$$

Substituting (10.79) into (10.80) yields the displacement equation of motion

$$\frac{\partial^2 u_r}{\partial r^2} + \frac{2}{r}\frac{\partial u_r}{\partial r} - \frac{2u_r}{r^2} + \frac{B_r}{\rho C_D^2} = \frac{1}{C_D^2}\frac{\partial^2 u_r}{\partial t^2} \tag{10.81}$$

where $C_D = \sqrt{\frac{\lambda+2\mu}{\rho}}$ is called the dilatational wave speed. Let R_I and R_o denote the inner and outer radii, respectively, of the concentric spherical surfaces which form the boundary of the shell. Then, at $r=R_I$, $\hat{n}=-\hat{e}_r$ and therefore $T_r=\tau_{rr}n_r=-\tau_{rr}$. Similarly, at $r=R_o$, $\hat{n}=\hat{e}_r$ and $T_r=\tau_{rr}n_r=\tau_{rr}$. The boundary conditions may thus be stated as follows:

At $r=R_I$, $u_r(R_I,t) = u_I^*(t)$ (10.82a)

or

$-\tau_{rr}(R_I,t) = T_I^*(t)$ (10.82b)

or

$-\tau_{rr} + \alpha_I u_r = T_I^*(t)$. (10.82c)

At $r=R_o$, $u_r(R_o,t) = u_o^*(t)$ (10.82d)

or

$$\tau_{rr}(R_o,t) = T_o^*(t) \tag{10.82e}$$

or

$$\tau_{rr} + \alpha_o u_r = T_o^*(t) \tag{10.82f}$$

where $\alpha_I > 0$ and $\alpha_o > 0$ are elastic foundation moduli.

The first step in our solution of the forced motion problem is to obtain the eigenvalues and eigenfunctions from the associated free vibration problem. In the present case the free vibration problem is characterized by the differential equation

$$\frac{d^2u_r}{dr^2} + \frac{2}{r}\frac{du_r}{dr} - \frac{2u_r}{r^2} = \frac{-\omega^2}{C_D^2}u_r \quad . \tag{10.81'}$$

The corresponding boundary conditions may be enumerated as follows:

If $u_r(R_I,t) = u_I^*(t)$ then $u_r(R_I) = 0$. (10.82'a)

If $-\tau_{rr}(R_I,t) = T_I^*(t)$ then $\tau_{rr}(R_I) = 0$. (10.82'b)

If $-\tau_{rr}(R_I,t) + \alpha_I u_r(R_I,t) = T_I^*(t)$

then $\tau_{rr}(R_I) - \alpha_I u_r(R_I) = 0$. (10.82'c)

If $u_r(R_o,t) = u_o^*(t)$ then $u_r(R_o) = 0$. (10.82'd)

If $\tau_{rr}(R_o,t) = T_o^*(t)$ then $\tau_{rr}(R_o) = 0$. (10.82'e)

If $\tau_{rr}(R_o,t) + \alpha_o u_r(R_o,t) = T_o^*(t)$

then $\tau_{rr}(R_o) + \alpha_o u_r(R_o) = 0$. (10.82'f)

By letting $z = \frac{\omega r}{C_D}$, (10.81') may be written as follows:

$$\frac{d^2u_r}{dz^2} + \frac{2}{z}\frac{du_r}{dz} + \left(1 - \frac{2}{z^2}\right)u_r = 0 \quad . \tag{10.83}$$

The solution of (10.83) is

$$u_r(z) = Aj_1(z) - By_1(z) \tag{10.84}$$

where

$$j_1(z) = \frac{\sin z}{z^2} - \frac{\cos z}{z}$$

and

$$y_1(z) = -\frac{\cos z}{z^2} - \frac{\sin z}{z}$$

are called spherical Bessel functions of the first and second kinds, respectively, of order one (see for example, M. Abramowitz and I.A. Stegun, "Handbook of Mathematical Functions," Dover, 1965, pp. 437-438). The arbitrary constants A and B will be determined so as to satisfy the boundary conditions (10.82') and the normalization condition. With the aid of (10.79), the stresses may be written as follows:

$$\left.\begin{aligned} \tau_{rr} &= \rho\omega C_D\left[\frac{du_r}{dz} + 2\,(1-2\gamma^2)\,\frac{u_r}{z}\right] \\ \tau_{\Theta\Theta} = \tau_{\phi\phi} &= \rho\omega C_D\left[(1-2\gamma^2)\,\frac{du_r}{dz} + 2\,(1-\gamma^2)\,\frac{u_r}{z}\right] \end{aligned}\right\} \tag{10.86}$$

where $\gamma^2 = \dfrac{C_S^2}{C_D^2} = \dfrac{\mu}{\lambda+2\mu} = \dfrac{1-2\upsilon}{2(1-\upsilon)}$. (10.87)

We shall call γ the wave speed ratio ($C_S = \sqrt{\frac{\mu}{\rho}}$ is the shear wave speed). By differentiating (10.85) with respect to z we find that

$$\frac{dj_1}{dz} = j_o(z) - \frac{2}{z}j_1(z)$$

and

$$\frac{dy_1}{dz} = y_o(z) - \frac{2}{z}y_1(z) \tag{10.88}$$

where

$$j_o(z) = \frac{\sin z}{z} \quad\text{and}\quad y_o(z) = -\frac{\cos z}{z}$$

are the spherical Bessel functions of order zero. Substituting (10.84) into (10.86) yields, with the aid of (10.88),

$$\tau_{rr}(z) = \rho\omega C_D\left[(j_o - \frac{4\gamma^2}{z}\,j_1)A - (y_o - \frac{4\gamma^2}{z}\,y_1)B\right] . \tag{10.89}$$

In order to proceed further and determine a set of natural frequencies for the shell we must specify a particular set of boundary conditions. Because of the flexibility they provide we shall consider the boundary conditions

$$\tau_{rr} - \alpha_I u_r = 0 \qquad \text{at } r = R_I \tag{10.90a}$$

and

$$\tau_{rr} + \alpha_o u_r = 0 \qquad \text{at } r = R_o . \tag{10.90b}$$

By extending the domain of the constants α_I and α_o to include zero and infinity we may obtain all the various possible boundary conditions from (10.90). For example, by setting $\alpha_I = 0$ we obtain the condition $\tau_{rr}(R_I) = 0$. On the other hand, if we divide (10.90a) by α_I and then let α_I approach infinity we obtain the condition $u_r(R_I) = 0$. Similar arguments apply at R_o as α_o varies between zero and infinity. If we let

$$a = \frac{\omega R_I}{C_D} , \quad b = \frac{\omega R_o}{C_D} , \quad \bar{\alpha}_I = \frac{\alpha_I R_I}{\rho C_D^2} \quad \text{and} \quad \bar{\alpha}_o = \frac{\alpha_o R_o}{\rho C_D^2} \tag{10.91}$$

then equations (10.90) may be written as follows:

$$\left.\begin{aligned} \tau_{rr}(a) - \rho\omega C_D \frac{\bar{\alpha}_I}{a} u_r(a) &= 0 \\ \tau_{rr}(b) + \rho\omega C_D \frac{\bar{\alpha}_o}{b} u_r(b) &= 0 \quad . \end{aligned}\right\} \tag{10.90'}$$

By substituting (10.84) and (10.89) into (10.90') we obtain the following set of equations for A and B.

$$\left.\begin{aligned} \left[j_o(a) - \frac{4\gamma^2 + \bar{\alpha}_I}{a} j_1(a)\right] A - \left[y_o(a) - \frac{4\gamma^2 + \bar{\alpha}_I}{a} y_1(a)\right] B &= 0 \\ \left[j_o(b) - \frac{4\gamma^2 - \bar{\alpha}_o}{b} j_1(b)\right] A - \left[y_o(b) - \frac{4\gamma^2 - \bar{\alpha}_o}{b} y_1(b)\right] B &= 0 \end{aligned}\right\} \tag{10.92}$$

Equations (10.92) will possess a nontrivial solution if and only if the determinant of their coefficient matrix vanishes, i.e.,

$$\left[j_o(a) - \frac{4\gamma^2 + \bar{\alpha}_I}{a} j_1(a)\right]\left[y_o(b) - \frac{4\gamma^2 - \bar{\alpha}_o}{b} y_1(b)\right] - \left[j_o(b) - \frac{4\gamma^2 - \bar{\alpha}_o}{b} j_1(b)\right]\left[y_o(a) - \frac{4\gamma^2 + \bar{\alpha}_I}{a} y_1(a)\right] = 0 \quad . \tag{10.93}$$

Equation (10.93) will be recast into a more familiar form with the aid of the following identities which were deduced from (10.85) and (10.88).

$$j_o(a)y_o(b) - j_o(b)y_o(a) = \frac{\sin(b-a)}{ab} \tag{10.94a}$$

$$j_o(a)y_1(b) - j_1(b)y_o(a) = \frac{\sin(b-a)}{ab^2} - \frac{\cos(b-a)}{ab} \tag{10.94b}$$

$$j_1(a)y_o(b) - j_o(b)y_1(a) = \frac{\sin(b-a)}{a^2 b} + \frac{\cos(b-a)}{ab} \tag{10.94c}$$

$$j_1(a)y_1(b) - j_1(b)y_1(a) = \left(\frac{1+ab}{a^2b^2}\right)\sin(b-a) - \left(\frac{b-a}{a^2b^2}\right)\cos(b-a) \tag{10.94d}$$

In view of (10.94), (10.93) may be written as

$$\tan (b-a) = \frac{4\gamma^2 ab(b-a)+ab(b\bar{\alpha}_I+a\bar{\alpha}_o)+(b-a)(4\gamma^2+\bar{\alpha}_I)(4\gamma^2-\bar{\alpha}_o)}{a^2b^2-4\gamma^2(a^2+b^2)+a^2\bar{\alpha}_o-b^2\bar{\alpha}_I+(1+ab)(4\gamma^2+\bar{\alpha}_I)(4\gamma^2-\bar{\alpha}_o)} \tag{10.95}$$

Equation (10.95) is easily modified to account for the various boundary conditions enumerated in (10.82'). For example, the frequency equation for a spherical shell with both surfaces rigidly fixed ($u_r=0$ at $r=R_I$ and $r=R_o$) is obtained by dividing both numerator and denominator on the right-hand side of (10.95) by $\bar{\alpha}_I\bar{\alpha}_o$ and then letting $\bar{\alpha}_I\to\infty$ and $\bar{\alpha}_o\to\infty$. The result is

$$\tan(b-a) = \frac{b-a}{1+ab} . \tag{10.96}$$

On the other hand, by setting $\bar{\alpha}_I=\bar{\alpha}_o=0$ in (10.95) we obtain the frequency equation for a spherical shell with both surfaces stress free ($\tau_{rr}= 0$ at $r=R_I$ and $r=R_o$), and this results in

$$\tan(b-a) = \frac{4\gamma^2(b-a)(ab+4\gamma^2)}{a^2b^2-4\gamma^2(a^2+b^2)+16\gamma^4(1+ab)} . \tag{10.97}$$

If we define a dimensionless frequency

$$\Omega=b-a= (R_o-R_I) \frac{\omega}{c_D} \tag{10.98}$$

then the frequency equation (10.95) or (10.96) or (10.97) assumes the form

$$\tan \Omega=f(\Omega) \tag{10.99}$$

where f is a real rational function of Ω. For example, (10.96) becomes

$$\tan \Omega = \frac{(1-\beta)^2\Omega}{\beta\Omega^2+(1-\beta)^2} = \left[f(\Omega)\right]_{\text{fixed}} \tag{10.96'}$$

and (10.97) becomes

$$\tan \Omega = \frac{4\gamma^2(1-\beta)^2\left[\beta\Omega^2+4\gamma^2(1-\beta)^2\right]\Omega}{\beta^2\Omega^4-4\gamma^2(1-\beta)^2(1+\beta^2-4\gamma^2\beta)\Omega^2+16\gamma^4(1-\beta)^4} = \left[f(\Omega)\right]_{\text{free}} \tag{10.97'}$$

where

$$0 < \beta = \frac{R_I}{R_o} < 1 .$$

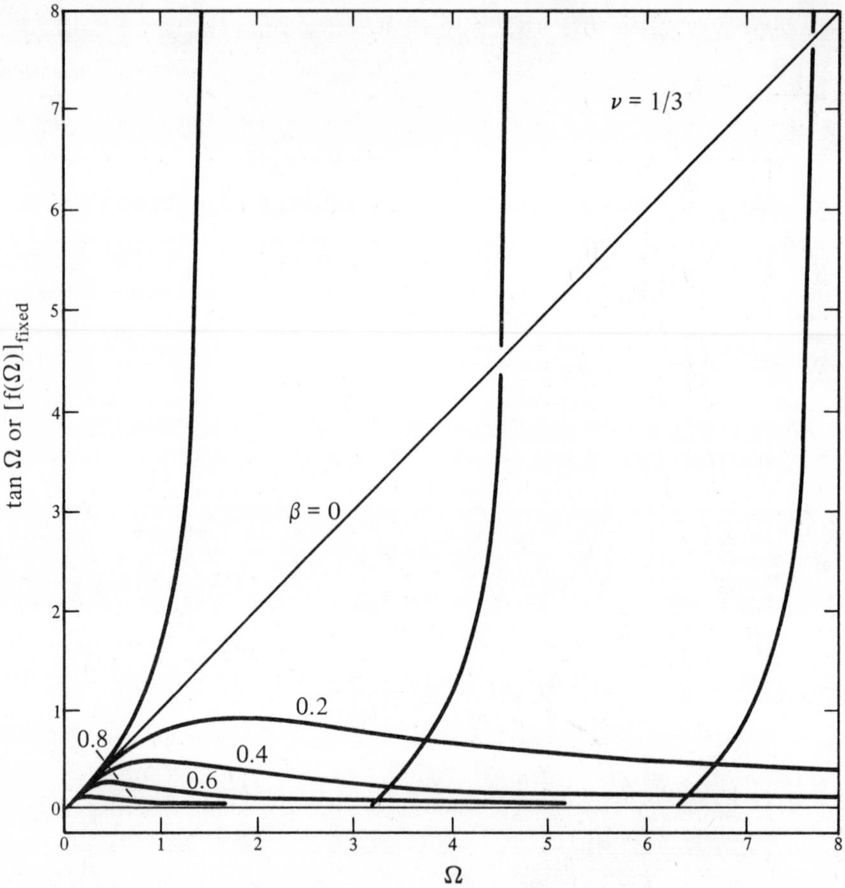

Figure 10.1

In Figs. 10.1 and 10.2 we show the graphs of tan Ω, $\left[f(\Omega)\right]_{fixed}$ and $\left[f(\Omega)\right]_{free}$ to illustrate the distribution of the roots of (10.96') and (10.97') for various values of β,the radius ratio. The value of γ^2 in (10.97') was computed from (10.87) with $\upsilon=\frac{1}{3}$. It follows from (10.99) and it is evident from Figs. 10.1 and 10.2 that, for any of the boundary conditions (10.82') and for any value of β between zero and one, the frequency equation has a denumerable infinity of positive real roots. If we denote these roots by Ω_n,n=1,2,3,... then

$$0<\Omega_1<\Omega_2<\Omega_3<\ldots$$

and

$$\lim_{n\to\infty}\Omega_n=\infty\ .$$

The natural frequencies of the shell are now given by

$$\omega_n = \frac{C_D \Omega_n}{R_o - R_I} \quad , \; n=1,2,3,\ldots \tag{10.100}$$

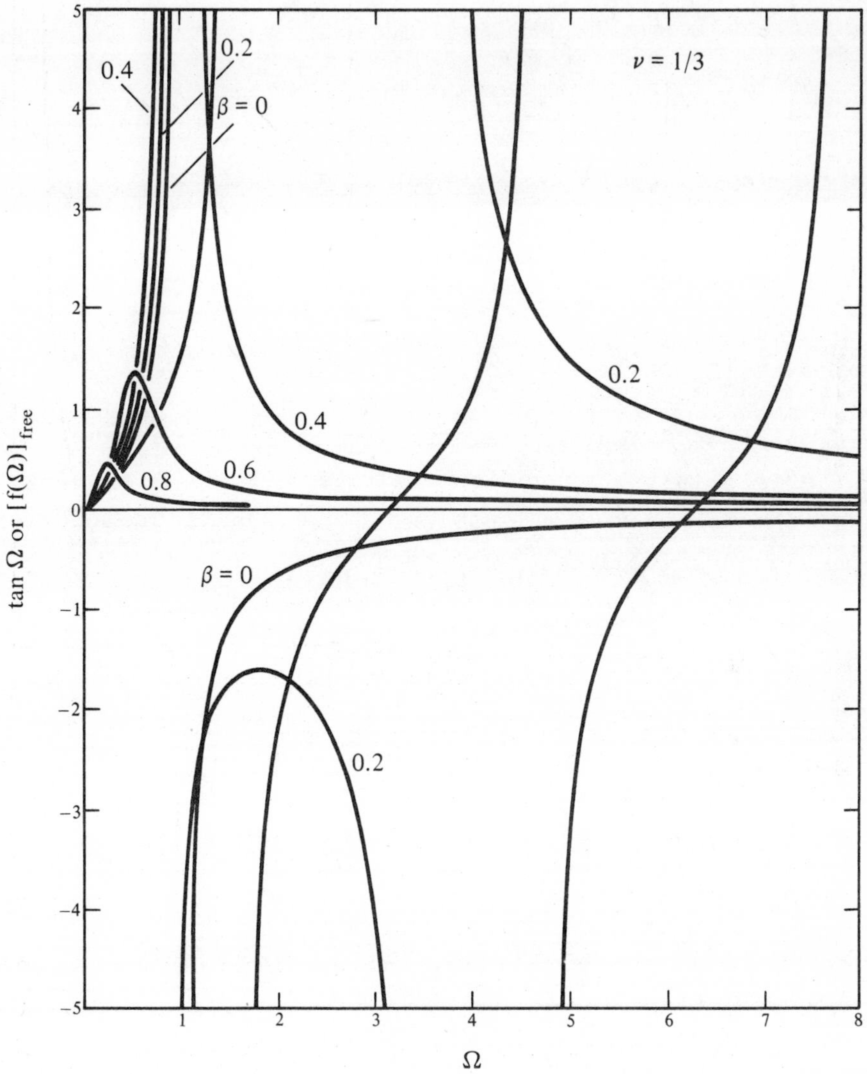

Figure 10.2

Corresponding to each natural frequency, there is a nontrivial solution of (10.92) which may be written as

$$\frac{B_n}{A_n} = \frac{j_o(a_n) - \dfrac{4\gamma^2 + \bar{\alpha}_I}{a_n} j_1(a_n)}{y_o(a_n) - \dfrac{4\gamma^2 + \bar{\alpha}_I}{a_n} y_1(a_n)} = \frac{j_o(b_n) - \dfrac{4\gamma^2 - \bar{\alpha}_o}{b_n} j_1(b_n)}{y_o(b_n) - \dfrac{4\gamma^2 - \bar{\alpha}_o}{b_n} y_1(b_n)} \tag{10.101}$$

where $a_n = \dfrac{R_I \omega_n}{C_D}$ and $b_n = \dfrac{R_o \omega_n}{C_D}$.

In order to uniquely determine the constants A_n and B_n we impose the normalization condition

$$\left[\int_{R_I}^{R_o} \rho u_r^{(n)2} r^2 dr\right]^{\frac{1}{2}} = 1 \quad . \tag{10.102}$$

Because of the point symmetry inherent in this example,

$$\int_V f(r)dV = 4\pi \int_{R_I}^{R_o} f(r)r^2 dr$$

and

$$\int_S f(r)dS = 4\pi\left[R_o^2 f(R_o) + R_I^2 f(R_I)\right] \quad .$$

Thus, the factor of 4π may be deleted from all of the integral relations derived in the preceding sections without affecting the solution of the forced motion problem. If we substitute (10.84) into (10.102) and then solve for A_n we obtain

$$A_n = \left\{\int_{R_I}^{R_o} \rho r^2 \left[j_1\left(\frac{\omega_n r}{C_D}\right) - \frac{B_n}{A_n} y_1\left(\frac{\omega_n r}{C_D}\right)\right]^2 dr\right\}^{-\frac{1}{2}} \tag{10.103}$$

where $\dfrac{B_n}{A_n}$ is given by (10.101). The integrations required to evaluate A_n by means of (10.103) are not difficult but they are rather tedious. However, with the aid of the reciprocal relation (10.33), the normalization integral in (10.102) may be evaluated by a process of differentiation rather than integration. To illustrate this we let the states (1) and (2) in (10.33) correspond to the eigenstates (n) and (m) respectively. Thus, (10.33) becomes

$$\int_V \left(\tau_{ij,j}^{(m)} u_i^{(n)} - \tau_{ij,j}^{(n)} u_i^{(m)}\right) dV = \int_S \left(T_i^{(m)} u_i^{(n)} - T_i^{(n)} u_i^{(m)}\right) dS \ .$$

According to (10.42),

$$\tau_{ij,j}^{(m)} = -\rho\omega_m^2 u_i^{(m)} \text{ and } \tau_{ij,j}^{(n)} = -\rho\omega_n^2 u_i^{(n)} .$$

Thus, the reciprocal relation becomes

$$(\omega_n^2 - \omega_m^2)\int_V \rho u_i^{(n)} u_i^{(m)} dV = \int_S (\tau_{ij}^{(m)} u_i^{(n)} - \tau_{ij}^{(n)} u_i^{(m)}) n_j dS \quad .$$

If we make use of the point symmetry involved in the present problem, the preceding result simplifies to

$$(\omega_n^2-\omega_m^2)\int_{R_I}^{R_o}\rho u_r^{(m)}u_r^{(n)}r^2dr = \left[r^2(\tau_{rr}^{(m)}u_r^{(n)}-\tau_{rr}^{(n)}u_r^{(m)})\right]_{R_I}^{R_o}$$

or

$$\int_{R_I}^{R_o}\rho u_r^{(n)}u_r^{(m)}r^2dr = \frac{\left[r^2(\tau_{rr}^{(m)}u_r^{(n)}-\tau_{rr}^{(n)}u_r^{(m)})\right]_{R_I}^{R_o}}{\omega_n^2-\omega_m^2} \quad . \tag{10.104}$$

If we let m=n in (10.104) the left-hand side of the equation becomes the desired normalization integral, however, the right-hand side assumes the indeterminate form $\frac{0}{0}$. This difficulty may be overcome by applying L'Hospital's rule to the right-hand side of (10.104). Thus, we differentiate both numerator and denominator with respect to ω_m and then take the limit as ω_m approaches ω_n to obtain

$$\int_{R_I}^{R_o}\rho u_r^{(n)^2}r^2dr = \lim_{\omega_m\to\omega_n}\left\{-\frac{1}{2\omega_m}\left[r^2\left(\frac{d\tau_{rr}^{(m)}}{d\omega_m}u_r^{(n)}-\tau_{rr}^{(n)}\frac{du_r^{(m)}}{d\omega_m}\right)\right]_{R_I}^{R_o}\right\}$$

$$= \frac{1}{2\omega_n}\left[r^2\left(\tau_{rr}^{(n)}\frac{du_r^{(n)}}{d\omega_n} - u_r^{(n)}\frac{d\tau_{rr}^{(n)}}{d\omega_n}\right)\right]_{R_I}^{R_o} \quad . \tag{10.105}$$

This expression can be simplified further by making use of the applicable boundary conditions. For example, if $u_r^{(n)}=0$ at $r=R_I$ and $r=R_o$, then

$$\int_{R_I}^{R_o}\rho u_r^{(n)^2}r^2dr = \frac{1}{2\omega_n}\left[r^2\tau_{rr}^{(n)}\frac{du_r^{(n)}}{d\omega_n}\right]_{R_I}^{R_o} \quad . \tag{10.106}$$

If $\tau_{rr}^{(n)}=0$ at $r=R_I$ and $r=R_o$, then

$$\int_{R_I}^{R_o}\rho u_r^{(n)^2}r^2dr = \frac{-1}{2\omega_n}\left[r^2u_r^{(n)}\frac{d\tau_{rr}^{(n)}}{d\omega_n}\right]_{R_I}^{R_o} \quad . \tag{10.107}$$

If $\tau_{rr}^{(n)}-\alpha_I u_r^{(n)}=0$ at $r=R_I$ and $\tau_{rr}^{(n)}+\alpha_o u_r^{(n)}=0$ at $r=R_o$, then

$$\int_{R_I}^{R_o}\rho u_r^{(n)^2}r^2dr = -\frac{R_o^2}{2\omega_n}\left[u_r^{(n)}\frac{d}{d\omega_n}(\tau_{rr}^{(n)}+\alpha_o u_r^{(n)})\right]_{r=R_o}$$

$$+\frac{R_I^2}{2\omega_n}\left[u_r^{(n)}\frac{d}{d\omega_n}(\tau_{rr}^{(n)}-\alpha_I u_r^{(n)})\right]_{r=R_I} \tag{10.108}$$

In every case the normalization integral is evaluated without actually performing any integrations.

The specific forced motion example we wish to examine is that of a spherical shell which is suddenly loaded at t=0 by internal and external pressures P_I and P_o, respectively. We shall assume the shell to be free of any body forces, i.e., $B_r=0$, and initially at rest in its reference configuration so that

$$u_r(r,0) = \dot{u}_r(r,0) = 0 \quad . \tag{10.109}$$

The surface tractions for this case are $T^*_o = -P_oH(t)$ and $T^*_I = P_IH(t)$. The proper boundary conditions are, according to (10.82)

$$\left.\begin{aligned} &\tau_{rr}(R_I,t) = -P_IH(t) \\ &\text{and} \\ &\tau_{rr}(R_o,t) = -P_oH(t) \quad . \end{aligned}\right\} \tag{10.110}$$

The eigenfunctions of the free vibration problem satisfy the corresponding homogeneous boundary conditions

$$\tau_{rr}^{(n)}(R_I) = \tau_{rr}^{(n)}(R_o) = 0 \quad . \tag{10.111}$$

The natural frequencies for this case are obtained by finding the roots of (10.97'). These roots are illustrated in Fig. 10.2 for various values of β and for $\upsilon = \frac{1}{3}$. The normalization integral for this case is given by (10.107). By differentiating the first of (10.86) with respect to ω we obtain, with the aid of (10.85) and (10.88)

$$\frac{d\tau_{rr}^{(n)}}{d\omega_n} = -\rho C_D\left\{\left[(4\gamma^2-1)j_0(z_n)+(z_n-\frac{8\gamma^2}{z_n})j_1(z_n)\right]A_n - \left[(4\gamma^2-1)y_0(z_n)+(z_n-\frac{8\gamma^2}{z_n})y_1(z_n)\right]B_n\right\}$$

where $z_n = \frac{r\omega_n}{C_D}$. Substituting this result, together with (10.84) into (10.107) yields

$$\int_{R_I}^{R_o} \rho u_r^{(n)2} r^2 dr = \frac{\rho C_D R_o^2}{2\omega_n} A_n^2 I_n \tag{10.112}$$

where

$$I_n = \left\{ \frac{r^2}{R_o^2} (4\gamma^2-1)\left[j_o(z_n) - \frac{B_n}{A_n} y_o(z_n)\right]\left[j_1(z_n) - \frac{B_n}{A_n} y_1(z_n)\right] + \frac{r^2}{R_o^2}\left[z_n - \frac{8\gamma^2}{z_n}\right]\left[j_1(z_n) - \frac{B_n}{A_n} y_1(z_n)\right]^2 \right\}_{R_I}^{R_o}$$

Let

$$f_o(z_n) \equiv j_o(z_n) - \frac{B_n}{A_n} y_o(z_n)$$

and

$$f_1(z_n) \equiv j_1(z_n) - \frac{B_n}{A_n} y_1(z_n) \quad . \tag{10.113}$$

In view of (10.92), we conclude that

$$f_o(z_n) = \frac{4\gamma^2}{z_n} f_1(z_n) \qquad \text{at } z_n = (a_n \text{ or } b_n) \quad . \tag{10.114}$$

Thus,

$$I_n = \left\{ \frac{r^2}{R_o^2}\left[z_n + \frac{4\gamma^2(4\gamma^2-3)}{z_n}\right] f_1^2(z_n) \right\}_{R_I}^{R_o}$$

or

$$I_n = \beta^2 b_n f_1^2(a_n) D_n$$

where

$$D_n \equiv \alpha_n^2\left[1 + \frac{4\gamma^2(4\gamma^2-3)}{b_n^2}\right] - \beta\left[1 + \frac{4\gamma^2(4\gamma^2-3)}{a_n^2}\right]$$

and

$$\alpha_n \equiv \frac{b_n}{a_n}\frac{f_1(b_n)}{f_1(a_n)} \quad . \tag{10.115}$$

In view of (10.114), (10.85), (10.88) and (10.101) we conclude that

$$f_1(z_n) = -\frac{1}{z_n}\left[\left(1 - \frac{4\gamma^2}{z_n^2}\right)\cos z_n - \frac{4\gamma^2}{z_n}\sin z_n\right]^{-1} \quad \text{at } z_n = a_n \text{ or } b_n$$

Thus,

$$\alpha_n = \frac{\left(1 - \frac{4\gamma^2}{a_n^2}\right)\cos a_n - \frac{4\gamma^2}{a_n}\sin a_n}{\left(1 - \frac{4\gamma^2}{b_n^2}\right)\cos b_n - \frac{4\gamma^2}{b_n}\sin b_n} \quad . \tag{10.116}$$

By substituting (10.112) into the normalization condition (10.102) we obtain

$$A_n = \sqrt{\frac{2\omega_n}{\rho C_D R_o^2 I_n}} \quad . \tag{10.117}$$

The normalized eigenfunctions and modal stresses are thus given by

$$u_r^{(n)}(r) = A_n f_1(z_n)$$

$$\tau_{rr}^{(n)}(r) = \rho C_D \omega_n A_n \left[f_o(z_n) - \frac{4\gamma^2}{z_n} f_1(z_n) \right]$$

$$\tau_{\Theta\Theta}^{(n)}(r) = \tau_{\phi\phi}^{(n)}(r) = \rho C_D \omega_n A_n \left[(1-2\gamma^2) f_o(z_n) + \frac{2\gamma^2}{z_n} f_1(z_n) \right] \tag{10.118}$$

where $z_n = \dfrac{r\omega_n}{C_D}$ and f_o, f_1 and A_n are defined by (10.113) and (10.117) respectively.

The static solution for the present case satisfies the differential equation

$$\frac{\partial^2 u_r^{(s)}}{\partial r^2} + \frac{2}{r}\frac{\partial u_r^{(s)}}{\partial r} - \frac{2u_r^{(s)}}{r^2} = 0 \tag{10.119}$$

and the boundary conditions

$$\left.\begin{aligned} &\tau_{rr}^{(s)}(R_I, t) = -P_I H(t) \\ &\text{and} \\ &\tau_{rr}^{(s)}(R_o, t) = -P_o H(t) \end{aligned}\right\} \tag{10.120}$$

where $\tau_{rr}^{(s)}$, $\tau_{\Theta\Theta}^{(s)}$ and $\tau_{\phi\phi}^{(s)}$ are determined by substituting $u_r^{(s)}$ into (10.79). It is not difficult to show that (10.119) and (10.120) are satisfied by the displacement and stress fields

$$\left.\begin{aligned} u_r^{(s)}(r,t) &= \frac{R_I H(t)}{1-\beta^3}\left[\frac{\beta^3 P_I - P_o}{3\lambda+2\mu}\frac{r}{R_I} + \frac{P_I - P_o}{4\mu}\frac{R_I^2}{r^2}\right] \\ \tau_{rr}^{(s)}(r,t) &= \frac{H(t)}{1-\beta^3}\left[\beta^3 P_I - P_o + (P_o - P_I)\frac{R_I^3}{r^3}\right] \\ \tau_{\Theta\Theta}^{(s)} = \tau_{\phi\phi}^{(s)} &= \frac{H(t)}{1-\beta^3}\left[\beta^3 P_I - P_o + \frac{P_I - P_o}{2}\frac{R_I^3}{r^3}\right] \end{aligned}\right\} \tag{10.121}$$

The normal coordinates of the shell $q_n(t)$ are obtained as follows. In the present case, (10.65) becomes

$$\omega_n^2 Q_n(t) = -R_o^2 T_o^* u_r^{(n)}(R_o) - R_I^2 T_I^* u_r^{(n)}(R_I)$$

or

$$Q_n(t) = \frac{1}{\omega_n^2}\left[R_o^2 P_o u_r^{(n)}(R_o) - R_I^2 P_I u_r^{(n)}(R_I)\right] H(t) \qquad (10.122)$$

By substituting (10.122) and (10.109) into (10.63) or (10.64) we obtain

$$q_n(t) = \left[R_o^2 P_o u_r^{(n)}(R_o) - R_I^2 P_I u_r^{(n)}(R_I)\right] \frac{\cos \omega_n t}{\omega_n^2}$$

or

$$q_n(t) = -R_I^2 P_I A_n f_1(a_n)\left(1 - \frac{\alpha_n}{\beta}\frac{P_o}{P_I}\right)\frac{\cos \omega_n t}{\omega_n^2} . \qquad (10.123)$$

The complete solution of the forced motion problem is obtained by substituting (10.123), (10.121) and (10.118) into (10.66). With the aid of (10.115), (10.116) and (10.117) this result may be expressed as follows:

$$\tilde{u}_r(r,t) = \tilde{u}_r^{(s)}(r,t) - 2(1-\beta)^2 \sum_{n=1}^{\infty}\left(1 - \frac{\alpha_n}{\beta}\frac{P_o}{P_I}\right)\frac{f_1(z_n)}{f_1(a_n)} \cdot \frac{\cos \Omega_n \tau}{\Omega_n^2 D_n}$$

$$\tilde{\tau}_{rr}(r,t) = \tilde{\tau}_{rr}^{(s)}(r,t) - 2(1-\beta)\sum_{n=1}^{\infty}\left(1 - \frac{\alpha_n}{\beta}\frac{P_o}{P_I}\right)\frac{f_0(z_n) - \frac{4\gamma^2}{z_n} f_1(z_n)}{f_1(a_n)} \cdot \frac{\cos \Omega_n \tau}{\Omega_n D_n}$$

$$\tilde{\tau}_{\phi\phi}(r,t) = \tilde{\tau}_{\phi\phi}^{(s)}(r,t) - 2(1-\beta)\sum_{n=1}^{\infty}\left(1 - \frac{\alpha_n}{\beta}\frac{P_o}{P_I}\right)\frac{(1-2\gamma^2) f_0(z_n) + \frac{2\gamma^2}{z_n} f_1(z_n)}{f_1(a_n)} \cdot \frac{\cos \Omega_n \tau}{\Omega_n D_n} \qquad (10.124)$$

wnere

$$\tilde{u}_r = \frac{\rho C_D^2}{P_I}\frac{u_r}{R_o} = \frac{\lambda + 2\mu}{P_I}\frac{u_r}{R_o}$$

$$\tilde{\tau}_{rr} = \frac{\tau_{rr}}{P_I} \text{ and } \tilde{\tau}_{\theta\theta} = \tilde{\tau}_{\phi\phi} = \frac{\tau_{\phi\phi}}{P_I} .$$

The dimensionless variables z_n and τ are defined by

$$\left.\begin{aligned} z_n &= \frac{r\omega_n}{C_D} = \frac{r}{R_o}\,\frac{\Omega_n}{(1-\beta)} \\ \text{and} \quad & \\ \tau &= \frac{\omega_n}{\Omega_n}\,t = \frac{C_D t}{R_o(1-\beta)}\,. \end{aligned}\right\} \tag{10.125}$$

The dimensionless static solutions are

$$\left.\begin{aligned} \tilde{u}_r^{(s)}(r,t) &= \frac{\beta H(t)}{1-\beta^3}\left[\frac{\beta^3-\frac{P_o}{P_I}}{3-4\gamma^2}\,\frac{r}{R_I} + \frac{1}{4\gamma^2}\left(1-\frac{P_o}{P_I}\right)\frac{R_I^2}{r^2}\right] \\ \tilde{\tau}_{rr}^{(s)}(r,t) &= \frac{H(t)}{1-\beta^3}\left[\beta^3-\frac{P_o}{P_I}-\left(1-\frac{P_o}{P_I}\right)\frac{R_I^3}{r^3}\right] \\ \tilde{\tau}_{\phi\phi}^{(s)}(r,t) &= \frac{H(t)}{1-\beta^3}\left[\beta^3-\frac{P_o}{P_I}+\tfrac{1}{2}\left(1-\frac{P_o}{P_I}\right)\frac{R_I^3}{r^3}\right]. \end{aligned}\right\} \tag{10.126}$$

The displacement and stress at the inner surface of the sphere $r=R_I$ are shown in Figs. 10.3 and 10.4 for the case $\frac{P_o}{P_I} = 0$ and $\upsilon = \frac{1}{3}$ (which yields $\gamma = \frac{1}{2}$). With the aid of (10.114) these results may be expressed analytically as

$$\tilde{u}_r(R_I,t) = \frac{\beta(2+\beta^3)}{2(1-\beta^3)} - 2(1-\beta)^2 g(\tau)$$

$$\tilde{\tau}_{\phi\phi}(R_I,t) = \frac{1+2\beta^3}{2(1-\beta^3)} - \frac{2}{\beta}(1-\beta)^2 g(\tau)$$

where

$$g(\tau) = \sum_{n=1}^{\infty} \frac{\cos\Omega_n\tau}{\Omega_n^2 D_n}$$

$$D_n = \alpha_n^2\left(1-\frac{2}{b_n^2}\right)-\beta\left(1-\frac{2}{a_n^2}\right)$$

$$\alpha_n = \frac{\left(1-\frac{1}{a_n^2}\right)\cos a_n - \frac{1}{a_n}\sin a_n}{\left(1-\frac{1}{b_n^2}\right)\cos b_n - \frac{1}{b_n}\sin b_n}$$

$$a_n = \frac{\beta\Omega_n}{1-\beta}\,, \quad b_n = \frac{\Omega_n}{1-\beta}$$

$$\tan\Omega_n = \frac{(1-\beta)^2\left[\beta\Omega_n^2+(1-\beta)^2\right]\Omega_n}{\beta^2\Omega_n^4-(1-\beta)^2(1-\beta+\beta^2)\Omega_n^2+(1-\beta)^4}\,.$$

The sudden application of the pressure P_I to the internal surface of the shell at $\tau=0$ produces a dilatational wave which propagates through the shell to the outer surface from which it is reflected at

$$\tau = \frac{C_D t}{R_o - R_I} = 1.$$

At $\tau=2$ this reflected wave arrives back at the inner surface where it is again reflected thus beginning another cycle. The arrival of these dilatational waves at the inner surface of the shell at the times $\tau=2,4,6,8$, etc. is clearly evident in Figs. 10.3 and 10.4.

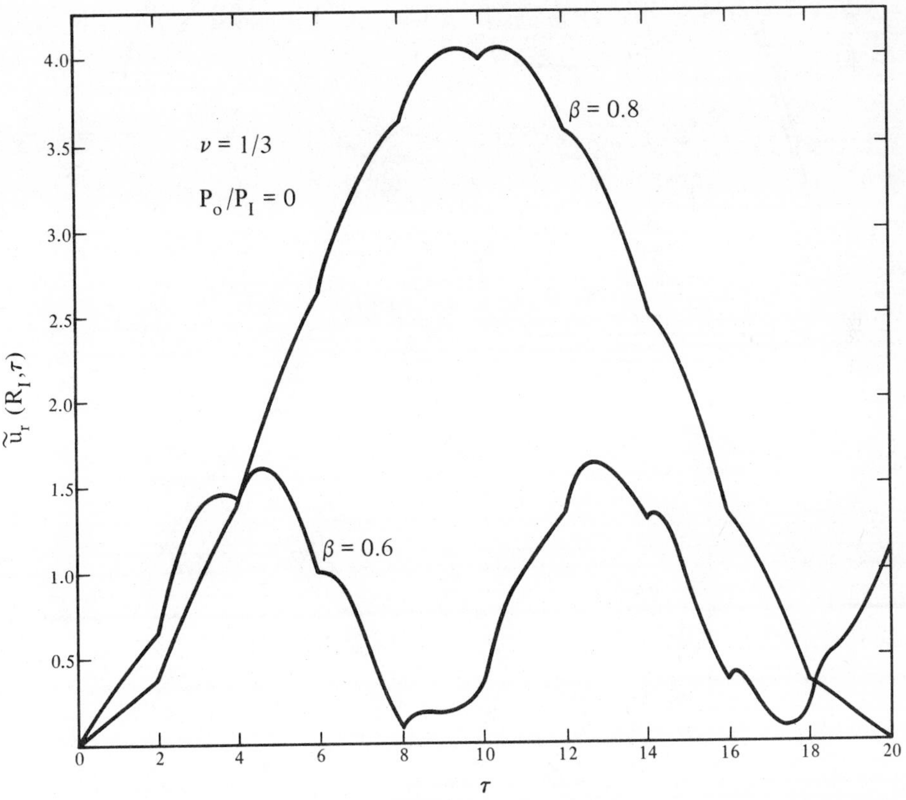

Figure 10.3

(a)

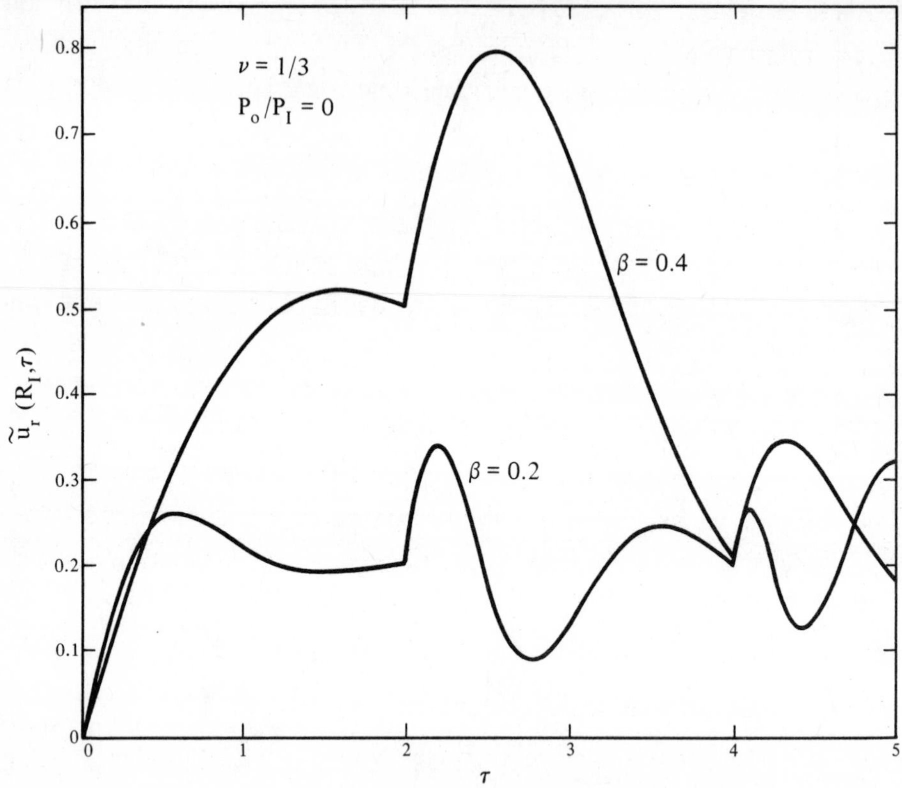

Fig. 10.3 (b)

Figure 10.4
(a)

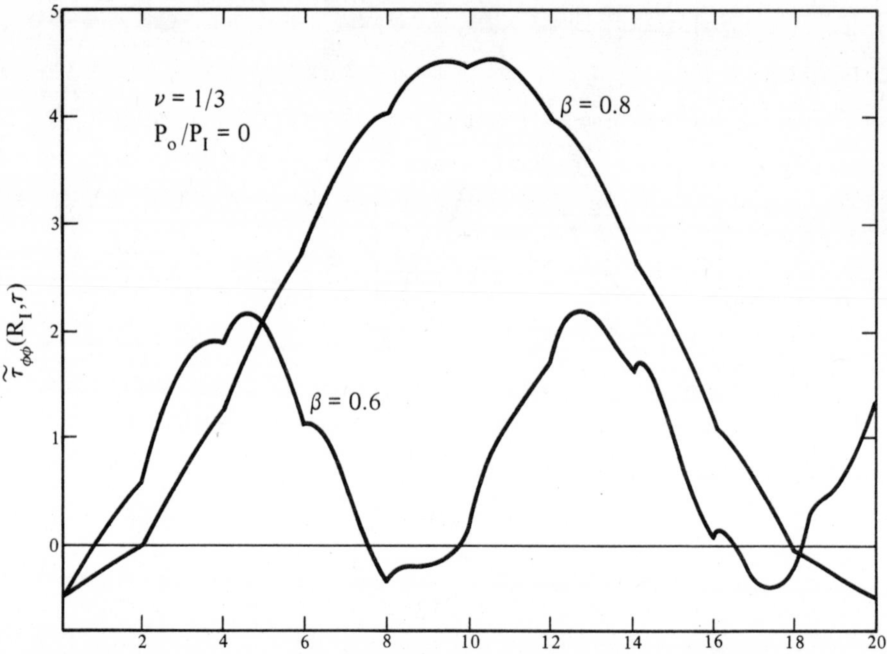

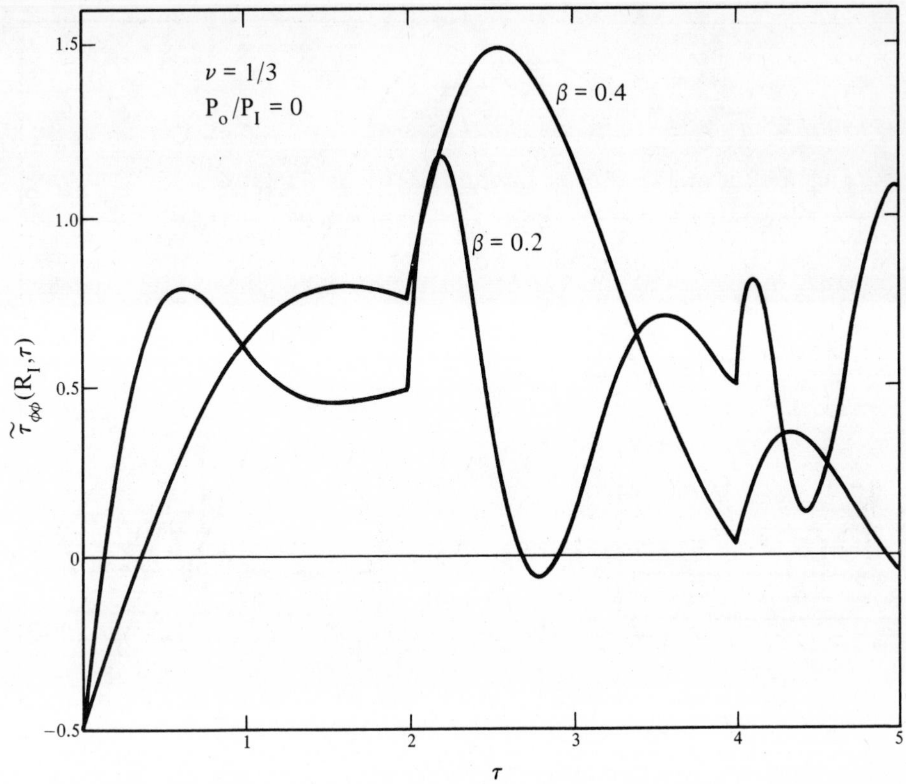

Fig. 10.4
(b)

EXERCISES

10.1 In the case of small deformations, show that the strains satisfy the following compatibility conditions

$$\varepsilon_{ij,kl}+\varepsilon_{kl,ij}=\varepsilon_{ik,jl}+\varepsilon_{jl,ik} \quad .$$

10.2 (a) Show that the general solution of the differential equations $\varepsilon_{ij}(x,t)=0$ is given by

$$u_i(x,t)=a_i(t)+e_{i\alpha\beta}b_\alpha(t)x_\beta .$$

(b) Show that small deformations imply small rigid body rotations, i.e., $b_\alpha \sim \delta$.

10.3 Show that the boundary conditions (10.24) satisfy the variational condition (10.11b).

10.4 The accompanying figure depicts two different loaded states of the same weightless, cantilevered beam. The beam is composed of a linearly elastic material. With the aid of the reciprocal relation (10.36), show that for $t > Max(t_1, t_2)$

$$F^{(2)}u_2^{(1)}(P_2, t-t_2) = F^{(1)}u_2^{(2)}(P_1, t-t_1)$$

where $u_2^{(1)}$ is the displacement produced by $F^{(1)}$ and $u_2^{(2)}$ is the displacement produced by $F^{(2)}$. Assume the beam to be initially undeformed and at rest in each case. Note,

$$H(t) = \begin{cases} 0 \text{ for } t<0 \\ 1 \text{ for } t>0 \end{cases} .$$

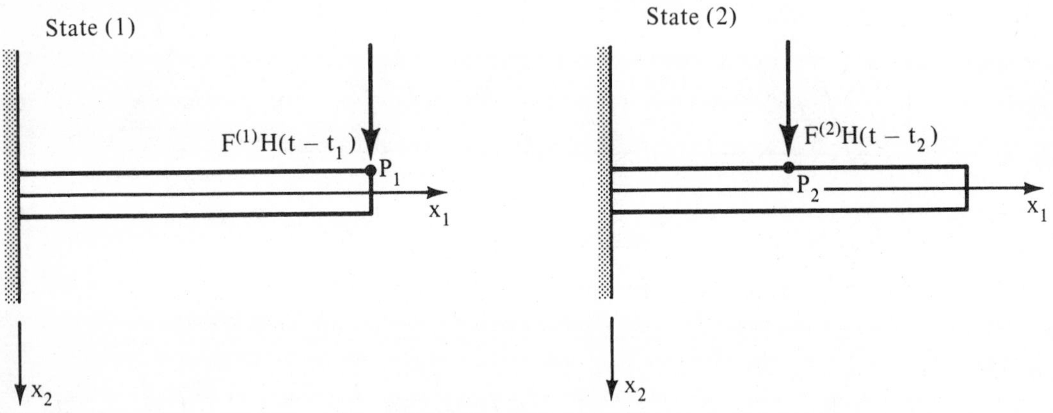

Figure: Exercise 10.4

10.5 Imagine a homogeneous, isotropic, linearly elastic medium filling the region

$$V = \{(x_1, x_2, x_3) \mid 0<x_1<L, \ -\infty<x_2<\infty, \ -\infty<x_3<\infty\} .$$

whose boundary is the surface $S=S^{(o)}+S^{(L)}$ where

$$S^{(o)} = \{(x_1, x_2, x_3) \mid x_1=0\} \text{ and } S^{(L)} = \{(x_1, x_2, x_3) \mid x_1=L\} .$$

Assume $u_1=u_1(x_1,t)$, $u_2=u_3\equiv 0$. This is called a transversely constrained medium. Find the eigenvalues, normalized eigenfunctions and modal stresses for the following cases:

(a) $u_1(0) = u_1(L) = 0$

(b) $\tau_{11}(0) = \tau_{11}(L) = 0$

(c) $u_1(0) = 0$, $\tau_{11}(L) = 0$

(d) $u_1(0) = 0$, $\tau_{11}(L) + \alpha u_1(L) = 0$, $\alpha>0$

Use the normalization condition

$$\left[\int_0^L \rho u_1^2 dx_1\right]^{\frac{1}{2}} = 1 \quad .$$

10.6 Find the displacement and stress fields produced in the transversely constrained medium described in Exercise 10.5 by the following loads.

(a) $u_1(0,t) = 0, \; T_1(L,t) = PH(t)$

(b) $u_1(0,t) = 0, \; T_1(L,t) = PH(t) - \alpha u_1(L,t)$

Describe the effect on the stress $\tau_{11}(x_1,t)$ of varying α between zero and infinity.

(c) $u_1(0,t) = 0, \; u_1(L,t) = u \sin \Omega t \cdot H(t)$

What happens in part (c) when Ω approaches one of the natural frequencies of the medium. Assume $B_1(x_1,t)=0$ and $u_1(x_1,0)=\dot{u}_1(x_1,0)=0$ in parts (a), (b) and (c).

10.7 A normal section of a long, homogeneous isotropic, circular cylindrical shell which is in a state of plane strain is shown below. Find the eigenvalues and eigenfunctions corresponding to the axi-symmetric (purely radial) vibrations of the shell for the following boundary conditions:

(a) $u_r(R_I) = u_r(R_o) = 0$

(b) $\tau_{rr}(R_I) = \tau_{rr}(R_o) = 0$

(c) $\tau_{rr}(R_I) - \alpha_I u_r(R_I) = 0$

$\tau_{rr}(R_o) + \alpha_o u_r(R_o) = 0$

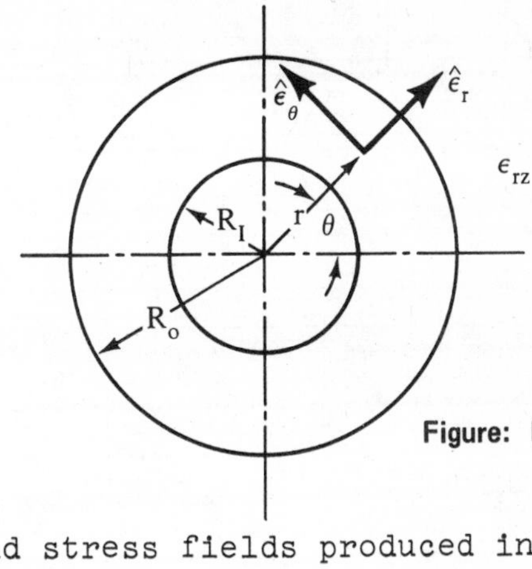

$\epsilon_{rz} = \epsilon_{\theta z} = \epsilon_{zz} = 0$

Figure: Exercise 10.7

10.8 Find the displacement and stress fields produced in the cylindrical shell of Exercise 10.7 by the suddenly applied pressure loading $T_r(R_I,t)=P_IH(t)$, $T_r(R_o,t)=-P_oH(t)$. Assume the shell to be initially undeformed and at rest and also free of body forces.

10.9 Assume that the outer surface of the cylinder of Exercise 10.7 is elastically restrained. Find the stresses which result from the internal pressure loading $T_r(R_I,t)=P_IH(t)$.

10.10 In each of the preceding forced motion problems involving a suddenly applied load, how does the maximum stress obtained compare with the maximum stress obtained from the corresponding statically loaded case?

10.11 The block shown in the accompanying figure is composed of a homogeneous, isotropic, linearly elastic material. Show that the eigenvalues and eigenfunctions corresponding to the case $S=S_{u_n}$ (see (10.40)) are given by

$$u_1^{(n_1n_2n_3)} = A_1^{(n_1n_2n_3)} \sin \frac{n_1\pi x_1}{\ell_1} \cos \frac{n_2\pi x_2}{\ell_2} \cos \frac{n_3\pi x_3}{\ell_3}$$

$$u_2^{(n_1n_2n_3)} = A_2^{(n_1,n_2n_3)} \cos \frac{n_1\pi x_1}{\ell_1} \sin \frac{n_2\pi x_2}{\ell_2} \cos \frac{n_3\pi x_3}{\ell_3}$$

$$u_3^{(n_1n_2n_3)} = A_3^{(n_1n_2n_3)} \cos \frac{n_1\pi x_1}{\ell_1} \cos \frac{n_2\pi x_2}{\ell_2} \sin \frac{n_3\pi x_3}{\ell_3}$$

$$\omega^2_{n_1n_2n_3} = c_D^2 \left[\left(\frac{n_1\pi}{\ell_1}\right)^2 + \left(\frac{n_2\pi}{\ell_2}\right)^2 + \left(\frac{n_3\pi}{\ell_3}\right)^2\right]$$

and

$$\omega^2_{n_1n_2n_3} = c_S^2 \left[\left(\frac{n_1\pi}{\ell_1}\right)^2 + \left(\frac{n_2\pi}{\ell_2}\right)^2 + \left(\frac{n_3\pi}{\ell_3}\right)^2\right]$$

where

$$c_D^2 = \frac{\lambda+2\mu}{\rho} \quad \text{and} \quad c_S^2 = \frac{\mu}{\rho}$$

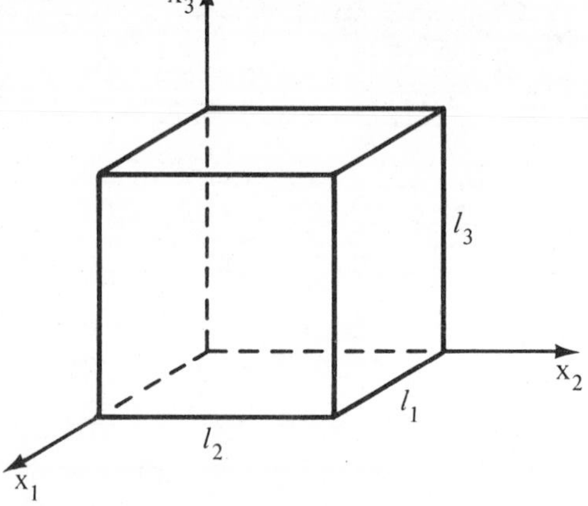

Figure: Exercise 10.11

10.12 A spherical shell is surrounded by a rigid matrix. An explosion within the shell causes the internal pressure at the shell surface to vary as follows:

$$T_r(R_I,t) = P_I(1+e^{-\kappa t})H(t),\ \kappa>0.$$

Assuming the shell is homogeneous, isotropic and linearly elastic, find the resulting stress and displacement fields in the shell.

10.13 The plate shown in the accompanying figure is simply supported at $x_1=(0,1)$ and at $x_2=(0,\beta)$. A simple support may be simulated by a boundary condition of the type S_{T_n}. The surfaces $x_3=\pm\frac{\Theta}{2}$ are stress free. Find the mode shape functions and the frequency equations for this plate. Hint: Try

$$u_1^{(n_1n_2)}= f_1^{(n_1n_2)}(x_3)\cos n_1\pi x_1 \sin\frac{n_2\pi x_2}{\beta}$$

$$u_2^{(n_1n_2)}= f_2^{(n_1n_2)}(x_3)\sin n_1\pi x_1 \cos\frac{n_2\pi x_2}{\beta}$$

$$u_3^{(n_1n_2)}= f_3^{(n_1n_2)}(x_3)\sin n_1\pi x_1 \sin\frac{n_2\pi x_2}{\beta}$$

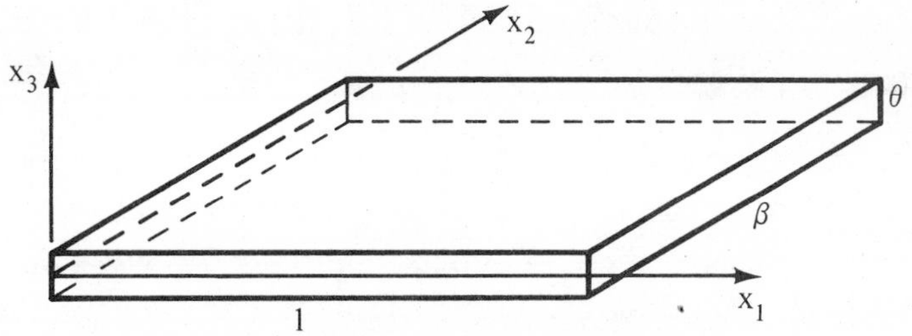

Figure: Exercise 10.13

The frequency equations together with a forced motion example for this plate may be found in the paper by Y.-C Lee and H. Reismann, "Dynamics of Rectangular Plates," International Journal of Engineering Science, Vol. 7, 1969, pp. 93-113.

10.14 Find the eigenfunctions and the frequency equation for the nonaxysymmetric vibrations of the cylindrical shell described in Exercise 10.7 for the case $\tau_{rr}(R_I,\Theta)=\tau_{rr}(R_o,\Theta)=0$. A forced motion example for this cylindrical shell may be found in the paper by P.S. Pawlik and H. Reismann, "Forced Plane Strain Motion of Cylindrical Shells - A Comparison of Shell Theory with Elasticity Theory," AFOSR Scientific Report, AFOSR 70-1377 TR, June 1970.

10.15 Find the eigenfunctions and the frequency equation for the simply supported beam shown in the accompanying figure. Assume a state of plane stress, i.e.,

$$\tau_{13} = \tau_{23} = \tau_{33} = 0, \; u_1 = u_1(x_1,x_2), \; u_2 = u_2(x_1,x_2) \quad .$$

Also assume the beam to be homogeneous, isotropic and linearly elastic.

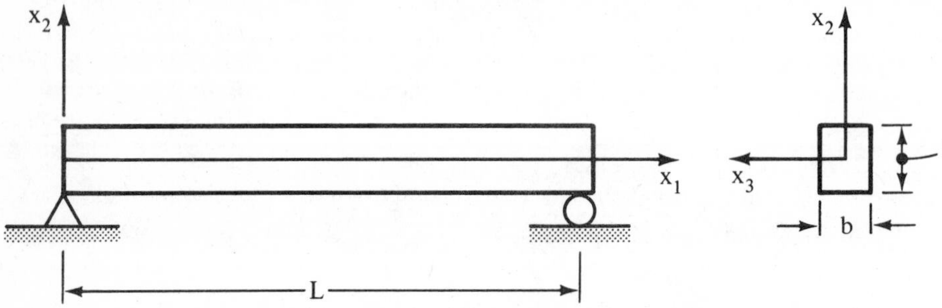

Figure: Exercise 10.15

10.16 Find the stress and displacement fields produced in the beam of Exercise 10.15 by a suddenly applied, uniformly distributed pressure load on the top surface, i.e., $T_2(x_1,\frac{h}{2},t) = -PH(t)$. Assume that the beam is weightless and is initially at rest in its undeformed state.

10.17 Consider an elastic body which satisfies homogeneous boundary conditions, i.e., $T_i u_i = 0$ on S. Use the separation of variables technique to show that the motion which results from the body force $B_i(x,t)$ and the initial conditions $u_i^{(o)}(x)$ and $\dot{u}_i^{(o)}(x)$ can be expressed in the form

$$u_i(x,t) = \sum_{n=1}^{\infty} u_i^{(n)}(x) q_n(t)$$

where $u_i^{(n)}(x)$ are the eigenfunctions of the body. Determine $q_n(t)$ in terms of $B_i(x,t)$, $u_i^{(o)}(x)$ and $\dot{u}_i^{(o)}(x)$.

Chapter

11

WAVE MOTION IN ELASTIC MEDIA

In modern observations of the tremors due to distant earthquakes three phases of the disturbance are often recognized. The first is interpreted as due to the arrival of the dilatational waves, propagated directly through the substance of the earth, the second as due to that of the distortional waves, also propagated directly, and the third to that of the Rayleigh waves, which have traveled over the surface and are therefore delayed more than in proportion to the difference of wave velocity.

Horace Lamb, "The Dynamical Theory of Sound," 1925.

11.1 INTRODUCTION

In this chapter we shall investigate the properties of traveling waves in elastic media which are unbounded in the direction of propagation of the wave. The purpose of this investigation is twofold. First, it will give us an insight into the manner by which disturbances propagate through bounded bodies and second it will provide us with a set of qualitative limits on the applicability of the approximate theories of torsion, extension and flexure to be presented in the succeeding chapters.

The traveling waves to be studied in the present chapter are not unrelated to the standing waves which we employed to construct solutions of forced motion problems for bounded bodies in the previous chapter. For example, consider the standing wave (normal mode of vibration)

$$u(x,t) = 2A \sin kx \cos \omega t$$

This standing wave can be represented as the sum of two traveling waves as follows:

$$u(x,t) = A \sin (kx+\omega t) + A \sin (kx-\omega t) \quad .$$

Thus, by studying the properties of the traveling waves we can also obtain information about the properties of the standing waves which are used to construct solutions of forced motion problems in bounded bodies.

11.2 WAVES IN UNBOUNDED MEDIA

The equations which characterize the motion (assuming sufficiently small deformations) of an isotropic, linearly elastic body are, according to (10.23),

$$(\lambda u_{j,j})_{,i} + [\mu(u_{i,j} + u_{j,i})]_{,j} + B_i = \rho\ddot{u}_i$$

If the body is uniform and homogeneous, i.e., if ρ, λ and μ are constants, and if we neglect the body force B_i, then the preceding equation can be written as

$$\mu u_{i,jj} + (\lambda+\mu)u_{j,ji} = \rho\ddot{u}_i \tag{11.1a}$$

or, in vector notation,

$$\mu\nabla^2\vec{u} + (\lambda+\mu)\vec{\nabla}(\vec{\nabla}\cdot\vec{u}) = \rho\ddot{\vec{u}} \quad . \tag{11.1b}$$

These are often called Navier's equations of motion for an elastic body. The solution of (11.1) can be written in the form

$$\vec{u} = \vec{\nabla}\phi+\vec{\nabla}\times\vec{\psi} \tag{11.2}$$

provided that the potentials ϕ and $\vec{\psi}$ satisfy

$$\vec{\nabla}\cdot\vec{\psi} = 0 \tag{11.3}$$

and the wave equations

$$c_D^2\nabla^2\phi - \ddot{\phi} = 0 \quad , \quad c_D^2 = \frac{\lambda+2\mu}{\rho} \tag{11.4a}$$

$$c_S^2\nabla^2\vec{\psi} - \ddot{\vec{\psi}} = 0 \quad , \quad c_S^2 = \frac{\mu}{\rho} \tag{11.4b}$$

This is Lamé's solution of Navier's equations. If we substitute (11.2) into (11.1) we obtain, with the aid of some vector identities (see Exercises 8.13 and 8.14),

$$\vec{\nabla}\,[(\lambda+2\mu)\nabla^2\phi-\rho\ddot{\phi}]+ \vec{\nabla}\times[\mu\nabla^2\vec{\psi}-\rho\ddot{\vec{\psi}}]= 0$$

In view of (11.4) it is now clear that the displacement field given by (11.2) is indeed a solution of (11.1). Moreover, it was shown by P. Duhem (see E. Sternberg, "On the Integration of the Equations of Motion in the Classical Theory of Elasticity," Archive for Rational Mechanics and Analysis, v.6, 1960, pp. 34-50) that every sufficiently regular solution of (11.1) may be expressed in the form (11.2) provided that ϕ and $\vec{\psi}$ satisfy (11.3) and (11.4). This is true even when the region V under consideration is unbounded and/or multiply connected.

The dilatation and rotation associated with the displacement field $\vec{u}(x,t)$ are defined as

$$\varepsilon \equiv \vec{\nabla}\cdot\vec{u} = \nabla^2\phi \qquad \text{(dilatation)} \tag{11.5}$$

and

$$\vec{\omega} \equiv \tfrac{1}{2}\vec{\nabla}\times\vec{u} = -\tfrac{1}{2}\nabla^2\vec{\psi} \qquad \text{(rotation)} \tag{11.6}$$

respectively. For sufficiently small deformations, the dilatation describes the manner in which the local capacity and density of the body change. To illustrate this assertion we write (9.60) and (9.67) in the following form

$$\frac{dV - dV_0}{dV_0} = J-1 \text{ and } \frac{\rho(a,t) - \rho(a,0)}{\rho(a,0)} = -\left(\frac{J-1}{J}\right).$$

The substitution of (9.5a) into the first of (9.59b) reveals that, for small deformations,

$$J-1 = u_{i,i} = \vec{\nabla}\cdot\vec{u} = \varepsilon \quad .$$

Thus, $\varepsilon(x)>0$ corresponds to an increase in volume and consequently a decrease in density or a rarefaction of the medium at the point x_i. On the other hand, $\varepsilon(x)<0$ corresponds to a decrease in volume which causes an increase in density or a condensation of the medium at the point x_i. For small deformations, the rotation vector may be interpreted physically as follows. In view of (9.5a), the body's motion will carry the infinitesimal vector da_i into the vector

$$dx_i = da_i + u_{i,j}da_j = (\delta_{ij} + \varepsilon_{ij} + \omega_{ij})da_j$$

where

$$\omega_{ij} = u_{i,j} - \varepsilon_{ij} = \tfrac{1}{2}(u_{i,j} - u_{j,i}) \quad .$$

If there is no deformation ($\varepsilon_{ij}=0$) then the change in the vector must be due to a pure rotation and therefore ω_{ij} is called the rotation tensor. Since the rotation tensor is skew-symmetric, it can, according to (8.42) and (8.43), be expressed in terms of a dual vector as follows:

$$\omega_{ij} = -\, e_{ijk}\omega_k$$

where

$$\omega_k = -\,\tfrac{1}{2}e_{ijk}\omega_{ij} = \tfrac{1}{2}e_{kji}u_{i,j}$$

Thus, $\vec{\omega}=\frac{1}{2}\vec{\nabla}\times\vec{u}$ is the dual vector of ω_{ij} and is therefore called the rotation vector. We now form the divergence and the curl of both sides of (11.1b). With the aid of (11.5) and (11.6) the following results are obtained:

$$c_D^2\nabla^2\varepsilon - \ddot{\varepsilon} = 0 \qquad (11.7a)$$

$$c_S^2\nabla^2\vec{\omega} - \ddot{\vec{\omega}} = 0 \qquad (11.7b)$$

$$\vec{\nabla}\cdot\vec{\omega} = 0 \qquad (11.8)$$

A comparison of (11.7) and (11.8) with (11.4) and (11.3) reveals that the dilatation and rotation are characterized by the same system of partial differential equations as the displacement potentials. Thus, although the following discussion refers to the dilatation and rotation the conclusions will also apply to the displacement potentials. Since equation (11.7a) does not contain $\vec{\omega}$ and (11.7b) does not contain ε we conclude that, in an unbounded elastic medium, it is possible for motions to be purely irrotational ($\vec{\omega}(x,t) \equiv 0$) or purely equivoluminal ($\varepsilon(x,t) \equiv 0$) or for the two to coexist independently of one another. Furthermore, since equations (11.7a) and (11.7b) are both of the hyperbolic type, commonly called wave equations, we conclude that a discontinuity in the dilatation will propagate through the medium with constant speed C_D and a discontinuity in the rotation will propagate through the medium with constant speed C_S. Thus, a disturbance which originates in some restricted region of the medium will require a finite amount of time to reach any part of the medium which is a finite distance from the source and an infinite amount of time to reach those parts of the medium which are infinitely far removed from the source. We wish to emphasize that the preceding discussion applies to unbounded media. If the body contains boundaries then ε and $\vec{\omega}$ can be coupled through the boundary conditions in which case wave speeds different than C_D and C_S can exist as they do, for example, in the case of Rayleigh surface waves (see Section 11.4).

The ratio of the two wave speeds $C_S:C_D$ occurs frequently in computations and will therefore be given the special symbol γ, i.e.,

$$\gamma \equiv \frac{C_S}{C_D} = \sqrt{\frac{\mu}{\lambda+2\mu}} = \sqrt{\frac{1-2\upsilon}{2(1-\upsilon)}} \quad . \tag{11.9}$$

For most materials, $0<\upsilon<\frac{1}{2}$. Thus, $\frac{1}{2}>\gamma^2>0$ and therefore $C_D>C_S$. Some typical values of the elastic constants and the wave speeds are given in Table 11.1. The Poisson ratios $\upsilon=(\frac{1}{4},\frac{1}{3})$ yield $\gamma^2=(\frac{1}{3},\frac{1}{4})$, respectively. These are reasonable and convenient values to use in illustrative examples.

To illustrate the preceding discussions we will examine some specific examples of wave propagation in an unbounded medium.

Table 11.1 Representative Values of Material Properties

Material	ν	E $\frac{lb}{in^2}$	$G=\mu$ $\frac{lb}{in^2}$	λ $\frac{lb}{in^2}$	$\rho g^{\dagger}$ $\frac{lb}{in^3}$	ρ $\frac{lb\text{-}sec^2}{in^4}$	C_D $\frac{ft}{sec}$	C_S $\frac{ft}{sec}$	γ
Aluminum	0.34	10×10^6	3.73×10^6	7.92×10^6	0.095	2.46×10^{-4}	20,850	10,200	.493
Concrete	0.20	4×10^6	1.67×10^6	1.11×10^6	0.0868	2.25×10^{-4}	11,700	7,200	.614
Copper	0.34	13×10^6	4.85×10^6	10.3×10^6	0.322	8.34×10^{-4}	12,900	6,360	.493
Glass	0.25	10×10^6	4.00×10^6	4×10^6	0.088	2.28×10^{-4}	19,100	11,000	.578
Nylon	0.40	4.1×10^5	1.46×10^5	5.84×10^5	0.0412	1.07×10^{-4}	7,550	3,080	.409
Rubber	0.499	285	95.1	47,500	0.033	0.855×10^{-4}	1,970	88	.045
Steel	0.29	30×10^6	11.6×10^6	16×10^6	0.283	7.34×10^{-4}	19,300	10,500	.545

$^{\dagger}g=32.2\ \frac{ft}{sec^2}$

(a) Irrotational (or Dilatational) Plane Waves.

Consider the motion characterized by the potentials

$$\phi=\phi(\vec{k}\cdot\vec{r}-\omega t),\vec{\psi}=0 \tag{11.10}$$

where $\vec{r}=x_i\hat{\varepsilon}_i$ is the position vector and $\vec{k}=k_i\hat{\varepsilon}_i$ is called the propagation vector. The corresponding dilatation and rotation are obtained by substituting (11.10) into (11.5) and (11.6). The result is,

$$\varepsilon=k^2\phi''(\vec{k}\cdot\vec{r}-\omega t),\ \vec{\omega}=0 \tag{11.11}$$

where

$$k=|\vec{k}|=\sqrt{k_1^2+k_2^2+k_3^2}$$

and a prime denotes differentiation with respect to the argument (or phase) $\alpha=\vec{k}\cdot\vec{r}-\omega t$. We shall assume $\phi(\alpha)$ and $\phi'(\alpha)$ to be continuous and $\phi''(\alpha)$ to be at least piecewise continuous for $-\infty<\alpha<\infty$. To an observer moving through space with velocity $\vec{v}=\frac{\omega\vec{k}}{k^2}=\frac{\omega}{k}\hat{k}$, the phase of the function ϕ appears to be stationary, i.e.,

$$\frac{d\alpha}{dt}=\vec{k}\cdot\frac{d\vec{r}}{dt}-\omega=\vec{k}\cdot\vec{v}-\omega=\omega-\omega=0 \quad .$$

Thus, this moving observer sees a stationary wave form $\phi(\alpha)$ and consequently a stationary observer would see the wave form $\phi(\alpha)$ advancing in the $\vec{k}$ direction with speed (phase velocity) $v=|\vec{v}|=\frac{\omega}{k}$. By substituting (11.10) into (11.3) and (11.4) we obtain

$$(C_D^2k^2-\omega^2)\phi''(\alpha)=0 \ .$$

Thus,

$$\omega^2=k^2C_D^2 \text{ or } v=\frac{\omega}{k}=C_D. \tag{11.12}$$

These are the dispersion relations for a plane, irrotational wave in an unbounded, elastic medium. We note that the phase velocity $v=C_D$ is independent of the wave number k and/or the frequency ω. In the light of this observation and the discussion in Appendix B we conclude that there is no dispersion of the wave form $\phi(\alpha)$ by the medium, i.e., it passes, undistorted, through the medium with speed C_D. The displacement field produced by this disturbance is obtained by substituting (11.10) into (11.2). This operation yields the result

$$\vec{u}=\vec{\nabla}\phi=\vec{k}\phi'(\vec{k}\cdot\vec{r}-\omega t) \ . \tag{11.13}$$

We note that the displacement vector is parallel to the propagation vector. For this reason the motion is said to be longitudinal. We also note that every point on the plane $\vec{k}\cdot\vec{r}=k_1x_1+k_2x_2+k_3x_3=a$, where a is an arbitrary constant, has the same motion. Thus, (11.10) characterizes a plane, longitudinal, wave motion. Since the vector $\vec{k}$ is perpendicular to the plane $\vec{k}\cdot\vec{r}=a$, we conclude that the motion of the medium and the direction of propagation of the wave are both perpendicular to this plane.

The stresses and strains produced by this disturbance are obtained by substituting (11.13) into (10.21) and (10.22). After noting that $u_{i,j}=k_ik_j\phi''(\alpha)$ one obtains the result

$$\left.\begin{aligned} \varepsilon_{ij}&=k_ik_j\phi''(\vec{k}\cdot\vec{r}-\omega t)\\ \tau_{ij}&=(\lambda k^2\delta_{ij}+2\mu k_ik_j)\phi''(\vec{k}\cdot\vec{r}-\omega t) \end{aligned}\right\} \qquad (11.14)$$

The stress vector acting on the plane $\vec{k}\cdot\vec{r}=a$ is given by

$$T_i=\tau_{ij}n_j=\tau_{ij}\,\frac{k_j}{k} = (\lambda+2\mu)kk_i\phi''(a-\omega t)$$

or, in vector form,

$$\vec{T}=\hat{k}\rho\omega^2\phi''(a-\omega t) \qquad (11.15)$$

where $\hat{k}\equiv\frac{\vec{k}}{k}$ is the unit vector in the $\vec{k}$ direction. It is evident from (11.15) that the stress vector is parallel to the propagation vector and therefore perpendicular to the plane $\vec{k}\cdot\vec{r}=a$. Hence, the normal stress (σ) on this plane is equal in magnitude to the stress vector,

$$\sigma \equiv \vec{T}\cdot\hat{n} = \vec{T}\cdot\hat{k} = \rho\omega^2\phi''(a-\omega t)$$

and the shear stress (τ) is zero,

$$\tau \equiv \sqrt{\vec{T}\cdot\vec{T}-\sigma^2} = 0.$$

By substituting (11.13) and (11.14) into (9.93) we obtain, with the aid of (11.12), the following expression for the energy flux vector:

$$\left.\begin{aligned} S_i&= -\tau_{ij}\dot{u}_j=\rho\omega^3k_i(\phi''(\alpha))^2\\ &\text{or in vector form}\\ \vec{S} &= \vec{k}[\rho\omega^3(\phi''(\alpha))^2]\ . \end{aligned}\right\} \qquad (11.16)$$

As one would expect, the energy flows in the direction of propagation of the plane wave.

Two specific wave forms which are of particular interest are the harmonic disturbance

$$\phi(\alpha) = A \sin \alpha$$

and the discontinuous stress wave

$$\phi(\alpha) = H(-\alpha) \int_0^{\alpha}\int_0^{\beta} f(\delta)\,d\delta\,d\beta$$

where $f(\alpha)$ is continuous on the interval $(-\infty<\alpha<\infty)$

and

$$H(-\alpha) = \begin{Bmatrix} 0 \text{ if } \alpha>0 \\ 1 \text{ if } \alpha<0 \end{Bmatrix} .$$

For the harmonic disturbance,

$$\varepsilon = -k^2 A \sin (\vec{k}\cdot\vec{r} - \omega t) .$$

Thus, the material lying on the plane $\vec{k}\cdot\vec{r} = a$ is alternately compressed and rarefied, the variation being harmonic in time with period $\frac{2\pi}{\omega}$. This disturbance is appropriately called a wave of condensation-rarefaction. For the discontinuous stress wave, the stress vectors on the planes $\vec{k}\cdot\vec{r}$ = constant are given by

$$\vec{T} = \hat{k}\rho\omega^2 f(\vec{k}\cdot\vec{r}-\omega t)H(\omega t-\vec{k}\cdot\vec{r})$$

or

$$\vec{T} = \hat{k}\rho\omega^2 \begin{Bmatrix} f(\vec{k}\cdot\vec{r}-\omega t) & \text{if } \vec{k}\cdot\vec{r}<\omega t \\ 0 & \text{if } \vec{k}\cdot\vec{r}>\omega t \end{Bmatrix}$$

If $f(0) \neq 0$ then we see that there will be a discontinuity in the stress across the plane $\vec{k}\cdot\vec{r}=\omega t$. For example, the time history of the normal stress on the plane $\vec{k}\cdot\vec{r}=a$ when $f(\alpha)=e^{\alpha}$ is given by

$$\sigma = 0 \qquad \text{for } \omega t<a$$

and

$$\sigma = \rho\omega^2 \exp(a-\omega t) \qquad \text{for } \omega t>a .$$

The discontinuity in this normal stress is illustrated in Fig. 11.1.

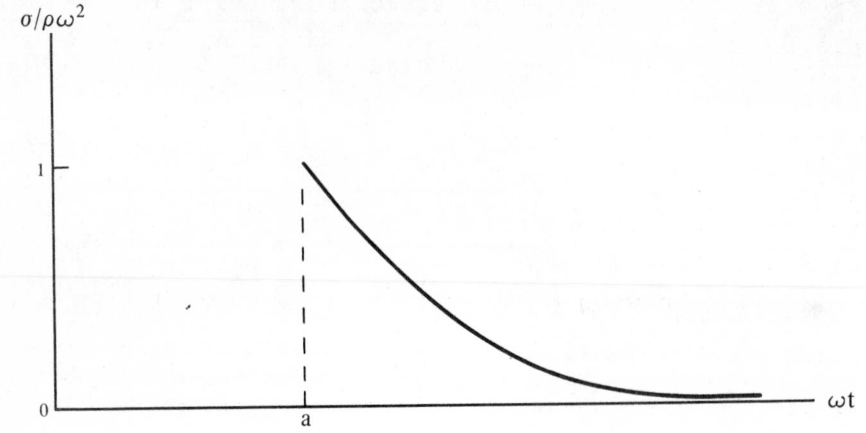

Figure 11.1

(b) Equivoluminal (or Solenoidal) Plane Waves.

Consider the motion characterized by the potentials

$$\phi = 0, \; \vec{\psi} = \vec{\psi}(\vec{k}\cdot\vec{r}-\omega t) \quad . \tag{11.17}$$

The potential $\vec{\psi}(\vec{k}\cdot\vec{r}-\omega t)$ represents a plane, vector wave propagating in the $\vec{k}$ direction with speed $v = \frac{\omega}{k}$. By substituting (11.17) into (11.3) and (11.4) we conclude that

$$\vec{k}\cdot\vec{\psi}'(\alpha) = 0, \; \alpha = \vec{k}\cdot\vec{r}-\omega t \tag{11.18}$$

and

$$(C_S^2k^2-\omega^2)\vec{\psi}''(\alpha) = 0$$

or

$$\omega^2 = k^2C_S^2 \text{ and } v = \frac{\omega}{k} = C_S \quad . \tag{11.19}$$

Equation (11.18) indicates that the vector $\vec{\psi}'$ must be orthogonal to the propagation vector $\vec{k}$. Since the phase velocity $v = C_S$ is independent of the wave number and/or the frequency, there is no dispersion, i.e., the wave form $\vec{\psi}(\alpha)$ propagates, undistorted, through the medium with speed C_S. By substituting (11.17) into (11.2) we obtain the displacement field

$$\vec{u} = \vec{\nabla}\times\vec{\psi} = \vec{k}\times\vec{\psi}'(\vec{k}\cdot\vec{r}-\omega t) \quad . \tag{11.20}$$

It is evident from (11.20) that the displacement vector is perpendicular to the propagation vector and therefore parallel to the planes $\vec{k}\cdot\vec{r}$ = constant. The disturbance characterized by (11.17) thus causes the plane $\vec{k}\cdot\vec{r} = a$ to displace, rigidly, parallel to itself or perpendicular to the direction of propagation of the disturbance and it is therefore called a transverse wave. The plane $\vec{k}\cdot\vec{r} = a$ is called the plane of vibrations and the plane defined by the two vectors $\vec{k}$ and $\vec{u}$ is called the plane of polarization. As shown in Fig. 11.2 these two planes are perpendicular to one another and the displacement vector lies along their line of intersection. This is referred to as plane polarization. In view of (11.5) and (11.6) we conclude that the disturbance characterized by (11.17) is equivoluminal, i.e.,

$$\varepsilon = 0, \; \vec{\omega} = \frac{-1}{2}\nabla^2\vec{\psi} = -\frac{1}{2}k^2\vec{\psi}''(\alpha) \; . \qquad (11.21)$$

The stress and strain fields associated with the disturbance 11.17) are

$$\left.\begin{aligned} \varepsilon_{ij} &= \frac{1}{2}\,(k_i e_{j\alpha\beta} + k_j e_{i\alpha\beta})k_\alpha \psi''_\beta \\ \tau_{ij} &= 2\mu\varepsilon_{ij} \quad . \end{aligned}\right\} \qquad (11.22)$$

The stress vector acting on the plane $\vec{k}\cdot\vec{r} = a$ (a is a constant) may be computed with the aid of (9.80) and (11.22). The result is,

$$T_i = \tau_{ij}n_j = \tau_{ij}\,\frac{k_j}{k} = \mu k e_{i\alpha\beta}k_\alpha \psi''_\beta(a-\omega t)$$

or, in vector form,

$$\vec{T} = \rho\omega^2\hat{k}\times\vec{\psi}''(a-\omega t) \quad . \qquad (11.23)$$

We note that the stress vector is perpendicular to the propagation vector or, in other words, parallel to the plane $\vec{k}\cdot\vec{r} = a$. Thus, the normal stress on this plane is zero,

$$\sigma \equiv \vec{T}\cdot\hat{n} = \vec{T}\cdot\left(\frac{\vec{k}}{k}\right) = \rho\omega^2\hat{k}\cdot\hat{k}\times\vec{\xi} = 0,$$

and the shear stress is just the magnitude of the stress vector,

$$\tau \equiv \sqrt{\vec{T}\cdot\vec{T}-\sigma^2} = \sqrt{T_i T_i} = \rho\omega^2\sqrt{\xi_i\xi_i - \frac{k_i k_j}{k^2}\,\xi_i\xi_j}$$

where $\xi_i \equiv \psi''_i(a-\omega t)$. For this reason the disturbance characterized by (11.7) is often called a shear wave. By substituting (11.20) and

(11.22) into (9.93) we obtain the following expression for the energy flux vector:

$$\left.\begin{aligned} S_i &= -\tau_{ij}\dot{u}_j = \rho\omega^3\left[\psi''_\alpha\psi''_\alpha - \frac{k_\alpha k_\beta}{k^2}\psi''_\alpha\psi''_\beta\right] k_i \\ &\text{or, in vector form,} \\ \vec{S} &= \rho\omega^3 |\hat{k}\times\vec{\psi}''|^2 \vec{k} \quad . \end{aligned}\right\} \qquad (11.24)$$

Once again we see that the energy flows in the direction of propagation of the plane disturbance.

In the case of a harmonic disturbance, $\vec{\psi}=\vec{B}\sin\alpha$, (11.18) requires $\vec{k}\cdot\vec{B} = 0$. According to (11.20) and (11.21), the displacement and rotation for this case are

$$\vec{u} = \vec{k}\times\vec{B}\cos\alpha \quad \text{and} \quad \vec{\omega} = \frac{1}{2}k^2\vec{B}\sin\alpha .$$

These are illustrated in Fig. 11.2.

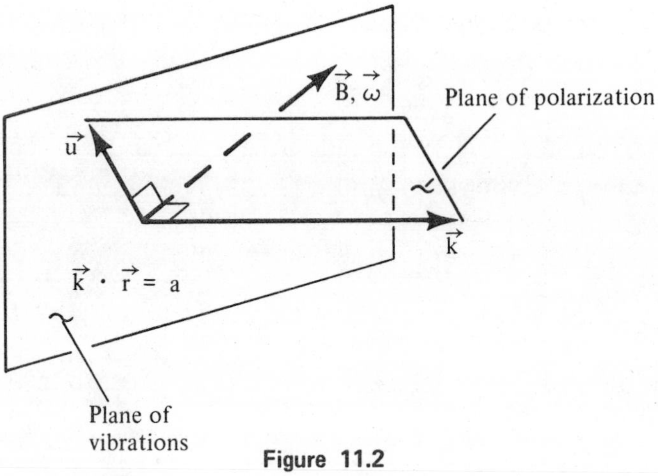

Figure 11.2

A discontinuous stress wave can be constructed by letting

$$\vec{\psi} = H(-\alpha)\int_0^\alpha\int_0^\beta \vec{f}(\delta)\,d\delta\,d\beta$$

where $\vec{f}(\alpha)$ is continuous on the interval $(-\infty<\alpha<\infty)$ and $\vec{k}\cdot\vec{f} = 0$. In this case, the stress vectors acting on the planes $\vec{k}\cdot\vec{r}$ = constant are given by

$$\vec{T} = \rho\omega^2[\hat{k}\times\vec{f}(\vec{k}\cdot\vec{r}-\omega t)]\; H(\omega t-\vec{k}\cdot\vec{r}) \quad .$$

If, for example, $\vec{f}(\alpha) = \vec{B}e^\alpha$ where $\vec{k}\cdot\vec{B} = 0$, then the shear stress on the plane $\vec{k}\cdot\vec{r} = a$ is given by

$$\tau = |\vec{T}| = \rho\omega^2|\hat{k}\times\vec{B}|e^{-(\omega t-a)}H(\omega t-a) \quad .$$

(c) Point Symmetric Dilatational Waves.

The potentials

$$\phi = \frac{1}{r}\,\Phi(\vec{k}\cdot\vec{r}-\omega t)\ ,\vec{\psi} = 0 \tag{11.25}$$

where $\vec{r} = r\hat{\varepsilon}_r$ is the position vector in spherical coordinates and $\vec{k} = k_r\hat{\varepsilon}_r$ is a radially directed propagation vector, characterize a point symmetric (or spherical) disturbance which propagates radially outward from the source r=0 with speed $v = \frac{\omega}{k}$. The amplitude of the disturbance is diminished by the factor $\frac{1}{r}$ as it progresses away from its source. Substitution of (11.25) into (11.3) and (11.4) yields the dispersion relations

$$k^2C_D^2 = \omega^2 \text{ and therefore } v = \frac{\omega}{k} = C_D \quad . \tag{11.26}$$

Since the phase velocity is independent of the wave number and/or the frequency, there is no dispersion, i.e., the wave form $\Phi(\alpha)$ is undistorted by the medium. The dilatation and rotation associated with this disturbance are

$$\varepsilon = \nabla^2\phi = \frac{k^2}{r}\,\Phi''(\alpha) \text{ and } \vec{\omega} = 0 \quad . \tag{11.27}$$

The corresponding displacement field is

$$\vec{u} = \vec{\nabla}\phi = \frac{\vec{k}}{r}\left[\Phi'(\alpha) - \frac{\Phi(\alpha)}{kr}\right] \tag{11.28}$$

where

$$\alpha = \vec{k}\cdot\vec{r} - \omega t = kr-\omega t = k(r-C_Dt) \quad .$$

Note that the displacement vector is parallel to the propagation vector $\vec{k}$ which in turn is perpendicular to the spherical surfaces given by $\vec{k}\cdot\vec{r} = kr = a$, where a is a constant. The stress vector acting on the spherical surface kr = a is given by

$$\vec{T} = \tau_{rr}\hat{\varepsilon}_r = \rho\omega^2\,\frac{k\hat{\varepsilon}_r}{a}\left[\Phi''(\alpha_a) - \frac{4\gamma^2}{a}\left(\Phi'(\alpha_a) - \frac{\Phi(\alpha_a)}{a}\right)\right] \tag{11.29}$$

where

$$\alpha_a = a-\omega t \quad .$$

The energy flux across the surface kr = a is given by

$$\vec{S} = S_r\hat{\varepsilon}_r = -\tau_{rr}\dot{u}_r\hat{\varepsilon}_r = \rho\omega^3\,\frac{k^3}{a^2}\left[\Phi''(\alpha_a) - \frac{4\gamma^2}{a}\left(\Phi'(\alpha_a) - \frac{\Phi(\alpha_a)}{a}\right)\right]\left[\Phi''(\alpha_a) - \frac{\Phi'(\alpha_a)}{a}\right]\hat{\varepsilon}_r \quad . \tag{11.30}$$

11.3 WAVE REFLECTION FROM A PLANE BOUNDARY

We shall now consider the problem of determining the reflected field which results when a plane, irrotational, wave impinges on a plane boundary as shown in Fig. 11.3.

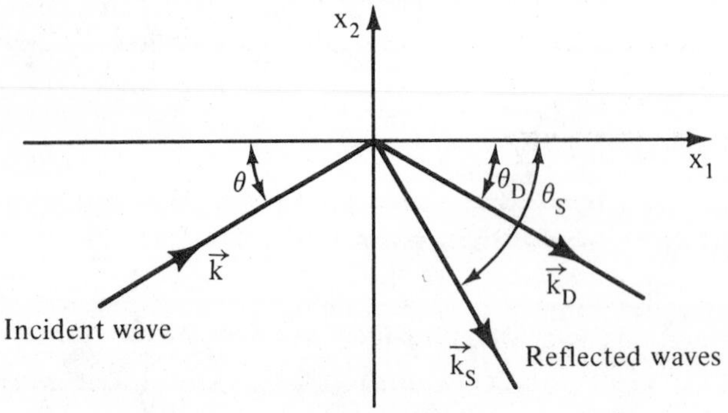

Figure 11.3

The boundary of the medium is the plane $x_2=0$ and, as shown in the figure, we assume that the propagation vector of the incident wave is parallel to the x_1x_2 plane, i.e.,

$$\vec{k}=k_1\hat{\varepsilon}_1 + k_2\hat{\varepsilon}_2 = k\ (\cos\Theta\hat{\varepsilon}_1+ \sin\Theta\hat{\varepsilon}_2) \tag{11.31}$$

where Θ is called the angle of incidence. We shall also assume that there is no motion in the x_3 direction ($u_3=0$) so that the medium is in a state of plane strain, i.e., $\varepsilon_{13} = \varepsilon_{23} = \varepsilon_{33}=0$. The boundary conditions to be considered are the following:

at $x_2 = 0$,

$$\left.\begin{aligned} &\text{either } u_1 \text{ or } \tau_{21} = 0 \\ \text{and}\quad &\\ &\text{either } u_2 \text{ or } \tau_{22} = 0 \ . \end{aligned}\right\} \tag{11.32}$$

The displacement potentials of the incident wave are

$$\phi^{(I)}= \phi(\vec{k}\cdot\vec{r}-\omega t) \text{ and } \vec{\psi}^{(I)}= 0 \quad . \tag{11.33}$$

It is assumed that the wave form $\phi(\alpha)$, and its first derivative $\phi'(\alpha)$, are continuous for $-\infty<\alpha<\infty$ and that its second derivative $\phi''(\alpha)$ is at least piecewise continuous in this interval. When we attempt to satisfy the boundary conditions (11.32) we will find that the reflected field may not be irrotational and that we must, in general, assume the displacement potentials of the reflected field to be of the form

$$\left.\begin{aligned}\phi^{(R)} &= A\phi\ (\vec{k}_D\cdot\vec{r}-\omega t)\\ \vec{\psi}^{(R)} &= B\phi\ (\vec{k}_S\cdot\vec{r}-\omega t)\hat{\varepsilon}_3\end{aligned}\right\} \tag{11.34}$$

where A and B are constants which are determined from the boundary conditions and where

$$\left.\begin{aligned}\vec{k}_D &= k_1^D\hat{\varepsilon}_1 + k_2^D\hat{\varepsilon}_2 = k_D(\cos\Theta_D\hat{\varepsilon}_1 - \sin\Theta_D\hat{\varepsilon}_2)\\ \vec{k}_S &= k_1^S\hat{\varepsilon}_1 + k_2^S\hat{\varepsilon}_2 = k_S(\cos\Theta_S\hat{\varepsilon}_1 - \sin\Theta_S\hat{\varepsilon}_2)\end{aligned}\right\} \tag{11.35}$$

are the propagation vectors of the reflected dilatational and shear waves, respectively. The total potential of the displacement field thus assumes the following form:

$$\left.\begin{aligned}\phi &= \phi^{(I)} + \phi^{(R)} = \phi(\vec{k}\cdot\vec{r}-\omega t) + A\phi(\vec{k}_D\cdot\vec{r}-\omega t)\\ \vec{\psi} &= \vec{\psi}^{(I)} + \vec{\psi}^{(R)} = B\phi(\vec{k}_S\cdot\vec{r}-\omega t)\hat{\varepsilon}_3\end{aligned}\right\} \tag{11.36}$$

The substitution of (11.36) into (11.4) yields

$$(k^2C_D^2-\omega^2)\phi''(\vec{k}\cdot\vec{r}-\omega t)+(k_D^2C_D^2-\omega^2)A\phi''(\vec{k}_D\cdot\vec{r}-\omega t) = 0$$

$$(k_S^2C_S^2-\omega^2)B\phi''(\vec{k}_S\cdot\vec{r}-\omega t) = 0 \quad .$$

These equations can be satisfied, in a nontrivial manner ($A\neq 0$, $B\neq 0$), by setting

$$\omega^2 = k^2C_D^2 = k_D^2C_D^2 = k_S^2C_S^2$$

or, in other words,

$$k_D = k \quad\text{and}\quad k_S = k\,\frac{C_D}{C_S} = \frac{k}{\gamma} \quad . \tag{11.37}$$

The displacement field is obtained by substituting (11.36) into (11.2). The result is,

$$\left.\begin{aligned}&\vec{u} = \vec{k}\phi'(\vec{k}\cdot\vec{r}-\omega t)+\vec{k}_DA\phi'(\vec{k}_D\cdot\vec{r}-\omega t)+(\vec{k}_S\times\hat{\varepsilon}_3)B\phi'(\vec{k}_S\cdot\vec{r}-\omega t)\\ &\text{or, in component form, } u_3=0,\\ &u_\alpha = k_\alpha\phi'(\vec{k}\cdot\vec{r}-\omega t)+k_\alpha^DA\phi'(\vec{k}_D\cdot\vec{r}-\omega t)+e_{\alpha\beta}k_\beta^SB\phi'(\vec{k}_S\cdot\vec{r}-\omega t)\end{aligned}\right\} \tag{11.38}$$

where $\alpha=1,2$, $\beta=1,2$ and $e_{\alpha\beta}\equiv e_{\alpha\beta 3}$ is the two-dimensional permutation tensor, i.e.,

$$e_{\alpha\beta}=\begin{bmatrix} 0 & 1 \\ -1 & 0 \end{bmatrix} .$$

In view of (11.14), (11.22) and our plane strain assumption, we conclude that the stress field in the medium can be expressed as follows:

$$\begin{aligned}\tau_{\alpha\beta} &= (\lambda k^2\delta_{\alpha\beta}+2\mu k_\alpha k_\beta)\phi''(\vec{k}\cdot\vec{r}-\omega t)\\ &+(\lambda k_D^2\delta_{\alpha\beta}+2\mu k_\alpha^D k_\beta^D)A\phi''(\vec{k}_D\cdot\vec{r}-\omega t)\\ &+\mu(k_\alpha^S e_{\beta\gamma}k_\gamma^S+k_\beta^S e_{\alpha\gamma}k_\gamma^S)B\phi''(\vec{k}_S\cdot\vec{r}-\omega t)\\ \tau_{\alpha 3} &= \tau_{3\alpha}= 0, \tau_{33}= \upsilon\tau_{\alpha\alpha}\end{aligned} \tag{11.39}$$

where $\alpha=1,2$, $\beta=1,2$ and $\gamma=1,2$.

$$\delta_{\alpha\beta}=\begin{bmatrix} 1 & 0 \\ 0 & 1 \end{bmatrix}$$

is the two-dimensional Kronecker delta. An examination of (11.38) and (11.39) reveals that, in order to satisfy the boundary conditions enumerated in (11.32), we must require

$$\vec{k}\cdot\vec{r} = \vec{k}_D\cdot\vec{r} = \vec{k}_S\cdot\vec{r} \quad \text{at } x_2= 0 \text{ and for all } x_1.$$

Thus,

$$\left.\begin{aligned} &k_1= k_1^D = k_1^S \\ &\text{or} \\ &k \cos \Theta = k_D \cos \Theta_D= k_S \cos \Theta_S \quad . \end{aligned}\right\} \tag{11.40}$$

This result, when combined with (11.37), yields

$$\left.\begin{aligned} &\cos \Theta_D= \cos \Theta \quad \text{and} \quad \cos \Theta_S= \gamma \cos \Theta \\ &\text{or} \\ &\Theta_D= \Theta \quad \text{and} \quad \Theta_S= \text{arc cos } (\gamma \cos \Theta) \quad . \end{aligned}\right\} \tag{11.41}$$

A graph of Θ_S vs. Θ is shown in Fig. 11.4.

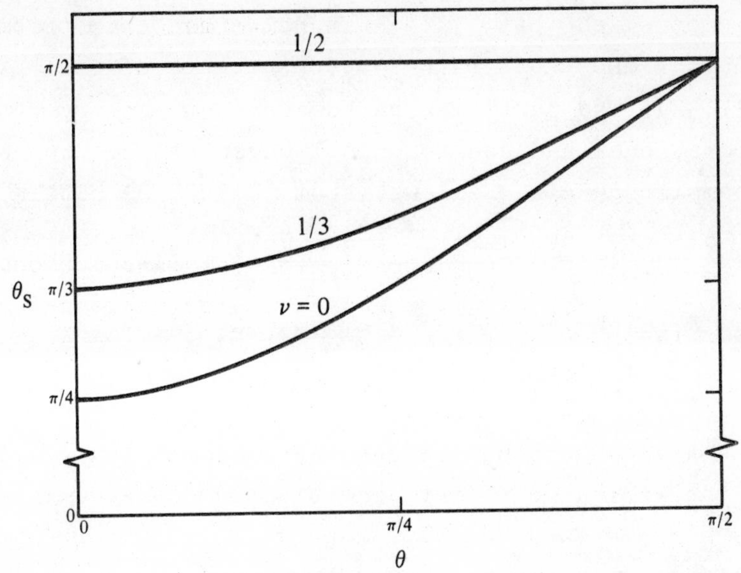

Figure 11.4

The components of the energy flux vector are, in the present case,

$$S_\alpha = -\tau_{\alpha\beta}\dot{u}_\beta, \; S_3 = 0 \quad .$$

In particular, the flux across the plane $x_2 = 0$ is given by

$$S_2 = -\tau_{21}\dot{u}_1 - \tau_{22}\dot{u}_2 \qquad \text{evaluated at } x_2=0 \quad .$$

An examination of the boundary conditons (11.32) reveals that $S_2=0$ at $x_2=0$, i.e., no energy is transmitted across the boundary of the medium. The flux S_2 at $x_2=0$ is a sum of the flux S_2^I due to the incident wave and the fluxes S_2^D and S_2^S due to the reflected waves. In view of (11.16), (11.24), (11.33), (11.34) and (11.40),

$$\left.\begin{aligned} S_2^I &= k_2\rho\omega^3\left[\phi''(k_1x_1-\omega t)\right]^2 \\ S_2^D &= k_2^D\rho\omega^3A^2\left[\phi''(k_1x_1-\omega t)\right]^2 \\ S_2^S &= k_2^S\rho\omega^3B^2\left[\phi''(k_1x_1-\omega t)\right]^2 \quad . \end{aligned}\right\} \tag{11.42}$$

Since $S_2 = S_2^I + S_2^D + S_2^S = 0$ for all x_1 and t we conclude that

$$\rho\omega^3\left[k_2 + k_2^DA^2 + k_2^SB^2\right] = 0$$

or

$$A^2 + \frac{\sin\Theta_S}{\gamma\sin\Theta}B^2 = 1 \quad . \tag{11.43}$$

It should be emphasized that (11.43) is not an independent restriction on the constants A and B but rather a consequence of the boundary conditions (11.32). The partition of the incident energy between the reflected irrotational and equivoluminal waves can be expressed in terms of the amplitude A as follows. From (11.42) we obtain

$$\frac{S_2^D}{S_2^I} = \frac{k_2^D}{k_2} A^2 = -A^2 \text{ and therfore } \frac{S_2^S}{S_2^I} = -\frac{S_2^I + S_2^D}{S_2^I} = -(1-A^2). \text{ Hence}$$

$$\left|\frac{S_2^D}{S_2^I}\right| = A^2, \quad \left|\frac{S_2^S}{S_2^I}\right| = 1-A^2 \quad . \tag{11.44}$$

The amplitudes A and B can be computed by substituting (11.38) and (11.39) into the various boundary conditions enumerated in (11.32). The results are summarized below.

(a) Free Boundary: $\tau_{21} = \tau_{22} = 0$ at $x_2 = 0$

$$\left.\begin{aligned} A &= \frac{-\cos^2 2\Theta_S + \gamma^2 \sin 2\Theta \sin 2\Theta_S}{\cos^2 2\Theta_S + \gamma^2 \sin 2\Theta \sin 2\Theta_S} \\ B &= \frac{2\gamma^2 \sin 2\Theta \cos 2\Theta_S}{\cos^2 2\Theta_S + \gamma^2 \sin 2\Theta \sin 2\Theta_S} \end{aligned}\right\} \tag{11.45}$$

The amplitudes of the reflected waves and the partition of energy for the case of a free boundary are shown in Fig. 11.5a and Fig. 11.5b, respectively for $\upsilon = \frac{1}{3}$.

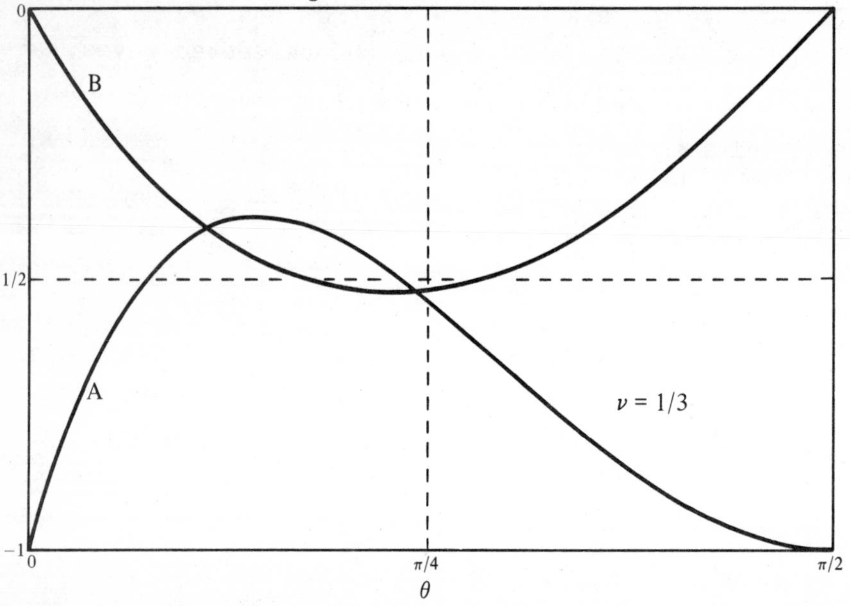

Figure 11.5
(a)

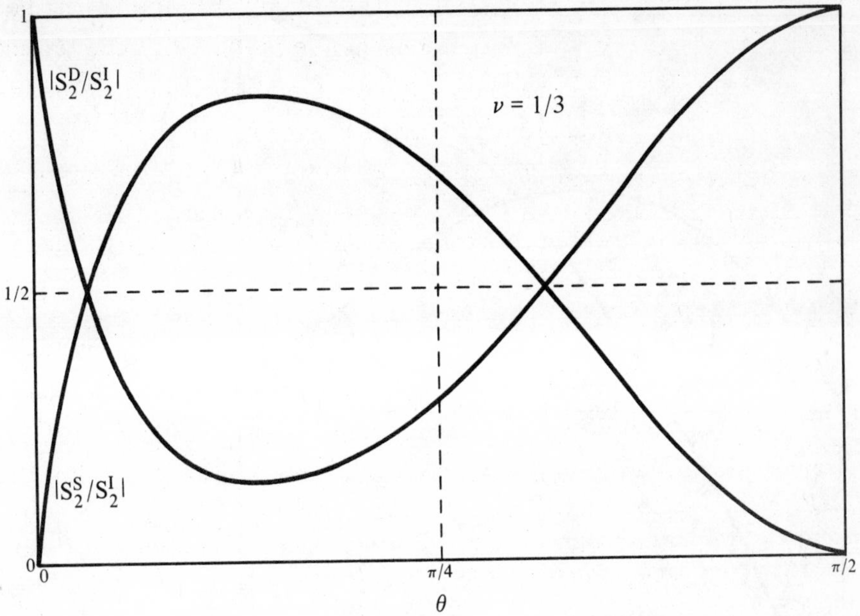

Fig. 11.5

(b)

(b) Fixed Boundary: $u_1 = u_2 = 0$ at $x_2 = 0$

$$\left.\begin{aligned} A &= \frac{-\cos^2\Theta_S+\gamma\sin\Theta\sin\Theta_S}{\cos^2\Theta_S+\gamma\sin\Theta\sin\Theta_S} \\ B &= \frac{2\gamma\sin\Theta\cos\Theta_S}{\cos^2\Theta_S+\gamma\sin\Theta\sin\Theta_S} \end{aligned}\right\} \tag{11.46}$$

The amplitudes of the reflected waves and the partition of energy for the case of a fixed boundary are shown in Fig. 11.6a and Fig. 11.6b, respectively for $\upsilon = \frac{1}{3}$.

(c) Frictionless Wall: $\tau_{21} = u_2 = 0$ at $x_2 = 0$

$$A = 1, \ B = 0 \tag{11.47}$$

(d) Simple Support: $u_1 = \tau_{22} = 0$ at $x_2 = 0$

$$A = -1, \ B = 0 \tag{11.48}$$

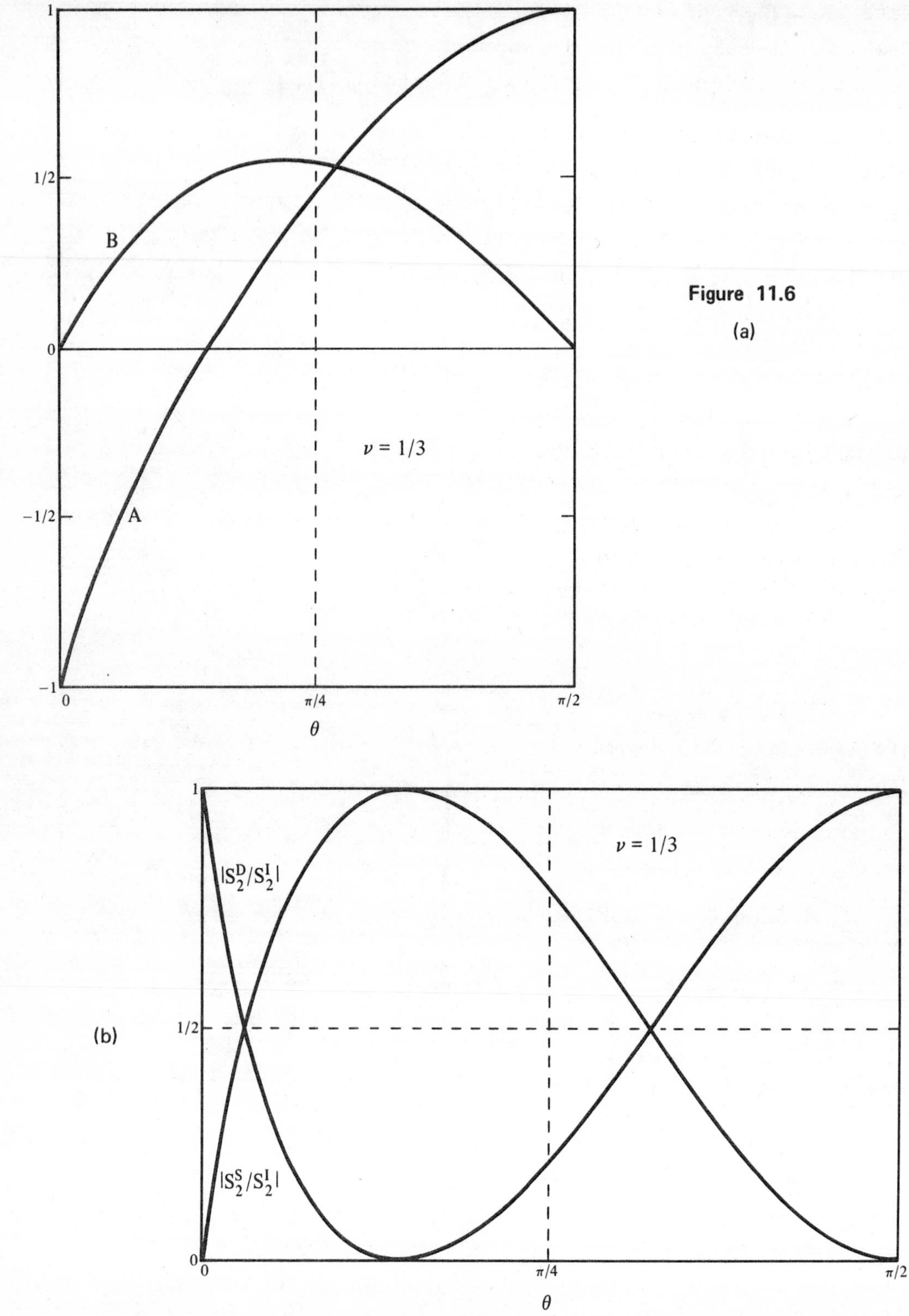

Figure 11.6

(a)

Note that the reflected field is irrotational in cases (c) and (d). Case (d) is called a simple support since this type of boundary condition (referred to as S_{T_n} in Chapter 10) may be used to simulate the simply supported end condition which occurs in classical beam and plate theories. In cases (a) and (b) the reflected field contains both dilatation and rotation for all angles of incidence except $\Theta = 0$ and $\Theta = \frac{\pi}{2}$. In the case of normal incidence, $\Theta = \frac{\pi}{2}$, we obtain $A=-1$, $B=0$ in case (a) and $A=1$, $B=0$ in case (b). Thus, a normally incident $\begin{pmatrix}\text{compressional}\\ \text{rarefactional}\end{pmatrix}$ wave is reflected as a $\begin{pmatrix}\text{rarefactional}\\ \text{compressional}\end{pmatrix}$ wave from a free boundary and as a $\begin{pmatrix}\text{compressional}\\ \text{rarefactional}\end{pmatrix}$ wave from a fixed boundary, i.e., a free boundary reverses the sense of the incoming wave while a fixed boundary does not. For the limiting value $\Theta = 0$ we obtain $A=-1$, $B = 0$ in both cases (a) and (b) which implies $\phi\equiv 0$ and $\vec{\psi}\equiv 0$. Thus, a plane wave cannot propagate parallel to a free or a fixed boundary. This conclusion is also seen to apply to the simply supported boundary, case (d). However, a plane wave can propagate parallel to a frictionless wall as indicated by (c). In the next section we will investigate another type of wave called a surface wave which can propagate parallel to a free boundary.

11.4 RAYLEIGH SURFACE WAVES

In the preceding section we found that the plane wave form $\phi(kx_1-\omega t)$ cannot propagate near a free boundary of an elastic medium. There is, however, another type of disturbance, called the Rayleigh surface wave, which does propagate parallel to a free boundary. Consider the displacement potentials

$$\left.\begin{aligned} \phi &= f(x_2)g(x_1-vt) \\ \vec{\psi} &= F(x_2)g(x_1-vt)\hat{e}_3 \end{aligned}\right\} \qquad (11.49)$$

for the elastic medium bounded by the plane $x_2=0$ as shown in Fig. 11.7.

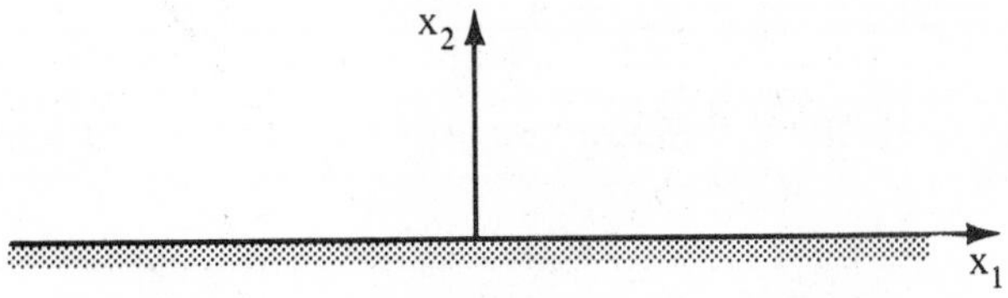

Figure 11.7

Substitution of (11.49) into (11.3) and (11.4) yields the equations

$$c_D^2\nabla^2\phi-\ddot{\phi}=(c_D^2-v^2)fg''+c_D^2f''g=0$$

$$c_S^2\nabla^2\psi_3-\ddot{\psi}_3=(c_S^2-v^2)Fg''+c_S^2F''g=0$$

or

$$\frac{g''}{g} = \frac{-c_D^2}{c_D^2-v^2}\frac{f''}{f} = \frac{-c_S^2}{c_S^2-v^2}\frac{F''}{F} = -k^2$$

where k is a separation constant (the wave number in this case). We thus obtain three differential equations for g,f and F. They are,

$$g''+k^2g = 0$$

$$f''-k^2\alpha_D^2f = 0, \quad \alpha_D^2 = 1 - \frac{v^2}{c_D^2}$$

$$F''-k^2\alpha_S^2F = 0, \quad \alpha_S^2 = 1 - \frac{v^2}{c_S^2} \quad .$$

The solutions of these equations are

$$g = \exp[\pm ik(x_1-vt)] , \quad f = \exp(\pm\alpha_Dkx_2), \quad F = \exp(\pm\alpha_Skx_2) \quad .$$

Since the solution must remain bounded as $x_2\to-\infty$ we choose $f=A\exp(\alpha_Dkx_2)$ and $F=B\exp(\alpha_Skx_2)$. As a matter of convenience we will use the exponential with positive exponent for g. The potentials of the form (11.49) which satisfy (11.3) and (11.4) are thus given by

$$\left.\begin{aligned}\phi &= A\exp(\alpha_Dkx_2)\exp[ik(x_1-vt)]\\ \vec{\psi} &= B\exp(\alpha_Skx_2)\exp[ik(x_1-vt)]\hat{e}_3\end{aligned}\right\} \tag{11.50}$$

where

$$\alpha_D = \sqrt{1-\frac{v^2}{c_D^2}} \quad \text{and} \quad \alpha_S = \sqrt{1-\frac{v^2}{c_S^2}} \quad . \tag{11.51}$$

The corresponding displacement field is obtained by substituting (11.50) into (11.2). The result is,

$$\left.\begin{aligned}u_1 &= \frac{\partial\phi}{\partial x_1} + \frac{\partial\psi_3}{\partial x_2} = [ikA\exp(\alpha_Dkx_2)+\alpha_SkB\exp(\alpha_Skx_2)]\exp[ik(x_1-vt)]\\ u_2 &= \frac{\partial\phi}{\partial x_2} - \frac{\partial\psi_3}{\partial x_1} = [\alpha_DkA\exp(\alpha_Dkx_2)-ikB\exp(\alpha_Skx_2)]\exp[ik(x_1-vt)]\end{aligned}\right\} \tag{11.52}$$

The stress and strain fields are obtained next by substituting (11.52) into (10.21) and (10.22). They are,

$$\left.\begin{aligned}\varepsilon_{11}&=k^2\left[-A\exp(\alpha_D kx_2)+i\alpha_S B\exp(\alpha_S kx_2)\right]\exp\left[ik(x_1-vt)\right]\\ \varepsilon_{22}&=k^2\left[\alpha_D^2 A\exp(\alpha_D kx_2)-i\alpha_S B\exp(\alpha_S kx_2)\right]\exp\left[ik(x_1-vt)\right]\\ \varepsilon_{12}&=\frac{k^2}{2}\left[2i\alpha_D A\exp(\alpha_D kx_2)+(1+\alpha_S^2)B\exp(\alpha_S kx_2)\right]\exp\left[ik(x_1-vt)\right]\end{aligned}\right\} \quad (11.53)$$

$$\tau_{11}=\rho C_S^2 k^2\left\{-\left[2+(1-2\gamma^2)\beta^2\right]A\exp(\alpha_D kx_2) + 2i\alpha_S B\exp(\alpha_S kx_2)\right\}\exp\left[ik(x_1-vt)\right] \quad (11.54a)$$

$$\tau_{22}=\rho C_S^2 k^2\left[(2-\beta^2)A\exp(\alpha_D kx_2) - 2i\alpha_S B\exp(\alpha_S kx_2)\right]\exp\left[ik(x_1-vt)\right] \quad (11.54b)$$

$$\tau_{12}=\rho C_S^2 k^2\left[2i\alpha_D A\exp(\alpha_D kx_2) +(2-\beta^2)B\exp(\alpha_S kx_2)\right]\exp\left[ik(x_1-vt)\right] \quad (11.54c)$$

where

$$\beta^2=\frac{v^2}{C_S^2}$$

In the case of a free boundary,

$$\tau_{12}=\tau_{22}=0 \text{ at } x_2=0 \quad .$$

Thus,

$$\left.\begin{aligned}(2-\beta^2)A-2i\alpha_S B&=0\\ 2i\alpha_D A+(2-\beta^2)B&=0 \quad .\end{aligned}\right\} \quad (11.55)$$

These equations will possess a nontrivial solution if and only if the determinant of their coefficient matrix vanishes, i.e.,

$$(2-\beta^2)^2=4\alpha_S\alpha_D=4\sqrt{(1-\beta^2)(1-\gamma^2\beta^2)} \quad . \quad (11.56a)$$

After squaring both sides of the preceding equation we obtain, after some rearrangement of the terms, the following polynomial equation for β^2:

$$\beta^2\left[\beta^6-8\beta^4+8(3-2\gamma^2)\beta^2-16(1-\gamma^2)\right]=0 \quad . \quad (11.56b)$$

The root $\beta^2=0$ implies v=0 and thus does not represent a traveling wave. For values of υ between 0 and $\frac{1}{2}$, the cubic equation in β^2 enclosed in parenthesis always has one real root which satisfies the inequalities

$$0<v^2<C_S^2<C_D^2$$

or

$$0<\beta^2<1<\frac{1}{\gamma^2}\ .$$

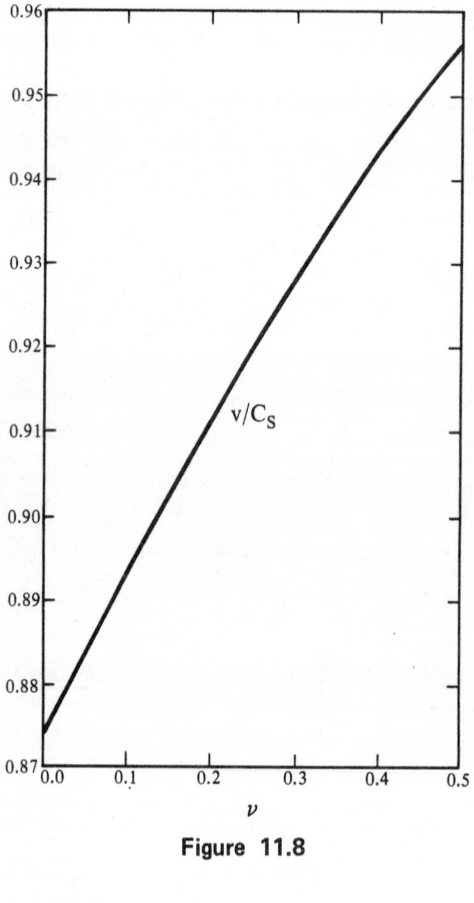

Figure 11.8

This root, which is illustrated in Fig. 11.8, yields real values of α_D and α_S when substituted into (11.51). Note that we have here an example of a wave which propagates in an elastic medium with a speed which is different from either C_D or C_S. An examination of (11.56) reveals that the phase velocity v is independent of the wave number k. Thus, there is no dispersion of the wave as it travels through the medium. With the aid of (11.55) the displacement and stress fields corresponding to a free boundary can be written as follows:

$$u_1 = \frac{ikA}{2}[2\exp(\alpha_D kx_2) - (2-\beta^2)\exp(\alpha_S kx_2)] \exp[ik(x_1 - vt)] \qquad (11.57a)$$

$$u_2 = \frac{\alpha_D kA}{2-\beta^2}[(2-\beta^2)\exp(\alpha_D kx_2) - 2\exp(\alpha_S kx_2)] \exp[ik(x_1 - vt)] \qquad (11.57b)$$

$$\left.\begin{aligned} \tau_{11} &= \rho C_S^2 k^2 A\{-[2+(1-2\gamma^2)\beta^2]\exp(\alpha_D kx_2) \\ &\qquad + (2-\beta^2)\exp(\alpha_S kx_2)\} \exp[ik(x_1 - vt)] \\ \tau_{22} &= \rho C_S^2 k^2 (2-\beta^2) A[\exp(\alpha_D kx_2) - \exp(\alpha_S kx_2)] \exp[ik(x_1 - vt)] \\ \tau_{12} &= \rho C_S^2 k^2 2i\alpha_D A[\exp(\alpha_D kx_2) - \exp(\alpha_S kx_2)] \exp[ik(x_1 - vt)] \end{aligned}\right\} \qquad (11.58)$$

If we assume a real disturbance, i.e.,

$$\phi = A\exp(\alpha_D kx_2) \cos k(x_1 - vt)$$

where A is a real constant, then the displacements at the free surface are found by taking the real part of (11.57) and then setting $x_2 = 0$. The result is

$$\left.\begin{aligned} u_1 &= -a \sin k(x_1 - vt) \\ u_2 &= -b \cos k(x_1 - vt) \end{aligned}\right\} \qquad (11.59)$$

where

$$a = \frac{kA\beta^2}{2} \quad \text{and} \quad b = \frac{kA\beta^2\alpha_D}{2-\beta^2} .$$

Thus, the particles on the free surface of the medium move in eliptical paths, i.e.,

$$\left(\frac{u_1}{a}\right)^2 + \left(\frac{u_2}{b}\right)^2 = 1 .$$

The period of this motion is $\frac{2\pi}{kv}$.

If we let ℓ represent the wavelength of the surface wave then $k = \frac{2\pi}{\ell}$. An examination of (11.57) and (11.58) reveals that this disturbance is attenuated by the factor $\exp(2\pi\alpha \frac{x_2}{\ell})$ as one moves away from the free surface, i.e., the shorter the wavelength the greater is the rate of attenuation. Hence, the effects of a disturbance which consists of very short wavelengths will be largely confined to a narrow strip adjacent to the free surface. This is sometimes called a boundary layer or skin effect.

11.5 PROGRESSIVE WAVES IN PLATES (THE RAYLEIGH-LAMB PROBLEM)

We now wish to examine the possibility of propagating a harmonic disturbance parallel to the free surfaces of an infinite plate. The x_1x_2 plane is chosen as the mid-plane of the plate whose free surfaces are the parallel planes $x_3 = \pm \frac{h}{2}$ as shown in Fig. 11.9. If we

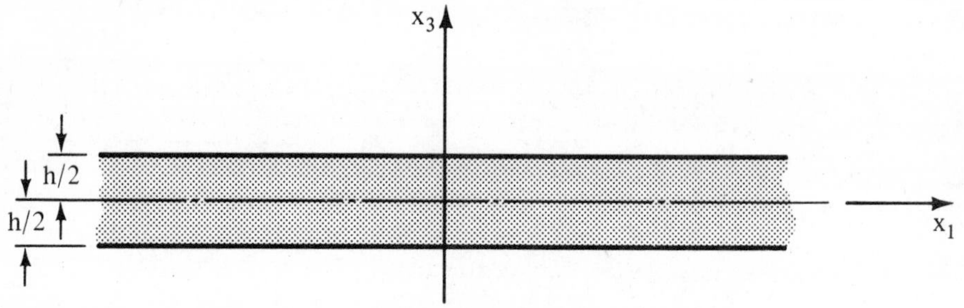

Figure 11.9

assume that the disturbance propagates in the positive x_1 direction then, in view of the results obtained in the preceding section, the displacement potentials which satisfy (11.3) and (11.4) may be written as follows:

$$\left.\begin{aligned} \phi &= f(x_3)\, \exp\left[ik(x_1 - vt)\right] \\ \vec{\psi} &= F(x_3)\, \exp\left[ik(x_1 - vt)\right] \hat{e}_3 \end{aligned}\right\} \qquad (11.60a)$$

where

$$\left.\begin{aligned} f(x_3) &= A_1 \cosh\alpha_D k x_3 + A_2 \sinh\alpha_D k x_3 \\ F(x_3) &= B_1 \sinh\alpha_S k x_3 + B_2 \cosh\alpha_S k x_3 \end{aligned}\right\} \qquad (11.60b)$$

and

$$\left.\begin{aligned} \alpha_D &= \sqrt{1-\gamma^2\beta^2} \quad , \quad \alpha_S = \sqrt{1-\beta^2} \\ \beta^2 &= \frac{v^2}{C_S^2} \quad . \end{aligned}\right\} \qquad (11.61)$$

By substituting (11.60) into (11.2) we find that the displacement field may be separated into two components as follows:

$$\vec{u} = \vec{u}^{(1)} + \vec{u}^{(2)} \qquad (11.62)$$

where

$$\left.\begin{aligned} u_1^{(1)} &= ik\left[A_1\cosh\alpha_D kx_3 + i\alpha_S B_1\cosh\alpha_S kx_3\right]\exp\left[ik(x_1-vt)\right] \\ u_2^{(1)} &= 0 \\ u_3^{(1)} &= k\left[\alpha_D A_1\sinh\alpha_D kx_3 + iB_1\sinh\alpha_S kx_3\right]\exp\left[ik(x_1-vt)\right] \end{aligned}\right\} \qquad (11.63)$$

and

$$\left.\begin{aligned} u_1^{(2)} &= ik\left[A_2\sinh\alpha_D kx_3 + i\alpha_S B_2\sinh\alpha_S kx_3\right]\exp\left[ik(x_1-vt)\right] \\ u_2^{(2)} &= 0 \\ u_3^{(2)} &= k\left[\alpha_D A_2\cosh\alpha_D kx_3 + iB_2\cosh\alpha_S kx_3\right]\exp\left[ik(x_1-vt)\right] \end{aligned}\right\} \qquad (11.64)$$

The displacement field $\vec{u}^{(1)}$ is characterized by the fact that the normal component of displacement $u_3^{(1)}$ is an odd function of x_3. Hence the deformed geometry of the free surfaces is symmetric about the plane $x_3=0$ as illustrated in Fig. 11.10. This is referred to as a longitudinal motion.

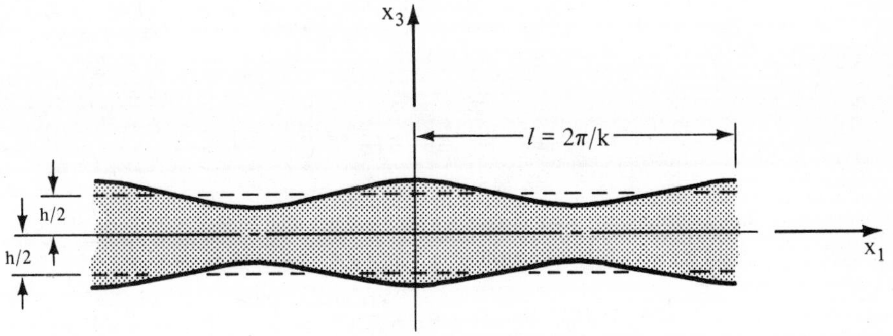

Figure 11.10

On the other hand, the displacement field $\vec{u}^{(2)}$ is characterized by the fact that the normal displacement $u_3^{(2)}$ is an even function of x_3. Hence the deformed geometry is antisymmetric with respect to the plane $x_3=0$ as illustrated in Fig. 11.11 and the motion is said to be transverse or flexural.

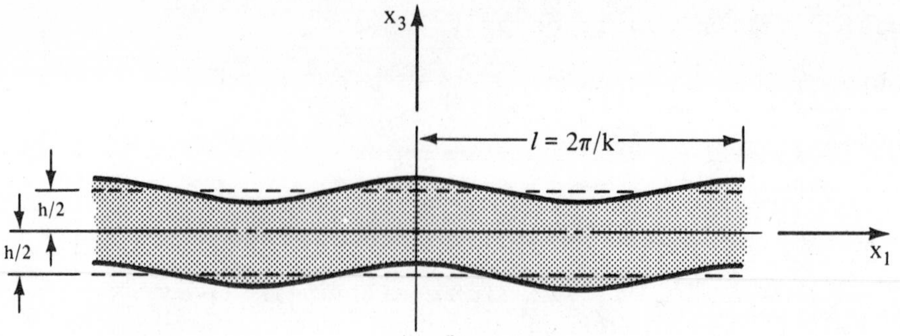

Figure 11.11

As a result of these symmetry properties the two motions are independent of one another and may be treated separately.

(a) Symmetric (Longitudinal) Motion.

The stress field associated with the longitudinal motion is obtained by substituting (11.63) into (10.21) and (10.22). The results are,

$$\tau_{11}^{(1)} = -\rho C_S^2 k^2 \left\{ \left[2+(1-2\gamma^2)\beta^2\right] A_1 \cosh\alpha_D kx_3 + 2i\alpha_S B_1 \cosh\alpha_S kx_3 \right\} \exp\left[ik(x_1-vt)\right] \tag{11.65a}$$

$$\tau_{33}^{(1)} = \rho C_S^2 k^2 \left[(2-\beta^2) A_1 \cosh\alpha_D kx_3 + 2i\alpha_S B_1 \cosh\alpha_S kx_3\right] \exp\left[ik(x_1-vt)\right] \tag{11.65b}$$

$$\tau_{13}^{(1)} = -\rho C_S^2 k^2 \left[-2i\alpha_D A_1 \sinh\alpha_D kx_3 + (2-\beta^2) B_1 \sinh\alpha_S kx_3\right] \exp\left[ik(x_1-vt)\right] \tag{11.65c}$$

The boundary conditions on the free surfaces are $\tau_{13}^{(1)}=\tau_{33}^{(1)}=0$ at $x_3= \pm \frac{h}{2}$. These conditions will be satisfied if we set

$$\left.\begin{aligned} &(2-\beta^2) A_1 \cosh\alpha_D \frac{kh}{2} + 2i\alpha_S B_1 \cosh\alpha_S \frac{kh}{2} = 0 \\ &\text{and} \\ &-2i\alpha_D A_1 \sinh\alpha_D \frac{kh}{2} + (2-\beta^2) B_1 \sinh\alpha_S \frac{kh}{2} = 0 \, . \end{aligned}\right\} \tag{11.66}$$

In order to obtain a nontrivial solution of (11.66) we must require that the determinant of its coefficient matrix vanish, i.e.,

$$(2-\beta^2)^2 \cosh\alpha_D \frac{kh}{2} \sinh\alpha_S \frac{kh}{2} = 4\alpha_S\alpha_D \cosh\alpha_S \frac{kh}{2} \sinh\alpha_D \frac{kh}{2} \, .$$

This equation may be rewritten as follows:

$$\frac{\tanh\pi\alpha_S \frac{h}{\ell}}{\tanh\pi\alpha_D \frac{h}{\ell}} = \frac{4\alpha_S\alpha_D}{(2-\beta^2)^2} \qquad (11.67)$$

where $\ell = \frac{2\pi}{k}$ is the wavelength of the harmonic wave propagating in the x_1 direction. Equation (11.67) is the dipsersion relation for this wave, it relates the dimensionless phase velocity $\beta = \frac{v}{c_S}$ to the dimensionless wavelength $\frac{\ell}{h}$. Since the phase velocity depends on the wavelength, the medium is dispersive. For any fixed value of the wavelength ratio $\frac{\ell}{h}$, equation (11.67) will possess a denumerable infinity of positive, real roots β_1, β_2, β_3, etc., each corresponding to a different mode of propagation. The limiting values of β for very long and for very short waves can be found from (11.67) as follows. For very short waves, $\frac{\ell}{h} \to 0$, $\tanh \pi\alpha \frac{h}{\ell} \to 1$ and equation (11.67) becomes

$$(2-\beta^2)^2 = 4\alpha_S\alpha_D$$

which is the dispersion relation for Rayleigh's surface waves (11.56a). For very long waves, $\frac{\ell}{h} \to \infty$, $\tanh \pi\alpha \frac{h}{\ell} \to \pi\alpha \frac{h}{\ell}$ and equation (11.67) becomes

$$\frac{\alpha_S}{\alpha_D} = \frac{4\alpha_S\alpha_D}{(2-\beta^2)^2}$$

which yields

$$\beta = \frac{v}{c_S} = 2\sqrt{1-\gamma^2} \quad \text{or} \quad v = \sqrt{\frac{E}{\rho(1-\upsilon^2)}} \ .$$

This is the dispersion relation which results from the classical plane stress theory for the extension of thin plates. Hence,this theory is applicable when the wavelength of the disturbance is very large compared to the thickness of the plate. The complete dispersion curve for the first mode of propagation when $\upsilon = 0.29$ is shown in Fig. 11.12 (after H. Kolsky, Stress Waves in Solids, Dover, 1963, p. 83, Fig. 20).

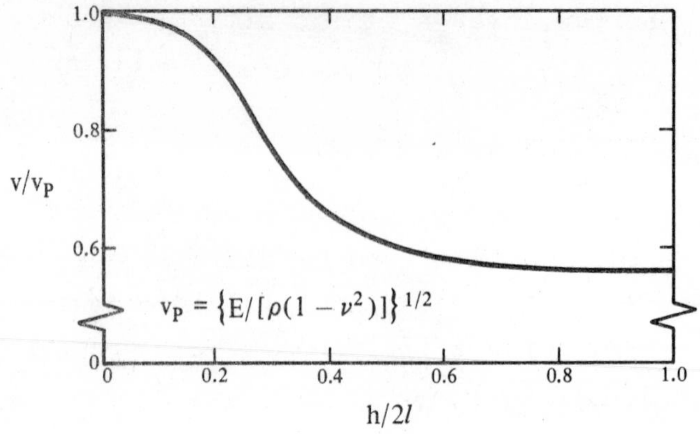

Figure 11.12

(b) Antisymmetric (Transverse or Flexural) Motion.

The stress field associated with the transverse motion is obtained by substituting (11.64) into (11.21) and (11.22). The results are,

$$\tau_{11}^{(2)} = -\rho c_S^2 k^2 \Big\{ \left[2+(1-2\gamma^2)\beta^2\right] A_2 \sinh \alpha_D k x_3 + 2i\alpha_S B_2 \sinh \alpha_S k x_3 \Big\} \exp\left[ik(x_1-vt)\right] \qquad (11.68a)$$

$$\tau_{33}^{(2)} = \rho c_S^2 k^2 \left[(2-\beta^2) A_2 \sinh \alpha_D k x_3 + 2i\alpha_S B_2 \sinh \alpha_S k x_3\right] \exp\left[ik(x_1-vt)\right] \qquad (11.68b)$$

$$\tau_{13}^{(2)} = -\rho c_S^2 k^2 \left[-2i\alpha_D A_2 \cosh \alpha_D k x_3 + (2-\beta^2) B_2 \cosh \alpha_S k x_3\right] \exp\left[ik(x_1-vt)\right] \qquad (11.68c)$$

The boundary conditions on the free surfaces, $\tau_{33}^{(2)}=\tau_{13}^{(2)}=0$ at $x_3=\pm \frac{h}{2}$ are satisfied by setting

$$(2-\beta^2) A_2 \sinh \alpha_D \frac{kh}{2} + 2i\alpha_S B_2 \sinh \alpha_S \frac{kh}{2} = 0$$

and

$$-2i\alpha_D A_2 \cosh \alpha_D \frac{kh}{2} + (2-\beta^2) B_2 \cosh \alpha_S \frac{kh}{2} = 0 \quad . \qquad (11.69)$$

The existence of a nontrivial solution of (11.69) is insured by setting

$$(2-\beta^2)^2 \sinh \alpha_D \frac{kh}{2} \cosh \alpha_S \frac{kh}{2} = 4\alpha_S \alpha_D \sinh \alpha_S \frac{kh}{2} \cosh \alpha_D \frac{kh}{2}$$

or

$$\frac{\tanh \pi\alpha_D \frac{h}{\ell}}{\tanh \pi\alpha_S \frac{h}{\ell}} = \frac{4\alpha_S\alpha_D}{(2-\beta^2)^2} \tag{11.70}$$

where $\ell = \frac{2\pi}{k}$ is the wavelength of the harmonic wave propagating in the x_1 direction. The dispersion relation (11.70) indicates that the dimensionless phase velocity $\beta = \frac{v}{C_S}$ depends on the dimensionless wavelength $\frac{\ell}{h}$, hence, the medium is dispersive. For a fixed $\frac{h}{\ell}$, equation (11.70) possesses a denumerable infinity of positive real roots $\beta_1, \beta_2, \beta_3$, etc., each corresponding to a different mode of propagation. The limiting values of β for very long and for very short waves can be found from (11.70) as follows. For very short waves, $\frac{\ell}{h} \to 0$, $\tanh \pi\alpha \frac{h}{\ell} \to 1$ and (11.70) becomes

$$(2-\beta^2)^2 = 4\alpha_S\alpha_D \tag{11.71}$$

which is the dispersion relation for Rayleigh's surface waves (11.56a). For very long waves, $\frac{\ell}{h} \to \infty$, we obtain the following power series expansion for the left-hand side of (11.70):

$$\frac{\tanh \pi\alpha_D \frac{h}{\ell}}{\tanh \pi\alpha_S \frac{h}{\ell}} = \frac{\alpha_D}{\alpha_S}\left[1+\left(\frac{\alpha_S^2-\alpha_D^2}{3}\right)\left(\frac{\pi h}{\ell}\right)^2 - \left(\frac{\alpha_S^4+5\alpha_S^2\alpha_D^2-6\alpha_D^4}{45}\right)\left(\frac{\pi h}{\ell}\right)^4 + \ldots\right]$$

By equating the above expression to the right-hand side of (11.70) we obtain, with the aid of (11.61),

$$0=\beta^4-\left(\frac{\pi h}{\ell}\right)^2\left(\frac{1-\gamma^2}{3}\right)\beta^2(2-\beta^2)^2+\left(\frac{\pi h}{\ell}\right)^4\frac{\beta^2}{45}(2-\beta^2)^2\left[7(1-\gamma^2)-\gamma^2\beta^2(5-6\gamma^2)\right]+\ldots$$

It is evident from the preceding equation that $\beta \to 0$ as $\frac{h}{\ell} \to 0$, therefore, let us assume that $\beta = \left(\frac{\pi h}{\ell}\right) x$. Substituting this expression for β into the preceding equation yields

$$0=\left(\frac{\pi h}{\ell}\right)^4\left[x^4-4\left(\frac{1-\gamma^2}{3}\right)x^2\right]+0\left(\frac{\pi h}{\ell}\right)^6$$

Thus,

$$x^2 = \frac{4}{3}(1-\gamma^2)$$

or

$$\beta^2 = \frac{v^2}{C_S^2} = \frac{4}{3}(1-\gamma^2)\left(\frac{\pi h}{\ell}\right)^2 = \frac{1-\gamma^2}{3}k^2h^2$$

and therefore

$$v = \sqrt{\frac{Eh^2}{12(1-\upsilon^2)\rho}} \quad k = \sqrt{\frac{D}{\rho h}} \quad k \tag{11.72}$$

where $D = \dfrac{Eh^3}{12(1-\upsilon^2)}$ is called the plate flexural rigidity. This is the dispersion relation which one obtains from the classical theory for the bending of thin plates. Thus, we should expect this theory to yield reasonably good results for disturbances whose wavelength is large compared to the plate thickness. The complete dispersion curve corresponding to the first mode of propagation of flexural waves together with the dispersion curves for several approximate theories have been plotted by R.D. Mindlin, ("Influence of Rotatory Inertia and Shear on Flexural Motions of Isotropic, Elastic Plates," Journal of Applied Mechanics, Vol. 18, No. 1; Trans. ASME 73, 1951, pp. 31-38) and are shown in Fig. 11.13.

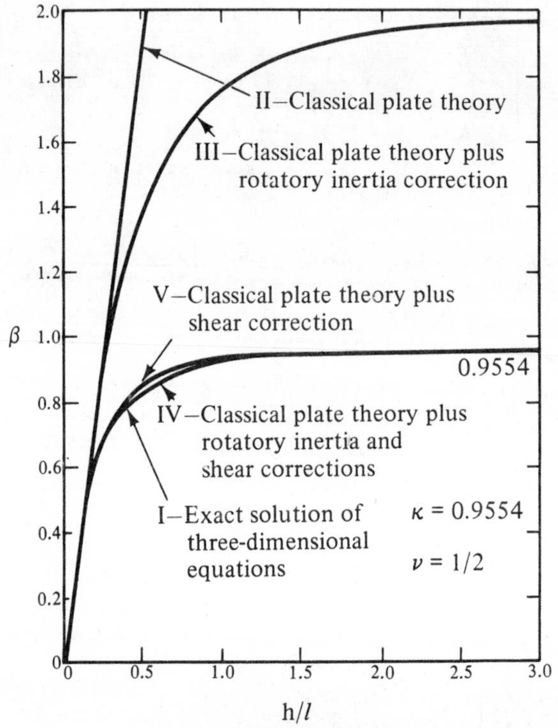

Figure 11.13

11.6 PROGRESSIVE WAVES IN CYLINDERS

The last example we shall consider is that of a harmonic disturbance propagating parallel to the axis of an infinitely long, circular cylinder whose boundary r=R is stress free. The geometry of the cylinder is shown in Fig. 11.14.

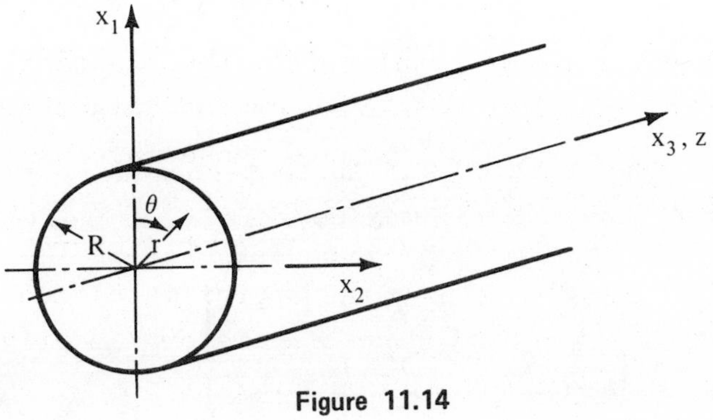

Figure 11.14

The displacement potentials for this example are assumed to have the following form

$$\left.\begin{aligned} \phi &= \Phi(r,\Theta)\exp\left[ik(z-vt)\right] \\ \vec{\psi} &= \vec{\Psi}(r,\Theta)\exp\left[ik(z-vt)\right] \end{aligned}\right\} \tag{11.73}$$

where

$$\vec{\Psi} = \Psi_r\hat{e}_r + \Psi_\Theta\hat{e}_\Theta + \Psi_z\hat{e}_z \quad .$$

By substituting (11.73) into (11.3) and (11.4) we obtain, with the aid of (8.107), (8.109) and Exercise 8.18, the following equations for Φ and $\vec{\Psi}$.

$$\frac{\partial\Psi_r}{\partial r} + \frac{\Psi_r}{r} + \frac{1}{r}\frac{\partial\Psi_\Theta}{\partial\Theta} + ik\,\Psi_z = 0 \tag{11.74}$$

$$\nabla_2^2\Phi + k^2\alpha_D^2\Phi = 0 \tag{11.75a}$$

$$\left.\begin{aligned} &\nabla_2^2\Psi_r - \frac{\Psi_r}{r^2} - \frac{2}{r^2}\frac{\partial\Psi_\Theta}{\partial\Theta} + k^2\alpha_S^2\Psi_r = 0 \\ &\nabla_2^2\Psi_\Theta - \frac{\Psi_\Theta}{r^2} + \frac{2}{r^2}\frac{\partial\Psi_r}{\partial\Theta} + k^2\alpha_S^2\Psi_\Theta = 0 \\ &\nabla_2^2\Psi_z + k^2\alpha_S^2\Psi_z = 0 \end{aligned}\right\} \tag{11.75b}$$

where

$$\alpha_D^2 = \frac{v^2}{c_D^2} - 1, \quad \alpha_S^2 = \frac{v^2}{c_S^2} - 1 \tag{11.76}$$

and

$$\nabla_2^2 \equiv \frac{\partial^2}{\partial r^2} + \frac{1}{r}\frac{\partial}{\partial r} + \frac{1}{r^2}\frac{\partial^2}{\partial \Theta^2} \quad . \tag{11.77}$$

By employing the separation of variables technique one finds that the solution of (11.74) and (11.75) may be written as follows:

$$\left.\begin{aligned} \Phi(r,\Theta) &= f(r)\left[a_n \cos n\Theta + b_n \sin n\Theta\right] \\ \Psi_r(r,\Theta) &= g_r(r)\left[a_n \sin n\Theta - b_n \cos n\Theta\right] \\ \Psi_\Theta(r,\Theta) &= g_\Theta(r)\left[a_n \cos n\Theta + b_n \sin n\Theta\right] \\ \Psi_z(r,\Theta) &= g_z(r)\left[a_n \sin n\Theta - b_n \cos n\Theta\right] \end{aligned}\right\} \tag{11.78}$$

where

$$\left.\begin{aligned} f(r) &= \tilde{A}_n J_n(\alpha_D kr) + A_n' Y_n(\alpha_D kr) \\ g_r(r) &= \tilde{B}_n J_{n-1}(\alpha_S kr) + \tilde{C}_n J_{n+1}(\alpha_S kr) \\ &\qquad + B_n' Y_{n-1}(\alpha_S kr) + C_n' Y_{n+1}(\alpha_S kr) \\ g_\Theta(r) &= \tilde{B}_n J_{n-1}(\alpha_S kr) - \tilde{C}_n J_{n+1}(\alpha_S kr) \\ &\qquad + B_n' Y_{n-1}(\alpha_S kr) - C_n' Y_{n+1}(\alpha_S kr) \\ g_z(r) &= i\alpha_S(\tilde{C}_n - \tilde{B}_n) J_n(\alpha_S kr) + i\alpha_S(C_n' - B_n') Y_n(\alpha_S kr) \end{aligned}\right\} \tag{11.79}$$

for $n=0,1,2,\ldots$.

Since we are considering a complete cylinder we must set $A_n'=B_n'=C_n'=0$ in order to obtain a solution which is bounded at $r=0$. This leaves three arbitrary constants $\tilde{A}_n$, $\tilde{B}_n$ and $\tilde{C}_n$ which we shall determine from the three boundary conditions $\tau_{rr}=\tau_{r\Theta}=\tau_{rz}=0$ at $r=R$. By substituting (11.73) into (11.2) we obtain, with the aid of (8.105), (8.108), (11.78) and (11.79),

$$\left.\begin{aligned} u_r &= \frac{\partial\phi}{\partial r} + \frac{1}{r}\frac{\partial\psi_z}{\partial\Theta} - \frac{\partial\psi_\Theta}{\partial z} = U_n\left[a_n\cos n\Theta + b_n\sin n\Theta\right]\exp\left[ik(z-vt)\right] \\ u_\Theta &= \frac{1}{r}\frac{\partial\phi}{\partial\Theta} + \frac{\partial\psi_r}{\partial z} - \frac{\partial\psi_z}{\partial r} = V_n\left[a_n\sin n\Theta - b_n\cos n\Theta\right]\exp\left[ik(z-vt)\right] \\ u_z &= \frac{\partial\phi}{\partial z} + \frac{\partial\psi_\Theta}{\partial r} + \frac{\psi_\Theta}{r} - \frac{1}{r}\frac{\partial\psi_r}{\partial\Theta} = W_n\left[a_n\cos n\Theta + b_n\sin n\Theta\right]\exp\left[ik(z-vt)\right] \end{aligned}\right\} \tag{11.80}$$

where

$$\left.\begin{aligned} U_n &= k\left[\alpha_D A_n J_n'(r_D) - B_n J_n'(r_S) + C_n\frac{n}{r_S}J_n(r_S)\right] \\ V_n &= -k\left[\alpha_D\frac{n}{r_D}A_n J_n(r_D) - B_n\frac{n}{r_S}J_n(r_S) + C_n J_n'(r_S)\right] \\ W_n &= ik\left[A_n J_n(r_D) + \alpha_S B_n J_n(r_S)\right] \end{aligned}\right\} \tag{11.81}$$

and

$$A_n \equiv \tilde{A}_n,\quad B_n \equiv i(\tilde{C}_n + \tilde{B}_n),\quad C_n \equiv i\beta^2(\tilde{C}_n - \tilde{B}_n)$$

$$r_D = \alpha_D kr,\quad r_S = \alpha_S kr,\quad \beta^2 = \frac{v^2}{c_S^2} \quad .$$

A prime denotes differentiation with respect to the argument of the function. With the aid of equations (8.110) and (10.22) the following stress-displacement relations are obtained:

$$\left.\begin{aligned} \tau_{rr} &= \lambda\varepsilon + 2\mu\frac{\partial u_r}{\partial r} \\ \tau_{r\Theta} &= \mu\left[\frac{1}{r}\frac{\partial u_r}{\partial\Theta} + \frac{\partial u_\Theta}{\partial r} - \frac{u_\Theta}{r}\right] \\ \tau_{rz} &= \mu\left[\frac{\partial u_z}{\partial r} + \frac{\partial u_r}{\partial z}\right] \end{aligned}\right\} \tag{11.82}$$

where

$$\varepsilon = \nabla^2\phi = \frac{1}{c_D^2}\ddot{\phi} = -k^2\frac{v^2}{c_D^2}\phi = -k^2\gamma^2\beta^2\phi \quad .$$

In view of (11.80) the stresses can be expressed as follows:

$$\left.\begin{aligned} \tau_{rr} &= \tau_{rr}^{(n)}\left[a_n\cos n\Theta + b_n\sin n\Theta\right]\exp\left[ik(z-vt)\right] \\ \tau_{r\Theta} &= \tau_{r\Theta}^{(n)}\left[a_n\sin n\Theta - b_n\cos n\Theta\right]\exp\left[ik(z-vt)\right] \\ \tau_{rz} &= \tau_{rz}^{(n)}\left[a_n\cos n\Theta + b_n\sin n\Theta\right]\exp\left[ik(z-vt)\right] \end{aligned}\right\} \tag{11.83}$$

where

$$\left.\begin{aligned}
\tau_{rr}^{(n)} &= \mu\left[-k^2\beta^2(1-2\gamma^2)f+2\,\frac{\partial U_n}{\partial r}\right]\\
\tau_{r\Theta}^{(n)} &= \mu\left[-\frac{n}{r}\,U_n+\frac{\partial V_n}{\partial r}-\frac{V_n}{r}\right]\\
\tau_{rz}^{(n)} &= \mu\left[\frac{\partial W_n}{\partial r}+ik\,U_n\right]
\end{aligned}\right\} \tag{11.84}$$

By substituting (11.79) and (11.81) into (11.84) we obtain

$$\left.\begin{aligned}
\tau_{rr}^{(n)} = \mu k^2\Bigg\{&-A_n\left[\frac{2\alpha_D^2}{r_D}\,J_n'(r_D)+\left(\alpha_S^2-1-2\alpha_D^2\,\frac{n^2}{r_D^2}\right)J_n(r_D)\right]\\
&+2\alpha_S B_n\left[\frac{1}{r_S}\,J_n'(r_S)+\left(1-\frac{n^2}{r_S^2}\right)J_n(r_S)\right]\\
&+2\alpha_S C_n\,\frac{n}{r_S}\left[J_n'(r_S)-\frac{1}{r_S}\,J_n(r_S)\right]\Bigg\}\\
\tau_{r\Theta}^{(n)} = \mu k^2\Bigg\{&-2\alpha_D^2 A_n\,\frac{n}{r_D}\left[J_n'(r_D)-\frac{1}{r_D}\,J_n(r_D)\right]\\
&+2\alpha_S B_n\,\frac{n}{r_S}\left[J_n'(r_S)-\frac{1}{r_S}\,J_n(r_S)\right]\\
&+\alpha_S C_n\left[\frac{2}{r_S}\,J_n'(r_S)+\left(1-\frac{2n^2}{r_S^2}\right)J_n(r_S)\right]\Bigg\}
\end{aligned}\right\} \tag{11.85}$$

$$\tau_{rz}^{(n)} = i\mu k^2\left\{2\alpha_D A_n J_n'(r_D)+(\alpha_S^2-1)B_n J_n'(r_S)+C_n\,\frac{n}{r_S}\,J_n(r_S)\right\}$$

We can now determine the constants A_n, B_n and C_n by setting

$$\tau_{rr}^{(n)}(R) = \tau_{r\Theta}^{(n)}(R) = \tau_{rz}^{(n)}(R) = 0$$

for each integer value of n. The n=0 and n=1 cases are especially interesting because of their physical significance.

(a) n = 0: Longitudinal and Torsional Waves.

With the aid of (11.85) we obtain the following expression of the boundary conditions when n=0.

$$\tau_{rr}^{(o)}(R) = \mu k^2 \left\{ A_o \left[\frac{2\alpha_D^2}{R_D} J_1(R_D) - (\alpha_S^2 - 1) J_o(R_D) \right] + 2\alpha_S B_o \left[J_o(R_S) - \frac{J_1(R_S)}{R_S} \right] \right\} = 0$$

$$\tau_{rz}^{(o)}(R) = -i\mu k^2 \left\{ 2\alpha_D A_o J_1(R_D) + (\alpha_S^2 - 1) B_o J_1(R_S) \right\} = 0$$

$$\tau_{r\Theta}^{(o)}(R) = \mu k^2 \alpha_S C_o \left[J_o(R_S) - \frac{2}{R_S} J_1(R_S) \right] = 0$$

where

$$R_D = \alpha_D kR \text{ and } R_S = \alpha_S kR \quad . \tag{11.86}$$

These boundary conditions may be satisfied in either of the following two ways.

(1) $A_o = B_o = 0$

and

$$R_S J_o(R_S) = 2J_1(R_S) \tag{11.87}$$

This solution corresponds to a torsional wave motion, i.e., if we choose $b_o = -1$ then,

$$u_r = u_z = 0, \ \tau_{rr} = \tau_{rz} = 0$$

$$u_\Theta = kC_o J_1(r_S) \exp[ik(z-vt)] \tag{11.88}$$

$$\tau_{r\Theta} = \mu k^2 \alpha_S C_o \left[J_o(r_S) - \frac{2}{r_S} J_1(r_S) \right] \exp[ik(z-vt)] \quad .$$

The smallest root of (11.87) is $R_S=0$ which implies $\alpha_S=0$ or, in other words, $v^2=C_S^2$. If we replace $\frac{1}{2}k^2\alpha_S C_o$ with $\overline{C}_o$ then (11.88) becomes

$$u_\Theta = \overline{C}_o r \exp[ik(z-C_S t)]$$

$$\tau_{r\Theta} = 2\mu\overline{C}_o [0] \exp[ik(z-C_S t)] = 0 \quad .$$

Thus, the motion associated with the phase velocity $v=C_S$ corresponds to a rigid body rotation of each normal section about the cylinder's axis. A normal section is the intersection of the cylinder with a plane whose normal is parallel to the cylinder's axis. Equation (11.87) possesses a denumerable infinity of positive real roots which can be ordered as follows: $0<R_S^{(1)}<R_S^{(2)}<R_S^{(3)}<\ldots$. For example, $R_S^{(1)}\cong 5.136$, $R_S^{(2)}\cong 8.417$, etc. The dispersion relation corresponding to the root $R_S^{(i)}$ is $\alpha_S kR = R_S^{(i)}$ or

$$\frac{v^2}{c_S^2} = 1 + \alpha_S^2 = 1 + \left(\frac{R_S^{(i)}}{kR}\right)^2 = 1 + \left(\frac{\ell R_S^{(i)}}{2\pi R}\right)^2 \tag{11.88}$$

where $\ell = \frac{2\pi}{k}$ is the wavelength of the harmonic wave propagating in the z direction.

(2) $C_o = 0$

$$\left.\begin{aligned}&\left[\frac{2\alpha_D^2}{R_D} J_1(R_D)-(\alpha_S^2-1)J_o(R_D)\right] A_o+2\alpha_S\left[J_o(R_S)-\frac{J_1(R_S)}{R_S}\right]B_o= 0\\ &2\alpha_D J_1(R_D)A_o+(\alpha_S^2-1)J_1(R_S)B_o= 0\end{aligned}\right\} \tag{11.89}$$

This solution corresponds to a longitudinal wave motion, i.e., if we set $a_o=1$ then,

$$\left.\begin{aligned}&u_\Theta=0, \quad \tau_{r\Theta}=0\\ &u_r=k\left[-\alpha_D A_o J_1(r_D)+B_o J_1(r_S)\right] \exp\left[ik(z-vt)\right]\\ &u_z=ik\left[A_o J_o(r_D)+\alpha_S B_o J_o(r_S)\right] \exp\left[ik(z-vt)\right] .\end{aligned}\right\} \tag{11.90}$$

In order to obtain a nontrivial solution for A_o and B_o we must set

$$(\alpha_S^2-1)J_1(R_S)\left[\frac{2\alpha_D^2}{R_D} J_1(R_D)-(\alpha_S^2-1)J_o(R_D)\right]= =4\alpha_D\alpha_S J_1(R_D)\left[J_o(R_S)-\frac{J_1(R_S)}{R_S}\right] \tag{11.91}$$

Equation (11.91) can be rewritten in a more convenient form as follows:

$$\beta^2\alpha_D^2-(\beta^2-2)^2F(R_D)-4\alpha_D^2F(R_S)= 0 \tag{11.92}$$

where

$$F(x) \equiv \frac{xJ_o(x)}{2J_1(x)}$$

$$\alpha_D^2= \gamma^2\beta^2-1, \quad R_D= kR\sqrt{\gamma^2\beta^2-1}, \quad R_S= kR\sqrt{\beta^2-1}$$

and $\beta^2= \frac{v^2}{c_S^2}$. Equation (11.92) is the dispersion relation for longitudinal motion of the cylinder. It relates the dimensionless phase velocity β to the dimensionless wave number $kR=2\pi(\frac{R}{\ell})$ for any given value of Poisson's ratio $\left(\gamma^2=\frac{1-2\upsilon}{2-2\upsilon}\right)$. If the wavelength of the harmonic wave propagating in the z direction is very long compared to the radius of the cylinder, i.e., if $kR=2\pi(\frac{R}{\ell})<<1$, then we can expand F(x) into a power series, the first two terms of which are,

$$F(x)\cong 1-\frac{1}{2}(\frac{x}{2})^2 \quad \text{for } x^2<<1 \quad .$$

If we substitute the above approximation for F(x) into (11.92) we obtain, after neglecting terms of order $\left(\frac{kR}{2}\right)^4$ and smaller,

$$-(1-\gamma^2)\beta^2+(3-4\gamma^2)+\frac{1}{2}\beta^2(\gamma^2\beta^2-1)(\frac{kR}{2})^2 = 0 \quad .$$

In keeping with the order of the approximation we assume that

$$\beta^2\cong a+b(\frac{kR}{2})^2$$

and then substitute this expression for β^2 into the preceding equation to obtain

$$a = \frac{3-4\gamma^2}{1-\gamma^2} = \frac{E}{\mu} \qquad \text{(E is Young's modulus)}$$

and

$$b=\frac{a}{2}\left(\frac{\gamma^2 a-1}{1-\gamma^2}\right) = -2\upsilon^2 a \quad .$$

Thus,

$$\beta^2\cong \frac{E}{\mu}\left[1-2\upsilon^2(\frac{kR}{2})^2\right]$$

or $\qquad$ for $(\frac{kR}{2})^2<<1$

$$v^2\cong \frac{E}{\rho}\left[1-2\upsilon^2(\frac{kR}{2})^2\right]$$

and therefore

$$v\cong\sqrt{\frac{E}{\rho}}\left[1-\upsilon^2(\frac{kR}{2})^2\right] \quad . \tag{11.93}$$

This result was first obtained by L. Pochhammer (J.f. Math. (Crelle), Bd. 81(1876), p.324) and was later rediscovered, independently, by C. Chree (Quart. J. of Math., Vol. 21(1886)). It indicates that very long waves will propagate parallel to the cylinder's axis with a speed very nearly equal to $\sqrt{\frac{E}{\rho}}$ which is the value that is used in approximate one-dimensional theories.

For fixed values of υ and kR, equation (11.92) will yield a denumerable infinity of positve real roots β_1^2, β_2^2, etc., each corresponding to a different mode of propagation. By varying kR one obtains the dispersion curves for the various modes of propagation. The dispersion curves corresponding to the lowest mode of propagation for various values of Poisson's ratio have been computed by D. Bancroft (Physical Review, Vol. 59, 1941, pp. 588-593) and are shown in Fig. 11.15. For very short wavelengths, i.e., as $kR=2\pi(\frac{R}{\ell})\to\infty$, we obtain the following asymptotic values for F(x).

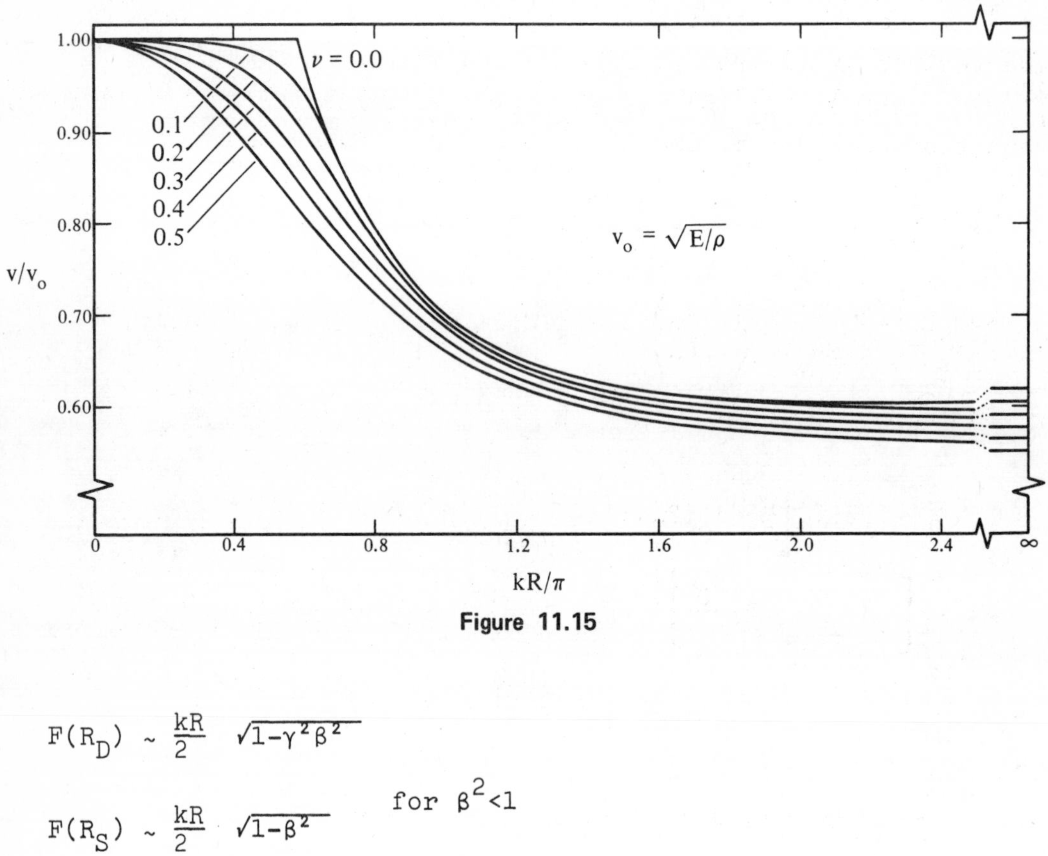

Figure 11.15

$$F(R_D) \sim \frac{kR}{2} \sqrt{1-\gamma^2\beta^2}$$

$$F(R_S) \sim \frac{kR}{2} \sqrt{1-\beta^2} \qquad \text{for } \beta^2<1$$

By substituting these expressions for $F(R_D)$ and $F(R_S)$ into (11.92) we obtain, after neglecting terms of order (2/kR),

$$(\beta^2-2)^2 = 4\sqrt{1-\gamma^2\beta^2}\ \sqrt{1-\beta^2}$$

which is the equation for Rayleigh surface waves (see (11.56a)).

(b) n = 1: Flexural Waves.

By letting $a_1=1$, $b_1=0$ in (11.80) we find that the displacement field assumes the following form for this case:

$$
\begin{aligned}
u_r &= U_1(r)\cos\Theta\exp[ik(z-vt)]\\
u_\Theta &= V_1(r)\sin\Theta\exp[ik(z-vt)] \qquad (11.94)\\
u_z &= W_1(r)\cos\Theta\exp[ik(z-vt)]\ .
\end{aligned}
$$

Referring to Fig. 11.14 and equation (11.94), we observe that the motion of particles lying on the x_1x_3 plane ($\Theta=0,\pi$) is confined to that plane while the motion of particles lying on the x_2x_3 plane ($\Theta=\pm\frac{\pi}{2}$) is normal to that plane, i.e., vertical. Thus, the motion characterized by (11.94) will be called flexural. By setting $\tau_{rr}^{(1)}(R)=\tau_{r\Theta}^{(1)}(R)=\tau_{rz}^{(1)}(R)=0$ we obtain from (11.85) the following set of equations for A_1, B_1 and C_1.

$$
-A_1\left[\frac{2\alpha_D^2}{R_D}J_o(R_D)+\left(\alpha_S^2-1-\frac{4\alpha_D^2}{R_D^2}\right)J_1(R_D)\right]
$$
$$
+2\alpha_S B_1\left[\frac{J_o(R_S)}{R_S}+\left(1-\frac{2}{R_S^2}\right)J_1(R_S)\right]+2\alpha_S C_1\left[\frac{J_o(R_S)}{R_S}-\frac{2}{R_S^2}J_1(R_S)\right]=0
$$
$$
-A_1\frac{2\alpha_D^2}{R_D}\left[J_o(R_D)-\frac{2}{R_D}J_1(R_D)\right]+B_1\frac{2\alpha_S}{R_S}\left[J_o(R_S)-\frac{2}{R_S}J_1(R_S)\right]
$$
$$
+\alpha_S C_1\left[\frac{2}{R_S}J_o(R_S)+\left(1-\frac{4}{R_S^2}\right)J_1(R_S)\right]=0
$$
$$
2\alpha_D A_1\left[J_o(R_D)-\frac{J_1(R_D)}{R_D}\right]+(\alpha_S^2-1)B_1\left[J_o(R_S)-\frac{J_1(R_S)}{R_S}\right]
$$
$$
+C_1\frac{J_1(R_S)}{R_S}=0
$$

If we let

$$
A_1=\frac{R_D}{2J_1(R_D)}\frac{A}{\alpha_D},\quad B_1=\frac{R_S B}{2J_1(R_S)}\quad\text{and}\quad C_1=\frac{R_S C}{2J_1(R_S)}
$$

then the preceding equations can be written in the following convenient form:

$$
\left.
\begin{aligned}
&-A[G(R_D)+(\alpha_S^2-1)K^2]+B[G(R_S)+2\alpha_S^2K^2]+CG(R_S)=0\\
&-AG(R_D)+BG(R_S)+C[G(R_S)+\alpha_S^2K^2]=0\\
&2A[2G(R_D)+1]+(\alpha_S^2-1)B[2G(R_S)+1]+C=0
\end{aligned}
\right\}\qquad (11.95)
$$

where $K = (\frac{kR}{2})$ is a dimensionless wave number and

$$G(x) \equiv F(x) - 1 = \frac{xJ_o(x)}{2J_1(x)} - 1 \quad . \qquad (11.96)$$

In order to obtain a nontrivial solution of (11.95) we must set the determinant of its coefficient matrix equal to zero, i.e.,

$$\begin{vmatrix} -\left[G(R_D)+(\alpha_S^2-1)K^2\right] & \left[G(R_S)+2\alpha_S^2K^2\right] & G(R_S) \\ -G(R_D) & G(R_S) & \left[G(R_S)+\alpha_S^2K^2\right] \\ 2\left[2G(R_D)+1\right] & (\alpha_S^2-1)\left[2G(R_S)+1\right] & 1 \end{vmatrix} = 0 \qquad (11.97a)$$

This is the dispersion relation for flexural waves. It relates the dimensionless phase velocity β to the dimensionless wave number K with Poisson's ratio appearing as a parameter ($\gamma^2 = \frac{1-2\upsilon}{2-2\upsilon}$). The determinant (11.97a) may be simplified by replacing the first row by the difference of the first row minus the second row. After dividing both sides of the resulting equation by K^2 we obtain

$$\begin{vmatrix} -(\alpha_S^2-1) & 2\alpha_S^2 & -\alpha_S^2 \\ -G(R_D) & G(R_S) & \left[G(R_S)+\alpha_S^2K^2\right] \\ 2\left[2G(R_D)+1\right] & (\alpha_S^2-1)\left[2G(R_S)+1\right] & 1 \end{vmatrix} = 0 \qquad (11.97b)$$

where

$$\alpha_S^2 = \beta^2-1, \quad \beta^2 = \frac{v^2}{c_S^2}$$

$$R_D = 2K\sqrt{\gamma^2\beta^2-1} \quad \text{and} \quad R_S = 2K\sqrt{\beta^2-1}.$$

After expanding (11.97b) we obtain the following expression for the dispersion relation:

$$2(\beta^2-2)^2G^2(R_S)+2(\beta^2-1)(\beta^2+4)G(R_S)G(R_D)$$
$$+\left[\beta^2(\beta^2+1)+2(\beta^2-1)(\beta^2-2)^2K^2\right]G(R_S)$$
$$+(\beta^2-1)\left[\beta^2+8(\beta^2-1)K^2\right]G(R_D)+\beta^4(\beta^2-1)K^2=0 \quad . \qquad (11.97c)$$

We note that $\beta^2=1$ or $\alpha_S^2=0$ is a root of (11.97c). However, an examination of (11.95) and (11.81) reveals that the corresponding solution

$A=0$, $B=C$ yields the trivial case $U_1=V_1=W_1=0$. We shall therefore disregard this root. If the wavelength of the harmonic wave propagating in the z direction becomes very small, i.e., if $K=\frac{kR}{2}=\pi(\frac{R}{\ell})\to\infty$, then $G(R_D)\to K\sqrt{1-\gamma^2\beta^2}$ and $G(R_S)\to K\sqrt{1-\beta^2}$, assuming $\beta^2<1$. If these asymptotic values for G are substituted into (11.97c) we obtain, after neglecting terms of order $\frac{1}{K}$ and smaller,

$$-2(1-\beta^2)^{\frac{3}{2}}\left[(\beta^2-2)^2-4\sqrt{1-\beta^2}\sqrt{1-\gamma^2\beta^2}\right]=0$$

or

$$(\beta^2-2)^2=4\sqrt{1-\beta^2}\sqrt{1-\gamma^2\beta^2}\quad .$$

This is the equation for Rayleigh surface waves which we obtained earlier (see (11.56a)). If, on the other hand, the wavelength is very large, i.e., if

$$K=\frac{kR}{2}=\pi(\frac{R}{\ell})\to 0,\ \text{then}$$

$$G(R_S)\to -\frac{\alpha_S^2K^2}{2}-\frac{\alpha_S^4K^4}{12}\quad\text{and}\quad G(R_D)\to -\frac{\alpha_D^2K^2}{2}-\frac{\alpha_D^4K^4}{12}\ .$$

If we assume $\beta^2\cong aK^2$ and then substitute this value for β^2 together with the preceding asymptotic values for $G(R_S)$ and $G(R_D)$ into (11.97c) we find that $a=\frac{3-4\gamma^2}{1-\gamma^2}=\frac{E}{\mu}$.

Thus,

$$\beta^2=\frac{v^2}{c_S^2}\cong\frac{E}{\mu}K^2\qquad\text{for } K^2<<1$$

or

$$v\cong\sqrt{\frac{E}{\rho}}\ K=\sqrt{\frac{E}{\rho}}\ (\frac{kR}{2})=\sqrt{\frac{E}{\rho}}\ \pi(\frac{R}{\ell}). \tag{11.98}$$

This is precisely the value which one would obtain from the classical, Euler-Bernoulli beam theory, i.e., $v=\sqrt{\frac{EI}{\rho A}}\,k=\sqrt{\frac{E}{\rho}}\ \frac{kR}{2}\ (\frac{I}{A}=\frac{R^2}{4}$ for a beam of circular cross section). Hence, we should expect Euler-Bernoulli beam theory to yield reasonable results for disturbances the wavelengths of which are large compared with the maximum chord of the beam's cross section.

Data for the complete dispersion curve for flexural waves has been computed by G.E. Hudson (Phys. Rev., Vol. 63, 1943, pp.46-51) for various values of Poisson's ratio between zero and one-half. R.M. Davies (Phil. Trans. Roy. Soc. (London), A, Vol. 240, 1948, p.454, Fig. 37) has interpolated from this data to obtain the dispersion curve for υ=0.29 which is shown, together with the dispersion curves for several approximate theories, in Fig. 11.16.

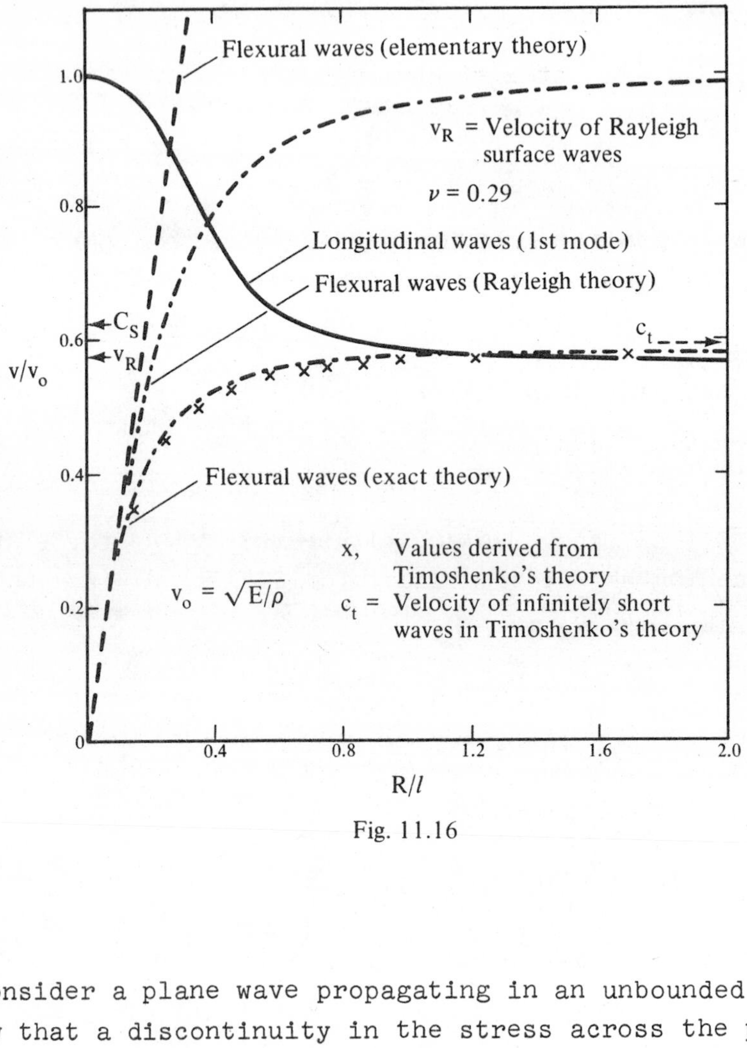

Fig. 11.16

EXERCISES

11.1 Consider a plane wave propagating in an unbounded elastic medium. Show that a discontinuity in the stress across the plane $\vec{k}\cdot\vec{r}=\omega t$ implies a discontinuity in the velocity across that plane.

11.2 Examine the properties of a time harmonic, irrotational wave which emanates from the line source r=0 in an unbounded elastic medium. Use cylindrical coordinates and assume

$$\phi = \phi(kr)\ \exp\ (i\omega t) \text{ and } \vec{\psi} = \vec{0}\ .$$

11.3 Determine the reflected field which results when a plane equivoluminal wave impinges on a plane boundary as shown in Fig. 11.3. Follow the procedure used in Section 11.3 with

$$\phi^{(I)} = 0, \ \vec{\psi}^{(I)} = \psi(\vec{k}\cdot\vec{r}-\omega t)\hat{e}_3 \ .$$

11.4 Find the dispersion relations for the surface waves of Section 11.4 for the following boundary conditions:

(a) $u_1 = u_2 = 0$ at $x_2 = 0$

(b) $u_1 = \tau_{22} = 0$ at $x_2 = 0$

(c) $u_2 = \tau_{21} = 0$ at $x_2 = 0$

11.5 Under what other boundary conditions can the symmetric and antisymmetric motions of an infinite plate be treated separately as they were in Section 11.5?

11.6 Find the dispersion relations for an infinite plate with both surfaces rigidly fixed. Discuss the limiting cases of very long and very short wavelengths.

11.7 Consider an infinite plate with one surface fixed and the other free. Find the limiting form of the dispersion relation as the ratio of the wavelength to thickness approaches infinity.

11.8 Find the dispersion relation for torsional waves in a hollow cylinder whose boundaries $r=R_I$ and $r=R_o$ are stress free.

11.9 Find the dispersion relation for longitudinal waves in a hollow cylinder whose boundaries $r=R_I$ and $r=R_o$ are stress free. Discuss the limiting cases of very long and very short wavelengths.

11.10 Show that the group velocity may be determined in dimensionless form as follows: $\frac{U}{C_S} = \beta + kR\frac{d\beta}{d(kR)}$. Find the group velocity of torsional and longitudinal waves in a solid cylinder. Discuss the limiting cases of very long and very short wavelengths.

11.11 Consider the plane strain dynamic response of an unbounded elastic medium with a cylindrical cavity. A harmonically oscillating load is uniformly distributed and acts normal to the cavity boundary. Find the response of the surrounding medium.

11.12 Consider the point symmetric dynamic response of an unbounded elastic medium with a spherical cavity. A harmonically oscillating load is uniformly distributed and acts normal to the cavity boundary. Find the response of the surrounding medium.

Chapter

12

DYNAMICS OF ELASTIC RODS

There are a few partial differential equations which occur so frequently in physical problems that they may be called classical. The first of these is the simple wave-equation which occurs in the theory of a vibrating string and also in the theory of the propagation of plane waves which travel without change in form. These waves may be waves of sound, elastic waves of various kinds, waves of light, electromagnetic waves and waves on the surface of the water.

H. Bateman, Partial Differential Equations of Mathematical Physics, 1959.

12.1 TORSIONAL MOTION OF RODS

An exact solution for wave motion in unbounded, cylindrical rods with circular cross-section was obtained in Chapter 11 within the context of classical elasticity theory. Many technical applications require a detailed knowledge of the free and forced motion of rods of finite length subjected to a variety of boundary and loading conditions. An inspection of the pertinent elasticity equations in Chapter 10 should convince the reader that their solution for technically significant boundary and loading conditions would require a monumental computational effort, if such solutions can be obtained at all. It is possible to derive equations of motion of rods which are approximations of elasticity theory that are sufficiently accurate for a host of engineering applications. The structure of the approximate equations is considerably simpler than the field equations of elasticity theory, and for this reason useful solutions for free and forced motion are readily obtained. Most engineering calculations pertaining to the static and dynamic response of structures are carried out within the framework of "approximate" theories of the type discussed in the present and succeeding chapters.

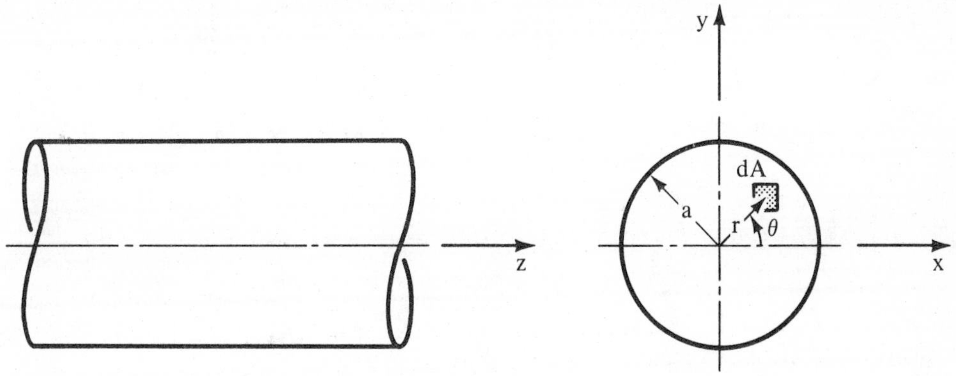

Figure 12.1

Let us consider the torsional motion of a rod of circular cross-section with radius a as shown in Fig. 12.1. We shall take the z-axis as the axis of revolution of the rod, and refer the displacement vector to cylindrical coordinates, r,θ,z as shown. It is now assumed that cross-sections of the rod originally plane remain plane, and radial lines originally straight remain radial and straight during deformation. In addition, we assume that there will be no radial displacements. Consequently, we postulate the displacement field

$$u_r = 0, \; u_\Theta = r\phi(z,t), \; u_z = 0 \tag{12.1}$$

where ϕ is the angle of twist. Upon substitution of (12.1) into the strain-displacement relations referred to cylindrical coordinates (8.110), we obtain

$$\begin{bmatrix} \varepsilon_{rr} & \varepsilon_{r\Theta} & \varepsilon_{rz} \\ \varepsilon_{\Theta r} & \varepsilon_{\Theta\Theta} & \varepsilon_{\Theta z} \\ \varepsilon_{zr} & \varepsilon_{z\Theta} & \varepsilon_{zz} \end{bmatrix} = \begin{bmatrix} 0 & 0 & 0 \\ 0 & 0 & \frac{1}{2} r \frac{\partial\phi}{\partial z} \\ 0 & \frac{1}{2} r \frac{\partial\phi}{\partial z} & 0 \end{bmatrix} \tag{12.2}$$

To derive the stress-equations of motion, we utilize Hamilton's principle in the form

$$\int_{t_1}^{t_2} (\delta T - \delta U + \delta W)dt = 0 \tag{12.3}$$

The kinetic energy of the rod extending from $z=z_1$ to $z=z_2$ is given by

$$T = \frac{1}{2} \int_V \rho(\dot{u}_r^2 + \dot{u}_\Theta^2 + \dot{u}_z^2)dV = \frac{1}{2} \int_{z_1}^{z_2} \int_A \rho r^2 \dot{\phi}^2 dA dz = \frac{1}{2} \int_{z_1}^{z_2} J\rho\dot{\phi}^2 dz \tag{12.4}$$

where $J = \frac{\pi a^4}{2}$ is the polar moment of inertia of the circular cross-section area with respect to the center of the circle. Consequently

$$\delta T = \int_{z_1}^{z_2} J\rho\dot{\phi} \frac{\partial}{\partial t} (\delta\phi)dz$$

and

$$\int_{t_1}^{t_2} \delta T dt = - \int_{t_1}^{t_2}\int_{z_1}^{z_2} J\rho\ddot{\phi}\delta\phi dz dt \quad . \tag{12.5}$$

To obtain the above result we have integrated by parts with respect to t and invoked the condition $\delta\phi=0$ for $t=t_1$ and $t=t_2$.

If the strain energy density is denoted by the symbol W*, then the total instantaneous strain energy in a rod segment extending from $z=z_1$ to $z=z_2$ is given by

$$U = \int_V W^* dV \tag{12.6}$$

so that

$$\delta U= \int_V \frac{\partial W^*}{\partial \varepsilon_{ij}} \delta\varepsilon_{ij} dV= \int_V \tau_{ij}\delta\varepsilon_{ij} dV= \int_V (\tau_{\Theta z}\delta\varepsilon_{\Theta z}+\tau_{z\Theta}\delta\varepsilon_{z\Theta}) dV$$

$$= \int_{z_1}^{z_2}\int_A \tau_{\Theta z} r \frac{\partial(\delta\phi)}{\partial z} dAdz= \int_{z_1}^{z_2} M \frac{\partial(\delta\phi)}{\partial z} dz=(M\delta\phi)\Big|_{z_1}^{z_2} - \int_{z_1}^{z_2} \frac{\partial M}{\partial z} \delta\phi dz \tag{12.7}$$

where M is the twisting moment (stress-resultant) defined by

$$M = \int_A r\tau_{\Theta z} dA \tag{12.8}$$

If body forces are neglected, then in accordance with (9.119) and (12.1), the virtual work of the surface tractions acting on a rod segment is given by

$$\delta W = \int_S T_i \delta u_i dS= \int_S T_\Theta \delta u_\Theta dS= \int_S rT_\Theta \delta\phi dS$$

where $S=A_1+A_2+\sigma$. In this expression, A_1 and A_2 denote the (flat) cross-sectional areas of the rod at $z=z_1$ and $z=z_2$, respectively, while σ is the (curved) surface area of the rod for $z_1<z<z_2$. With reference to (9.142a), we have

$$T_\Theta = \tau_{r\Theta}n_r + \tau_{\Theta\Theta}n_\Theta + \tau_{z\Theta}n_z$$

and the components of the unit outer normal $\hat{n}$ are

on A_1: $n_r = 0$, $n_\Theta = 0$, $n_z = -1$

on A_2: $n_r = 0$, $n_\Theta = 0$, $n_z = 1$

Consequently

$$\delta W= \int_\sigma rT_\Theta \delta\phi dA+\delta\phi \int_{A_2} r\tau_{z\Theta} dA-\delta\phi \int_{A_1} r\tau_{z\Theta} dA$$

But

$$\int_\sigma rT_\Theta \delta\phi dS= \int_{z_1}^{z_2} \delta\phi \oint_c rT_\Theta dc dz= \int_{z_1}^{z_2} m\delta\phi dz$$

where we let the line integral $\oint_c rT_\Theta dc=m$. Furthermore, we let

$$M^*(z_1,t)= \int_{A_1} r\tau_{z\Theta} dA$$

$$M^*(z_2,t)= \int_{A_2} r\tau_{z\Theta} dA$$

With these notations, the expression for the virtual work reduces to

$$\delta W= \int_{z_1}^{z_2} m\delta\phi dz+(M^*\delta\phi)\Big|_{z_1}^{z_2} \tag{12.9}$$

We note that the symbol m denotes the (external) distributed twisting moment intensity, while M* is the twisting moment acting on the end of the rod. Upon substitution of (12.5), (12.7), and (12.9) into (12.3), we obtain

$$\int_{t_1}^{t_2}\int_{z_1}^{z_2}(-J\rho\ddot{\phi}+\frac{\partial M}{\partial z}+m)\delta\phi dzdt+\int_{t_1}^{t_2}\Big[(M^*-M)\delta\phi\Big]_{z_1}^{z_2}dt=0$$

The time interval t_2-t_1, as well as the length of the rod segment z_2-z_1 are arbitrary. Therefore the integrands of the double and single integrals must vanish separately. For arbitrary $\delta\phi$ we have

$$-J\rho\frac{\partial^2\phi}{\partial t^2}+\frac{\partial M}{\partial z}+m=0 \text{ for } z_1<z<z_2 \tag{12.10a}$$

At $z=z_1$ and $z=z_2$ we must specify either $\phi=\phi(t)$ or $M=M^*(t)$ (12.10b)
Equation (12.10a) is the (approximate) stress equation of torsional motion of a rod, and (12.10b) are the associated, admissible boundary conditions.

We shall next derive an expression for energy flux in the rod. Differentiating (12.6) with respect to t, we obtain

$$\frac{dU}{dt}=\int_V\frac{\partial W^*}{\partial\varepsilon_{ij}}\frac{d\varepsilon_{ij}}{dt}dV=\int_V\tau_{ij}\dot{\varepsilon}_{ij}dV=\int_{z_1}^{z_2}\int_A\tau_{\theta z}r\frac{\partial\dot{\phi}}{\partial z}dzdA$$

$$=\int_{z_1}^{z_2}M\frac{\partial\dot{\phi}}{\partial z}dz$$

where we have used (12.2) and (12.8). Differentiating (12.4) with respect to t, we obtain

$$\frac{dT}{dt}=\int_{z_1}^{z_2}J\rho\dot{\phi}\ddot{\phi}dz=\int_{z_1}^{z_2}(\dot{\phi}m+\dot{\phi}\frac{\partial M}{\partial z})dz$$

where we have utilized the stress-equation of motion (12.10a). If we denote the instantaneous energy per unit length of the rod by ε, then

$$\int_{z_1}^{z_2}\dot{\varepsilon}dz=\frac{d}{dt}(T+U)=\int_{z_1}^{z_2}\Big[\dot{\phi}m+\frac{\partial}{\partial z}(\dot{\phi}M)\Big]dz$$

$$=\int_{z_1}^{z_2}(\dot{\phi}m-\frac{\partial S}{\partial z})dz$$

where the energy flux is defined by

$$S=-\dot{\phi}M \tag{12.11}$$

Since the length z_2-z_1 of the rod segment is arbitrary, we obtain the energy continuity equation (see 9.94).

$$\dot{\varepsilon} + \frac{\partial S}{\partial z} = m\dot{\phi} \tag{12.12}$$

If we set the twisting moment intensity m=0, we have

$$\frac{d}{dt}(T+U) = (-S)_{z_1}^{z_2} = S(z_1,t)-S(z_2,t) \tag{12.13}$$

i.e., the time rate of change of energy in the rod is equal to the energy flux at $z=z_1$ minus the energy flux at $z=z_2$.

In order to find an expression for momentum flux in the rod, we re-write (12.10a) in the form

$$\dot{\alpha} = \frac{\partial M}{\partial z} + m$$

where $\alpha=J\rho\dot{\phi}$ is the magnitude of the angular momentum per unit length of the rod. If we now define the (angular) momentum flux J^* by

$$J^* = -M \tag{12.14}$$

then we readily obtain the momentum continuity equation

$$\dot{\alpha} + \frac{\partial J^*}{\partial z} = m \tag{12.15}$$

If we set m=0, then

$$\int_{z_1}^{z_2}\dot{\alpha}dz = J^*(z_1,t) - J^*(z_2,t) \tag{12.16}$$

i.e., the rate of change of angular momentum contained in a rod segment extending from $z=z_1$ to $z=z_2$ is equal to the momentum flux at $z=z_1$ minus the momentum flux at $z=z_2$. Combining (12.11) and (12.14), we obtain a relationship between energy flux and momentum flux in the rod given by

$$S = \dot{\phi}J^* \tag{12.17}$$

The three-dimensional version of (12.17) is exhibited by (9.93b).

The stress-equation of motion (12.10a) cannot be integrated as it stands, and we require a constitutive relation to complete our mathematical model. Assuming a linearly elastic, isotropic material, Hooke's law, in conjunction with (12.2) yields $\tau_{rr}=\tau_{\theta\theta}=\tau_{zz}=\tau_{rz}=\tau_{r\theta}=0$. and

$$\tau_{\theta z}= 2G\varepsilon_{\theta z}= Gr\frac{\partial\phi}{\partial z} \tag{12.18}$$

where G is the shear modulus. Upon substitution of (12.18) into (12.8), we obtain

$$M = 2\pi G \frac{\partial\phi}{\partial z} \int_0^a r^3 dr = GJ \frac{\partial\phi}{\partial z} \tag{12.19}$$

Further substitution of (12.19) into (12.10a) results in the desired displacement equation of motion of the rod:

$$J\rho \frac{\partial^2\phi}{\partial t^2} = \frac{\partial}{\partial z}\left(JG \frac{\partial\phi}{\partial z}\right) + m \tag{12.20}$$

12.2 LONGITUDINAL MOTION OF RODS

Because of our assumptions with respect to displacement (12.1), the approximate theory in section 12.1 is restricted to torsional motion of rods. We shall now formulate an analogous theory valid for the longitudinal motion of rods. Because of the similarities with the analysis in section 12.1, the present derivation is presented in outline form only, and the reader can readily supply the missing details by referring to the preceding section.

The displacement vector is again referred to cylindrical coordinates, and with reference to Fig. 12.1 we now assume that all points on any plane cross-section normal to the rod axis deform the same amount in the axial direction. We also assume that radial and circumferential displacement components vanish. In view of these assumptions the displacement vector is taken as

$$u_r = 0, \; u_\Theta = 0, \; u_z = w(z,t) \tag{12.21}$$

Upon substitution of (12.21) into the strain displacement relations (8.110), we obtain the strain field

$$\begin{bmatrix} \varepsilon_{rr} & \varepsilon_{r\Theta} & \varepsilon_{rz} \\ \varepsilon_{\Theta r} & \varepsilon_{\Theta\Theta} & \varepsilon_{\Theta z} \\ \varepsilon_{zr} & \varepsilon_{z\Theta} & \varepsilon_{zz} \end{bmatrix} = \begin{bmatrix} 0 & 0 & 0 \\ 0 & 0 & 0 \\ 0 & 0 & \frac{\partial w}{\partial z} \end{bmatrix} \tag{12.22}$$

By methods entirely analogous to those of section 12.1, it can be shown, that based on the present assumptions, the kinetic energy of a rod segment extending from $z=z_1$ to $z=z_2$ is given by

$$T = \frac{1}{2} \int_{z_1}^{z_2} A\rho \dot{w}^2 dz$$

where $A=\pi a^2$ is the cross-sectional area of the rod. Consequently

$$\int_{t_1}^{t_2}\delta T dt=-\int_{t_1}^{t_2}\int_{z_1}^{z_2}A\rho\ddot{w}\delta w dz dt \tag{12.23}$$

The variation of the potential energy is

$$\delta U = (P\delta w)\Big|_{z_1}^{z_2}-\int_{z_1}^{z_2}\frac{\partial P}{\partial z}\delta w dz \tag{12.24}$$

where

$$P = \int_A \tau_{zz} dA$$

and the virtual work of all applied forces is given by

$$\delta W= \int_S T_i\delta u_i dS= \int_S T_z\delta w dS$$

or

$$\delta W= \int_{z_1}^{z_2} p\delta w dz+(P^*\delta w)\Big|_{z_1}^{z_2} \tag{12.25}$$

where

$$p = \oint_c T_z dc$$

and P^* is the value of P at either $z=z_1$ or at $z=z_2$. We note that P is the longitudinal force in the rod, while p is the intensity of the distributed axial force applied to the rod. Upon substitution of (12.23), (12.24), and (12.25) into (12.3), we obtain the appropriate stress-equation of motion and the associated, admissible boundary conditions:

$$-A\rho\frac{\partial^2 w}{\partial t^2}+\frac{\partial P}{\partial z}+p = 0 \quad \text{for } z_1<z<z_2 \tag{12.26a}$$

At $z=z_1$ and at $z=z_2$ we must specify either $w=w(t)$ or $P=P^*(t)$ (12.26b)
In the present case, the energy flux is in the z-direction and is given by $S=-P\dot{w}$, while the (linear) momentum flux in the z-direction is $J^*=-P$. To transform (12.26a) to a displacement equation of motion, we utilize only one of the six equations characterizing Hooke's law, and ignore the remaining five:

$$E\varepsilon_{zz}= \tau_{zz}-\nu(\tau_{rr}+\tau_{\theta\theta}) = E\frac{\partial w}{\partial z}$$

so that

$$P = \int_A \tau_{zz} dA = AE \frac{\partial w}{\partial z} + \int_A \nu(\tau_{rr}+\tau_{\theta\theta}) dA$$

The present (elementary) rod theory assumes that

$$\nu \int_A (\tau_{rr}+\tau_{\theta\theta}) dA << AE \frac{\partial w}{\partial z}$$

and it can be shown that this assumption is justified, provided that the wave lengths of axial disturbances are large compared to the rod diameter 2a. Consequently we shall use the approximation

$$P = AE \frac{\partial w}{\partial z} \qquad (12.27)$$

and upon substitution of (12.27) into (12.26a), we obtain the displacement equation of motion

$$A\rho \frac{\partial^2 w}{\partial t^2} = \frac{\partial}{\partial z} (AE \frac{\partial w}{\partial z}) + p \qquad (12.28)$$

By a simple change of notation, equations (12.28) and (12.20) are readily shown to be identical. An equation of this type is often referred to as a one-dimensional, non-homogeneous wave equation, and it becomes homogeneous when m (or p) vanishes. We have already encountered a homogeneous wave equation in section 11.2 and we shall encounter it again in section 12.4. A variety of physical situations can be characterized by the wave equation, such as wave motion in an unbounded solid, vibrations of a string, longitudinal and torsional motion of rods, transverse motion of a shear beam, etc. Because of its fundamental importance in the study of the motion of elastic solids, we shall give detailed consideration to the solution of the one-dimensional, non-homogeneous wave equation for bounded as well as unbounded regions. For convenience as well as consistency, we shall develop the theory within the context of torsional motion of rods, but applications to other physical situations characterized by the one-dimensional wave equation can be readily made by a simple change of symbolism.

12.3 THE ROD OF FINITE LENGTH

The present section is concerned with the solution of the rod problem for time-dependent (admissible) boundary conditions, as well as time-dependent, distributed torques. Although we shall treat the torsion model of the rod developed in section 12.1, the method will be seen to be applicable to any phenomenon characterized by the one-dimensional wave equation.

A properly posed problem may be stated as follows: find a solution of the equation

$$J\rho \frac{\partial^2 \phi}{\partial t^2} = \frac{\partial}{\partial z}\left(JG \frac{\partial \phi}{\partial z}\right) + m \tag{12.29}$$

where

$m=m(z,t)$ and $G=G(z)$.

are known functions of the axial coordinate z, and the boundary conditions are

$$\text{either } \phi(0,t) \text{ or } M(0,t) = f(t) \tag{12.30a}$$

and

$$\text{either } \phi(\ell,t) \text{ or } M(\ell,t) = g(t) \tag{12.30b}$$

where $z=0$ and $z=\ell$ denote the extremities of the rod of length ℓ. In addition, to obtain a unique solution it is sufficient to specify the initial conditions

$$\phi(z,0) = \phi_0(z) \tag{12.31a}$$

$$\dot{\phi}(z,0) = \dot{\phi}_0(z) \tag{12.31b}$$

Proceeding as in Chapters 3 and 10, we shall first consider the free vibration problem for homogeneous boundary conditions, i.e., we shall seek a solution of (12.29) with $m\equiv 0$ and boundary conditions

$$\phi(0,t) \text{ or } M(0,t) = 0 \tag{12.32a}$$

and

$$\phi(\ell,t) \text{ or } M(\ell,t) = 0 \tag{12.32b}$$

Assuming a product solution in the form

$$\phi(z,t) = \Phi(z) \begin{matrix} \cos \omega t \\ \sin \omega t \end{matrix} \tag{12.33}$$

and substituting (12.33) into the homogeneous form of (12.29), we obtain an equation which characterizes the eigenfunctions (mode shapes)

$$\frac{d}{dz}\left(JG \frac{d\Phi}{dz}\right) + J\rho\omega^2\Phi = 0 \tag{12.34}$$

In view of (12.32), (12.33), and (12.34), the boundary conditions on $\Phi(z)$ are either $\Phi=0$ or $\frac{d\Phi}{dz} = 0$ at $z=0$ and at $z = \ell$. (12.35)

Non-trivial solutions of (12.34) which also satisfy (12.35) exist only for certain values of ω called the natural frequencies or eigenvalues of the rod. There is a denumerable infinity of these natural frequencies ω_i, $i=1,2,3\ldots$. The solution Φ_i corresponding to the natural frequency ω_i is called the i'th normal mode or eigenfunction of the rod. Equations (12.34) and (12.35) specify each normal mode only to within an indeterminate constant multiplier. By imposing the normalization condition

$$\left(\int_0^\ell \rho J \Phi_i^2 dz\right)^{\frac{1}{2}} = 1 \tag{12.36}$$

unique expressions for the eigenfunctions are obtained. The form of (12.36) is selected because of its convenience in the subsequent solution of the forced motion problem.

Let us now consider two different eigenfunctions Φ_i and Φ_j, and their associated frequencies ω_i and ω_j, respectively. By definition, both sets (Φ_i,ω_i) and (Φ_j,ω_j) satisfy (12.34). We now form the expression

$$\int_0^\ell \left[\left(\frac{dM_i}{dz} + J\rho\omega_i^2\Phi_i\right)\Phi_j - \left(\frac{dM_j}{dz} + J\rho\omega_j^2\Phi_j\right)\Phi_i\right] dz = 0$$

where

$$M_i = GJ\,\frac{d\Phi_i}{dz}$$

and upon rearrangement we obtain

$$(\omega_i^2-\omega_j^2)\int_0^\ell \rho J\Phi_i\Phi_j dz = \int_0^\ell \left(\frac{dM_j}{dz}\,\Phi_i - \frac{dM_i}{dz}\,\Phi_j\right) dz \quad .$$

The right-hand side of this equation is readily integrated by parts with the result

$$(\omega_i^2-\omega_j^2)\int_0^\ell \rho J\Phi_i\Phi_j dz = \left(\Phi_i M_j - \Phi_j M_i\right)_0^\ell \tag{12.37}$$

If homogeneous boundary conditions (12.35) are assumed and substituted into (12.37), we obtain the orthogonality relation

$$\left.\begin{aligned} &\int_0^\ell \rho J \Phi_i \Phi_j \, dz = 0 \\ &\text{provided} \\ &\omega_i^2 \neq \omega_j^2 \end{aligned}\right\} \qquad (12.38)$$

We now proceed to the solution of the forced motion problem as characterized by (12.29), (12.30), and (12.31). It is our intention to obtain a solution in terms of an eigenfunction expansion, utilizing eigenfunctions with homogeneous boundary conditions. Toward this end we write the solution in the form

$$\phi(z,t) = \phi_s(z,t) + \sum_{i=1}^{\infty} \Phi_i(z) q_i(t) \qquad (12.39)$$

where $q_i(t)$, i=1,2,..., are functions of the time and where $\phi_s(z,t)$ is the quasi static solution which satisfies (12.29) with inertia terms deleted, i.e.,

$$\frac{\partial}{\partial z}\left(JG \frac{\partial \phi_s}{\partial z}\right) = -m(z,t) \qquad (12.40a)$$

In addition, the "static" solution $\phi_s(z,t)$ must satisfy the boundary conditions

$$\left.\begin{aligned} &\text{either } \phi_s(0,t) \text{ or } M_s(0,t) = f(t) \\ &\text{and} \\ &\text{either } \phi_s(\ell,t) \text{ or } M_s(\ell,t) = g(t) \end{aligned}\right\} \qquad (12.40b)$$

where

$$M_s = GJ \frac{\partial \phi_s}{\partial z} \qquad (12.40c)$$

The eigenfunctions $\Phi_i(z)$ in (12.39) satisfy (12.34) and the following homogeneous boundary conditions:

$$\left.\begin{aligned} &\text{If } \phi_s(0,t) = f(t), \text{ then } \Phi_i(0) = 0 \\ &\text{if } M_s(0,t) = f(t), \text{ then } M_i(0) = 0 \\ &\text{if } \phi_s(\ell,t) = g(t), \text{ then } \Phi_i(\ell) = 0 \\ &\text{if } M_s(\ell,t) = g(t), \text{ then } M_i(\ell) = 0 \end{aligned}\right\} \qquad (12.41)$$

where it is understood that

$$M_i = GJ \frac{d\Phi_i}{dz} \qquad (12.42)$$

We now substitute (12.39) into (12.29) and utilize (12.40a) and (12.34). The result is

$$\sum_{i=1}^{\infty} \rho J \Phi_i (\ddot{q}_i + \omega_i^2 q_i) = -\rho J \ddot{\phi}_s$$

Both sides of this equation are now multiplied by $\Phi_j(z)$, and the products so obtained are integrated over the length of the rod. With the help of the relations (12.36) and (12.38) we obtain

$$\ddot{q}_i + \omega_i^2 q_i = -\int_0^{\ell} \rho J \ddot{\phi}_s \Phi_i dz = \ddot{Q}_i \qquad (12.43a)$$

where

$$Q_i(t) = -\int_0^{\ell} \rho J \phi_s \Phi_i dz = \frac{1}{\omega_i^2} \int_0^{\ell} \phi_s \frac{dM_i}{dz} dz \qquad (12.43b)$$

Equation (12.43b) is now integrated by parts twice, and (12.40a), (12.40c), and (12.42) are applied. The final result is

$$\omega_i^2 Q_i(t) = (\phi_s M_i - M_s \Phi_i)\Big|_0^{\ell} - \int_0^{\ell} m \Phi_i dz \qquad (12.44)$$

The complete solution of (12.43a) is obtained with the aid of (1.50):

$$q_i(t) = q_i(0) \cos\omega_i t + \frac{1}{\omega_i} \dot{q}_i(0) \sin\omega_i t + \frac{1}{\omega_i} \int_0^t \ddot{Q}_i(\tau) \sin\omega_i (t-\tau) d\tau \qquad (12.45a)$$

and integrating by parts twice we obtain

$$q_i(t) = [q_i(0) - Q_i(0)] \cos\omega_i t + \frac{1}{\omega_i}[\dot{q}_i(0) - \dot{Q}_i(0)] \sin\omega_i t$$

$$+ Q_i(t) - \omega_i \int_0^t Q_i(\tau) \sin\omega_i (t-\tau) d\tau \qquad (12.45b)$$

We shall now proceed to determine appropriate initial conditions on $q_i(t)$. With reference to (12.31a) and (12.39) we have

$$\phi_s(z,0) + \sum_{i=1}^{\infty} \Phi_i q_i(0) = \phi_0(z)$$

If we now multiply both sides of this equation by $\rho J \Phi_j$, integrate the products so obtained over the length of the rod, and then utilize the relations (12.36), (12.38), and (12.43b), we obtain

$$q_i(0) = Q_i(0) + \int_0^{\ell} \rho J \phi_0 \Phi_i dz \qquad (12.46a)$$

By an entirely analogous analysis, it can be shown that

$$\dot{q}_i(0) = \dot{Q}_i(0) + \int_0^{\ell} \rho J \dot{\phi}_0 \Phi_i dz \qquad (12.46b)$$

This completes the formal solution of the forced motion problem.

The above resolution of the free and forced torsional motion problem will now be applied to three examples, each treating a uniform, homogeneous rod of length ℓ, i.e., the properties G, J and ρ are constant over the length of the rod.

(a) Free Torsional Vibrations of a Rod clamped at z = 0 and free at z = ℓ.

In this case the solution of (12.34) is

$$\Phi(z) = B\cos r\omega z + A\sin r\omega z, \text{ where } r^2 = \frac{\rho}{G}$$

with $\Phi(0)=0$ and $\left(\frac{d\Phi}{dz}\right)_{z=\ell}=0$. Consequently B=0, A$\neq$0, and $r\omega_i\ell = \frac{(2i-1)\pi}{2}$, $i=1,2,\ldots$, i.e., $\cos r\omega_i\ell=0$ and $\sin r\omega_i\ell=(-1)^{i-1}$. In order to normalize the eigenfunctions in accordance with (12.3b), it is necessary to integrate the square of the eigenfunctions over the length of the rod. In the present case, the integration is elementary and can be readily performed. The formal integration can be circumvented, however, by a method of normalization which becomes particularly useful in more complicated cases, and which will now be demonstrated for the present case. If we let $\Phi' \equiv \frac{d\Phi}{d(r\omega z)}$, then $\frac{d\Phi}{dz} = \Phi' r\omega$, and with the aid of (12.37) we obtain

$$\int_0^\ell \rho J\Phi_i\Phi_j dz = \frac{GJr(\Phi_j'\Phi_i\omega_j - \Phi_i'\Phi_j\omega_i)_0^\ell}{\omega_i^2-\omega_j^2}$$

The right-hand side of this expression is of the form $\frac{0}{0}$ when i=j. Consequently we may apply L'Hospital's rule to this fraction, i.e., we differentiate numerator and denominator with respect to ω_i and subsequently take the limit as $i\to j$. If, in addition we apply (12.35), the result is

$$\int_0^\ell \rho J\Phi_i^2 dz = \frac{\rho J\ell}{2}\left[(\Phi_i')^2 - \Phi_i\Phi_i''\right]_{z=\ell}$$

If the rod is free at z=ℓ, this expression reduces to

$$\int_0^\ell \rho J\Phi_i^2 dz = -\frac{\ell\rho J}{2}(\Phi_i\Phi_i'')_{z=\ell}$$

and if the rod is fixed at z=ℓ, we obtain

$$\int_0^\ell \rho J\Phi_i^2 dz = \frac{\rho J\ell}{2}(\Phi_i')^2_{z=\ell} .$$

In the present case the rod is free at $z=\ell$, and we have

$$(\Phi_i)_{z=\ell} = A_i \sin r\omega_i \ell = (-1)^{i-1} A_i$$

$$(\Phi_i'')_{z=\ell} = -A_i \sin r\omega_i \ell = -(-1)^{i-1} A_i$$

so that

$$\int_0^\ell \rho J \Phi_i^2 dz = \frac{1}{2} \ell \rho J A_i^2 = 1$$

and

$$A_i = \sqrt{\frac{2}{\ell \rho J}}$$

Our results can be summarized as follows: For a rod clamped at $z=0$ and free at $z=\ell$, the normalized eigenfunctions are

$$\Phi_i(z) = \sqrt{\frac{2}{\rho J \ell}} \sin r\omega_i z, \quad \omega_i = \frac{(2i-1)\pi}{2r\ell}, \quad i=1,2,\dots. \tag{12.47}$$

The corresponding modal moment is

$$M_i = \omega_i \sqrt{\frac{2GJ}{\ell}} \cos r\omega_i z$$

and

$$\Phi_i(0) = 0, \quad \Phi_i(\ell) = \sqrt{\frac{2}{\rho J \ell}} (-1)^{i-1}$$

$$M_i(0) = \omega_i \sqrt{\frac{2GJ}{\ell}}, \quad M_i(\ell) = 0 \quad .$$

(b) Forced Torsional Motion of a Rod Caused by a Distributed Torque.

We assume that the rod is clamped at $z=0$ and free at $z=\ell$. In the region $0<z<\ell$ it is acted upon by a uniformly distributed, time-dependent torque of intensity $m=m_0 h(t)$, where m_0 is a constant. It is assumed that at $t=0$, the rod is in equilbrium in its natural state, i.e.,

$$\phi_0(z) = \dot{\phi}_0(z) = 0$$

With reference to (12.40), the static solution can be characterized by

$$JG \frac{\partial^2 \phi_s}{\partial z^2} = -m_0 h(t)$$

$$\phi_s(0,t) = 0, \quad M_s(\ell,t) = 0 \ .$$

Consequently

$$\frac{JG}{m_o \ell^2} \phi_s = \frac{z}{\ell}\left(1-\frac{z}{2\ell}\right) h(t)$$

$$M_s = m_o(\ell - z)\, h(t)$$

and

$$\phi_s(\ell,t) = \frac{m_o \ell^2 h(t)}{2JG} \quad , \quad M_s(0,t) = m_o \ell h(t) \quad .$$

In view of (12.41), the eigenfunctions with homogeneous boundary conditions appropriate to the present forced motion case, are given by (12.47). We now substitute all required terms into (12.44) and (12.46), and obtain

$$Q_i(t) = -\frac{m_o}{\rho}\sqrt{\frac{2G}{J\ell}}\,\frac{h(t)}{\omega_i^3}$$

$$q_i(0) = Q_i(0) = -\frac{m_o}{\rho}\sqrt{\frac{2G}{J\ell}}\,\frac{h(0)}{\omega_i^3}$$

$$\dot{q}_i(0) = \dot{Q}_i(0) = -\frac{m_o}{\rho}\sqrt{\frac{2G}{J\ell}}\,\frac{\dot{h}(0)}{\omega_i^3}$$

These quantities, in turn, are substituted into (12.45a) resulting in the desired forced motion solution

$$\frac{JG}{m_o \ell^2}\phi(z,t) = \left(\frac{z}{\ell}\right)\left(1-\frac{1}{2}\frac{z}{\ell}\right)h(t) - \frac{16}{\pi^3}\sum_{i=1}^{\infty}\frac{\sin r\omega_i z}{(2i-1)^3}\left[h(0)\cos\omega_i t + \frac{\dot{h}(0)}{\omega_i}\sin\omega_i t + \frac{1}{\omega_i}\int_0^t \ddot{h}(\tau)\sin\omega_i(t-\tau)d\tau\right] \qquad (12.48)$$

(c) Dynamic Response of a Rod Subjected to a Boundary Torque.

The rod is clamped at z=0 and free from distributed torques along its length $0<z<\ell$. A time-dependent torque $M_o g(t)$ is applied to the end at $z=\ell$, where M_o is a constant. It is assumed that the rod is in equilibrium in its natural state at t=0, i.e.,

$$\phi_o(z) = \dot{\phi}_o(z) = 0 \quad .$$

With reference to (12.40), the static solution is characterized by

$$JG\,\frac{\partial^2\phi_s}{\partial z^2} = 0$$

where $\phi_s(0,t)=0$ and $M_s(\ell,t)=M_o g(t)$.
Therefore

$$JG\phi_s(z,t) = M_o z g(t)$$
$$M_s(z,t) = M_o g(t)$$

and

$$\phi_s(0,t)=0,\ \phi_s(\ell,t)=M_o \ell g(t)$$
$$M_s(0,t)=M_s(\ell,t)=M_o g(t) \quad .$$

In view of (12.41), the eigenfunctions with homogeneous boundary conditions appropriate to the present forced motion case are given by (12.47). We now substitute all required terms into (12.44) and (12.46), and obtain

$$\omega_i^2 Q_i(t) = (-1)^i \sqrt{\frac{2}{\rho J\ell}}\ M_o g(t)$$
$$q_i(0) = Q_i(0) = \frac{(-1)^i}{\omega_i^2}\sqrt{\frac{2}{\rho J\ell}}\ M_o g(0)$$
$$\dot{q}_i(0) = \dot{Q}_i(0) = \frac{(-1)^i}{\omega_i^2}\sqrt{\frac{2}{\rho J\ell}}\ M_o \dot{g}(0)$$

We now have all quantities required for substitution into (12.45a). Thus the forced motion solution is

$$\frac{GJ}{M_o\ell}\,\phi(z,t) = \frac{z}{\ell}g(t) + \frac{8}{\pi^2}\sum_{i=1}^{\infty}\frac{(-1)^i}{(2i-1)^2}\sin r\omega_i z \ \cdot$$

$$\left[g(0)\cos\omega_i t + \frac{1}{\omega_i}\,\dot{g}(0)\sin\omega_i t + \frac{1}{\omega_i}\int_0^t \ddot{g}(\tau)\sin\omega_i(t-\tau)d\tau\right] \qquad (12.49)$$

Let us consider the special case of a suddenly applied, constant torque at $z=\ell$, i.e., $g(t)=H(t)$ where $H(t)$ is the Heaviside unit step function. In this case (12.49) reduces to

$$\frac{GJ}{M_o\ell}\,\phi(z,t) = \frac{z}{\ell} + \frac{8}{\pi^2}\sum_{i=1}^{\infty}\frac{(-1)^i}{(2i-1)^2}\ \sin r\omega_i z\cos\omega_i t,\ t\geq 0 \qquad (12.50a)$$

and the corresponding twisting moment is

$$\frac{M(z,t)}{M_o} = 1 + \frac{4}{\pi} \sum_{i=1}^{\infty} \frac{(-1)^i}{(2i-1)} \cos r\omega_i z \cos\omega_i t, \ t \geq 0 \tag{12.50b}$$

To obtain the time-history of the angle of twist at the free end of the rod, we set $z=\ell$ in (12.50a):

$$\frac{GJ}{M_o \ell} \phi(\ell,t) = 1 - \frac{8}{\pi^2} \sum_{i=1}^{\infty} \frac{1}{(2i-1)^2} \cos \omega_i t, \ t \geq 0 \tag{12.51}$$

The right-hand side of (12.51) is recognized as the Fourier series expansion of the saw-tooth function shown in Fig. 12.2a. When $t=0$,

$$\frac{GJ}{M_o \ell} \phi(\ell,0) = 1 - \frac{8}{\pi^2} \sum_{i=1}^{\infty} \frac{1}{(2i-1)^2} = 1-1=0$$

because $\sum_{i=1}^{\infty} \frac{1}{(2i-1)^2} = \frac{\pi^2}{8}$, and this provides a check on our solution.

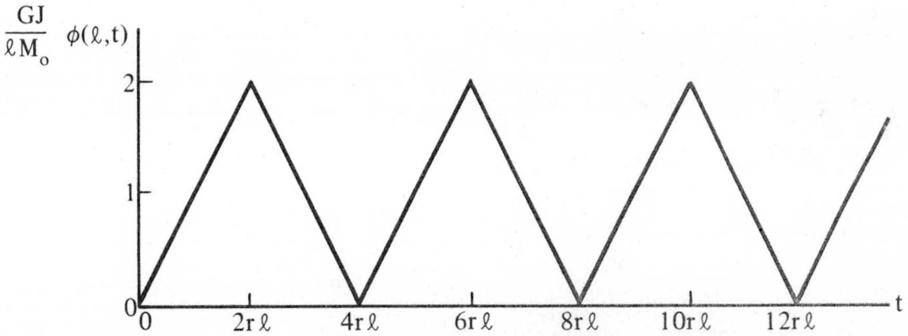

Figure 12.2 (a)

The maximum rotation angle occurs when $\cos \omega_i t=-1$, or for $t=t^*=2r\ell$, $6r\ell$, $10r\ell$, etc. Consequently

$$\frac{GJ}{\ell M_o} \phi(\ell,t^*) = 1 + \frac{8}{\pi^2} \sum_{i=1}^{\infty} \frac{1}{(2i-1)^2} = 1+1=2,$$

and we arrive at the conclusion that for the present case, a suddenly applied torque produces twice as great a rotation than one which is statically applied. To obtain the time history of the twisting moment at the fixed end, we set $z=0$ in (12.50b):

$$\frac{M(0,t)}{M_o} = 1 + \frac{4}{\pi} \sum_{i=1}^{\infty} \frac{(-1)^i}{(2i-1)} \cos \omega_i t, \ t \geq 0 \tag{12.52}$$

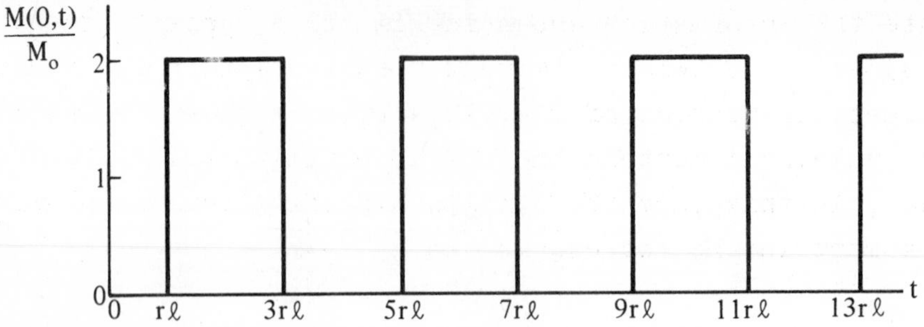

Figure 12.2 (b)

The right-hand side of (12.52) is recognized as the Fourier series expansion of the function shown in Fig. 12.2b. When t=0,

$$\frac{M(0,0)}{M_o} = 1 + \frac{4}{\pi}\sum_{i=1}^{\infty}\frac{(-1)^i}{(2i-1)} = 1-1=0$$

because $\sum_{i=1}^{\infty}\frac{(-1)^i}{(2i-1)} = -\frac{\pi}{4}$, and this provides an additional check on our solution. The present solution can also be interpreted as the superposition of traveling waves. Toward this end we re-write (12.50b) in the form

$$\frac{M(z,t)}{M_o} = 1 + \frac{2}{\pi}\sum_{i=1}^{\infty}\frac{(-1)^i}{(2i-1)}\cos(\omega_i t-k_i z)$$

$$+ \frac{2}{\pi}\sum_{i=1}^{\infty}\frac{(-1)^i}{(2i-1)}\cos(\omega_i t+k_i z) \qquad (12.53)$$

where $k_i = r\omega_i = \frac{(2i-1)\pi}{2\ell}$. The Fourier series expansion $\frac{2}{\pi}\sum_{i=1}^{\infty}\frac{(-1)^i}{(2i-1)}\cdot\cos(\omega_i t-k_i z)$ characterizes the function shown in Fig. 12.3a. It is a wave which travels without change of shape, in the direction of the positive z-direction with constant speed $v = \frac{\omega_i}{k_i} = \frac{1}{r}$. Similarly the Fourier series expansion $\frac{2}{\pi}\sum_{i=1}^{\infty}\frac{(-1)^i}{(2i-1)}\cos(\omega_i t+k_i z)$ characterizes the function shown in Fig. 12.3b. It is a wave which travels, without change of shape, in the direction of the negative z-axis with constant speed v. Thus the sum of the three terms on the right-hand side of (12.53) can be visualized as the superposition of Figs. 12.3a, 12.3b, and 12.3c, resulting in the (instantaneous) distribution of twisting moment in the rod as shown in Fig. 12.3d. This procedure can be used to generate the information shown in Fig. 12.4, depicting the (discontinuous) state of twisting moment distribution in the rod for various times. Inspection of Fig. 12.4 reveals that the maximum dynamical twisting moment in the rod is equal to twice its static value, i.e., in this case the dynamic amplification factor for the twisting moment in the rod equals two.

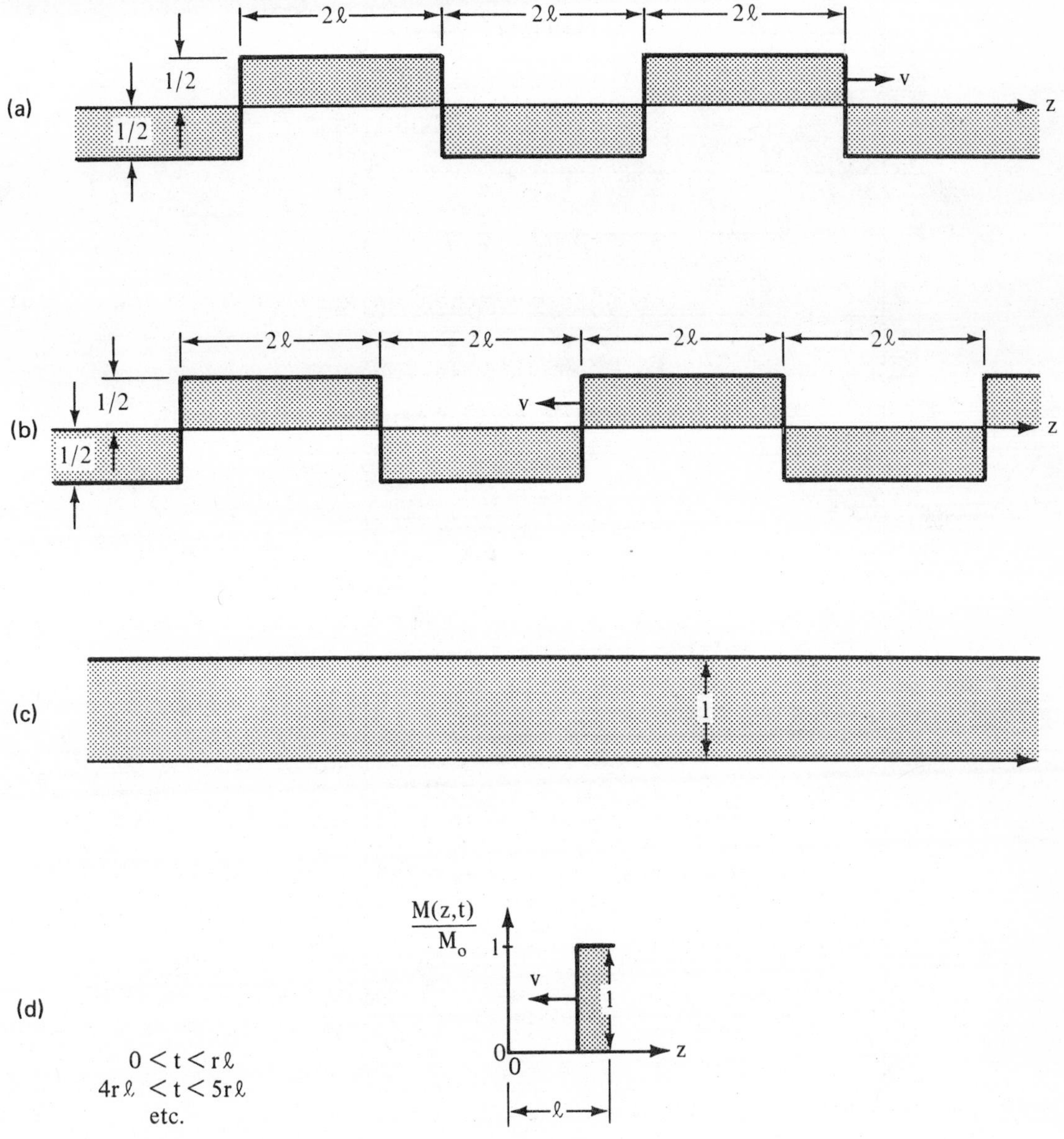

Figure 12.3

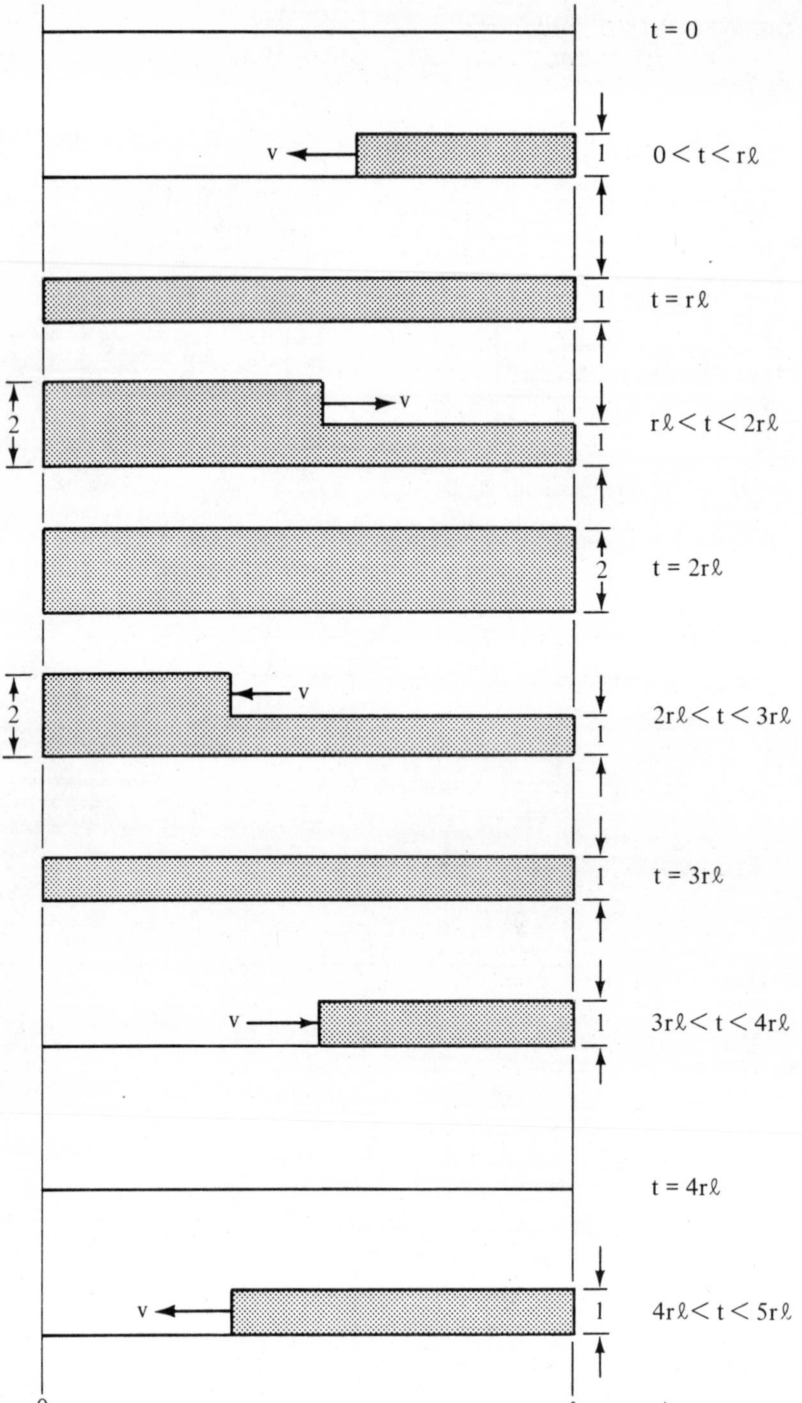

t = 0
v
1
0 < t < rℓ
1
t = rℓ
2
v
1
rℓ < t < 2rℓ
2
t = 2rℓ
2
v
1
2rℓ < t < 3rℓ
1
t = 3rℓ
v
1
3rℓ < t < 4rℓ
t = 4rℓ
v
1
4rℓ < t < 5rℓ
0
ℓ
etc.

Figure 12.4

12.4 THE ROD OF UNBOUNDED LENGTH

In example (c) of section 12.3 it was shown that the solution for a rod of finite length, originally obtained by the method of eigenfunctions, can be interpreted as the superposition of traveling waves in rods of infinite extent. This indicates that solutions for rods of finite length can be built up with the aid of solutions valid for rods of unbounded length. For these as well as other reasons, it will be useful to consider a series of specific examples exhibiting the important aspects of torsional wave motion (propagation, reflection, etc.) in elastic rods with circular cross-section.

(a) Time-harmonic Waves in a Homogeneous Rod.

We consider the possibility of a propagating wave (see Appendix B) of the type

$$\phi(z,t) = A\exp i(kz-\omega t) \tag{12.54}$$

where k, ω, and A are the wave number, frequency, and amplitude of the wave, respectively. The torsional motion of the rod is characterized by

$$\frac{\partial^2\phi}{\partial t^2} = \frac{1}{r^2}\frac{\partial^2\phi}{\partial z^2} \tag{12.55}$$

where $\frac{1}{r^2} = \frac{G}{\rho}$ = constant. Upon substitution of (12.54), into (12.55), we obtain the phase velocity

$$v = \frac{\omega}{k} = \frac{1}{r}$$

i.e., for the present case the phase velocity is a constant which is independent of frequency (or wave number). Within the framework of the terminology developed in Appendix B, we can say that the elastic rod is a non-dispersive medium with respect to torsional wave propagation.

(b) The Effect of Damping on Wave Motion.

We now modify the situation treated in example (a) by the addition of a viscous damping term (in the sense of Chapter 6). In this case (12.20) is modified to read

$$J\rho\frac{\partial^2\phi}{\partial t^2} + b\frac{\partial\phi}{\partial t} = JG\frac{\partial^2\phi}{\partial z^2} \tag{12.56a}$$

where b>0 is the coefficient of viscous damping. It will be convenient to write (12.56a) in the form

$$\frac{\partial^2\phi}{\partial t^2} + \beta\frac{\partial\phi}{\partial t} = \frac{1}{r^2}\frac{\partial^2\phi}{\partial z^2} \tag{12.56b}$$

where

$$\beta = \frac{b}{J\rho} > 0 \text{ and } \frac{1}{r^2} = \frac{G}{\rho}.$$

Assuming a propagating wave in the form

$$\phi(z,t) = A\exp i(\kappa z-\omega t) \tag{12.57}$$

and substituting (12.57) into (12.56b), we obtain

$$\omega^2 + i\beta\omega = \frac{\kappa^2}{r^2} \tag{12.58}$$

If we set $\kappa=k+i\mu$, substitute into (12.58), and separate real and imaginary parts, we obtain

$$r^2\omega^2=k^2-\mu^2,\quad \beta\omega = \frac{2\mu k}{r^2},$$

and these equations are readily combined, resulting in

$$k = \frac{r\omega}{\sqrt{2}}\left(1+\sqrt{1+\frac{\beta^2}{\omega^2}}\right)^{\frac{1}{2}} > 0 \tag{12.59}$$

$$\mu = \frac{r\omega}{\sqrt{2}}\left(-1+\sqrt{1+\frac{\beta^2}{\omega^2}}\right)^{\frac{1}{2}} > 0 \tag{12.60}$$

We note that k>0 for a wave traveling in the direction of the positive z-axis, while μ>0 to insure that the medium is dissipative. Consequently the wave motion can be characterized by

$$\phi(z,t) = \exp(-\mu z)\exp i(kz-\omega t) \tag{12.61}$$

With reference to (12.59) and Appendix B, the phase velocity is given by

$$v = \frac{\omega}{k} = \frac{\sqrt{2}}{r\left(1+\sqrt{1+\frac{\beta^2}{\omega^2}}\right)^{\frac{1}{2}}}$$

The phase velocity is seen to be a function of frequency, and therefore the wave motion characterized by (12.61) is dispersive. We also note that the wave is exponentially attenuated as it propagates in the positive z-direction. It is now noted that

$$\sqrt{1+\frac{\beta^2}{\omega^2}} = 1 + \frac{1}{2}\frac{\beta^2}{\omega^2} + O\left(\frac{\beta^4}{\omega^4}\right)$$

so that for a sufficiently small damping coefficient β we have the approximation

$$\sqrt{2}\ k \cong \sqrt{2}\ r\omega\left(1 + \frac{1}{4}\frac{\beta^2}{\omega^2}\right)^{\frac{1}{2}} \cong \sqrt{2}r\omega\left(1 + \frac{1}{8}\frac{\beta^2}{\omega^2}\right)$$

$$\sqrt{2}\ \mu \cong r\omega\left(\frac{1}{2}\frac{\beta^2}{\omega^2}\right)^{\frac{1}{2}} = \frac{r\beta}{\sqrt{2}}$$

Consequently, to the first order in β,

$$k = r\omega \text{ and } \mu = \frac{r\beta}{2} .$$

Thus for sufficiently small damping, we can write (12.61) as

$$\phi(z,t) = \exp\left(-\frac{1}{2}\, r\beta z\right)\exp\ ik(z-vt)$$

where $v= \frac{\omega}{k} = \frac{1}{r}$ = constant. In this case the motion is non-dispersive, but it is spatially attenuated. We can summarize our conclusions as follows: When (linear) damping is present, the rod is a dispersive medium with respect to torsional wave propagation. For sufficiently small damping, wave motion will be non-dispersive but spatially attenuated.

(c) The Elastically Restrained Rod. (An example of a mechanical high-pass filter)

We shall now consider an unbounded, elastic rod which is embedded in a continuous restraining medium. If the torsional restraint is linearly proportional to local rod rotation ϕ, the equation of motion (12.20) must be modified to read

$$J\rho\frac{\partial^2\phi}{\partial t^2} - JG\frac{\partial^2\phi}{\partial z^2} + c\phi = 0 \qquad (12.62)$$

where $c>0$ is the restraining constant. If we assume the possibility of traveling, time-harmonic waves

$$\phi(z,t) = A\exp i(kz-\omega t) \qquad (12.63)$$

and substitute (12.63) into (12.62), we obtain the dispersion relation

$$\omega = \sqrt{\frac{G}{\rho} k^2 + \frac{c}{J\rho}} \quad \text{or} \quad k = \sqrt{\frac{c}{JG}} \sqrt{\frac{J\rho}{c}\omega^2 - 1} \tag{12.64}$$

The phase velocity is

$$\sqrt{\frac{\rho}{G}}\, v = \frac{\rho}{G}\frac{\omega}{k} = \frac{\sqrt{\frac{J\rho}{c}}}{\sqrt{\frac{J\rho}{c}\omega^2 - 1}} = \sqrt{1 + \frac{1}{\frac{JG}{c}k^2}} \tag{12.65}$$

The corresponding group velocity is given by (B.43) in Appendix B. With reference to (12.64) we note that the wave number will be real for $\sqrt{\frac{c}{J\rho}} < \omega$. When $0 < \omega < \sqrt{\frac{c}{J\rho}}$, the number will be imaginary (complex, with zero real part), and in this range we set $k = i\mu$, where $\mu > 0$ to insure attenuation. In this case

$$\sqrt{\frac{JG}{c}}\, \mu = \sqrt{1 - \frac{J\rho}{c}\omega^2} \tag{12.66}$$

and (12.63) becomes

$$\phi(z,t) = A\exp(-\mu z)\exp(-i\omega t) \tag{12.67}$$

We can draw the following conclusions: When $0 < \omega < \sqrt{\frac{c}{J\rho}}$, there is a standing attenuated wave characterized by (12.67), with attenuation number μ given by (12.66). When $\sqrt{\frac{c}{J\rho}} < \omega$, we have traveling waves characterized by (12.63), with wave number k given by (12.64). Plots of (12.65) are shown in Fig. B.2. A plot of wave number (or attenuation number) versus frequency is shown in Fig. 12.5. With reference to Fig. 12.5, and in view of the above analysis, we can conclude that the elastically restrained rod acts as a mechanical high-pass filter with cut-off frequency $\omega = \sqrt{\frac{c}{J\rho}}$.

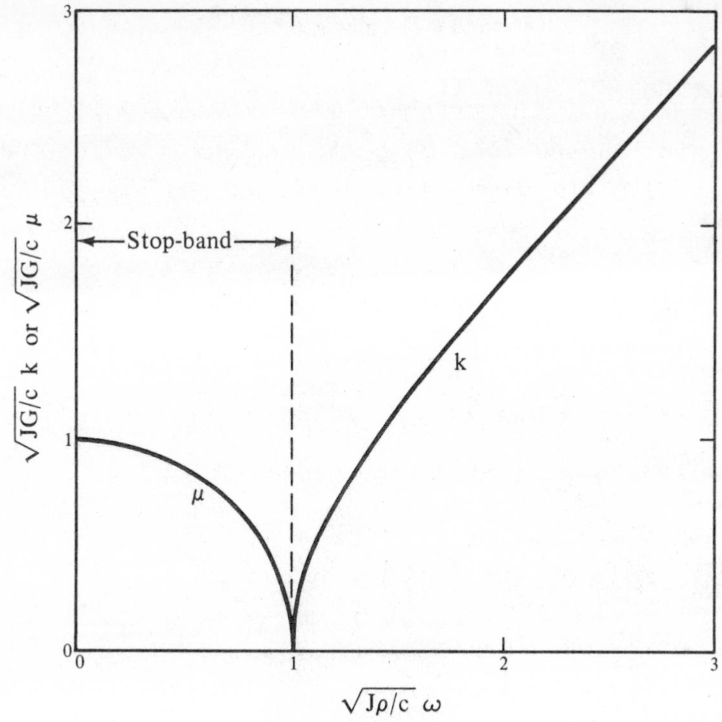

Figure 12.5

(d) Wave Motion in Two Bonded, Semi-infinite Rods.

We now consider the propagation of time-harmonic torsional waves in two dissimilar, bonded, semi-infinite rods shown in Fig. 12.6.

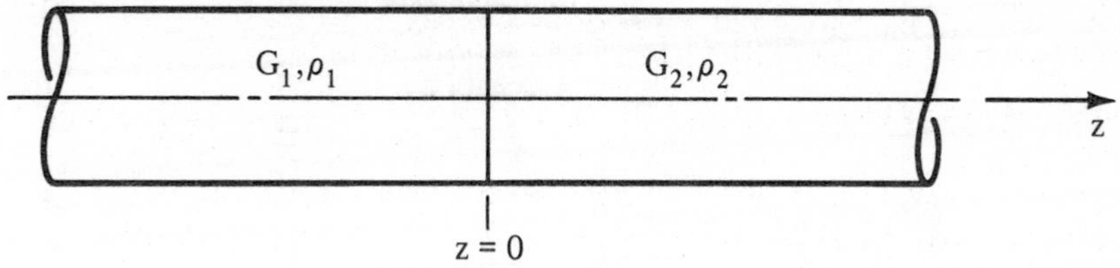

Figure 12.6

Quantitites with subscripts 1 and 2 will pertain to regions $-\infty<z<0$ and $0<z<\infty$, respectively. The (free) wave motion is characterized by

$$\frac{\partial^2 \phi_1}{\partial t^2} = v_1^2 \frac{\partial^2 \phi_1}{\partial z^2} \text{ for } -\infty<z<0$$

and

$$\frac{\partial^2 \phi_2}{\partial t^2} = v_2^2 \frac{\partial^2 \phi_2}{\partial z^2} \text{ for } 0<z<\infty$$

where

$$v_1^2 = \frac{G_1}{\rho_1} \text{ and } v_2^2 = \frac{G_2}{\rho_2}$$

An incident, fully specified wave characterized by

$$\phi_i = A\exp i\omega\left(t - \frac{z}{v_1}\right), \quad -\infty<z<0$$

gives rise to a reflected wave

$$\phi_r = B\exp i\omega\left(t + \frac{z}{v_1}\right) \quad -\infty<z<0$$

and results in a transmitted wave

$$\phi_t = C\exp i\omega\left(t - \frac{z}{v_2}\right) \quad 0<z<\infty$$

Thus,

$$\phi_1 = \phi_i + \phi_r = A\exp\, i\omega\left(t - \frac{z}{v_1}\right) + B\exp i\omega\left(t + \frac{z}{v_1}\right) \tag{12.68}$$

$$\phi_2 = \phi_t = C\exp\, i\omega\left(t - \frac{z}{v_2}\right) \tag{12.69}$$

The real quantities A, ω, v_1 and v_2 are to be considered as known. To find the constants B and C, we impose the requirement of continuity on the rotation angle ϕ and twisting moment M, i.e.,

$$\phi_1(0,t) = \phi_2(0,t) \tag{12.70}$$

$$M_1(0,t) = M_2(0,t) \tag{12.71}$$

Upon substitution of (12.68) and (12.69) into (12.7) and (12.71), and with the aid of (12.19), we readily obtain

$$A + B = C, \quad A - B = \lambda C$$

where $\lambda = \dfrac{G_2 v_1}{G_1 v_2} = \sqrt{\dfrac{G_2 \rho_2}{G_1 \rho_1}}$. Consequently

$$\frac{C}{A} = \frac{2}{1+\lambda} \quad , \quad \frac{B}{A} = \frac{1-\lambda}{1+\lambda} \tag{12.72}$$

We shall next consider the flux of energy in the composite rod. Toward this end we write the component waves in real form:

$$\phi_i = A\cos\omega(t - \frac{z}{v_1})$$

$$\phi_r = B\cos\omega(t + \frac{z}{v_1})$$

$$\phi_t = C\cos\omega(t - \frac{z}{v_2})$$

With the aid of (12.11) and (12.19), we compute the energy flux at z=0:

$$S_i(0,t) = A^2 \frac{G_1J_1}{v_1} \omega^2\sin^2\omega t$$

$$S_r(0,t) = -B^2 \frac{G_1J_1}{v_1} \omega^2\sin^2\omega t$$

$$S_t(0,t) = C^2 \frac{G_2J_2}{v_2} \omega^2\sin^2\omega t$$

We now utilize (7.28a) to compute the time averages of energy flux (energy flux per cycle):

$$\langle S_i(0,t)\rangle = \frac{1}{2} A^2\omega^2 \frac{G_1J_1}{v_1} \tag{12.73a}$$

$$\langle S_r(0,t)\rangle = -\frac{1}{2} B^2\omega^2 \frac{G_1J_1}{v_1} \tag{12.73b}$$

$$\langle S_t(0,t)\rangle = \frac{1}{2} C^2\omega^2 \frac{G_2J_2}{v_2} \tag{12.73c}$$

No energy is stored at z=0. Therefore the following energy flux balance must hold:

$$\langle S_i(0,t)\rangle + \langle S_r(0,t)\rangle = \langle S_t(0,t)\rangle \tag{12.74}$$

Upon substitution of (12.73) into (12.74) we obtain

$$A^2 - B^2 = \lambda C^2$$

and with the aid of (12.72), this reduces to

$$1 - \left(\frac{1-\lambda}{1+\lambda}\right)^2 = \frac{4\lambda}{(1+\lambda^2)}$$

which is an identity in λ, thus providing a check on our solution. It is customary to define

$$T^* = \text{coefficient of transmission} = \left|\frac{\text{transmitted energy per cycle}}{\text{incident energy per cycle}}\right|$$

$$R^* = \text{coefficient of reflection} = \left|\frac{\text{reflected energy per cycle}}{\text{incident energy per cycle}}\right|$$

so that

$$T^* = \left|\frac{\langle S_t(0,t)\rangle}{\langle S_i(0,t)\rangle}\right| = \lambda\left(\frac{C}{A}\right)^2 = \frac{4\lambda}{(1+\lambda)^2}$$

$$R^* = \left|\frac{\langle S_r(0,t)\rangle}{\langle S_i(0,t)\rangle}\right| = \left(\frac{B}{A}\right)^2 = \left(\frac{1-\lambda}{1+\lambda}\right)^2$$

and we note that $T^*+R^*=1$. A plot of the transmission coefficient T^* and the reflection coefficient R^* vs. $\lambda=\sqrt{\frac{G_2\rho_2}{G_1\rho_1}}$ is shown in Fig. 12.7.

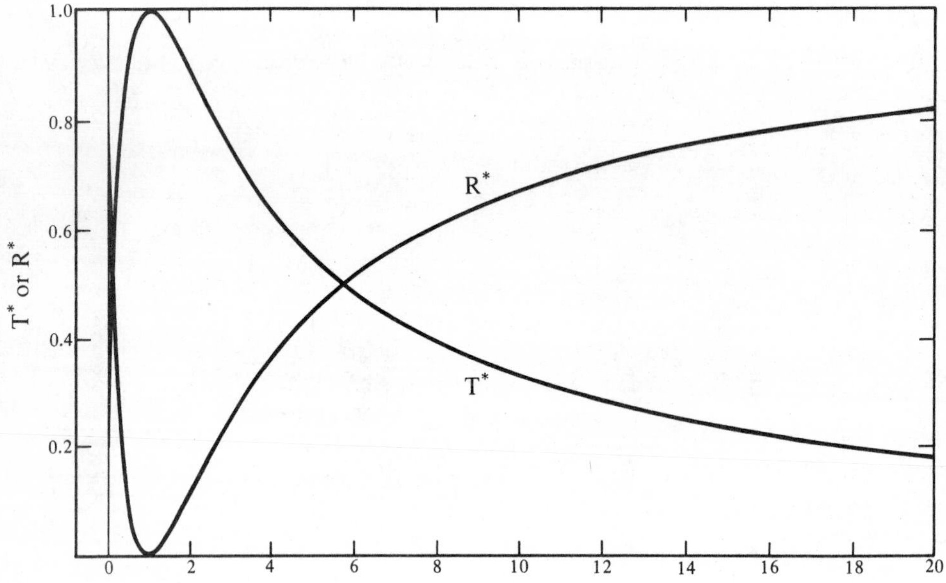

Figure 12.7

There are three cases of special interest: When $\lambda=0$, we have $T^*=0$, $R^*=1$, and $B=A$, corresponding to the case of a free end at $z=0$. When $\lambda=1$ we have $T^*=1$, $R^*=0$, $B=0$, and $C=A$, correspondong to a homogeneous rod for $-\infty<z<\infty$. When $\lambda\to\infty$, $T^*=0$, $R^*=1$, $B=-A$, and we have the case of a rigidly fixed rod at $z=0$.

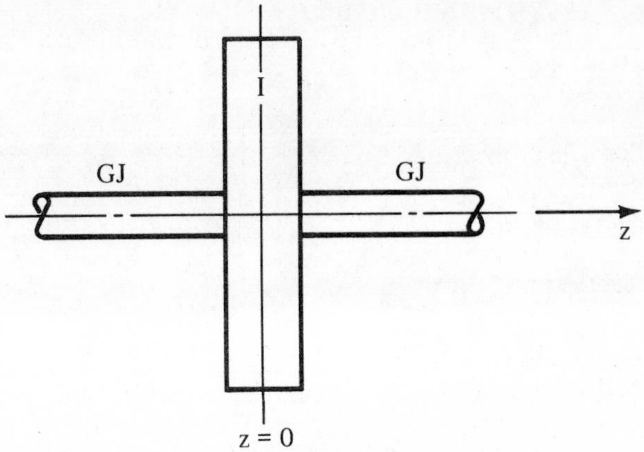

Figure 12.8

(e) Rigid Pulley Mounted on an Elastic Shaft.

With reference to Fig. 12.8, we now consider an unbounded, elastic rod $-\infty<z<\infty$ to which a rigid disc with moment of inertia I is attached at z=0. The equation of torsional motion of the shaft is

$$\frac{\partial^2\phi}{\partial t^2} = v^2 \frac{\partial^2\phi}{\partial z^2}, \quad z\neq 0, \text{ where } v=\sqrt{\frac{G}{\rho}}$$

and the equation of motion of the disc is

$$M(0^+,t) - M(0^-,t) = I\left(\frac{\partial^2\phi}{\partial t^2}\right)_{z=0} \tag{12.75}$$

The incoming wave is fully specified as

$$\phi_i = A\exp i\omega(t-\frac{z}{v}), \quad -\infty<z<0 \tag{12.76}$$

where the real constants A, ω, and v are to be considered as known. The incoming wave ϕ_i gives rise to a reflected wave

$$\phi_r = B\exp i\omega(t+\frac{z}{v}), \quad -\infty<z<0 \tag{12.77}$$

and a transmitted wave

$$\phi_t = C\exp i\omega(t-\frac{z}{v}), \quad 0<z<\infty \tag{12.78}$$

To determine the unknown constants B anc C, we require continuity in the angular coordinate ϕ at z=0, i.e.,

$$\phi(0^+,t) = \phi(0^-,t) \tag{12.79}$$

and we also impose (12.75) in conjunction with (12.19). Thus

$$\phi(z,t) = \phi_i + \phi_r \text{ for } -\infty<z<0 \tag{12.80}$$

$$\phi(z,t) = \phi_t \text{ for } 0<z<\infty \tag{12.81}$$

and upon substitution of (12.80) and (12.81) into (12.79) and (12.75) we obtain

$$A + B = C$$
$$A - B = (1+i2\lambda)C$$

where

$$\lambda = \frac{1}{2}\frac{I}{J}\frac{\omega}{\sqrt{\rho G}}$$

is a convenient frequency parameter. Solving,

$$-\frac{B}{A} = \frac{(i+\lambda)\lambda}{(1+\lambda^2)} \tag{12.82a}$$

$$\frac{C}{A} = \frac{1-i\lambda}{(1+\lambda^2)} \tag{12.82b}$$

We shall next investigate the flux of energy in the rod. Toward this end we utilize (12.82), (12.76), (12.77), and (12.78), and subsequently take real parts:

$$\phi_i = A\cos\omega(t - \frac{z}{v})$$

$$\phi_r = \frac{A\lambda}{1+\lambda^2}\sin\omega(t+\frac{z}{v}) - \frac{A\lambda^2}{1+\lambda^2}\cos\omega(t+\frac{z}{v})$$
$$= -\frac{A\lambda}{\sqrt{1+\lambda^2}}\cos\left[\omega(t+\frac{z}{v}) + \mu_r\right]$$

where $\tan\mu_r = \frac{1}{\lambda}$, $\sin\mu_r = \frac{1}{\sqrt{1+\lambda^2}}$, $\cos\mu_r = \frac{\lambda}{\sqrt{1+\lambda^2}}$

$$\phi_t = \frac{A}{1+\lambda^2}\cos\omega(t-\frac{z}{v}) + \frac{A\lambda}{1+\lambda^2}\sin\omega(t-\frac{z}{v})$$
$$= \frac{A}{\sqrt{1+\lambda^2}}\cos\left[\omega(t-\frac{z}{v}) - \mu_t\right]$$

where $\tan\mu_t = \lambda$, $\sin\mu_t = \frac{\lambda}{\sqrt{1+\lambda^2}}$, $\cos\mu_t = \frac{1}{\sqrt{1+\lambda^2}}$.

With the aid of (12.11) and (12.19), we can now compute the energy flux at z=0:

$$S_i(0,t) = A^2 \frac{GJ}{v} \omega^2 \sin^2 \omega t$$
$$S_r(0,t) = -\frac{A^2\lambda^2}{1+\lambda^2} \frac{GJ}{v} \omega^2 \sin^2(\omega t+\mu_r)$$
$$S_t(0,t) = \frac{A^2}{1+\lambda^2} \frac{GJ}{v} \omega^2 \sin^2(\omega t-\mu_t) \quad .$$

To find the time average of energy flux, we utilize (7.28a):

$$\langle S_i(0,t)\rangle = \frac{1}{2} A^2 \frac{GJ}{v} \omega^2 \tag{12.83a}$$

$$\langle S_r(0,t)\rangle = -\frac{1}{2} \frac{A^2\lambda^2}{(1+\lambda^2)} \frac{GJ}{v} \omega^2 \tag{12.83b}$$

$$\langle S_t(0,t)\rangle = \frac{1}{2} \frac{A^2}{(1+\lambda^2)} \frac{GJ}{v} \omega^2 \tag{12.83c}$$

The time average of energy stored at z=0 vanishes. Consequently

$$\langle S_i(0,t)\rangle = \langle S_t(0,t)\rangle - \langle S_r(0,t)\rangle \tag{12.84}$$

and upon substitution of (12.83) into (12.84), we obtain

$$\frac{1}{2} A^2 \frac{GJ}{v} \omega^2 = \frac{1}{2} \frac{A^2}{1+\lambda^2} \frac{GJ}{v} \omega^2 + \frac{1}{2} \frac{A^2\lambda^2}{1+\lambda^2} \frac{GJ}{v} \omega^2$$

which is seen to be an identity in λ, thus providing a check upon our calculations. We can again define a transmission and reflection coefficient as in example (d):

$$T^* = \left|\frac{\langle S_t(0,t)\rangle}{\langle S_i(0,t)\rangle}\right| = \frac{1}{1+\lambda^2}$$
$$R^* = \left|\frac{\langle S_r(0,t)\rangle}{\langle S_i(0,t)\rangle}\right| = \frac{\lambda^2}{1+\lambda^2}$$

We note that $T^*+R^*=1$. A plot of the transmission and reflection coefficient as a function of λ is shown in Fig. 12.9. We note the following limiting cases: When $\lambda=0$, $I=0$ and we have an unbounded, continuous shaft without a disc as in example (a). When $\lambda\to\infty$, $I\to\infty$, and this case corresponds to a rigidly fixed end at z=0.

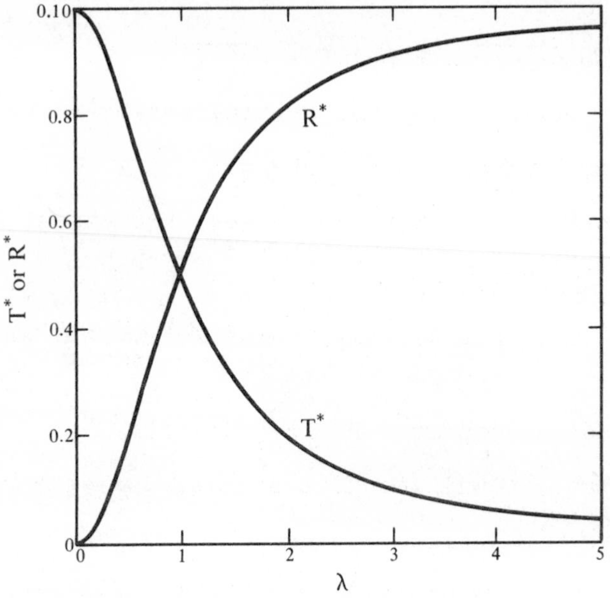

Figure 12.9

(f) Forced, Time-harmonic Motion of a Semi-infinite Rod.

We now consider an elastically restrained, semi-infinite rod $0<z<\infty$ which is excited by a time-harmonic twisting moment at $z=0$. The motion of the rod is characterized by (12.62) which is now written as

$$\frac{\partial^2\phi}{\partial t^2} - \frac{1}{r^2}\frac{\partial^2\phi}{\partial z^2} + \gamma^2\phi = 0, \quad 0<z<\infty \tag{12.85}$$

where

$$\frac{1}{r^2} = \frac{G}{\rho}, \quad \gamma^2 = \frac{c}{J\rho}$$

The moment at $z=0$ is given by

$$M(0,t) = GJ\left(\frac{\partial\phi}{\partial z}\right)_{(0,t)} = -M_0 \exp i\omega t \tag{12.86}$$

We shall neglect any transients, and concentrate on the steady state solution. Thus, if we assume

$$\phi(z,t) = \Phi(z)\exp i\omega t \tag{12.87}$$

and substitute (12.87) into (12.85), we obtain

$$\frac{d^2\Phi}{dz^2} + r^2(\omega^2-\gamma^2)\Phi = 0$$

so that

$$\Phi(z) = A\exp(ir\sqrt{\omega^2-\gamma^2}\,z) + B\exp(-ir\sqrt{\omega^2-\gamma^2}\,z) \tag{12.88}$$

Combining (12.88) and (12.87), and setting $k=r\sqrt{\omega^2-\gamma^2}$, we obtain

$$\phi(z,t) = A\exp i(\omega t+kz) + B\exp i(\omega t-kz) \tag{12.89}$$

It will be necessary to distinguish between two cases.

Case I: $0<\omega<\gamma$. In this case k is imaginary (complex, with real part equal to zero), and we set $k=i\mu$, where $\mu=r\sqrt{\gamma^2-\omega^2}>0$. Therefore (12.89) becomes

$$\phi(z,t) = A\exp(-\mu z)\exp i\omega t + B\exp(\mu z)\exp i\omega t$$

We now require that ϕ remain bounded as $z\to\infty$. Consequently B=0, and

$$\phi(z,t) = A\exp(-\mu z)\exp i\omega t \tag{12.90}$$

To determine A, we substitute (12.90) into (12.86), and readily obtain

$$A = \frac{M_o}{\mu JG} = \frac{M_o}{rJG\sqrt{\gamma^2-\omega^2}}$$

and the solution is

$$\phi(z,t) = \frac{M_o}{JGr\sqrt{\gamma^2-\omega^2}}\exp(-\mu z)\exp i\omega t, \quad 0<\omega<\gamma \tag{12.91}$$

Case II: $\gamma<\omega$. In this case k is real. We now observe that the energy input occurs at z=0, and the average energy flux must be away from z=0, in the direction of the positive z-axis. We also note, that in the present case, the group velocity is equal to the velocity of energy transport (see Exercise 12.9). In addition, as shown in Fig. B.2a, the sign of the group velocity is the same as the sign of the phase velocity, i.e., the average energy flux vector and the phase velocity vector point in the same direction. We thus conclude that only outgoing waves are possible, and we must discard incoming waves, i.e., A=0 in (12.89) and

$$\phi(z,t) = B\exp i(\omega t-kz) \tag{12.92}$$

To determine B, we substitute (12.92) into (12.86), resulting in

$$B = -\frac{iM_o}{kGJ}$$

so that the solution is

$$\phi(z,t) = -\frac{iM_o}{kGJ}\,\mathrm{expi}(\omega t-kz), \quad \gamma<\omega \tag{12.93}$$

We may summarize our conclusions as follows: If the forcing frequency ω is below the cut off frequency $\gamma=\sqrt{\frac{c}{J\rho}}$, we obtain standing, spatially attenuated waves characterized by (12.91). If the forcing frequency is above the cut off frequency γ, we obtain traveling waves characterized by (12.93).

(g) Forced, Transient Motion of an Elastically Restrained Rod.

An infinite, elastically restrained rod $-\infty<z<\infty$ is acted upon by a concentrated angular impulse of magnitude M_o at $z=0$, $t=0$. It is required to find the ensuing, transient dynamic response of the rod. The equation characterizing the motion is given by (12.62), and it is now written as

$$\frac{\partial^2\phi}{\partial t^2}-\frac{1}{r^2}\frac{\partial^2\phi}{\partial z^2}+\gamma^2\phi = \frac{M_o}{J\rho}\,\delta(z)\delta(t) \tag{12.94}$$

where the symbol δ denotes the Dirac delta function and where $\frac{1}{r^2}=\frac{G}{\rho}$, $\gamma^2=\frac{c}{J\rho}$. We shall assume that

$$\lim_{z\to\pm\infty}\phi(z,t) = \lim_{z\to\pm\infty}\left(\frac{\partial\phi}{\partial z}\right)_{(z,t)} = 0 \tag{12.95}$$

and that initially the rod is at rest in its reference state, i.e.,

$$\phi(z,0) = \dot{\phi}(z,0) = 0 \tag{12.96}$$

We shall employ both Fourier and Laplace transforms to solve our problem. The Laplace transform will be denoted by a bar over the transformed variable (see Chapter 1), while the Fourier transform is denoted by a star. Thus

$$\phi^*(k,t)= \frac{1}{\sqrt{2\pi}}\int_{-\infty}^{\infty}\phi(z,t)e^{ikz}dz=F^*\{\phi(z,t)\}$$

$$\phi(z,t) = \frac{1}{\sqrt{2\pi}}\int_{-\infty}^{\infty}\phi^*(k,t)e^{-izk}dk$$

$$\left(\frac{\partial^2\phi}{\partial z^2}\right)^* = -k^2\phi^*=F^*\left\{\frac{\partial^2\phi}{\partial z^2}\right\}, \quad F^*\{\delta(z)\} = \frac{1}{\sqrt{2\pi}}$$

Taking Laplace transforms of (12.94) and observing (12.96), we obtain

$$p^2\bar{\phi} - \frac{1}{r^2}\frac{d^2\bar{\phi}}{dz^2}+\gamma^2\bar{\phi} = \frac{M_o}{J\rho}\,\delta(z) \tag{12.97}$$

Taking Fourier transforms of (12.97), and observing (12.95), we obtain

$$\bar{\phi}^*(k,p) = \frac{M_o r^2}{J\rho\sqrt{2\pi}} \frac{1}{[k^2+r^2(p^2+\gamma^2)]} \tag{12.98}$$

Inversion of (12.98) with respect to k results in

$$\bar{\phi}(z,p) = \frac{1}{2\pi}\frac{M_o}{JG}\int_{-\infty}^{\infty}\frac{e^{-izk}dk}{\left[k^2+r^2(\gamma^2+p^2)\right]} = \frac{1}{\pi}\frac{M_o}{JG}\int_0^{\infty}\frac{\cos zk dk}{\left[k^2+r^2(\gamma^2+p^2)\right]}$$

and this integral can be evaluated by contour integration, resulting in

$$\bar{\phi}(z,p) = \frac{M_o}{2JG}\frac{e^{-|z|r\sqrt{\gamma^2+p^2}}}{r\sqrt{\gamma^2+p^2}} \tag{12.99}$$

With the aid of entry (19) in Table 1.1 of Chapter 1, we can invert (12.99) with respect to p:

$$\begin{aligned}\phi(z,t) &= 0, \quad 0<t<r|z| \\ &= \frac{M_o}{2JGr} J_o\left(\gamma\sqrt{t^2-r^2z^2}\right), \quad r|z|\le t\end{aligned} \tag{12.100}$$

where J_o denotes the Bessel function of order zero. The solution (12.100) indicates that the disturbance travels away from the origin and has a definite wave front, which travels with speed $\frac{1}{r}$. The value of the rotation angle ϕ at the wave front is

$$\phi(\frac{t}{r},t) = \frac{M_o}{2JGr} J_o(0) = \frac{M_o}{2JGr}\ .$$

To obtain (12.100), we inverted first with respect to k, then with respect to p. It is possible to reverse this process, and toward this end we write (12.98) in the form

$$\bar{\phi}^*(k,p) = \frac{M_o}{J\rho}\frac{1}{\sqrt{2\pi}}\frac{1}{\sqrt{\gamma^2+\frac{k^2}{r^2}}}\frac{\sqrt{\gamma^2+\frac{k^2}{r^2}}}{\left[p^2+\left(\gamma^2+\frac{k^2}{r^2}\right)\right]}$$

Inversion with respect to p can now be accomplished with the aid of entry (11) in Table 1.1 of Chapter 1:

$$\phi^*(k,t) = \frac{M_o}{GJr\sqrt{2\pi}}\frac{\sin\sqrt{r^2\gamma^2+k^2}\,\frac{t}{r}}{\sqrt{r^2\gamma^2+k^2}}$$

and therefore

$$\phi(z,t) = \frac{M_o}{GJr}\frac{1}{2\pi}\int_{-\infty}^{\infty}\frac{\left(\sin\sqrt{r^2\gamma^2+k^2}\,\frac{t}{r}\right)\exp(-izk)dk}{\sqrt{r^2\gamma^2+k^2}}$$

$$= \frac{M_o}{GJr}\frac{1}{\pi}\int_{0}^{\infty}\frac{\left(\sin\sqrt{r^2\gamma^2+k^2}\,\frac{t}{r}\right)\cos zkdk}{\sqrt{r^2\gamma^2+k^2}} \qquad (12.101)$$

The value of the integral in (12.101) is known (see, for instance, A. Erdélyi, Ed., Tables of Integral Transforms, vol. I, p. 26, McGraw Hill Book Co., Inc., 1954), and upon its evaluation we again obtain (12.100). The form of the integral in (12.101) often arises in the study of wave motion in dispersive media, and its evaluation in closed form is usually not possible. When this is the case, an approximate evaluation is usually possible by Kelvin's method of stationary phase, and the integral in (12.101) is approximated by that method in Section 4 of Appendix B.

EXERCISES

12.1 Consider the torsional (or logitudinal) motion of an elastic rod of length ℓ, free at both ends. Find the natural frequencies and associated (normalized) mode shapes.

12.2 Consider the torsional (or longitudinal) motion of an elastic rod fixed at $z=0$ and at $z=\ell$. Find the natural frequencies and associated (normalized) mode shapes.

12.3 The rod of Exercise 12.2 is subjected to a uniformly distributed, suddenly applied twisting moment of intensity $m_o H(t)$, where $H(t)$ is the Heaviside unit step function. Find the transient response of the rod if it is in equilibrium in its reference position at $t=0$.

12.4 The rod of Exercise 12.2 is subjected to a pure (longitudinal) impulse of magnitude I_o at $z=\frac{\ell}{2}$, $t=0$. Find the transient response of the rod if it is in equilibrium in its reference position at $t=0$.

12.5 Find the general solution of the problem of torsional oscillations of a rod consisting of two rigidly fastened sections with different dimensions and physical properties as shown in the figure.

It is assumed that the ends of the rod are clamped and that the initial state of the rod is characterized by the conditions

$$\phi(z,0) = \phi_0(z); \quad \dot{\phi}(z,0) = \dot{\phi}_0(z).$$

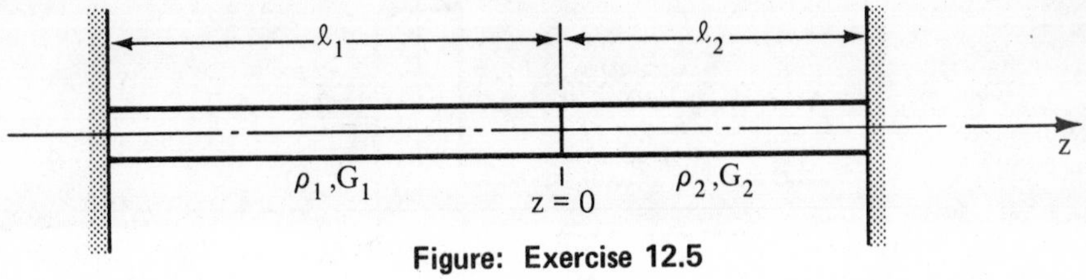

Figure: Exercise 12.5

12.6 An inhomogeneous rod consisting of two sections made from different materials is clamped at one end and is initially at rest in its reference configuration. Find the dynamic response if a constant force P_0 is suddenly applied to the free end at t=0.

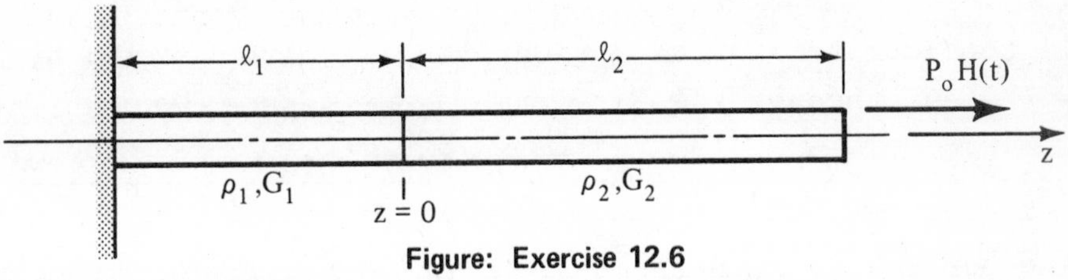

Figure: Exercise 12.6

12.7 A semi-infinite elastic rod $0<z<\infty$ is subjected to a twisting moment at z=0.

(a) Find the steady state dynamic response of the rod if the end moment at z=0 is given by $M(0,t)=-M_0 \exp i\omega t$. (b) Find the transient response if $M(0,t)=-M_0 H(t)$. In this case assume that $\phi(z,0)=\dot{\phi}(z,0)=0$.

12.8 A semi-infinite elastic rod $0<z<\infty$ is embedded in a restraining medium so that local axial deformation w is resisted by a force which is proportional to deformation. At t=0 a pure axial impulse of magnitude I_0 is applied to the end z=0. Find the response of the rod if $w(z,0)=\dot{w}(z,0)=0$.

12.9 Consider the torsional motion of an unbounded, homogeneous, elastic rod embedded in a restraining medium. The motion can be characterized by (12.62). Prove that the group velocity is equal to the velocity of energy transport (see (7.14)).

12.10 A semi-infinite elastic rod $-\infty<z<0$ is attached to a rigid disc with moment of inertia I at z=0. A torsional wave $\phi(z,t)=A\exp ik\cdot$

(vt−z) is incident upon the disc. Find the reflected wave, and discuss the present phenomenon with particular reference to the physical parameters of the problem.

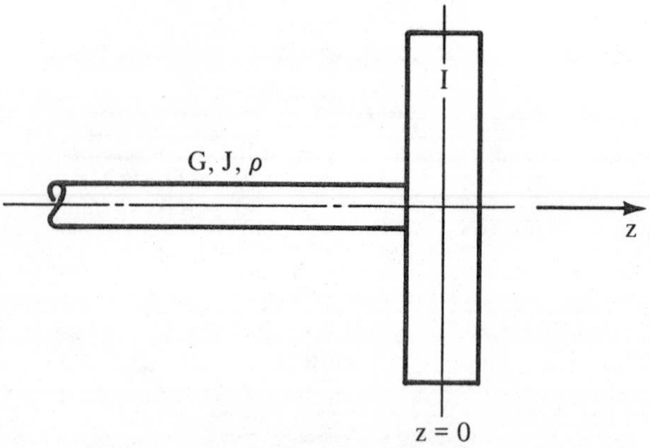

Figure: Exercise 12.10

12.11 Consider the forced motion problem characterized by equations (12.29) through (12.31) with f≡g≡0. Assume a solution of the form $\phi(z,t) = \sum_{i=1}^{\infty} \Phi_i(z)\cdot q_i(t)$, where $\Phi_i(z)$ are the eigenfunctions which satisfy the particular homogeneous boundary conditions of the problem Show that

$$\ddot{q}_i+\omega_i^2 q_i=Q_i(t) = \int_0^{\ell} m\Phi_i dz;\; q_i(0) = \int_0^{\ell} \rho J\phi_0\Phi_i dz,\; \dot{q}_i(0) = \int_0^{\ell} \rho J\dot{\phi}_0\Phi_i dz$$

and

$$q_i(t) = q_i(0)\cos\omega_i t + \frac{1}{\omega_i}\dot{q}_i(0)\sin\omega_i t + \frac{1}{\omega_i}\int_0^t Q_i(\tau)\sin\omega_i(t-\tau)d\tau.$$

Chapter

13

DYNAMICS OF ELASTIC BEAMS

The statement that every small motion of the system can be presented as the result of superposed motions in normal modes is equivalent to a theorem, viz.: that any arbitrary displacement (or velocity) can be represented as the sum of a finite or infinite series of normal functions. Such theorems concerning the expansions of functions are generalizations of Fourier's theorem, and, from the point of view of a rigorous analysis, they require independent proof. Every problem of free vibrations suggests such a theorem of expansion.

A.E.H. Love, A Treatise on the Mathematical Theory of Elasticity, 4th Ed., 1927.

13.1 EQUATIONS OF MOTION

A beam is usually defined as a transversely loaded, prismatic structural member, the length of which is large compared to its largest, linear cross-sectional dimension. The widespread utilization of beams in applications such as shafts in machines, girders in buildings and bridges, wingspars in airplanes, etc., makes it mandatory that engineers have methods of beam analysis at their disposal which are not overly cumbersome, yet are capable of yielding results of acceptable accuracy. We have already considered flexural wave motion, within the framework of three-dimensional elasticity theory, in an unbounded, free beam with circular cross section in Section 11.6. In the usual applications, beams are transversely loaded and supported, and come in a variety of cross-sectional shapes. To generate solutions of practical value within the framework of the field equations of classical elasticity theory is either impossible or results in complexities which are generally not warranted in terms of the accuracy which suffices for many problems. For this reason it is customary to use an approximate, technical beam theory, and we shall now proceed with the development of such a theory.

Let us consider the "small" motion of a beam segment of length x_2-x_1. We shall assume that the cross section of the beam has at least one plane of symmetry, which we shall take as the x-z plane of our Cartesian axis system, where x is the longitudinal axis of the beam (see Fig. 13.1). As a basis of the present beam model we take the (small) displacement assumptions

$$u_x=z\Psi(x,t),\ u_y=0,\ u_z=w(x,t) \tag{13.1}$$

which imply that

(a) Planes which are normal to the beam x-axis in the undeformed state remain plane in the deformed state.

(b) The beam is in a state of plane strain with respect to the x-z plane, i.e., displacements normal to the x-z plane vanish.

(c) The vertical displacements of all points on any normal cross section are the same, i.e., there is no "thickness stretch."

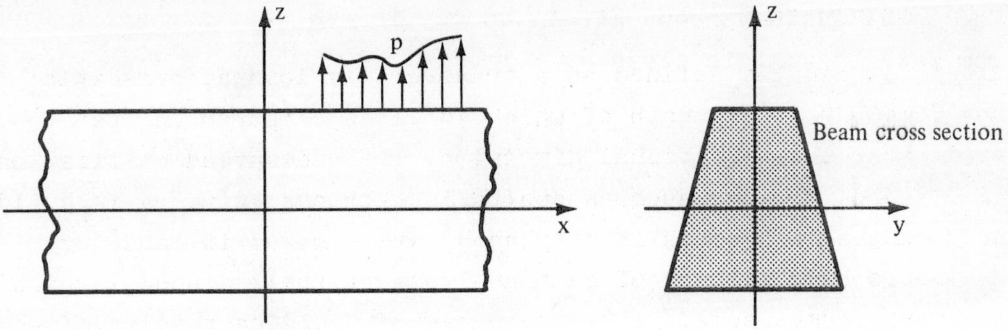

Figure 13.1

Upon substitution of (13.1) into the strain-displacement relations (10.21), we obtain the strain field

$$\begin{bmatrix} \varepsilon_{xx} & \varepsilon_{xy} & \varepsilon_{xz} \\ \varepsilon_{yx} & \varepsilon_{yy} & \varepsilon_{yz} \\ \varepsilon_{zx} & \varepsilon_{zy} & \varepsilon_{zz} \end{bmatrix} = \begin{bmatrix} z\,\frac{\partial \Psi}{\partial x} & 0 & \frac{1}{2}(\psi + \frac{\partial w}{\partial x}) \\ 0 & 0 & 0 \\ \frac{1}{2}(\psi + \frac{\partial w}{\partial x}) & 0 & 0 \end{bmatrix} \qquad (13.2)$$

The equations of motion will now be derived with the aid of Hamilton's principle. The kinetic energy of the beam segment extending from $x=x_1$ to $x=x_2$ is given by

$$T = \frac{1}{2}\int_V \rho(\dot{u}_x^2+\dot{u}_y^2+\dot{u}_z^2)dV = \frac{1}{2}\int_{x_1}^{x_2}\int_A \rho(z^2\dot{\psi}^2+\dot{w}^2)dA\,dx$$

$$= \frac{1}{2}\int_{x_1}^{x_2}(\rho I\dot{\psi}^2+\rho A\dot{w}^2)dx \qquad (13.3)$$

where $I = \int_A z^2 dA$ is the moment of inertia and A is the magnitude of the beam cross-sectional area.

Consequently

$$\delta T = \int_{x_1}^{x_2}(\rho I\dot{\psi}\delta\dot{\psi}+\rho A\dot{w}\delta\dot{w})dx$$

and

$$\int_{t_1}^{t_2}\delta T\,dt = -\int_{t_1}^{t_2}\int_{x_1}^{x_2}(\rho I\ddot{\psi}\delta\psi+\rho A\ddot{w}\delta w)dx\,dt \qquad (13.4)$$

To obtain the above result, we had to integrate by parts with respect to t and also invoked the conditions $\delta\psi=0$ and $\delta w=0$ for $t=t_1$ and $t=t_2$.

The total instantaneous strain energy in the beam segment extending from $x=x_1$ to $x=x_2$ is given by

$$U = \int_V W^* dV \tag{13.5}$$

where W^* denotes the strain energy density.
Consequently

$$\delta U = \int_V \frac{\partial W^*}{\partial \varepsilon_{ij}} \delta\varepsilon_{ij} dV = \int_V \tau_{ij}\delta\varepsilon_{ij} dV$$

$$= \int_{x_1}^{x_2}\int_A \left[\tau_{xx} z \frac{\partial}{\partial x}(\delta\psi) + \tau_{xz}\delta\psi + \tau_{xz}\frac{\partial}{\partial x}(\delta w)\right] dA\, dx$$

$$= \int_{x_1}^{x_2}\left[M \frac{\partial}{\partial x}(\delta\psi) + Q\delta\psi + Q\frac{\partial}{\partial x}(\delta w)\right] dx \tag{13.6}$$

where the stress-resultants M and Q are the bending moment and shear force, respectively, defined by

$$M = \int_A \tau_{xx} z dA \tag{13.7a}$$

$$Q = \int_A \tau_{xz} dA \tag{13.7b}$$

We can now integrate (13.6) by parts with respect to x, and readily obtain

$$\delta U = (M\delta\psi + Q\delta w)\Big|_{x_1}^{x_2} + \int_{x_1}^{x_2}\left[\left(Q - \frac{\partial M}{\partial x}\right)\delta\psi - \frac{\partial Q}{\partial x}\delta w\right] dx \tag{13.8}$$

If body forces are neglected, then in accordance with (9.119) and (13.1), the virtual work of the surface tractions acting on the beam segment is given by

$$\delta W = \int_S T_i \delta u_i dS = \int_S (T_x\delta u_x + T_z \delta u_z) dS$$

$$= \int_S (T_x z\delta\psi + T_z\delta w) dS$$

where $S=A_1+A_2+\sigma$. In this expression A_1 and A_2 denote the magnitudes of the areas of the (plane) cross sections of the beam at $x=x_1$ and $x=x_2$, respectively, while σ is the surface area of the beam for $x_1<x<x_2$. With

reference to (9.80) we have $T_i=\tau_{ij}n_j$, and since deformation is symmetrical with respect to the x-z plane, we require

$$T_x(x,y,z,t) = T_x(x,-y,z,t)$$
$$T_z(x,y,z,t) = T_z(x,-y,z,t)$$

The components of the unit outer normal $\hat{n}$ are

on A_1: $n_x=-1$, $n_y=0$, $n_z=0$

on A_2: $n_x=1$, $n_y=0$, $n_z=0$

so that

$$\delta W= \int_\sigma (T_x z\delta\psi+T_z\delta w)dS$$
$$+ \delta\psi \int_{A_2} z\tau_{xx}dA+\delta w \int_{A_2} \tau_{xz}dA$$
$$- \delta\psi \int_{A_1} z\tau_{xx}dA-\delta w \int_{A_1} \tau_{xz}dA$$

But

$$\int_\sigma T_x z\delta\psi dS= \int_{x_1}^{x_2}\oint_c T_x z\delta\psi\cdot dc\cdot dx = \int_{x_1}^{x_2} m\delta\psi dx$$

$$\int_\sigma T_z\delta w dS = \int_{x_1}^{x_2} \oint_c T_z \ \delta w\cdot dc\cdot dx = \int_{x_1}^{x_2} p\delta w \ dx$$

where

$$\oint_c T_x z dc = m, \quad \oint_c T_z dc = p$$

and where the symbol $\oint_c$ denotes the line integral taken over the circumference of cross-sectional area of the beam. The quantities p and m are readily identified as the intensities of the distributed transverse load and moment, respectively. In conformance with (13.7), we also set

$$Q^*(x_i,t) = \int_{A_i} \tau_{xz}dA$$
$$M^*(x_i,t) = \int_{A_i} z\tau_{xx}dA$$
$$i = 1,2.$$

With these notations, the expression for the virtual work reduces to

$$\delta W = \int_{x_1}^{x_2} (p\delta w + m\delta\psi)dx + (M^*\delta\psi + Q^*\delta w)\Big|_{x_1}^{x_2} \tag{13.9}$$

We note that M^* and Q^* are the bending moment and shear, respectively, acting upon the ends of the beam. Upon substitution of (13.4), (13.8), and (13.9) into (9.121), we obtain

$$\int_{t_1}^{t_2}\int_{x_1}^{x_2} \left[\left(-\rho A\ddot{w} + \frac{\partial Q}{\partial x} + p\right)\delta w + \left(-\rho I\ddot{\psi} + \frac{\partial M}{\partial x} - Q + m\right)\delta\psi\right] dxdt$$

$$+ \int_{t_1}^{t_2} \left[(M^* - M)\delta\psi + (Q^* - Q)\delta w\right]_{x_1}^{x_2} dt = 0$$

The time interval $t_2 - t_1$ as well as the length of the beam segment are arbitrary. Therefore the integrands of the double and single integrals must vanish separately. For arbitrary δw and $\delta\psi$ we have

$$-\rho A\ddot{w} + \frac{\partial Q}{\partial x} + p = 0 \quad \text{for } x_1 < x < x_2 \tag{13.10a}$$

$$-\rho I\ddot{\psi} + \frac{\partial M}{\partial x} - Q + m = 0 \quad \text{for } x_1 < x < x_2 \tag{13.10b}$$

At $x = x_1$ and $x = x_2$, one member of each of the pairs

$$\begin{Bmatrix} M(x,t) = M^*(t) \\ \psi(x,t) = \psi^*(t) \end{Bmatrix}, \quad \begin{Bmatrix} Q(x,t) = Q^*(t) \\ w(x,t) = w^*(t) \end{Bmatrix} \tag{13.10c}$$

must be specified. Equations (13.10a) and (13.10b) are the stress-equations of small motion of a beam, and (13.10c) are the associated, admissible boundary conditions.

We shall next derive an expression for energy flux in the beam. Differentiating (13.5) with respect to t, we obtain

$$\frac{dU}{dt} = \int_V \frac{\partial W^*}{\partial \varepsilon_{ij}} \frac{d\varepsilon_{ij}}{dt} dV = \int_V \tau_{ij}\dot{\varepsilon}_{ij} dV$$

$$= \int_{x_1}^{x_2}\int_A (\tau_{xx}\dot{\varepsilon}_{xx} + 2\tau_{xz}\dot{\varepsilon}_{xz}) dxdA$$

$$= \int_{x_1}^{x_2}\int_A \left[\tau_{xx} z \frac{\partial\dot{\psi}}{\partial x} + \tau_{xz}\left(\dot{\psi} + \frac{\partial\dot{w}}{\partial x}\right)\right] dxdA$$

$$= \int_{x_1}^{x_2} \left[M \frac{\partial\dot{\psi}}{\partial x} + Q\left(\dot{\psi} + \frac{\partial\dot{w}}{\partial x}\right)\right] dx$$

where we have used (9.101), (13.2), and (13.7). Differentiating (13.3) with respect to t, we obtain

$$\frac{dT}{dt} = \int_{x_1}^{x_2} (\rho I\dot{\psi}\ddot{\psi}+\rho A\dot{w}\ddot{w})dx$$

$$= \int_{x_1}^{x_2} \left[\dot{\psi}\left(\frac{\partial M}{\partial x} - Q + m\right) + \dot{w}\left(p + \frac{\partial Q}{\partial x}\right)\right] dx$$

where we have utilized the stress-equations of motion (13.10a) and (13.10b). If we denote the instantaneous energy per unit length of the beam by E*, then

$$\int_{x_1}^{x_2} \dot{E}^* dx = \frac{d}{dt}(T+U) = \int_{x_1}^{x_2} \left[\frac{\partial}{\partial x}(M\dot{\psi}+Q\dot{w})+m\dot{\psi}+p\dot{w}\right] dx$$

$$= \int_{x_1}^{x_2} (m\dot{\psi}+p\dot{w} - \frac{\partial S}{\partial x})dx$$

where the energy flux is defined by

$$S = -M\dot{\psi} - Q\dot{w} \tag{13.11}$$

Since the length $x_2 - x_1$ of the beam segment is arbitrary, we obtain the energy continuity equation (see 9.94)

$$\dot{E}^* = m\dot{\psi} + p\dot{w} - \frac{\partial S}{\partial x} \tag{13.12}$$

For vanishing force and moment intensity, p=m=0, and

$$\frac{d}{dt}(T+U) = (-S)\Big|_{x_1}^{x_2} = S(x_1,t) - S(x_2,t) \tag{13.13}$$

i.e., the time rate of change of energy in the beam is equal to the energy flux at $x=x_1$ minus the energy flux at $x=x_2$.

The stress equations of motion (13.10a) and (13.10b) cannot be integrated as they stand, and constitutive relations are required to complete the mathematical model of the beam. We assume a linearly elastic, isotropic medium, and utilize two of the six equations of Hooke's law (10.22):

$$\tau_{xx} = E\varepsilon_{xx} + \nu(\tau_{yy}+\tau_{zz})$$

$$= Ez\frac{\partial\psi}{\partial x} + \nu(\tau_{yy}+\tau_{zz}) \tag{13.14a}$$

$$\tau_{xz} = 2G\varepsilon_{xz} = G\left(\psi + \frac{\partial w}{\partial x}\right) \tag{13.14b}$$

We now substitute (13.14) into (13.7) and alter the results in two respects: (a) we suppress (drop) the term $\nu \int_A z(\tau_{yy}+\tau_{zz})dA$ and (b) we change G to $\kappa^2 G$, where κ^2 is a constant to be determined later. It is noted that added to the present approximation is the fact that we simply ignore the remaining four equations of Hooke's law. The results are

$$M = EI \frac{\partial\psi}{\partial x} \tag{13.15a}$$

$$Q = \kappa^2 AG\left(\frac{\partial w}{\partial x} + \psi\right) \tag{13.15b}$$

Equations (13.15) are the "stress-displacement" relations of the beam. We also note that the force on any cross-sectional area of the beam must vanish. Consequently

$$\int_A \tau_{xx} dA = E \frac{\partial\psi}{\partial x} \int_A z dA = 0$$

where we have dropped the term $\nu\int_A(\tau_{yy}+\tau_{zz})dA$. Therefore $\int_A z\, dA = 0$, and this implies that the x-axis is the centroidal axis of the beam. Upon substitution of (13.15) into (13.10a) and (13.10b) we obtain the displacement equations of motion:

$$-\rho A\ddot{w} + \frac{\partial}{\partial x}\left[\kappa^2 AG\left(\frac{\partial w}{\partial x} + \psi\right)\right] + p = 0 \tag{13.16a}$$

$$-\rho I\ddot{\psi} + \frac{\partial}{\partial x}\left(EI \frac{\partial\psi}{\partial x}\right) - \kappa^2 AG\left(\frac{\partial w}{\partial x} + \psi\right) + m = 0 \tag{13.16b}$$

Equations (13.16) are often referred to as the Timoshenko beam equations, and they account for transverse and rotatory inertia, and for deformations due to beam flexure and shear. The shear coefficient κ^2 was originally determined from static considerations, and was given as the ratio of the average shear strain on the beam cross section to the shear strain at the centroid. Other investigators calculate the phase velocity of flexural waves with the aid of the approximate beam equations (13.16), and perform a similar calculation within the framework of the exact, three-dimensional theory. The value of κ^2 is then adjusted to result in a good match of the phase velocity vs. wave number curve between the two theories, using the lowest mode of propagation in the case of elasticity theory. The different methods for evaluating κ^2 has given rise to a number of interesting papers in the literature, and a useful summary, as well as an additional method for evaluating κ^2 is contained in a paper by G.R. Cowper, "The Shear Coefficient in Timoshenko's Beam Theory," Journal of Applied Mechanics, June 1966, pp. 335-340. This paper also lists a series of formulae for κ^2 for the commonly used beam cross sections. The following values are taken from this paper:

TABLE 13.1 VALUES OF THE SHEAR COEFFICIENT κ^2

ν	Rectangular Cross Section	Circular Cross Section
0	0.833	0.857
0.3	0.850	0.886
0.5	0.870	0.900

The most commonly used beam theory for technical applications proceeds from the displacement equations (13.1) with two additional assumptions: (d) $\psi = -\frac{\partial w}{\partial x}$, i.e., planes which are normal to the beam axis in the undeformed state remain plane and normal in the deformed state. This assumption is equivalent to neglecting shear deformations, i.e., the beam is assumed to be rigid with respect to shear deformations, and all deformations are due to longitudinal fiber extension and compression alone.

(e) The effects of rotatory inertia are assumed to be negligible, and therefore we set $\rho I\ddot{\psi} = 0$.

If we now set m=0, and incorporate assumptions (d) and (e) into (13.10a) and (13.10b), we obtain the stress-equation of motion

$$-\rho A\ddot{w} + \frac{\partial Q}{\partial x} + p = 0$$

where

$$Q = \frac{\partial M}{\partial x}$$

and with the aid of (13.15a), the pertinent stress-displacement relation assumes the form

$$M = -EI\,\frac{\partial^2 w}{\partial x^2} .$$

Consequently, the displacement-equation of motion is

$$-\rho A\ddot{w} - \frac{\partial^2}{\partial x^2}\left(EI\,\frac{\partial^2 w}{\partial x^2}\right) + p = 0 \qquad (13.17a)$$

and the associated, admissible boundary conditions are now obtained by specifying one member of each of the following pairs at the ends of the beam

$$(M, \frac{\partial w}{\partial x}), \ (Q, w) \tag{13.17b}$$

Equations (13.17) characterize the well-known Euler-Bernoulli theory of beams, so-named after two early scientists who made major contributions to this branch of elasticity theory.

13.2 FORCED MOTION OF EULER-BERNOULLI BEAMS

A problem which is frequently encountered by engineers is the prediction of the dynamic response of beams subjected to time-dependent loads and/or boundary conditions, and for this reason the present section is concerned with the resolution of the forced motion problem of Euler-Bernoulli beams of finite length.

A properly posed problem may be stated as follows: find a solution of the equation

$$\rho A \frac{\partial^2 w}{\partial t^2} = \frac{\partial Q}{\partial x} + p \tag{13.18}$$

where

$$Q = \frac{\partial M}{\partial x} = - \frac{\partial}{\partial x} \left(EI \frac{\partial^2 w}{\partial x^2} \right), \quad M = - EI \frac{\partial^2 w}{\partial x^2} \tag{13.19}$$

In the above equations, ρ, A, E, and I are known functions of x and $p=p(x,t)$. The boundary conditions are given by

$$\left.\begin{array}{ll} \text{either } w(0,t) \text{ or } Q(0,t) = f_1(t) \\ \text{either } (\frac{\partial w}{\partial x})_{(0,t)} \text{ or } M(0,t) = f_2(t) \\ \text{either } w(\ell,t) \text{ or } Q(\ell,t) = g_1(t) \\ \text{either } (\frac{\partial w}{\partial x})_{(\ell,t)} \text{ or } M(\ell,t) = g_2(t) \end{array}\right\} \tag{13.20}$$

where $x=0$ and $x=\ell$ denote the extremities of the beam. Initial conditions which are sufficient to insure a unique solution (see Exercise 13.4) are given by

$$w(x,0) = w_o(x), \ \dot{w}(x,0) = \dot{w}_o(x) \tag{13.21}$$

To obtain a solution of the problem characterized by (13.18) through (13.21), we proceed as in the case of the rod (Section 12.3), and first consider the free vibration problem for homogeneous boundary conditions, i.e., $p \equiv 0$, and

$$\left.\begin{aligned} &w(0,t) \text{ or } Q(0,t) = 0 \\ &\left(\frac{\partial w}{\partial x}\right)_{(0,t)} \text{ or } M(0,t) = 0 \\ &w(\ell,t) \text{ or } Q(\ell,t) = 0 \\ &\left(\frac{\partial w}{\partial x}\right)_{(\ell,t)} \text{ or } M(\ell,t) = 0 \end{aligned}\right\} \tag{13.22}$$

Assume a product solution of the form

$$w(x,t) = W(x) \cos\omega t \tag{13.23}$$

and substitute (13.23) into the homogeneous form of (13.18), in conjunction with (13.19). The result is the differential equation

$$\frac{d^2}{dx^2}\left(EI \frac{d^2W}{dx^2}\right) - \rho A \omega^2 W = 0 \tag{13.24}$$

characterizing the eigenfunctions (or mode shapes) of the beam. In view of (13.22), (13.23), and (13.19), the appropriate boundary conditions on $W(x)$ are

$$\left.\begin{aligned} &\text{either } W(0) \text{ or } \left[\frac{d}{dx}\left(EI \frac{d^2W}{dx^2}\right)\right]_{(0)} = 0 \\ &\text{either } \left(\frac{dW}{dx}\right)_{(0)} \text{ or } \left(EI \frac{d^2W}{dx^2}\right)_{(0)} = 0 \\ &\text{either } W(\ell) \text{ or } \left[\frac{d}{dx}\left(EI \frac{d^2W}{dx^2}\right)\right]_{(\ell)} = 0 \\ &\text{either } \left(\frac{dW}{dx}\right)_{(\ell)} \text{ or } \left(EI \frac{d^2W}{dx^2}\right)_{(\ell)} = 0 \end{aligned}\right\} \tag{13.25}$$

Non-trivial solutions of (13.24) which also satisfy (13.25) exist only for certain discrete values of ω called the natural frequencies (or eigenvalues) of the beam. There is a denumerable infinity of these natural frequencies $\omega_i, i=1,2\ldots$. The solution $W_i(x)$ corresponding to the natural frequency ω_i is called the i'th normal mode (or eigenfunction) of the beam. Equations (13.24) and (13.25) specify each normal mode only to within an indeterminate, constant multiplier. By imposing the normalization condition

$$\left(\int_0^\ell \rho A W_i^2 dx\right)^{\frac{1}{2}} = 1 \tag{13.26}$$

unique expressions for the eigenfunctions are obtained. The form (13.26) was selected because of its convenience in the subsequent solution of forced motion problems.

Let us now consider two different eigenfunctions W_i and W_j, and their associated frequencies ω_i and ω_j. By definition, each of the sets (W_i,ω_i) and (W_j,ω_j) satisfies (13.24) and (13.25). We now form the expression

$$\int_0^\ell \left\{\left[\frac{d^2}{dx^2}\left(EI\,\frac{d^2W_i}{dx^2}\right) - \rho A\omega_i^2 W_i\right]W_j - \left[\frac{d^2}{dx^2}\left(EI\,\frac{d^2W_j}{dx^2}\right) - \rho A\omega_j^2 W_j\right]W_i\right\} dx = 0$$

and upon rearrangement we obtain

$$(\omega_i^2-\omega_j^2)\int_0^\ell \rho A W_i W_j dx = \int_0^\ell \left[\left(W_j\,\frac{d^2}{dx^2}\left(EI\,\frac{d^2W_i}{dx^2}\right) - W_i\,\frac{d^2}{dx^2}\left(EI\,\frac{d^2W_j}{dx^2}\right)\right] dx$$

The right-hand side of this equation can be reduced by successive integration by parts, resulting in

$$(\omega_i^2-\omega_j^2)\int_0^\ell \rho A W_i W_j dx = \left(Q_j W_i - Q_i W_j + M_i\,\frac{dW_j}{dx} - M_j\,\frac{dW_i}{dx}\right)_0^\ell \tag{13.27}$$

where

$$M_i = -EI\,\frac{d^2W_i}{dx^2}\,,\quad Q_i = -\frac{d}{dx}\left(EI\,\frac{d^2W_i}{dx^2}\right) \tag{13.28}$$

If homogeneous boundary conditions (13.25) are now assumed, (13.27) reduces to the orthogonality relation

$$\left.\begin{aligned} &\int_0^\ell \rho A W_i W_j dx = 0 \\ &\text{provided} \\ &\omega_i^2 \neq \omega_j^2 \end{aligned}\right\} \tag{13.29}$$

We now proceed to the solution of the forced motion problem as characterized by (13.18) through (13.21). It is our intention to obtain a solution in terms of an eigenfunction expansion, utilizing eigenfunctions with homogeneous boundary conditions. Toward this end we write the solution in the form

$$w(x,t) = w_s(x,t) + \sum_{i=1}^{\infty} W_i(x)\, q_i(t) \tag{13.30a}$$

and, in view of the linearity of the present problem

$$Q(x,t) = Q_s(x,t) + \sum_{i=1}^{\infty} Q_i(x)\, q_i(t) \tag{13.30b}$$

where $q_i(t)$, i=1,2..., are functions of time and where $w_s(x,t)$ and $Q_s(x,t)$ pertain to the quasi-static solution, i.e., they satisfy

$$-\frac{\partial^2}{\partial x^2}\left(EI\,\frac{\partial^2 w_s}{\partial x^2}\right) = \frac{\partial Q_s}{\partial x} = -p \tag{13.31a}$$

$$Q_s = \frac{\partial M_s}{\partial x} = -\frac{\partial}{\partial x}\left(EI\,\frac{\partial^2 w_s}{\partial x^2}\right), \quad M_s = -EI\,\frac{\partial^2 w_s}{\partial x^2} \tag{13.31b}$$

In addition, the "static" solution $w_s(x,t)$ must satisfy the boundary conditions

$$\left.\begin{array}{l}
\text{either } w_s(0,t) \text{ or } Q_s(0,t) = f_1(t) \\
\text{either } \left(\dfrac{\partial w_s}{\partial x}\right)_{(0,t)} \text{ or } M_s(0,t) = f_2(t) \\
\text{either } w_s(\ell,t) \text{ or } Q_s(\ell,t) = g_1(t) \\
\text{either } \left(\dfrac{\partial w_s}{\partial x}\right)_{(\ell,t)} \text{ or } M_s(\ell,t) = g_2(t)
\end{array}\right\} \tag{13.32}$$

The eigenfunctions $W_i(x)$ in (13.30a) satisfy (13.24) and the following homogeneous boundary conditions:

$$\left.\begin{array}{l}
\text{If } w_s(0,t) = f_1(t), \text{ then } W_i(0) = 0 \\
\text{if } Q_s(0,t) = f_1(t), \text{ then } Q_i(0) = 0 \\
\text{if } \left(\dfrac{\partial w_s}{\partial x}\right)_{(0,t)} = f_2(t), \text{ then } \left(\dfrac{dW_i}{dx}\right)_{(0)} = 0 \\
\text{if } M_s(0,t) = f_2(t), \text{ then } M_i(0) = 0 \\
\text{if } w_s(\ell,t) = g_1(t), \text{ then } W_i(\ell) = 0 \\
\text{if } Q_s(\ell,t) = g_1(t), \text{ then } Q_i(\ell) = 0 \\
\text{if } \left(\dfrac{\partial w_s}{\partial x}\right)_{(\ell,t)} = g_2(t), \text{ then } \left(\dfrac{dW_i}{dx}\right)_{(\ell)} = 0 \\
\text{if } M_s(\ell,t) = g_2(t), \text{ then } M_i(\ell) = 0
\end{array}\right\} \tag{13.33}$$

We now substitute (13.30) into (13.18), and utilize (13.24) and (13.31). The result is

$$\sum_{i=1}^{\infty} \rho A W_i(\ddot{q}_i+\omega_i^2 q_i) = -\rho A \ddot{w}_s$$

Both sides of this equation are now multiplied by $W_j(x)$, and the products so obtained are integrated over the length of the beam. With the aid of relations (13.26) and (13.29), we obtain

$$\ddot{q}_i+\omega_i^2 q_i = -\int_0^{\ell} \rho A \ddot{w}_s W_i dx = \ddot{P}_i \qquad (13.34a)$$

where

$$P_i(t) = -\int_0^{\ell} \rho A w_s W_i dx = -\frac{1}{\omega_i^2}\int_0^{\ell} w_s \frac{d^2}{dx^2}\left(EI\frac{d^2W_i}{dx^2}\right)dx \qquad (13.34b)$$

We now integrate by parts the right-hand side of (13.34b) four times in succession, and utilize (13.28) and (13.31), resulting in the convenient expression

$$\omega_i^2 P_i = (-W_i Q_s + w_s Q_i)\Big|_0^{\ell} + \left(-M_i\frac{\partial w_s}{\partial x} + M_s\frac{dW_i}{dx}\right)\Big|_0^{\ell} - \int_0^{\ell} pW_i dx \qquad (13.35)$$

The complete solution of (13.34a) is obtained with the aid of (1.50):

$$q_i(t) = q_i(0)\cos\omega_i t + \frac{1}{\omega_i}\dot{q}_i(0)\sin\omega_i t + \frac{1}{\omega_i}\int_0^t \ddot{P}_i(\tau)\sin\omega_i(t-\tau)d\tau \qquad (13.36a)$$

and upon integrating by parts twice we obtain

$$q_i(t) = \left[q_i(0)-P_i(0)\right]\cos\omega_i t + \frac{1}{\omega_i}\left[\dot{q}_i(0)-\dot{P}_i(0)\right]\sin\omega_i t + P_i(t) - \omega_i\int_0^t P_i(\tau)\sin\omega_i(t-\tau)d\tau \qquad (13.36b)$$

We now proceed to determine the appropriate initial conditions on $q_i(t)$. With reference to (13.21) and (13.30a) we have

$$w_0(x) = w_s(x,0) + \sum_{i=1}^{\infty} W_i(x)\, q_i(0)$$

If we now multiply both sides of this equation by $\rho A W_j$, integrate the products so obtained over the length of the beam, and then utilize the relations (13.26) and (13.29), we obtain

$$q_i(0) = P_i(0) + \int_0^{\ell} \rho A w_0 W_i dx \qquad (13.37a)$$

By an entirely analogous analysis, it can be shown that

$$\dot{q}_i(0) = \dot{P}_i(0) + \int_0^{\ell} \rho A\dot{w}_o W_i dx \qquad (13.37b)$$

This completes the formal solution of the forced motion problem.

The above method of solution will now be applied to three examples, each treating a uniform, homogeneous Euler-Bernoulli beam of length ℓ, i.e., the properties A,E,I, and ρ are assumed to be constant over the length of the beam.

(a) Free Vibrations of a Simply Supported Euler-Bernoulli Beam.

In this case the solution of (13.24) is

$$W(x) = C \sin \beta x + B\cos\beta x + D\sinh\beta x + F\cosh\beta x$$

where

$$\beta = \sqrt{\omega}\sqrt[4]{\frac{\rho A}{EI}}$$

with $W(0) = \left(\frac{d^2W}{dx^2}\right)_{x=0} = W(\ell) = \left(\frac{d^2W}{dx^2}\right)_{x=\ell} = 0$. It is readily shown that $B=D=F=0$ and $\beta_i\ell=i\pi$, $i=1,2,\ldots$, i.e., $\sin \beta_i\ell=0$ and $\cos \beta_i\ell=(-1)^i$. To normalize the eigenfunctions in accordance with (13.26), we follow the procedure of Section 12.3 and write (13.27) in the form

$$\int_0^{\ell} \rho AW_iW_j dx= \frac{\left(Q_jW_i-Q_iW_j+M_i\frac{dW_j}{dx} - M_j\frac{dW_i}{dx}\right)_0^{\ell}}{(\omega_i^2-\omega_j^2)}$$

If we let $W'= \frac{dW}{d(\beta x)}$, then $\frac{dW}{dx}= W'\beta$, $\frac{d^2W}{dx^2} = W''\beta^2$, etc. Thus with the aid of (13.28) we have

$$\int_0^{\ell} \rho AW_iW_j dx= \frac{\rho A\left[W_jW_i'''\beta_i^3-W_iW_j'''\beta_j^3+W_i'\beta_iW_j''\beta_j^2-W_j'\beta_jW_i''\beta_i^2\right]_0^{\ell}}{(\beta_i^4-\beta_j^4)} .$$

The right-hand side of this expression is of the form $\frac{0}{0}$ when $i=j$. Consequently, we can apply L'Hospital's rule to this fraction, i.e., we differentiate numerator denominator with respect to β_i, and subsequently take the limit as $i\to j$. If, in addition, we apply (13.24) and (13.25), the result is

$$\int_0^{\ell} \rho AW_i^2dx= \frac{\rho A\ell}{4}\left[W_i^2-2W_i'W_i'''+(W_i'')^2\right]_{x=\ell}$$

In the case of a simple support at $x=\ell$, $W_i''(\ell)=W_i(\ell)=0$ and

$$\int_0^\ell \rho A W_i^2 dx = -\frac{1}{2}\rho A \ell W_i'(\ell)\, W_i'''(\ell) \ .$$

If the end at $x=\ell$ is free, we have $W_i''(\ell)=W_i'''(\ell)=0$ and

$$\int_0^\ell \rho A W_i^2 dx = \frac{1}{4}\ell \rho A W_i^2(\ell) \ .$$

If the end at $x=\ell$ is rigidly clamped, $W_i(\ell)=W_i'(\ell)=0$ and

$$\int_0^\ell \rho A W_i^2 dx = \frac{1}{4}\rho A \ell \left[W_i''(\ell)\right]^2 \ .$$

In the present case the beam is simply supported at $x=\ell$, and we have

$$(W_i')_{x=\ell} = C_i \cos\beta_i \ell = (-1)^i C_i$$

$$(W_i''')_{x=\ell} = -C_i \cos\beta_i \ell = -(-1)^i C_i$$

so that

$$\int_0^\ell \rho A W_i^2 dx = \frac{1}{2}\rho A \ell C_i^2 = 1$$

and

$$C_i = \sqrt{\frac{2}{\rho A \ell}} \ .$$

Our results may be summarized as follows: For a beam simply supported at $x=0$ and $x=\ell$, the normalized eigenfunctions are

$$W_i(x) = \sqrt{\frac{2}{\rho A \ell}} \quad \sin \beta_i x \tag{13.38a}$$

$$\beta_i = \frac{i\pi}{\ell}\ , \ i=1,2... \tag{13.38b}$$

and the associated natural frequencies are

$$\omega_i = \frac{i^2\pi^2}{\ell^2}\sqrt{\frac{EI}{\rho A}} \tag{13.38c}$$

(b) The Simply Supported Beam Subjected to a Time-dependent End Moment.

We now consider a simple supported Euler-Bernoulli beam of length ℓ shown in Fig. 13.2.

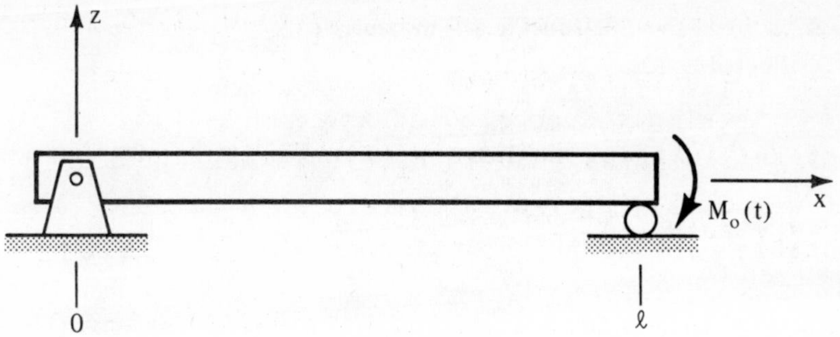

Figure 13.2

The beam is subjected to an end-moment $M_o(t)$ at the end $x=\ell$. We want to predict the dynamic response if the initial conditions are $w(x,0)=\dot{w}(x,0)=0$. With reference to (13.31) and (13.32), the quasi-static part of the solution can be characterized by

$$EI\frac{\partial^4 w_s}{\partial x^4}=0$$

$$w_s(0,t)=0,\ w_s(\ell,t)=0,\ M_s(0,t)=0,\ M_s(\ell,t)=M_o(t)$$

and the solution is obtained by elementary means:

$$w_s=\frac{M_o\ell^2}{6EI}\left(\frac{x}{\ell}-\frac{x^3}{\ell^3}\right),\quad t>0.$$

In view of (13.33), the eigenfunctions with homogeneous boundary conditions appropriate to the present forced motion case are given by (13.38). We now substitute all required terms into (13.35) and (13.37), and obtain

$$P_i(t)=\frac{(-1)^i C_i\beta_i}{\omega_i^2}M_o(t)$$

$$q_i(0)=P_i(0),\ \dot{q}_i(0)=\dot{P}_i(0)$$

These quantities, in turn, are substituted into (13.36b). Utilizing (13.30a) the desired forced motion solution is

$$w(x,t)=\left(\frac{x}{\ell}-\frac{x^3}{\ell^3}\right)\frac{\ell^2M_o(t)}{6EI}+\frac{2\ell^2}{\pi^3EI}\sum_{i=1}^{\infty}\frac{(-1)^i}{i^3}\left[M_o(t)-\omega_i\int_0^t M_o(\tau)\sin\omega_i(t-\tau)\,d\tau\right]\sin\frac{i\pi x}{\ell}\tag{13.39}$$

where

$$\omega_i=\frac{i^2\pi^2}{\ell^2}\sqrt{\frac{EI}{\rho A}}\ .$$

(c) The Simply Supported Beam Subjected to a Time-dependent, Uniformly Distributed Load.

The beam is simply supported at x=0 and at x=ℓ. The intensity of the uniformly distributed load is $p_0(t)$. We want to find the dynamic response if $w(x,0) = \dot{w}(x,0)=0$. With reference to (13.31) and (13.32), the given static part of the solution is characterized by

$$EI \frac{\partial^4 w_s}{\partial x^4} = p_0(t)$$

$$w_s(0,t) = w_s(\ell,t) = M_s(0,t) = M_s(\ell,t) = 0$$

and it is readily shown by elementary methods that

$$w_s = \frac{\ell^4 p_0(t)}{24EI} \left[\left(\frac{x}{\ell}\right)^4 -2 \left(\frac{x}{\ell}\right)^3 + \left(\frac{x}{\ell}\right)\right] , \; t>0 \; .$$

In view of (13.33), the eigenfunctions with homogeneous boundary conditions appropriate to the present problem are given by (13.38). Upon substitution of all required terms into (13.35) and (13.37), we obtain

$$P_i = -\frac{2C_i}{\omega_i^2 \beta_i} p_0(t), \; i=1,3,5,\ldots$$

$$= \quad 0, \qquad i=2,4,6,\ldots$$

$$q_i(0) = P_i(0), \; \dot{q}_i(0) = \dot{P}_i(0)$$

These quantities are now substituted into (13.36b), and the desired forced motion solution is now obtained with the aid of (13.30a):

$$w(x,t) = \frac{\ell^4 p_0(t)}{24EI} \left[\left(\frac{x}{\ell}\right)^4 -2 \left(\frac{x}{\ell}\right)^3 + \left(\frac{x}{\ell}\right)\right]$$

$$- \frac{4\ell^4}{\pi^5 EI} \sum_{i=1,3,5,..}^{\infty} \frac{1}{i^5} \sin \frac{i\pi x}{\ell}\left[p_0(t) - \omega_i \int_0^t p_0(\tau) \sin\omega_i (t-\tau) d\tau\right] \tag{13.40}$$

where

$$\omega_i = \frac{i^2\pi^2}{\ell^2} \sqrt{\frac{EI}{\rho A}}$$

We now specialize (13.40) to the case where the load is suddenly applied at t=0 and thereafter maintained constant, i.e., $p_o(t)=p^*H(t)$, where H(t) is the Heaviside unit step function. In this case

$$w(x,t)= \frac{\ell^4 p^*}{24EI}\left[\left(\frac{x}{\ell}\right)^4 -2\left(\frac{x}{\ell}\right)^3 + \left(\frac{x}{\ell}\right)\right]$$

$$-\frac{4\ell^4 p^*}{\pi^5 EI}\sum_{i=1,3,5,..}^{\infty} \frac{1}{i^5}\sin\frac{i\pi x}{\ell}\cos\omega_i t,\ t>0.$$

The center deflection is

$$w\left(\frac{\ell}{2},t\right) = \frac{5\ell^4 p^*}{384EI} - \frac{4\ell^4 p^*}{\pi^5 EI}\sum_{i=1,3,5,..}^{\infty} \frac{1}{i^5}\sin\frac{i\pi}{2}\cos\omega_i t$$

The maximum center deflection occurs when $t=t^*=\frac{\pi}{\omega_1}=\frac{\ell^2}{\pi}\sqrt{\frac{\rho A}{EI}}$. At that instant $\cos\omega_i t=-1$ for i=1,3,5,..., and

$$w\left(\frac{\ell}{2},t^*\right)= \frac{5\ell^4 p^*}{384EI} + \frac{4\ell^4 p^*}{\pi^5 EI}\left(1-\frac{1}{3^5}+\frac{1}{5^5}-\frac{1}{7^5}+\ldots\right)$$

But

$$1-\frac{1}{3^5}+\frac{1}{5^5}-\frac{1}{7^5}+\ldots = \frac{5\pi^5}{1536} = \frac{5\pi^5}{(4)(384)}$$

so that

$$w\left(\frac{\ell}{2},t^*\right) = \frac{5\ell^4 p^*}{384EI}(2)\quad .$$

We conclude that the dynamic load factor for center deflection of the beam is two.

Before we conclude the present section, we want to note that the eigenfunctions and associated eigenvalues can be used to construct static solutions of beam problems with homogeneous boundary conditions. The static beam problem is characterized by (13.18), (13.19), and (13.20) with inertia terms deleted:

$$\frac{d^2}{dx^2}\left(EI\,\frac{d^2w}{dx^2}\right)= p(x)$$

and we assume that at x=0 and at x=ℓ, either w or Q=0 and either $\frac{dw}{dx}$ or M=0. The solution is taken in the form of an eigenfunction expansion

$$w = \sum_{i=1}^{\infty} a_i W_i(x)$$

where a_i, i=1,2,..., are constants to be determined and the eigenfunctions are subject to the same, homogeneous boundary conditions as the static beam problem to be solved. Upon substitution of the assumed solution into the differential equation and utilization of (13.24), we obtain

$$\sum_{i=1}^{\infty} a_i \rho A \omega_i^2 W_i = p$$

We now multiply both sides of this equation by W_j and subsequently integrate both sides over the length of the beam. Upon application of (13.26) and (13.29), we obtain

$$a_i = \frac{1}{\omega_i^2} \int_0^{\ell} p W_i \, dx$$

Th s we can utilize the results of a free vibration analysis to obtain a series solution for static beam problems with homogeneous boundary conditions.

13.3 FORCED MOTION OF TIMOSHENKO BEAMS

Just as we were able to obtain a resolution of the forced motion of Euler-Bernoulli beams in Section 13.2, it is possible to find a general solution of the following well-posed problem of Timoshenko beam theory.

Stress-equations of motion:

$$\left.\begin{aligned} -\rho A\ddot{w} + \frac{\partial Q}{\partial x} + p &= 0 \\ -\rho I\ddot{\psi} + \frac{\partial M}{\partial x} - Q + m &= 0 \end{aligned}\right\} \tag{13.41}$$

Stress-displacement relations:

$$M = EI\,\frac{\partial \psi}{\partial x}, \quad Q = \kappa^2 AG\left(\frac{\partial w}{\partial x} + \psi\right) \tag{13.42}$$

Boundary conditions:

$$\left.\begin{aligned} &\text{either } w(0,t) \text{ or } Q(0,t) = f_1(t) \\ &\text{either } \psi(0,t) \text{ or } M(0,t) = f_2(t) \\ &\text{either } w(\ell,t) \text{ or } Q(\ell,t) = g_1(t) \\ &\text{either } \psi(\ell,t) \text{ or } M(\ell,t) = g_2(t) \end{aligned}\right\} \tag{13.43}$$

Initial conditions:

$$\begin{aligned} w(x,0) &= w_o(x),\ \dot{w}(x,0) = \dot{w}_o(x) \\ \psi(x,0) &= \psi_o(x),\ \dot{\psi}(x,0) = \dot{\psi}_o(x) \end{aligned} \tag{13.44}$$

It can be shown (see Exercise 13.5) that the forced motion problem as characterized by (13.41) through (13.44) has a unique solution.

We shall construct the forced motion solution with the aid of eigenfunctions which satisfy homogeneous boundary conditions. For this reason we consider the free vibration problem characterized by (13.41) when we set p=m=0, and the homogeneous boundary conditions

$$\left.\begin{aligned} &\text{either } w(0,t) \text{ or } Q(0,t) = 0 \\ &\text{either } \psi(0,t) \text{ or } M(0,t) = 0 \\ &\text{either } w(\ell,t) \text{ or } Q(\ell,t) = 0 \\ &\text{either } \psi(\ell,t) \text{ or } M(\ell,t) = 0 \end{aligned}\right\} \tag{13.45}$$

Assume a product solution of the form

$$\left.\begin{aligned} w(x,t) &= W_i(x)\cos\omega_i t, \\ \psi(x,t) &= \Psi_i(x)\cos\omega_i t \end{aligned}\right\} \tag{13.46}$$

and substitute into the homogeneous form of (13.41) and into (13.45). The result is

$$\left.\begin{aligned} &\rho A\omega_i^2 W_i + \frac{dQ_i}{dx} = 0 \\ &\rho I\omega_i^2 \Psi_i + \frac{dM_i}{dx} - Q_i = 0 \end{aligned}\right\} \tag{13.47}$$

where

$$Q_i = \kappa^2 AG\left(\frac{dW_i}{dx} + \Psi_i\right),\quad M_i = EI\,\frac{d\Psi_i}{dx} \tag{13.48}$$

and

$$\left.\begin{aligned} &\text{either } W_i(0) \text{ or } Q_i(0) = 0 \\ &\text{either } \Psi_i(0) \text{ or } M_i(0) = 0 \\ &\text{either } W_i(\ell) \text{ or } Q_i(\ell) = 0 \\ &\text{either } \Psi_i(\ell) \text{ or } M_i(\ell) = 0 \end{aligned}\right\} \tag{13.49}$$

In general, there will be a denumerable infinity of natural frequencies ω_i, i=1,2,..., each frequency being associated with a particular normal mode characterized by the functions $W_i(x)$ and $\Psi_i(x)$. Equations (13.47) through (13.49) specify each normal mode only to within an undetermined, constant multiplier common to each member of the set of two mode shape functions. By introducing a mode normalization condition, unique expressions for the eigenfunctions are obtained. In the present case, the form of this condition is selected because of its convenience in the subsequent solution of the forced motion problem, and it is given by

$$\left(\int_0^{\ell} (\rho A W_i^2 + \rho I \Psi_i^2)\,dx\right)^{\frac{1}{2}} = 1. \tag{13.50}$$

We now form the expression

$$\int_0^{\ell} \left[\left(\rho A\omega_i^2 W_i + \frac{dQ_i}{dx}\right)W_j - \left(\rho A\omega_j^2 W_j + \frac{dQ_j}{dx}\right)W_i\right.$$

$$\left. + \left(\rho I\omega_i^2 \Psi_i + \frac{dM_i}{dx} - Q_i\right)\Psi_j - \left(\rho I\omega_j^2 \Psi_j + \frac{dM_j}{dx} - Q_j\right)\Psi_i\right] dx = 0$$

Upon rearranging,

$$(\omega_i^2 - \omega_j^2)\int_0^{\ell} (\rho A W_i W_j + \rho I \Psi_i \Psi_j)\,dx$$

$$= \int_0^{\ell} \left[\left(\frac{dQ_j}{dx} W_i - \frac{dQ_i}{dx} W_j\right) + \left(\frac{dM_j}{dx} \Psi_i - \frac{dM_i}{dx} \Psi_j\right)\right.$$

$$\left. + (Q_i \Psi_j - Q_j \Psi_i)\right] dx$$

If we integrate the right-hand side of this equation by parts, and then utilize (13.48), we readily obtain

$$(\omega_i^2 - \omega_j^2)\int_0^{\ell} (\rho A W_i W_j + \rho I \Psi_i \Psi_j)\,dx$$

$$= (Q_j W_i - Q_i W_j + M_j \Psi_i - M_i \Psi_j)\Big|_0^{\ell} \tag{13.51}$$

Thus, for homogeneous boundary conditions (13.49) the right-hand side of (13.51) vanishes and

$$\left.\begin{aligned} &\int_0^{\ell} (\rho A W_i W_j + \rho I \Psi_i \Psi_j)\,dx = 0 \\ &\text{provided } \omega_i^2 \neq \omega_j^2 \end{aligned}\right\} \tag{13.52}$$

Equation (13.52) is the orthogonality relation for the eigenfunctions arising from the Timoshenko beam equations.

We now assume that the solution of the forced motion problem characterized by (13.41) through (13.44) can be written in the form

$$w(x,t) = w_s(x,t) + \sum_{i=1}^{\infty} W_i(x)\, q_i(t) \tag{13.53a}$$

and in view of the linearity of the Timoshenko beam theory,

$$Q(x,t) = Q_s(x,t) + \sum_{i=1}^{\infty} Q_i(x)\, q_i(t) \tag{13.53b}$$

$$\psi(x,t) = \psi_s(x,t) + \sum_{i=1}^{\infty} \Psi_i(x)\, q_i(t) \tag{13.53c}$$

$$M(x,t) = M_s(x,t) + \sum_{i=1}^{\infty} M_i(x)\, q_i(t) \tag{13.53d}$$

where

$$Q_s = \kappa^2 AG\left(\frac{\partial w_s}{\partial x} + \psi_s\right), \quad M_s = EI\,\frac{\partial \psi_s}{\partial x}. \tag{13.54}$$

The quasi-static solution (subscript s) is characterized by the equations

$$\left.\begin{aligned} &\frac{\partial Q_s}{\partial x} + p = 0 \\ &\frac{\partial M_s}{\partial x} - Q_s + m = 0 \end{aligned}\right\} \tag{13.55}$$

and boundary conditions

$$\left.\begin{aligned} &\text{either } w_s(0,t) \text{ or } Q_s(0,t) = f_1(t) \\ &\text{either } \psi_s(0,t) \text{ or } M_s(0,t) = f_2(t) \\ &\text{either } w_s(\ell,t) \text{ or } Q_s(\ell,t) = g_1(t) \\ &\text{either } \psi_s(\ell,t) \text{ or } M_s(\ell,t) = g_2(t) \end{aligned}\right\} \tag{13.56}$$

If we now substitute (13.53) into (13.41), and utilize (13.55) and (13.47), we obtain

$$\sum_{i=1}^{\infty} (\ddot{q}_i + \omega_i^2 q_i)\rho A W_i = -\rho A \ddot{w}_s \tag{13.57a}$$

$$\sum_{i=1}^{\infty} (\ddot{q}_i + \omega_i^2 q_i)\rho I \Psi_i = -\rho I \ddot{\psi}_s \tag{13.57b}$$

Equations (13.57a) and (13.57b) are now multiplied by W_j and Ψ_j, respectively, and the products so obtained are added. The resulting sum is then integrated over the length of the beam, and the relations (13.50) and (13.52) are applied. The result is

$$\ddot{q}_i + \omega_i^2 q_i = \ddot{P}_i(t) \tag{13.58}$$

where

$$P_i(t) = -\int_0^{\ell} (\rho A w_s W_i + \rho I \psi_s \Psi_i)dx \tag{13.59}$$

With the aid of (13.47), we can transform (13.59) into

$$P_i(t) = \frac{1}{\omega_i^2} \int_0^{\ell} \left[w_s \frac{dQ_i}{dx} + \psi_s \left(\frac{dM_i}{dx} - Q_i \right) \right] dx$$

The right-hand side of this equation is integrated by parts, and upon application of (13.54), (13.55), and (13.48) we obtain the convenient expression

$$\omega_i^2 P_i = (w_s Q_i - W_i Q_s + \psi_s M_i - \Psi_i M_s) \Big|_0^{\ell} - \int_0^{\ell} (\Psi_i m + W_i p)dx \tag{13.60}$$

The complete solution of (13.58) is given by (13.36).
Initial conditions on q_i are now determined. With reference to (13.44), (13.53a), (13.53c), we have

$$w_o(x) = w_s(x,0) + \sum_{i=1}^{\infty} W_i(x)\, q_i(0) \tag{13.61a}$$

$$\psi_o(x) = \psi_s(x,0) + \sum_{i=1}^{\infty} \Psi_i(x)\, q_i(0) \tag{13.61b}$$

Equations (13.61a) and (13.61b) are now multiplied by $\rho A W_j$ and $\rho I \Psi_j$, respectively, and the products so obtained are added. The sum is then integrated over the length of the beam, and equations (13.50), (13.52) and (13.59) are applied. The result is

$$q_i(0) = P_i(0) + \int_0^{\ell} (\rho A w_o W_i + \rho I \psi_o \Psi_i)dx \tag{13.62a}$$

An entirely analogous analysis leads to

$$\dot{q}_i(0) = \dot{P}_i(0) + \int_0^{\ell} (\rho A \dot{w}_o W_i + \rho I \dot{\psi}_o \Psi_i)dx \tag{13.62b}$$

This completes the resolution of the forced motion problem for Timoshenko beams.

We shall now illustrate the application of the above general theory by considering a homogeneous Timoshenko beam of uniform cross section, i.e., the quantities A, ρ, and I are taken to be constant over the length of the beam.

(a) Free Vibrations of a Simply Supported Timoshenko Beam.

In this case we seek a solution of (13.47) subject to the boundary conditions

$$W_i(0) = W_i(\ell) = M_i(0) = M_i(\ell) = 0 \tag{13.63}$$

Such a solution is provided by

$$\left.\begin{aligned} W_i(x) &= \ell B_i \sin\frac{i\pi x}{\ell} \\ \Psi_i(x) &= C_i \cos\frac{i\pi x}{\ell} \end{aligned}\right\} \tag{13.64}$$

Upon substitution of (13.64) into (13.47), we obtain

$$B_i(\Omega_i^2 - k^2 i^2 \pi^2) + C_i(-k^2 i\pi) = 0 \tag{13.65a}$$

$$B_i(-k^2 i\pi) + C_i(\alpha^2\Omega_i^2 - k^2 - \alpha^2 i^2 \pi^2) = 0 \tag{13.65b}$$

where

$$\alpha^2 = \frac{I}{A\ell^2}, \quad k^2 = \kappa^2\,\frac{G}{E}, \quad \Omega_i^2 = \frac{\ell^2 \rho\omega_i^2}{E} .$$

Upon substitution of (13.64) into the normalization condition (13.50), we obtain

$$A\ell^2 B_i^2 + IC_i^2 = \frac{2}{\rho\ell} \tag{13.66}$$

Equations (13.65) and (13.66) are now solved for the mode coefficients B_i and C_i, resulting in

$$\ell B_i = \sqrt{\frac{2}{A\rho\ell}}\;\frac{k^2 i\pi}{\sqrt{k^4 i^2\pi^2 + \alpha^2(\Omega_i^2 - k^2 i^2\pi^2)^2}} \tag{13.67a}$$

$$\ell C_i = \sqrt{\frac{2}{A\rho\ell}}\;\frac{(\Omega_i^2 - k^2 i^2\pi^2)}{\sqrt{k^4 i^2\pi^2 + \alpha^2(\Omega_i^2 - k^2 i^2\pi^2)^2}} \tag{13.67b}$$

In addition, to obtain a non-trivial solution for the constants B_i and C_i, it is necessary and sufficient that the coefficient determinant of (13.65) vanish, i.e.,

$$\begin{vmatrix} (\Omega_i^2-k^2i^2\pi^2) & (-k^2i\pi) \\ (-k^2i\pi) & (\alpha^2\Omega_i^2-k^2-\alpha^2i^2\pi^2) \end{vmatrix} = 0$$

This is an equation for Ω_i, which can be solved to yield

$$\left[\Omega_i^{(1)}\right]^2 = \frac{i^2\pi^2\alpha^2(1+k^2)+k^2-\sqrt{\alpha^4i^4\pi^4(1-k^2)^2+2i^2\pi^2\alpha^2k^2(1+k^2)+k^4}}{2\alpha^2} \tag{13.68a}$$

$$\left[\Omega_i^{(2)}\right]^2 = \frac{i^2\pi^2\alpha^2(1+k^2)+k^2+\sqrt{\alpha^4i^4\pi^4(1-k^2)^2+2i^2\pi^2\alpha^2k^2(1+k^2)+k^4}}{2\alpha^2} \tag{13.68b}$$

With reference to (13.68) we can see that there are two sets of frequencies for each i=1,2,... . Consequently, there will be two sets of eigenfunctions corresponding to these frequencies, and they will be denoted by

$$\left.\begin{aligned} W_i^{(r)} &= \ell B_i^{(r)}\sin\frac{i\pi x}{\ell} \\ \Psi_i^{(r)} &= C_i^{(r)}\cos\frac{i\pi x}{\ell} \end{aligned}\right\} \quad \begin{pmatrix}\text{corresponding to} \\ \omega_i^{(r)},\ r=1,2. \\ i=1,2,3,\ldots\ . \end{pmatrix} \tag{13.69}$$

Dimensionless graphs of $\frac{\Omega_i^{(r)}}{i}$ vs. i are shown in Figs. 13.3a and 13.3b for the values $\alpha=0.0288$, $k^2=0.3308$ and $\alpha=0.0866$, $k^2=0.3308$, respectively. In the notation of the present section, the frequencies of the corresponding Euler-Bernoulli beam are given by $\Omega_i=\alpha\pi^2i^2$, and a graph of $\frac{\Omega_i}{i}$ is also shown in Figs. 13.3. It can be concluded that for sufficiently low mode numbers, the first set of frequencies $\Omega_i^{(1)}$ is approximated by the Euler-Bernoulli theory, but the accuracy of the approximation deteriorates as i increases. Consequently, for the specific cases considered here, the effects of shear deformation and rotatory inertia must be considered if the higher modes of motion play an important role.

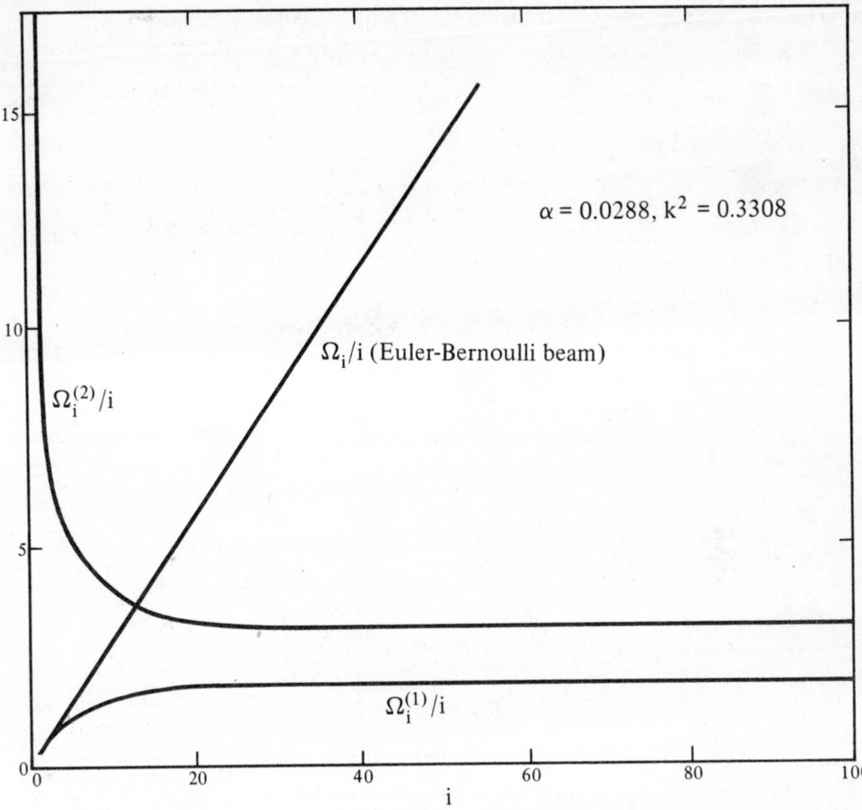

Figure 13.3 (a)

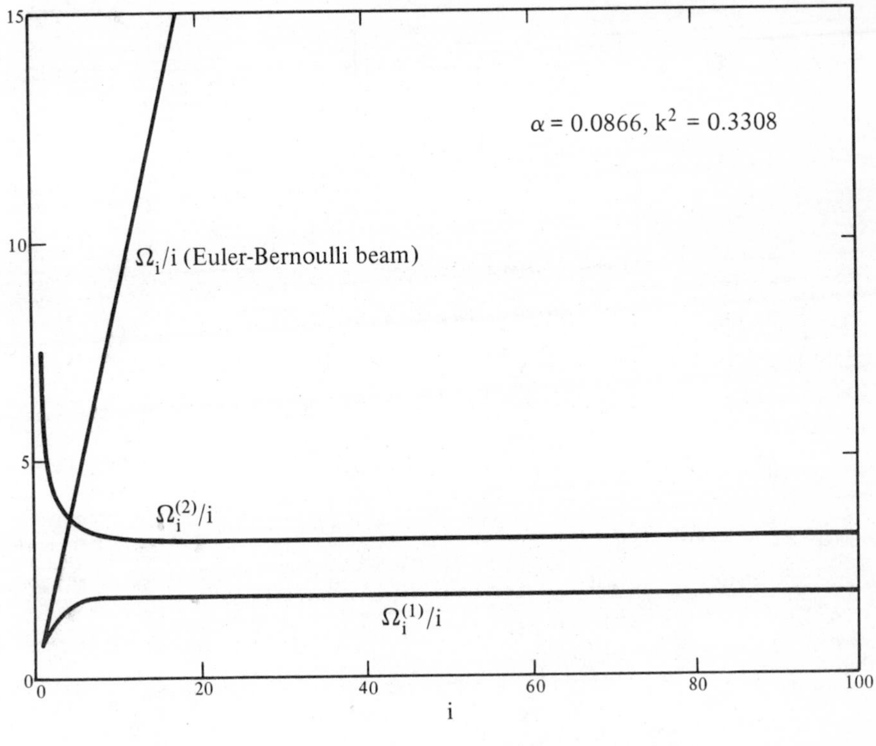

Figure 13.3 (b)

(b) The Simply Supported Timoshenko Beam Under Central Impact.

A simply supported Timoshenko beam of length ℓ is subjected to a suddenly applied concentrated load of magnitude F_o at $x=\frac{\ell}{2}$ and at t=0. The load remains in position for t>0. The beam is at rest and in its reference position at t=0 (see Fig. 13.4a). In the present case the quasi-static solution must satisfy (13.55) and (13.54) as well as the boundary conditions

$$w_s(0,t) = w_s(\ell,t) = M_s(0,t) = M_s(\ell,t) = 0 \tag{13.70}$$

By the usual elementary means, we obtain for $0<x<\frac{\ell}{2}$, t>0

$$w_s = \frac{F_o \ell^3}{2EI}\left(\frac{\alpha^2}{k^2}\frac{x}{\ell} + \frac{1}{8}\frac{x}{\ell} - \frac{1}{6}\frac{x^3}{\ell^3}\right) \tag{13.71a}$$

$$\psi_s = \frac{F_o \ell^2}{4EI}\left(\frac{x^2}{\ell^2} - \frac{1}{4}\right) \tag{13.71b}$$

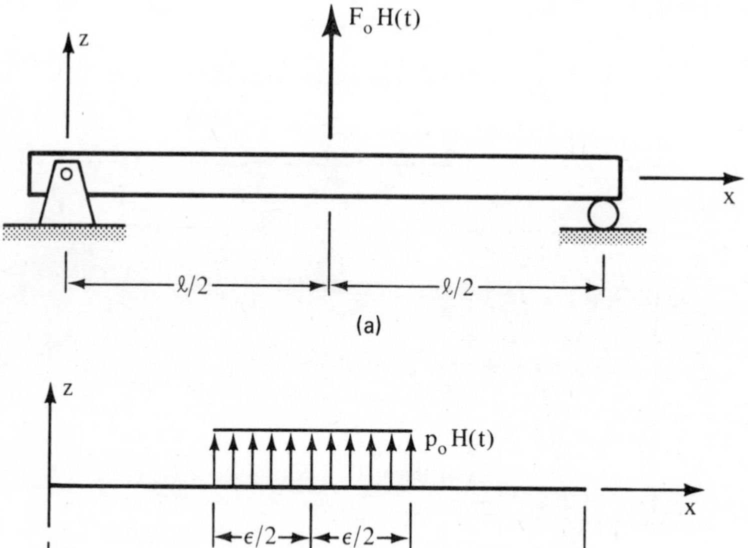

(a)

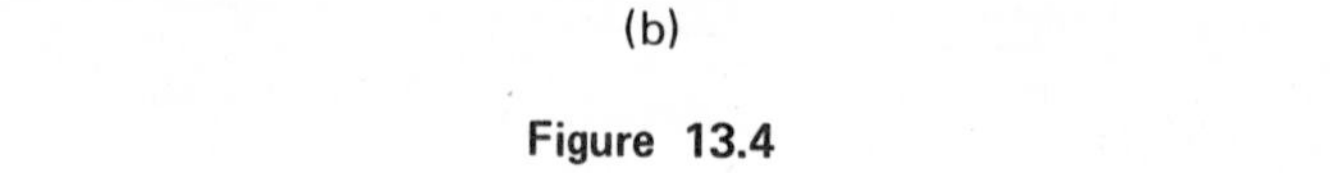

(b)

Figure 13.4

Similar expressions can be obtained for $\frac{\ell}{2}<x<\ell$ (see Exercise 13.6). We now proceed to adapt (13.60) to the present case. In view of (13.70) and (13.63) we arrive at

$$\omega_i^2 P_i(t) = -\int_0^\ell W_i(x)\ p(x,t)dx$$

where we define (see Fig. 13.4b) for t>0

$$p(x,t) = p_o \text{ for } \frac{\ell}{2} - \frac{\varepsilon}{2} < x < \frac{\ell}{2} + \frac{\varepsilon}{2}$$

$$= 0 \text{ for } \ell > \frac{\ell}{2} + \frac{\varepsilon}{2} < x < \frac{\ell}{2} - \frac{\varepsilon}{2} > 0.$$

such that $\varepsilon p_o = F_o$. Then

$$\omega_i^2 P_i(t) = -p_o \int_{\frac{\ell}{2}-\frac{\varepsilon}{2}}^{\frac{\ell}{2}+\frac{\varepsilon}{2}} \ell B_i \sin \frac{i\pi x}{\ell}\, dx = -\frac{2\ell^2 p_o B_i}{\pi i} \sin \frac{i\pi}{2} \cdot \sin \frac{i\pi\varepsilon}{2\ell}$$

or

$$\frac{\omega_i^2 P_i}{(p_o\varepsilon)\ell B_i} = -\sin \frac{i\pi}{2} \frac{\sin \frac{i\pi\varepsilon}{2\ell}}{\left(\frac{i\pi\varepsilon}{2\ell}\right)}$$

We now take the limit of this expression as $\varepsilon \to 0$ and $p_o\varepsilon = F_o$. Observing that

$$-\sin \frac{i\pi}{2} = (-1)^{\frac{i+1}{2}}, \quad i=1,3,5,\ldots$$

$$= 0 \quad , \quad i=2,4,6,\ldots$$

We obtain for t>0

$$P_i(t) = \frac{\ell B_i F_o}{\omega_i^2} (-1)^{\frac{i+1}{2}} \quad , \quad i=1,3,5,\ldots$$

$$= 0 \quad , \quad i=2,4,6,\ldots \tag{13.72}$$

In the present case we have

$$w_o(x) = \psi_o(x) = \dot{w}_o(x) = \dot{\psi}_o(x) = 0$$

and according to (13.62) this results in

$$q_i(0) = P_i(0), \quad \dot{q}_i(0) = \dot{P}_i(0) \tag{13.73}$$

Further substitution of (13.72) and (13.73) into (13.36b) results in

$$\left.\begin{aligned} q_i(t) &= \frac{(-1)^{\frac{i+1}{2}} \ell B_i F_o}{\omega_i^2} \cos\omega_i t, \quad t>0, \quad i=1,3,5,\ldots \\ &= 0, \quad i=2,4,6,\ldots \end{aligned}\right\} \tag{13.74}$$

Observing that there are two sets of frequencies $\omega_i^{(r)}$, r=1,2, we now substitute (13.74) and (13.69) and (13.71) into (13.53a) and (13.53c), resulting in the solution valid for $0<x<\frac{\ell}{2}$ and t>0:

$$w = \frac{F_o \ell^3}{2EI}\left(\frac{\alpha^2}{k^2}\frac{x}{\ell} + \frac{1}{8}\frac{x}{\ell} - \frac{1}{6}\frac{x^3}{\ell^3}\right)$$

$$+ \ell^2 F_o \sum_{i=1,3,5,\dots}^{\infty} (-1)^{\frac{i+1}{2}} \left(\frac{B_i^{(1)}}{\omega_i^{(1)}}\right)^2 \sin\frac{i\pi x}{\ell} \cos\omega_i^{(1)} t$$

$$+ \ell^2 F_o \sum_{i=1,3,5,\dots}^{\infty} (-1)^{\frac{i+1}{2}} \left(\frac{B_i^{(2)}}{\omega_i^{(2)}}\right)^2 \sin\frac{i\pi x}{\ell} \cos\omega_i^{(2)} t \qquad (13.75a)$$

$$\psi = \frac{F_o \ell^2}{4EI}\left(\frac{x^2}{\ell^2} - \frac{1}{4}\right) + \ell F_o \sum_{i=1,3,5,\dots}^{\infty} \frac{(-1)^{\frac{i+1}{2}}}{\left[\omega_i^{(1)}\right]^2} B_i^{(1)} C_i^{(1)} \cos\frac{i\pi x}{\ell} \cos\omega_i^{(1)} t$$

$$+ \ell F_o \sum_{i=1,3,5,\dots}^{\infty} \frac{(-1)^{\frac{i+1}{2}}}{\left[\omega_i^{(2)}\right]^2} B_i^{(2)} C_i^{(2)} \cos\frac{i\pi x}{\ell} \cos\omega_i^{(2)} t \qquad (13.75b)$$

The dimensionless graphs in Figs. 13.5 and 13.6 were obtained with the aid of equations (13.75) in conjunction with (13.67), (13.68), and (13.15a). All results correspond to the beam center span point $x=\frac{\ell}{2}$ and $k^2=0.3308$. The present problem can also be solved within the framework of Euler-Bernoulli beam theory (see Exercise 13.7), and the results of that analysis have been superimposed in Figs. 13.5 and 13.6 to provide a comparison between the two beam theories. An inspection of the results reveals that in the case of the beam with the lower slenderness ratio $\alpha=0.0288$, the forced motion results of the Euler-Bernoulli beam theory closely approximate the Timoshenko beam results. However, there are pronounced differences in the case of the beam with the higher slenderness ratio $\alpha=0.0866$ as shown in Fig. 13.6a.

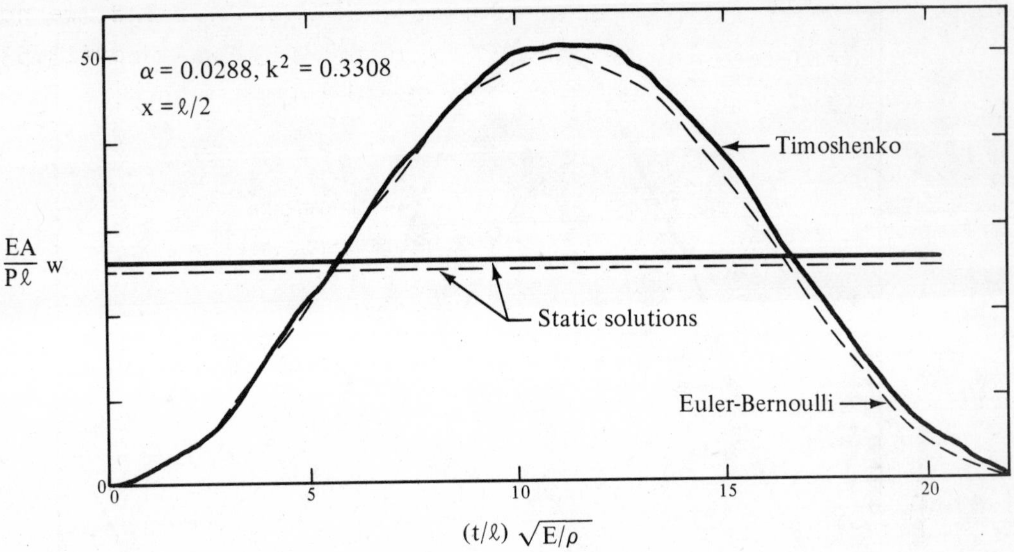

Figure 13.5 (a)

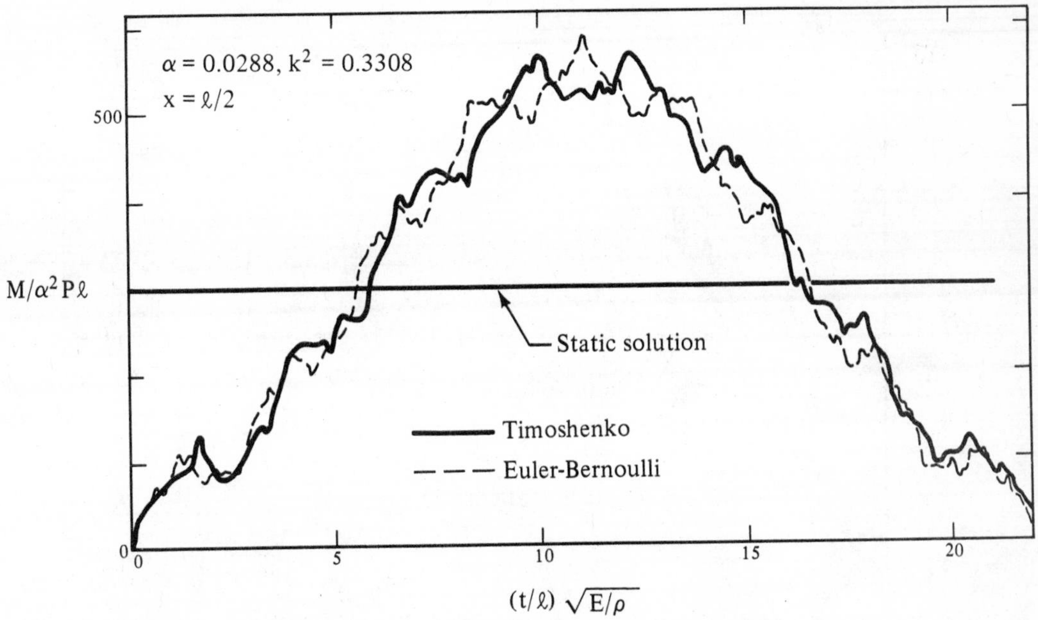

Figure 13.5 (b)

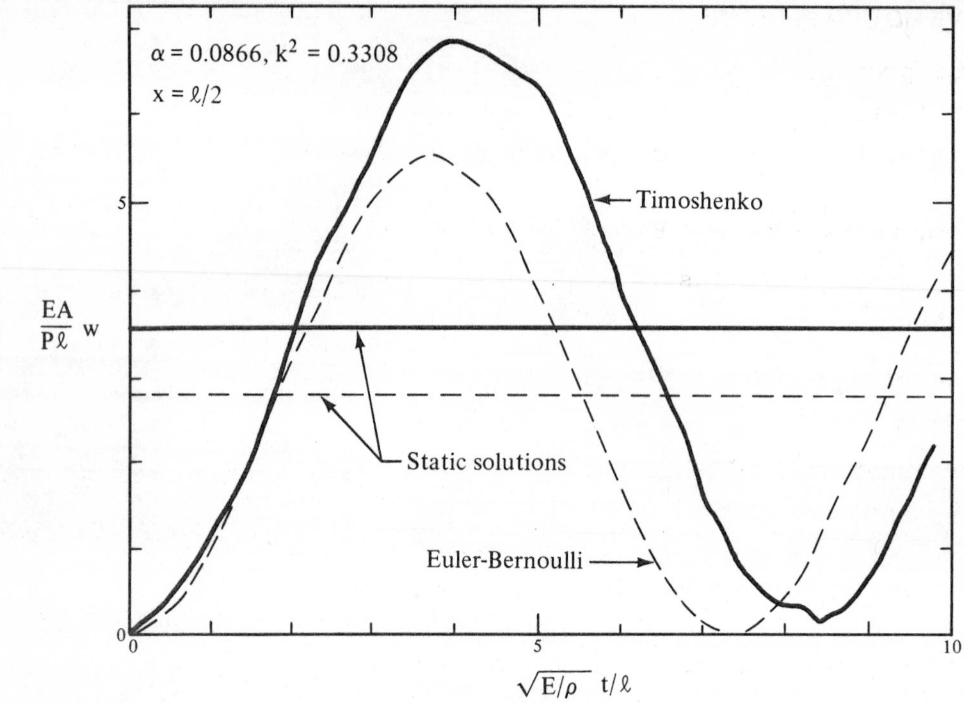

Figure 13.6 (a)

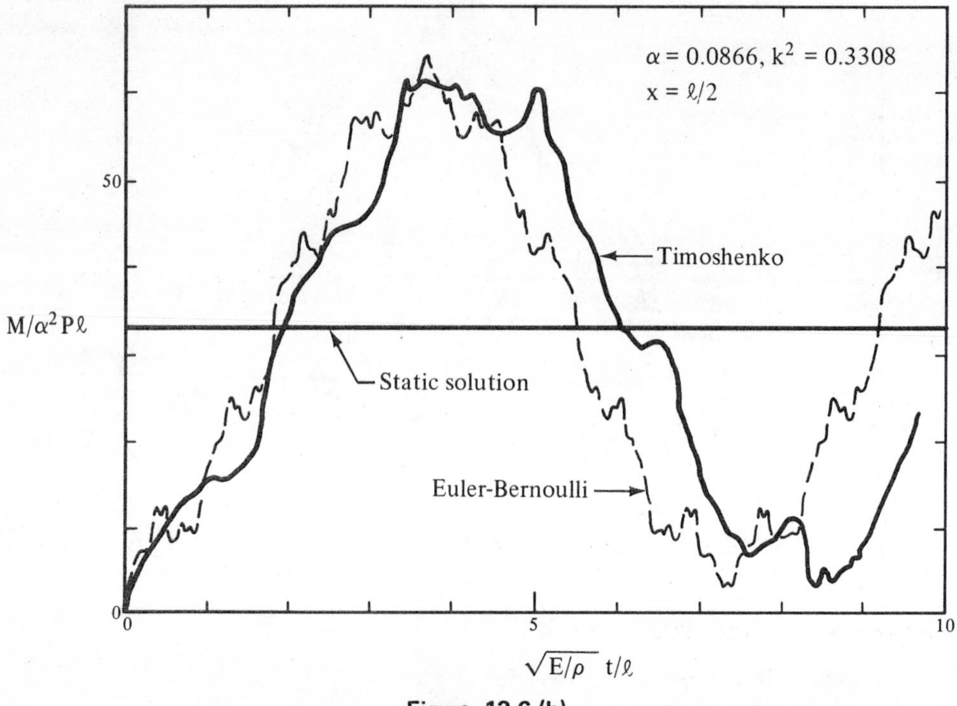

Figure 13.6 (b)

13.4 WAVE MOTION IN BEAMS

The present section is concerned with wave motion in infinite and semi-infinite, homogeneous beams of constant cross section, i.e., the quanitities E, I, and ρ are assumed to be constant.

(a) Time-harmonic Waves in an Euler-Bernoulli Beam of Unbounded Length.

We consider the possibility of a propagating wave (see Appendix B) of the type

$$w(x,t) = B\exp i(\omega t - kx) \tag{13.76}$$

where k, ω, and B are the wave number, frequency, and amplitude of the wave, respectively. The flexural motion of the beam is characterized by the equation

$$EI \frac{\partial^4 w}{\partial x^4} + \rho A \frac{\partial^2 w}{\partial t^2} = 0 \tag{13.77}$$

Upon substitution of (13.76) into (13.77), we obtain

$$EIk^4 = \rho A\omega^2 \tag{13.78}$$

In order to achieve a nondimensional representation, it will be convenient to define $\alpha_1^2 = \frac{E}{\rho}$, $k_o^2 = \frac{A}{I}$. Then with the aid of (13.78) we obtain the dispersion relation

$$\frac{k}{k_o} = \sqrt{\frac{\omega}{\alpha_1 k_o}} \tag{13.79}$$

the phase velocity

$$\frac{v}{\alpha_1} = \frac{1}{\alpha_1}\frac{\omega}{k} = \sqrt{\frac{\omega}{\alpha_1 k_o}} \tag{13.80}$$

and the group velocity

$$\frac{v_G}{\alpha_1} = \frac{1}{\alpha_1}\frac{d\omega}{dk} = 2\sqrt{\frac{\omega}{\alpha_1 k_o}} \tag{13.81}$$

A graph of wave number vs. frequency is shown in Fig. 13.7, while graphs of phase and group velocity vs. frequency are shown in Fig. 13.8. We note that the phase velocity is dependent upon frequency (or wave number) and therefore, within the framework of the terminology developed in Appendix B, we can say that the Euler-Bernoulli beam is a dispersive medium with respect to flexural wave propagation.

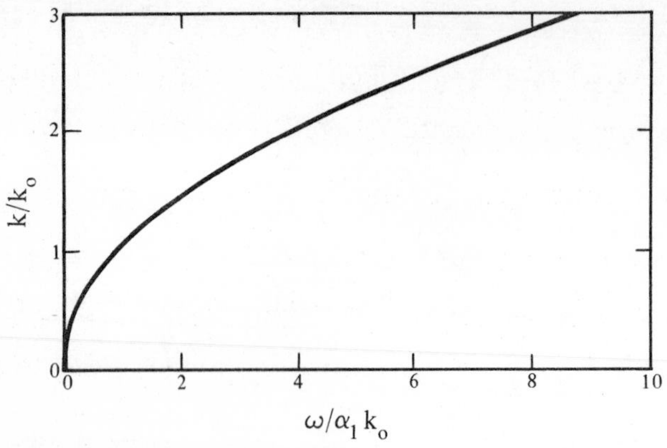

Figure 13.7

Figure 13.8

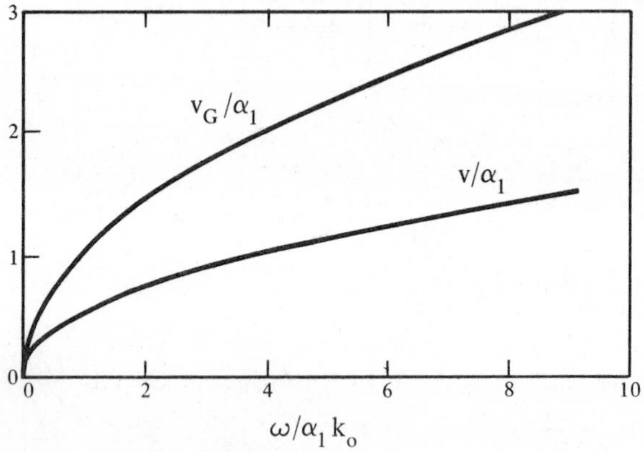

The average velocity at which energy is transported by a harmonic wave is defined to be the time average of the energy flux $\langle S\rangle$ divided by the time average of the energy density $\langle E^*\rangle$, i.e.,

$$\text{velocity of energy transport} = \frac{\langle S\rangle}{\langle E^*\rangle} \tag{13.82}$$

If (13.11) is adapted for the case of an Euler-Bernoulli beam where $\psi=-\frac{\partial w}{\partial x}$, then

$$S = M\frac{\partial \dot{w}}{\partial x} - Q\dot{w} \tag{13.83}$$

The total energy in a beam segment extending from $x=x_1$ to $x=x_2$ is

$$\int_{x_1}^{x_2} E^* dx = T + U$$

where

$$T = \frac{1}{2}\int_{x_1}^{x_2} \rho A\dot{w}^2 dx \quad \text{and} \quad U = \frac{1}{2}\int_{x_1}^{x_2} EI\left(\frac{\partial^2 w}{\partial x^2}\right)^2 dx$$

and therefore

$$E^* = \frac{1}{2}\rho A\dot{w}^2 + \frac{1}{2} EI\left(\frac{\partial^2 w}{\partial x^2}\right)^2 \tag{13.84}$$

In view of (10.176) and the present displacement assumptions, the strain energy density is given by

$$W^* = \frac{1}{2}\tau_{xx}e_{xx} = -\frac{1}{2}\tau_{xx}z\frac{\partial^2 w}{\partial x^2} = \frac{M}{2EI}z\tau_{xx}$$

Consequently

$$U = \int_V W^* dV = \int_{x_1}^{x_2}\int_A \frac{M}{2EI}z\tau_{xx}dAdx = \int_{x_1}^{x_2}\frac{M^2}{2EI}dx = \int_{x_1}^{x_2}\frac{1}{2}EI\left(\frac{\partial^2 w}{\partial x^2}\right)^2 dx$$

and the above result follows. In the present case it will be convenient to work with the real part of (13.76), i.e.,

$$w = B\cos(\omega t - kx) \tag{13.85}$$

Upon substitution of (13.85) into (13.83) and (13.84) we obtain

$$\langle S\rangle = \omega k^3 EIB^2, \quad \langle E^*\rangle = \frac{1}{2}\rho A\omega^2 B^2$$

and further substitution into (13.82) yields the result

$$\text{velocity of energy transport} = \frac{2EI}{A\rho}\frac{k^3}{\omega} = v_G$$

i.e., in the case of an Euler-Bernoulli beam, the group velocity (13.81) is equal to the velocity of energy transport.

(b) Time-harmonic Waves in a Semi-infinite, Clamped Euler-Bernoulli Beam.

A time-harmonic incident wave is reflected from the clamped end x=0 of a semi-infinite beam. We wish to find the reflected wave (or waves) in the beam. The equation of motion of the beam is given by (13.77). For time-harmonic motion we assume $w(x,t)=W(x)\cdot\exp i\omega t$, and substitute this expression into (13.77). We thus arrive at the ordinary differential equation

$$W^{IV} - k^4 W = 0 \tag{13.86}$$

where the wave number $k= \sqrt{\omega}\sqrt[4]{\frac{\rho A}{EI}}$.
Assuming a solution $W(x)=\exp i\lambda x$ and substituting into (13.86), we readily obtain $\lambda^4-k^4=0$, or $\lambda_1=k$, $\lambda_2=-k$, $\lambda_3=ik$, $\lambda_4=-ik$. Thus there are four possible solution types

$$\left.\begin{array}{l} w = \exp i\,(\omega t+kx) \\ w = \exp i\,(\omega t-kx) \end{array}\right\} \text{traveling waves}$$

$$\left.\begin{array}{l} w = \exp i\omega t \cdot \exp(kx) \\ w = \exp i\omega t \cdot \exp(-kx) \end{array}\right\} \text{standing waves}$$

With reference to Fig. 13.9, we now assume an incident wave characterized by

$$w_1= C_1 \exp i(\omega t+kx)$$

and in view of the above discussion, we must tentatively assume that the reflected waves are characterized by

$$w_2= C_2\exp i(\omega t-kx)$$

$$w_3= C_3\exp i\omega t \cdot \exp(-kx)$$

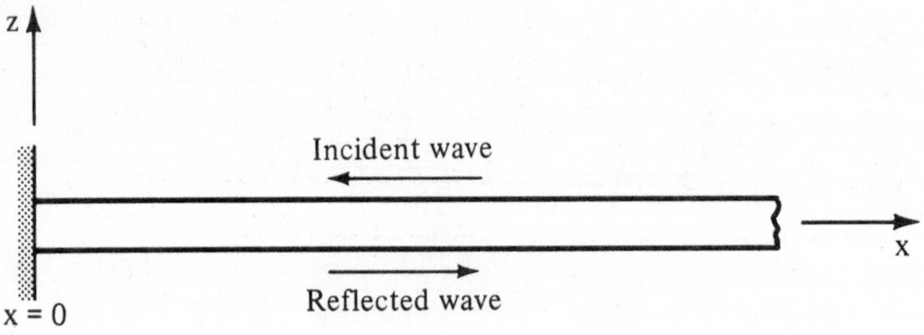

Figure 13.9

The total deflection of the beam at any instant of time is obtained by the superposition of w_1, w_2, and w_3, so that

$$w = C_1\exp i(\omega t+kx) + C_2\exp i(\omega t-kx) + C_3\exp i\omega t\cdot\exp(-kx) \quad (13.87)$$

The beam is rigidly clamped at x=0, so that

$$w(0,t) = \left(\frac{\partial w}{\partial x}\right)_{(0,t)} = 0 \quad (13.88)$$

Upon substitution of (13.87) into (13.88) we obtain two equations in the two unknowns C_2 and C_3. These equations are readily solved to yield

$$C_2 = C_1 \exp(-i\,\tfrac{\pi}{2}),\quad C_3 = -\sqrt{2}\;C_1 \exp(-i\,\tfrac{\pi}{4}) \quad .$$

and therefore the solution is

$$w_2 = C_1 \exp\left[i(\omega t - kx - \tfrac{\pi}{2})\right]$$

$$w_3 = -\sqrt{2}\;C_1 \exp(-kx) \cdot \exp\left[i(\omega t - \tfrac{\pi}{4})\right]$$

We thus conclude that there are two reflected waves, one of which is a traveling wave, while the other one is a standing, exponentially decaying wave.

(c) Forced, Time-harmonic Motion of an Euler-Bernoulli Beam.

We now consider the steady-state motion of an Euler-Bernoulli beam under the influence of a concentrated, time varying load of magnitude $F_0 \exp i\omega t$ applied at x=0, as shown in Fig. 13.10. The equation which characterizes the motion of the beam is given by (13.77).

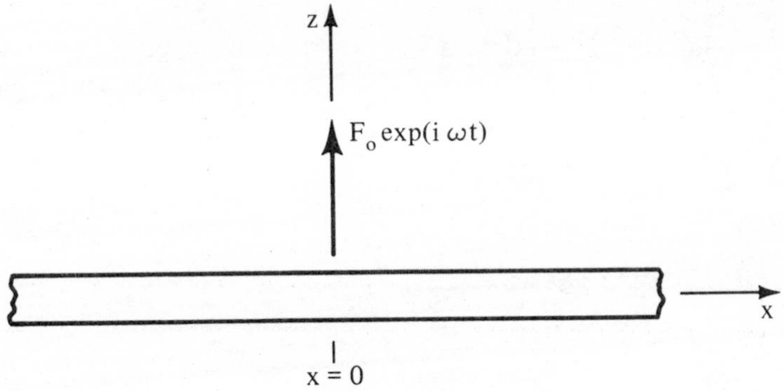

Figure 13.10

Since we are interested in the steady-state solution only, we assume

$$w(x,t) = W(x)\exp i\omega t \tag{13.89}$$

and upon substitution of (13.89) into (13.77), we obtain

$$W^{IV} - k^4 W = 0, \text{ where } k^4 = \omega^2 \frac{\rho A}{EI} \; .$$

If $W(x) = e^{\lambda x}$, then $\lambda_1 = k$, $\lambda_2 = -k$, $\lambda_3 = ik$, $\lambda_4 = -ik$ so that we can write

$$W(x) = C_1\exp(kx) + C_2\exp(-kx) + C_3\exp(ikx) + C_4\exp(-ikx) \quad .$$

The solution we are seeking must satisfy a number of conditions: (a) The deflection w must remain bounded as $x \to \pm\infty$. (b) The energy input occurs at x=0, and the average energy flux must be directed away from the point x=0. We have shown (Example (a)) that the group velocity is equal to the velocity of energy transport. In addition as shown in Fig. 13.8, the sign of the group velocity is the same as the sign of the phase velocity over the entire frequency range, and therefore the energy flux vector and phase velocity vector point in the same direction. We are thus forced to conclude that only outgoing traveling waves are physically admissible, and that incoming waves must be discarded. Consequently, we take

$$\text{for } x>0\text{: } W(x) = C_2\exp(-kx) + C_4\exp(-ikx) \tag{13.90a}$$

$$\text{for } x<0\text{: } W(x) = C_1\exp(kx) + C_3\exp(ikx) \tag{13.90b}$$

The four constants C_1, C_2, C_3, and C_4 in (13.90) are determined by invoking the following conditions: The deflection, slope, and moment must be continuous at x=0; i.e.,

$$[w(0^-,t)-w(0^+,t)] = 0 \tag{13.91a}$$

$$\left[\left(\frac{\partial w}{\partial x}\right)_{(0^-,t)} - \left(\frac{\partial w}{\partial x}\right)_{(0^+,t)}\right] = 0 \tag{13.91b}$$

$$[M(0^-,t)-M(0^+,t)] = 0 \tag{13.91c}$$

In addition, the shear force in the beam must have the appropriate discontinuity at x=0, i.e.,

$$[Q(0^-,t)-Q(0^+,t)] = F_o\exp i\omega t \tag{13.91d}$$

Upon substitution of (13.89) into (13.91) and utilization of (13.90) and (13.19), we obtain

$$(C_2+C_4) - (C_1+C_3) = 0$$

$$(-C_2-iC_4) - (C_1+iC_3) = 0$$

$$(C_2-C_4) - (C_1-C_3) = 0$$

$$(C_1-iC_3) - (-C_2+iC_4) = -\frac{F_o}{EIk^3}$$

Solving,

$$C_1 = C_2 = -\frac{F_o}{4EIk^3} \tag{13.92a}$$

$$C_3 = C_4 = iC_1 = -\frac{iF_o}{4EIk^3} \tag{13.92b}$$

If we now combine (13.92), (13.90), and (13.89), we readily obtain the steady-state part of the forced motion solution

for x>0:

$$w(x,t) = -\frac{F_o}{4EIk^3}\left[\exp(-kx)\exp(i\omega t) + i\exp i(\omega t - kx)\right] \tag{13.93a}$$

for x<0:

$$w(x,t) = -\frac{F_o}{4EIk^3}\left[\exp(kx)\exp(i\omega t) + i\exp i(\omega t + kx)\right] \tag{13.93b}$$

An inspection of (13.93) reveals that each half of the beam will have a traveling wave of constant amplitude, as well as an exponentially attenuated, standing wave.

(d) The Impulsively Loaded Euler-Bernoulli Beam.

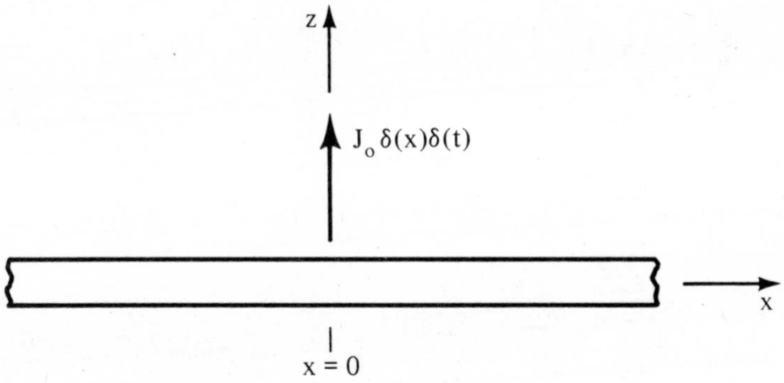

Figure 13.11

An Euler-Bernoulli beam of unbounded extent is subjected to a concentrated transverse impulse of magnitude J_o at x=0 and t=0 (see Fig. 13.11). The equation of motion is (see (13.17a))

$$\rho A \frac{\partial^2 w}{\partial t^2} + EI\frac{\partial^4 w}{\partial x^4} = J_o\,\delta(x)\,\delta(t) \tag{13.94}$$

where the symbol δ denotes the Dirac delta (or impulse) function. We shall assume that

$$\lim_{x\to\pm\infty} w(x,t) = \lim_{x\to\pm\infty}\left(\frac{\partial w}{\partial x}\right)_{(x,t)} = \lim_{x\to\pm\infty} M(x,t) = \lim_{x\to\pm\infty} Q(x,t) = 0 \tag{13.95}$$

and that

$$w(x,0) = \dot{w}(x,0) = 0 \quad . \tag{13.96}$$

We shall employ both Fourier and Laplace transforms to solve our problem. The Laplace transform will be denoted by a bar over the transformed variable (see Chapter 1), while the Fourier transform is denoted by a star. Thus

$$w^*(k,t) = \frac{1}{\sqrt{2\pi}} \int_{-\infty}^{\infty} w(x,t) \exp ikx dx = F^*\{w(x,t)\}$$

$$w(x,t) = \frac{1}{\sqrt{2\pi}} \int_{-\infty}^{\infty} w^*(k,t) \exp(-ixk) dk$$

$F^* \{\delta(x)\} = \frac{1}{\sqrt{2\pi}}$, and in the present case we have

$$\left(\frac{\partial^4 w}{\partial x^4}\right)^* = k^4 w^* = F^* \left\{\frac{\partial^4 w}{\partial x^4}\right\}$$

Taking Laplace transforms of (13.94) and observing (13.96), we obtain

$$\rho A p^2 \bar{w} + EI \frac{d^4 \bar{w}}{dx^4} = J_o \delta(x) \tag{13.97}$$

Taking Fourier transforms of (13.97) and observing (13.95), we obtain

$$\bar{w}^* = \frac{J_o}{\rho A \sqrt{2\pi}} \frac{1}{\omega} \frac{\omega}{p^2 + \omega^2} \tag{13.98}$$

where we have used the dispersion relation $\omega^2 = \lambda^4 k^4$, where $\lambda^4 = \frac{EI}{\rho A}$. Inversion of (13.98) with respect to p results in (see Table 1.1)

$$w^* = \frac{J_o}{\sqrt{2\pi}\, \rho A \lambda^2 k^2} \sin \lambda^2 k^2 t$$

and application of the inversion theorem for Fourier transforms results in

$$w(x,t) = \frac{J_o}{2\pi \rho A} \int_{-\infty}^{\infty} \frac{\sin \lambda^2 k^2 t}{\lambda^2 k^2} \exp(-ixk) dk \tag{13.99}$$

and we note that

$$\dot{w}(x,t) = \frac{J_o}{2\pi \rho A} \int_{-\infty}^{\infty} \cos \lambda^2 k^2 t \exp(-ixk) dk \tag{13.100}$$

To evaluate the improper integral in (13.100), we can use

$$\int_{-\infty}^{\infty} \cos ak^2 \exp(-ixk)\, dk = \frac{\sqrt{2\pi}}{2\sqrt{a}} \left(\cos \frac{x^2}{4a} + \sin \frac{x^2}{4a} \right)$$

(See: I.N. Sneddon, Fourier Transforms, McGraw-Hill Book Co., Inc., New York 1951, p. 112). Consequently, we have

$$\dot{w}(x,t) = \frac{J_o}{\sqrt{2\pi}\ \rho A \sqrt{4\lambda^2 t}} \left(\cos \frac{x^2}{4\lambda^2 t} + \sin \frac{x^2}{4\lambda^2 t} \right) \tag{13.101}$$

With regard to the improper integral in (13.99), it appears that it cannot be expressed in terms of elementary functions. For this reason, we shall approximate it for sufficiently large x and t by applying Kelvin's method of stationary phase as developed in Appendix B. Toward this end we rewrite (13.99) in the form

$$w(x,t) = \frac{J_o}{\pi \rho A} \int_0^{\infty} \frac{\sin \lambda^2 k^2 t}{\lambda^2 k^2} \cos xk \cdot dk$$

But

$$2 \cos kx \sin \lambda^2 k^2 t = \sin(kvt+kx) + \sin(kvt-kx)$$

where the phase velocity $v=\lambda^2 k$. Consequently, $w(x,t)=w_1+w_2$, where

$$w_1 = \frac{J_o}{2\pi\rho A} \int_0^{\infty} \frac{1}{\lambda^2 k^2} \sin k(vt+x)dk \tag{13.102a}$$

$$w_2 = \frac{J_o}{2\pi\rho A} \int_0^{\infty} \frac{1}{\lambda^2 k^2} \sin k(vt-x)dk \tag{13.102b}$$

With reference to Appendix B, it is now clear that w_1 represents a wave packet which progresses in the negative x-direction, while w_2 is a similar wave packet progressing in the positive x-direction. We shall now apply Kelvin's method to approximate w_2 as given by (13.102b). The phase is stationary when $x=v_G(k_o)t=2\lambda^2 k_o t$ so that the predominant wave number is $k_o = \frac{x}{2\lambda^2 t}$. Therefore $v_G'(k_o)=2\lambda^2$ and $v_o = \frac{\omega_o}{k_o} = \frac{x}{2t}$. We now compare the imaginary part of (B.44) with (13.102b) and readily identify

$$a(k) = \frac{J_o}{\sqrt{2\pi}\ \rho A} \frac{1}{\lambda^2 k^2}$$

so that

$$a(k_o) = \frac{J_o \lambda^2}{\sqrt{2\pi}\ \rho A} \cdot \frac{4t^2}{x^2} \quad .$$

We now substitute into the imaginary part of (B.49a) and obtain the approximation

$$w_2(x,t) \cong \frac{a(k_o)}{\sqrt{v_G' t}} \sin\left[k_o(v_o t - x) + \frac{\pi}{4}\right]$$

$$= \sqrt{\frac{2}{\pi}} \frac{J_o \lambda}{\rho A} \frac{\sqrt{t^3}}{x^2} \left(\cos \frac{x^2}{4\lambda^2 t} - \sin \frac{x^2}{4\lambda^2 t}\right) \tag{13.103}$$

The validity of the approximation (13.103) is restricted to values of x and t which are sufficiently large. We also note that the method of Kelvin can be used to approximate the integral in (13.100). It is noteworthy that in this case the resulting "approximation" is equal to the exact value as given by (13.101), (see Exercise 13.11).

(e) Moving Load on an Euler-Bernoulli Beam Resting on an Elastic Foundation.

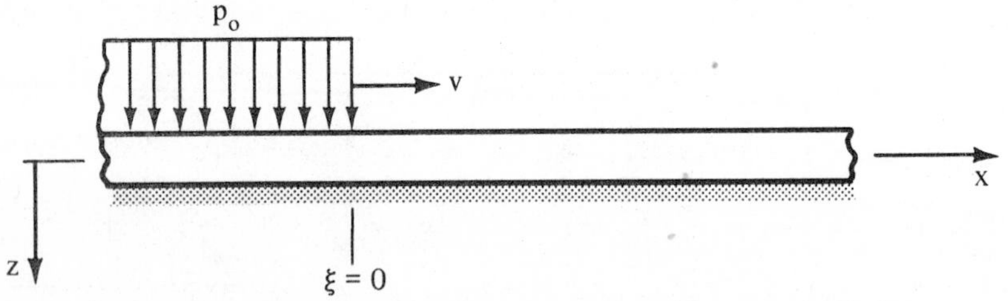

Figure 13.12

In the following we consider the dynamic response of a beam under the influence of a uniformly distributed load of intensity p_o. The load distribution has a distinct load front which moves with constant velocity v in the direction of the positive x-axis as shown in Fig. 13.12. It is assumed that the beam rests on an elastic foundation with foundation modulus c, i.e., local beam deflection is opposed by a force intensity which is linearly proportional to deflection, the constant of proportionality being $c>0$. To facilitate the subsequent analysis, we shall now derive some basic physical properties of the beam with respect to flexural wave propagation along the x-axis. Accordingly, we consider the (modified) Euler-Bernoulli beam equation

$$EI \frac{\partial^4 w}{\partial x^4} + \rho A \frac{\partial^2 w}{\partial t^2} + cw = 0 \tag{13.104}$$

and assume the possibility of a traveling flexural wave

$$w = \exp i(\omega t - kx) \tag{13.105}$$

where k and ω are the wave number and frequency, respectively. Upon substitution of (13.105) into (13.104) we readily obtain the dispersion relation, and from the latter we can calculate the phase and group velocity (see Appendix B). The results are

$$\frac{v}{v_o} = \sqrt{2}\ \Omega = \sqrt{\frac{\alpha^4+1}{\alpha^2}}$$

$$\frac{v_G}{v_o} = \frac{2\alpha^3}{\sqrt{\alpha^4+1}}$$

where

$$\alpha = \frac{k}{k_o}\ ;\ k_o = \sqrt[4]{\frac{c}{EI}}\ ,\ v_o = \frac{\sqrt[4]{cEI}}{\sqrt{\rho A}}\ .$$

Dimensionless graphs of phase and group velocity vs. wave number ratio α are shown in Fig. 13.13.

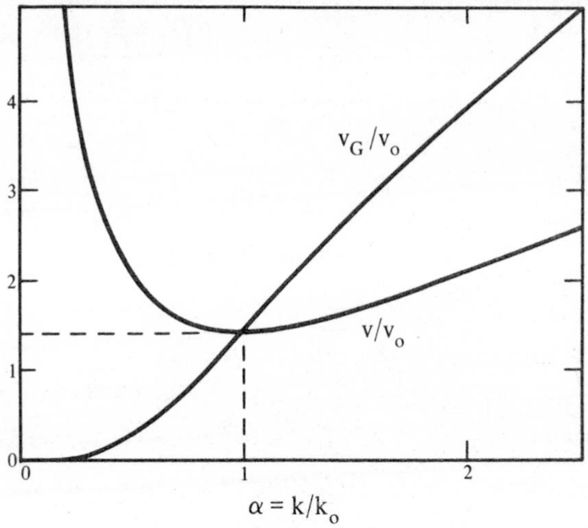

Figure 13.13

We now proceed to the forced motion problem which is characterized by the equation

$$EI\,\frac{\partial^4 w}{\partial x^4} + \rho A\,\frac{\partial^2 w}{\partial t^2} + cw = p(x,t) \tag{13.106}$$

In the following we are concerned with loads which propagate with constant speed v in the direction of the positive x-axis. Moreover, we shall concentrate on the steady-state solution and neglect transients. For these reasons it is permissible as well as convenient to describe the response of the beam in terms of the moving coordinate ξ. This is accomplished by the change of variable

$$\xi = x - vt \tag{13.107}$$

so that (13.106) is now transformed into

$$\frac{d^4w}{d\xi^4} + 2\Omega^2 k_o^2 \frac{d^2w}{d\xi^2} + k_o^4 w = \frac{p(\xi)}{EI} \tag{13.108}$$

The change of variable (13.107) can be given the following physical interpretation: An observer fixed with respect to the x-coordinate will see the distributed load advance in the direction of the positive x-axis, and to him the deflection of the beam will appear to be a function of x and t. However, an observer fixed with respect to the ξ-axis will move with the advancing load distribution, and to him the beam deflection profile will appear to be stationary, i.e., independent of t and a function of ξ only. We note that by neglecting the (damped) transients due to the starting of the motion, we have made the implicit assumption that the load has been moving for a sufficiently long period. Moreover, we assume that whatever damping exists in the system, it is sufficiently small so as not to affect the steady-state motion, as characterized by (13.108), in an important way.

We are now ready to consider the infinite beam $-\infty<\xi<\infty$, where ξ is the moving coordinate defined by (13.107). The uniformly distributed load of intensity p_o advancing in the positive x-direction is characterized by

$$\left.\begin{aligned} &p=p_o=\text{constant}, && \xi<0 \\ &p=0, && \xi>0 \end{aligned}\right\} \tag{13.109}$$

Thus, with reference to (13.108), we have

$$\frac{d^4w^{(1)}}{d\xi^4} + 2\Omega^2 k_o^2 \frac{d^2w^{(1)}}{d\xi^2} + k_o^4 w^{(1)} = 0, \quad \xi>0 \tag{13.110a}$$

$$\frac{d^4w^{(2)}}{d\xi^4} + 2\Omega^2 k_o^2 \frac{d^2w^{(2)}}{d\xi^2} + k_o^4 w^{(2)} = \frac{p_o}{EI}\ ; \quad \xi<0 \tag{13.110b}$$

where the superscripts (1) and (2) refer to the region $\xi>0$ and $\xi<0$, respectively. Solutions of (13.110) must be bounded for $\xi\to\pm\infty$, and at $\xi=0$ we require that the deflection, slope, moment and shear be continuous, i.e.,

$$\left.\begin{aligned} & w^{(1)}(0) = w^{(2)}(0), \left(\frac{dw^{(1)}}{d\xi}\right)_{\xi=0} = \left(\frac{dw^{(2)}}{d\xi}\right)_{\xi=0} \\ & \left(\frac{d^2w^{(1)}}{d\xi^2}\right)_{\xi=0} = \left(\frac{d^2w^{(2)}}{d\xi^2}\right)_{\xi=0}, \left(\frac{d^3w^{(1)}}{d\xi^3}\right)_{\xi=0} = \left(\frac{d^3w^{(2)}}{d\xi^3}\right)_{\xi=0} \end{aligned}\right\} \tag{13.111}$$

The characteristic equation obtained by assuming $w^{(i)}=e^{k\xi}$, $i=1,2$; k=constant, and substituting in the homogeneous equation (13.110) is

$$\alpha^4 + 2\Omega^2\alpha^2 + 1 = 0, \ \alpha = \frac{k}{k_o} \tag{13.112}$$

and its roots are

$$\sqrt{2}\ \alpha = \pm\ (\sqrt{-\Omega^2-1} \pm \sqrt{1-\Omega^2}\ ) \tag{13.113}$$

The roots of (13.112) are complex for $0<\Omega^2<1$ and pure imaginary for $1\leq\Omega^2$. Solutions of (13.110) subject to the conditions (13.111) and bounded for $\xi\to\pm\infty$ are found in the usual manner when the roots of the characteristic equation (13.113) are complex, and are given by

$$\frac{w^{(1)}}{w_o} = \frac{1}{2}\cdot\exp(-ak_o\xi)(\cos bk_o\xi + \frac{a^2-b^2}{2ab}\sin bk_o\xi),\ \xi\geq 0 \tag{13.114a}$$

$$\frac{w^{(2)}}{w_o} = \frac{1}{2}\cdot\exp(ak_o\xi)(-\cos bk_o\xi + \frac{a^2-b^2}{2ab}\sin bk_o\xi)+1,\ \xi\leq 0 \tag{13.114b}$$

where $w_o = \frac{p_o}{c}$ and

$$\sqrt{2a} = \sqrt{1-\Omega^2},\quad \sqrt{2b} = \sqrt{1+\Omega^2} \tag{13.114c}$$

In the case of imaginary roots ($1<\Omega^2$), the solutions of (13.110) are

$$w^{(1)}=C_1^{(1)}\cos ak_o\xi+C_2^{(1)}\sin ak_o\xi+C_3^{(1)}\cos bk_o\xi+C_4^{(1)}\sin bk_o\xi \tag{13.115a}$$

$$w^{(2)}=C_1^{(2)}\cos bk_o\xi+C_2^{(2)}\sin bk_o\xi+C_3^{(2)}\cos ak_o\xi+C_4^{(2)}\sin ak_o\xi+ \frac{p_o}{c} \tag{13.115b}$$

where

$$\sqrt{2}\ a=\sqrt{1+\Omega^2} +\sqrt{-1+\Omega^2} \ >\sqrt{2}\ ,\sqrt{2}\ b=\sqrt{1+\Omega^2} -\sqrt{-1+\Omega^2} \ <\sqrt{2}\ .$$

There are only four equations (13.111) to determine the eight constants of integration $C_i^{(j)}$, $i=1,2,3,4$; $j=1,2$. At this point we can use the concept of group velocity to determine the appropriate steady-state motion. The group velocity is the velocity of energy transport, and the physically appropriate solution requires a flow of energy away from the load front. With reference to Fig. 13.13, we note that corresponding to a given phase velocity v greater than $\sqrt{2}\, v_o$, there are two wave numbers, k_1 and k_2, one of them $k_1<k_o$, the other $k_o<k_2$. For k_2, the group velocity is always greater than the phase velocity, and for the present case this must pertain to waves which precede the load front at $\xi=0$. For the smaller wave number k_1, the group velocity is always smaller than the phase velocity, and therefore k_1 corresponds to waves behind the load front. Since a>b, we readily identify $k_1=k_o b$ and $k_2=k_o a$, and the argument presented enables us to set $C_3^{(1)}=C_4^{(1)}=C_3^{(2)}=C_4^{(2)}=0$ in (13.115). The remaining constants in (13.115) are now determined by the application of the continuity conditions (13.111) and the final results are

$$\frac{w^{(1)}}{w_o} = -\frac{b^2}{a^2-b^2}\cos ak_o\xi, \quad \xi\geq 0 \tag{13.117a}$$

$$\frac{w^{(2)}}{w_o} = -\frac{a^2}{a^2-b^2}\cos bk_o\xi+1, \quad \xi\leq 0 \tag{13.117b}$$

where $w_o = \frac{p}{c}$, $\Omega^2>1$, and

$$\sqrt{2}\,a=\sqrt{1+\Omega^2}+\sqrt{-1+\Omega^2}\,,\quad \sqrt{2}\,b=\sqrt{1+\Omega^2}-\sqrt{-1+\Omega^2} \tag{13.117c}$$

We note that when $\Omega^2=1$, $\frac{v}{v_o} = \sqrt{2}$ and the solutions (13.114) and (13.117) become unbounded. For this reason the velocity $v_{cr} = \sqrt{2}\, v_o = \sqrt{\frac{2}{\rho A}}\cdot\sqrt[4]{cEI}$ is called the critical speed. With the aid of (13.114) and (13.117), we can draw graphs of the deflection curves as shown in Figs. 13.14a and 13.14b, respectively. The wave length ratios pertaining to the motion are shown in Fig. 13.15. They were obtained with the aid of (13.114c) and (13.117c). The character of the solution for $v<v_{cr}$ is seen to be entirely different from the case $v_{cr}<v$. When $0<\Omega^2<1$ (or $0<v<v_{cr}$), the traveling wave is an attenuated sinuisoid as shown in Fig. 13.14a. In this case $w\to 0$ as $\xi\to\infty$, $w\to w_o$ as $\xi\to-\infty$, the deflection curve is anti-symmetrical with respect to the point

$(\xi, W)=(0, \frac{1}{2}w_o)$ and $w=\frac{1}{2}w_o$ at the load front $\xi=0$. When $1<\Omega^2$ (or $v_{cr}<v$) we obtain a sinusoidal deflection curve as shown in Fig. 13.14b. In this case the wavelength of waves ahead of the load front is smaller than that of waves behind the load front. Waves ahead of the load front oscillate about $w=0$, while waves behind the load front oscillate about $w=w_o$. We also note that the deflection at the load front $\xi=0$ is negative in this case. With reference to Fig. 13.15, we note that the wavelength in front and behind the load front are identical for $0<v<v_{cr}$, but for $v_{cr}<v$ the wavelength in front of the load is smaller than the wavelength behind it. The difference in wavelength increases substantially with an increase in v for $v_{cr}<v$. It is also worth noting that the solution (13.114) approaches the known, static solution when $v\to0$ (or $\Omega^2\to0$).

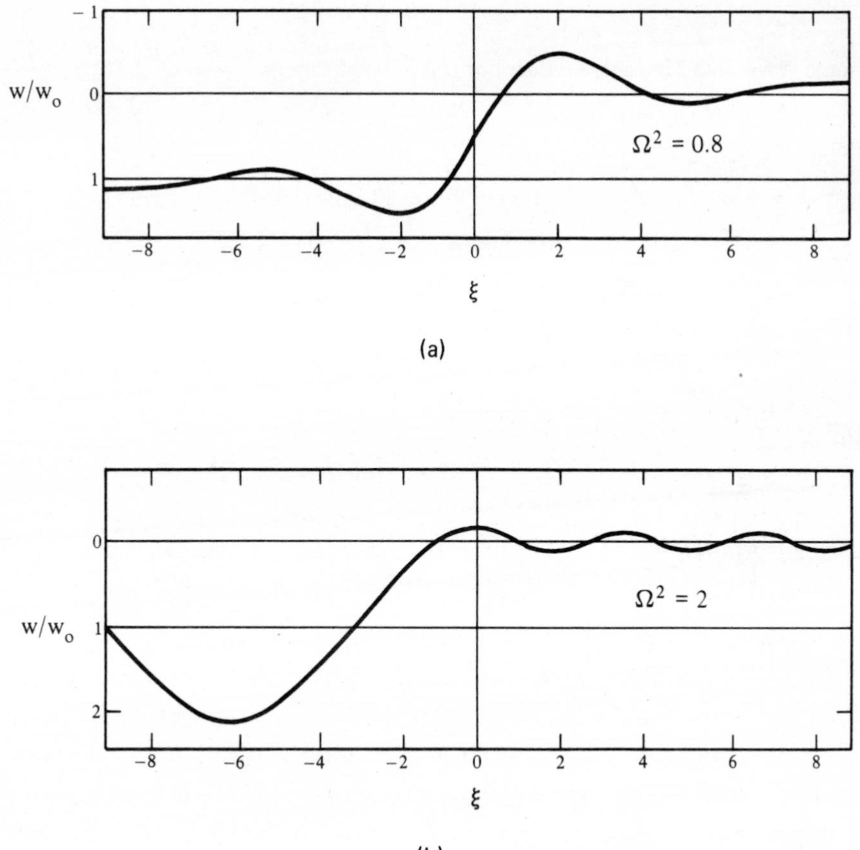

Figure 13.14

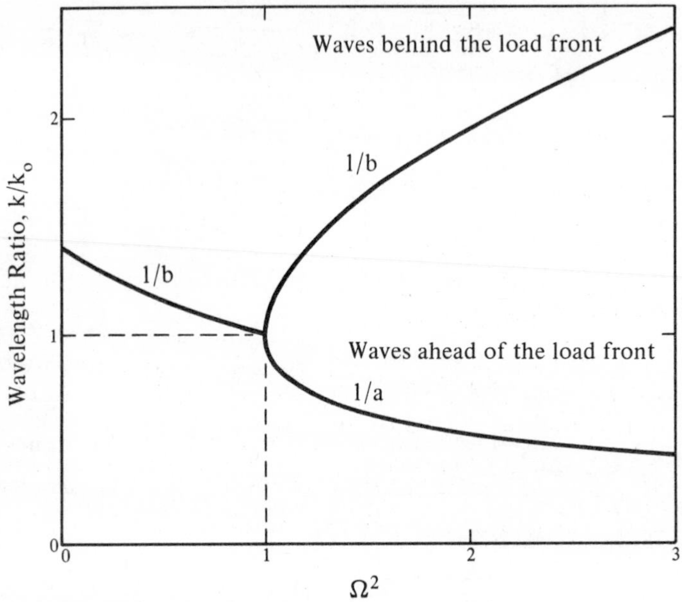

Figure 13.15

(f) Time-harmonic Waves in a Timoshenko Beam of Unbounded Extent.

In the following we are concerned with free wave motion in a Timoshenko beam of unbounded extent. For the present purpose, we write the (homogeneous) Timoshenko beam equations (13.41) and (13.42) in the form

$$\alpha_2^2\left(\frac{\partial^2 w}{\partial x^2}+\frac{\partial\psi}{\partial x}\right)=\frac{\partial^2 w}{\partial t^2} \tag{13.118a}$$

$$\alpha_1^2\frac{\partial^2\psi}{\partial x^2}-\alpha_2^2 k_o^2\left(\frac{\partial w}{\partial x}+\psi\right)=\frac{\partial^2\psi}{\partial t^2} \tag{13.118b}$$

where $\alpha_1^2=\frac{E}{\rho}$, $\alpha_2^2=\kappa^2\frac{G}{\rho}$, $k_o^2=\frac{A}{I}$.

To explore the possibility of the passage of a time-harmonic wave, we assume

$$w=\frac{B}{k_o}\,\mathrm{expi}(\omega t-kx) \tag{13.119a}$$

$$\psi=C\,\mathrm{expi}(\omega t-kx) \tag{13.119b}$$

where B, C are constants and ω and k are the frequency and wave number of the assumed, traveling wave, respectively. Upon substitution of (13.119) into (13.118), we obtain

$$B(\Omega^2-K^2)+C(-iK)=0 \tag{13.120a}$$

$$B(iK)+C(\Omega^2-\gamma^2K^2-1)=0 \tag{13.120b}$$

where $\Omega = \frac{\omega}{\alpha_2 k_o}$, $K = \frac{k}{k_o}$, and $\gamma^2 = \frac{\alpha_1^2}{\alpha_2^2} = \frac{E}{\kappa^2 G}$. To obtain a non-trivial solution for the constants A and B, we set the determinant of coefficients in (13.120) equal to zero, ultimately resulting in

$$\Omega^4 - \Omega^2(1+K^2+\gamma^2K^2) + \gamma^2K^4 = 0 \tag{13.121}$$

or

$$\frac{\gamma^2}{\Omega^2} K^4 - K^2(1+\gamma^2) + (\Omega^2 - 1) = 0 \tag{13.122}$$

With the aid of (13.121) and (13.122), we obtain dimensionless expressions for wave number, phase velocity, and group velocity in the usual manner (see Appendix B):

$$\frac{k^2}{k_o^2} = K^2 = \frac{\Omega^2}{2\gamma^2}\left(1+\gamma^2 \pm \sqrt{(1-\gamma^2)^2 + \frac{4\gamma^2}{\Omega^2}}\right) \tag{13.123}$$

$$\frac{\alpha_2^2}{v^2} = \frac{K^2}{\Omega^2} = \frac{1}{2\gamma^2}\left(1+\gamma^2 \pm \sqrt{(1-\gamma^2)^2 + \frac{4\gamma^2}{\Omega^2}}\right) \tag{13.124}$$

$$\frac{v_G}{\alpha_2} = \frac{d\Omega}{dK} = \frac{K}{\Omega}\,\frac{[2\gamma^2K^2 - (1+\gamma^2)\Omega^2]}{[(1+\gamma^2)K^2 - (2\Omega^2 - 1)]} \tag{13.125}$$

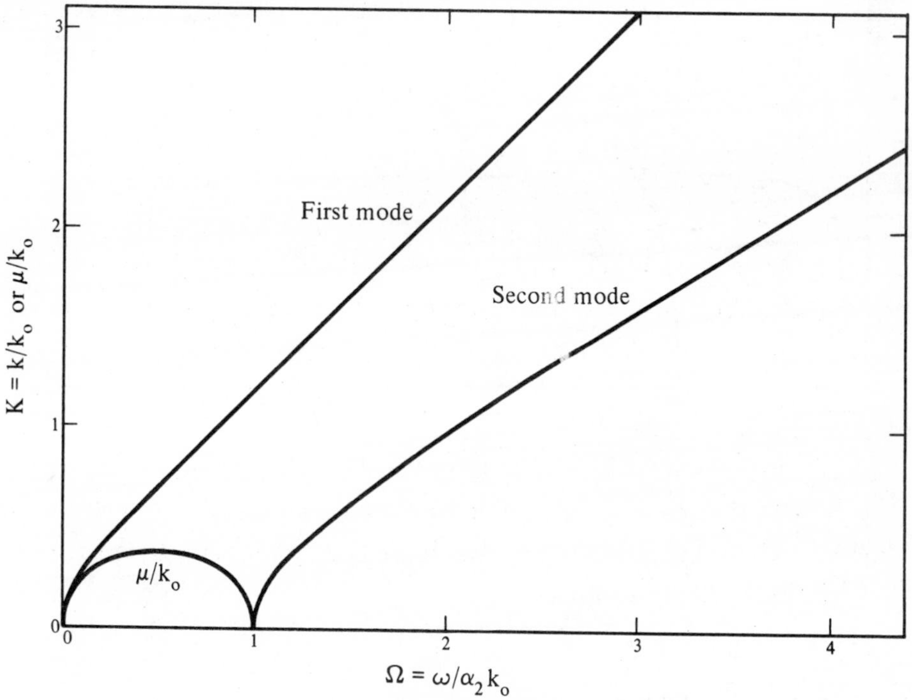

Figure 13.16

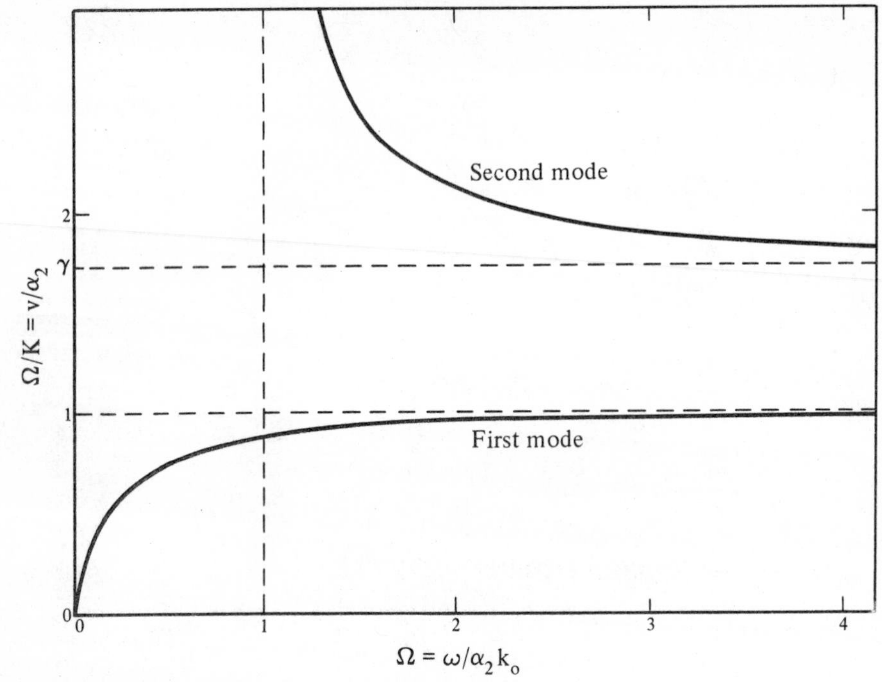

Figure 13.17

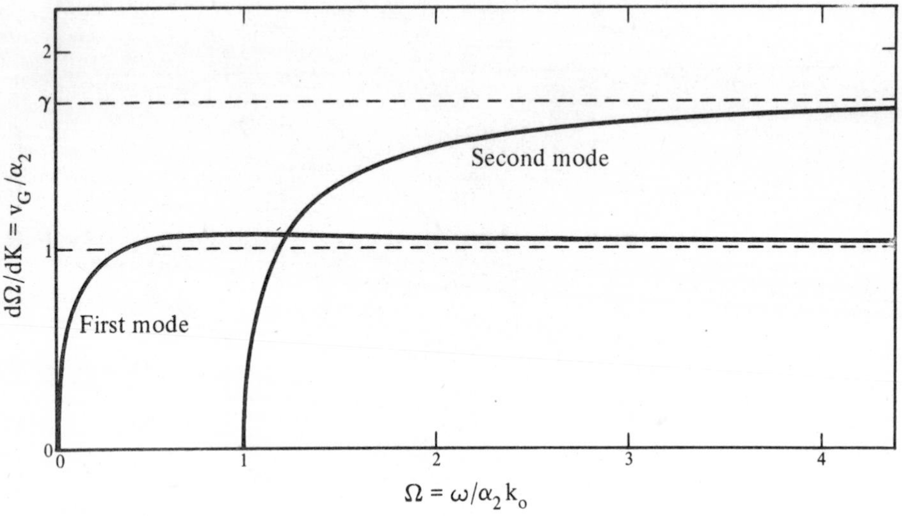

Figure 13.18

The relations (13.123), (13.124), and (13.125) have been used to construct the graphs in Figs. 13.16, 13.17, and 13.18, respectively, for the specific value γ=1.74. With reference to these figures, we note that, in general, there are two distinct modes of wave propagation in a Timoshenko beam. In the first mode K is real for the entire frequency spectrum $0<\Omega$. In the second mode of propagation, the wave number ratio K is real for $1<\Omega$, but K will be imaginary for $0<\Omega<1$. In this case we set $k=-i\mu$, where $i=\sqrt{-1}$ and $\mu>0$ is the attenuation number, so that for $0<\Omega<1$ we have a standing, attenuated wave in the second mode, which is characterized by

$$\left.\begin{aligned} w &= \frac{B}{k_o}\exp(-\mu x)\cdot \exp i\omega t \\ \psi &= C\exp(-\mu x)\cdot \exp i\omega t \end{aligned}\right\} \tag{13.126}$$

The constants B and C in (13.119) are not independent, but they are related by (see 13.120a)

$$\frac{C}{B} = -\frac{i(\Omega^2-K^2)}{K} \tag{13.127}$$

Because of the complexity of the present results, it will be informative to consider the motion for the extreme limits of the wave number spectrum $K\to 0$ and $K\to\infty$. When $K\to 0$ we have $\Omega=0$ in the first mode, and equations (13.120) reveal that $C=0$ and B is arbitrary. This motion is characterized by $w=\frac{B}{k_o}$ and $\psi\equiv 0$, i.e., we have a rigid body motion at zero frequency. The corresponding limit for the second mode is $\Omega=1$ (or $\omega=\alpha_2 k_o$), resulting in $B=0$ and C an arbitrary constant. Thus $w=0$, $\psi=C\exp i\alpha_2 k_o t$, and we have a "thickness shear" motion wherein all cross sections rotate back and forth in unison, with frequency $\omega=\alpha_2 k_o$. To obtain the limiting cases corresponding to $K\to\infty$, it is convenient to write (13.120) in the form

$$B\left(\frac{\Omega^2}{K^2}-1\right)+C\left(-i\,\frac{1}{K}\right)=0$$

$$B\left(\frac{i}{K}\right)+\left(C\,\frac{\Omega^2}{K^2}-\gamma^2-\frac{1}{K^2}\right)=0$$

Letting $K\to\infty$, we have $\Omega^2=K^2$ (or $\omega=\alpha_2 k$) in the first mode, and $C=0$, B is arbitrary, resulting in $w=\frac{B}{k_o}\exp i(\alpha_2 kt-kx)$, $\psi\equiv 0$. In this motion cross sections will not rotate, and beam deformation is in shear only. The corresponding limit for the second mode is $\Omega^2=\gamma^2K^2$ (or $\omega=\alpha_1 k$), and $B=0$, $C=$arbitrary constant. The resulting limiting motion is characterized by $w\equiv 0$, $\psi=C\exp i(\alpha_1 kt-kx)$. In this case beam cross sections rotate back and forth, alternately stretching and compressing the fibers at points in the cross section where $z\neq 0$.

(g) Time-harmonic Waves in a Semi-infinite, Clamped Timoshenko Beam.

As the final example of wave motion we reconsider example (b), except that this time we use the Timoshenko beam model to obtain a solution. With reference to both examples (b) and (f), we now characterize the incident wave (see Fig. 13.9) by

$$w_i = \frac{B_1^{(i)}}{k_o} \exp i(\omega t + k_1 x), \quad \psi = C_1^{(i)} \exp i(\omega t + k_1 x) \tag{13.128}$$

where, utilizing the notations of example (f),

$$\frac{C_1^{(i)}}{B_1^{(i)}} = i\beta_1, \quad \beta_1 = \frac{\Omega^2 - K_1^2}{K_1}$$

and where we have made the assumptions that only the first mode (subscript 1) is present in the incident wave. It is assumed that the quantities $B_1^{(i)}$ and ω are given. Since two modes of propagation exist in a Timoshenko beam, we must (tentatively) assume that both of these modes will be present in the reflected wave. Accordingly, we characterize the reflected wave by

$$\left.\begin{aligned} w_r &= \frac{B_1^{(r)}}{k_o} \exp i(\omega t - k_1 x) + \frac{B_2^{(r)}}{k_o} \exp i(\omega t - k_2 x) \\ \psi_r &= C_1^{(r)} \exp i(\omega t - k_1 x) + C_2^{(r)} \exp i(\omega t - k_2 x) \end{aligned}\right\} \tag{13.129}$$

where (see (13.127))

$$\frac{C_1^{(r)}}{B_1^{(r)}} = -i\beta_1, \quad \frac{C_2^{(r)}}{B_2^{(r)}} = -i\beta_2$$

and where $\beta_2 = \dfrac{\Omega^2 - K_2^2}{K_2}$. The total deformation of the beam is obtained by a superposition of the incident and reflected waves, i.e.,

$$w = w_i + w_r, \quad \psi_i + \psi_r = \psi \tag{13.130}$$

If the beam is rigidly clamped at x=0, we require

$$w(0,t) = \psi(0,t) = 0 \tag{13.131}$$

We now substitute (13.128) and (13.129) into (13.130) and apply (13.131). The result of this computation is

$$B_1^{(r)} + B_2^{(r)} = -B_1^{(i)}$$

$$\beta_1 B_1^{(r)} + \beta_2 B_2^{(r)} = \beta_1 B_1^{(i)}$$

and the solution for this system of equations is

$$\frac{B_1^{(r)}}{B_1^{(i)}} = \frac{\beta_1+\beta_2}{\beta_1-\beta_2}, \quad \frac{B_2^{(r)}}{B_1^{(i)}} = \frac{2\beta_1}{\beta_2-\beta_1}$$

so that

$$\frac{C_1^{(r)}}{C_1^{(i)}} = -\frac{\beta_1+\beta_2}{\beta_1-\beta_2}, \quad \frac{C_2^{(r)}}{C_1^{(i)}} = -\frac{2\beta_2}{\beta_2-\beta_1}$$

The constants in (13.129) are now fully evaluated in terms of the parameters of the incident wave, and they do not vanish in general. We are thus led to the conclusion that even though only the first mode of propagation is represented in the incident wave, both modes will appear in the reflected wave when the end of the beam is clamped.

EXERCISES

13.1 Consider the mathematical model of a shear beam, i.e., a beam which deforms in shear only, but remains rigid with respect to other types of deformation. Assume (and justify) the displacement field $u_1=0$, $u_2=0$, $u_3=w(x,t)$, and use Hamilton's principle to derive the stress equation of motion

$$-\rho A\ddot{w} + \frac{\partial Q}{\partial x} + p = 0$$

and the associated admissible boundary conditions: at $x=x_1$ and $x=x_2$ either w or Q must be specified. Justify the stress-displacement relation $Q = \int_A \tau_{zx} dA = \kappa^2 AG \frac{\partial w}{\partial x}$, where κ^2 is the shear coefficient. Formulate a well posed forced motion problem, and discuss its solution. Derive an expression for energy flux in the shear beam.

13.2 Starting with the displacement assumptions

$$u_1 = -z \frac{\partial w}{\partial x}, \quad u_2 = 0, \quad u_3 = w(x,t)$$

use Hamilton's principle to derive the stress equation of motion for a Rayleigh beam

$$\frac{\partial^2}{\partial x \partial t}\left(\rho I \frac{\partial^2 w}{\partial x \partial t}\right) - \rho A \frac{\partial^2 w}{\partial t^2} + \frac{\partial^2 M}{\partial x^2} + p = 0$$

where $M = \int_A z\tau_{xx} dA = -EI \frac{\partial^2 w}{\partial x^2}$. Show that the admissible boundary conditions at $x=x_1$ and $x=x_2$ are

either M or $\frac{\partial w}{\partial x}$ must be specified

and

either $\frac{\partial M}{\partial x} + \frac{\partial}{\partial t}\left(\rho I \frac{\partial^2 w}{\partial x \partial t}\right)$ or w must be specified.

Show that upon neglect of the rotatory inertia term $\rho I \frac{\partial^2 w}{\partial x \partial t}$, the Rayleigh beam degenerates to an Euler-Bernoulli beam. Derive a dispersion relation for the Rayleigh beam, and calculate phase and group velocities.

13.3 Derive a Timoshenko type mathematical model for the sandwich beam shown in the figure. Assume that the facings are monolithically bonded to the core of the beam.

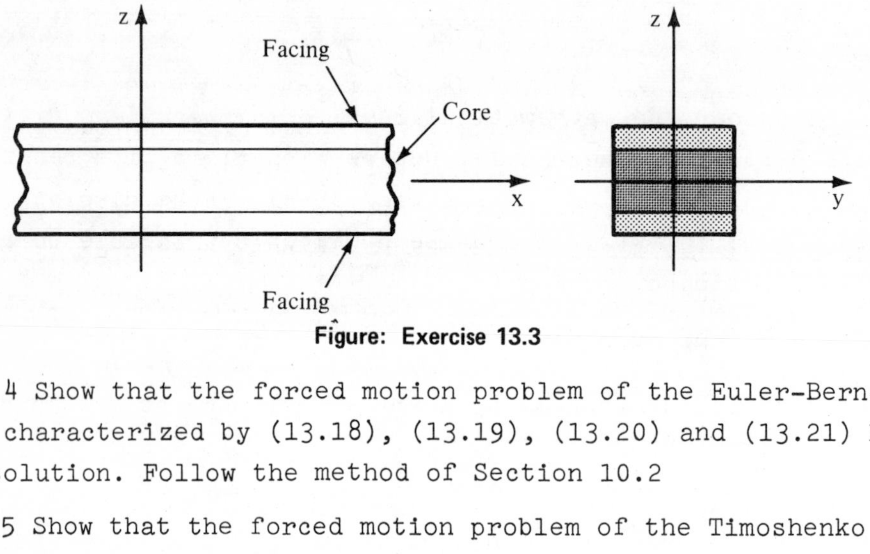

Figure: Exercise 13.3

13.4 Show that the forced motion problem of the Euler-Bernoulli beam as characterized by (13.18), (13.19), (13.20) and (13.21) has a unique solution. Follow the method of Section 10.2

13.5 Show that the forced motion problem of the Timoshenko beam as characterized by (13.41), (13.42), (13.43), and (13.44) has a unique solution. Follow the method of Section 10.2.

13.6 Find the quasi-static solution of example (b) in Section 13.3 valid for $\frac{\ell}{2} < x < \ell$ (see Fig. 13.4a).

13.7 Consider the simply supported Euler-Bernoulli beam shown in Fig. 13.4a. Find the dynamic response if a concentrated load of magnitude F_0 is suddenly applied at t=0 at the center of the beam. Assume that the beam is at rest in the reference configuration at t=0.

13.8 Consider the Euler-Bernoulli cantilever beam shown in the figure. An end load of magnitude F_0 is suddenly applied to the free end at t=0. The beam is at rest and in its reference configuration at t=0. (a) Find the free vibration frequencies and normalized mode shapes (eigenfunctions). (b) Obtain a solution of the forced motion problem.

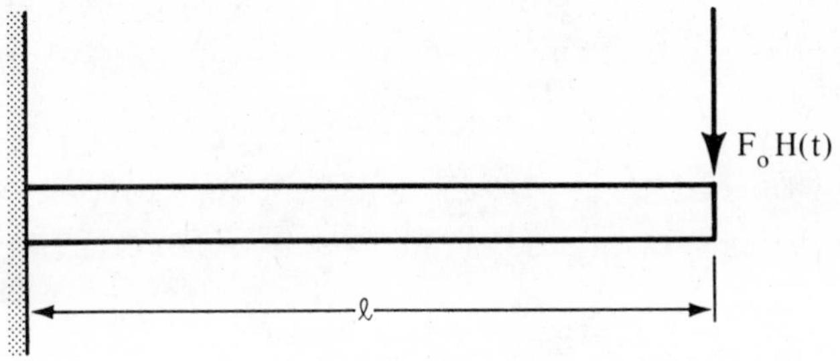

Figure: Exercise 13.8

13.9 Find the frequencies and normalized mode shapes (eigenfunctions) of a clamped-clamped Euler-Bernoulli beam of length ℓ.

13.10 A simply supported Euler-Bernoulli beam is excited by a concentrated, harmonically oscillating load of magnitude $F_0 \sin\omega t \cdot H(t)$ as shown in the figure. The beam is at rest and in its reference configuration at t=0. Find the dynamic response w=w(x,t). What happends if ω is very close to a natural frequency? What happens if ω is equal to one of the frequencies?

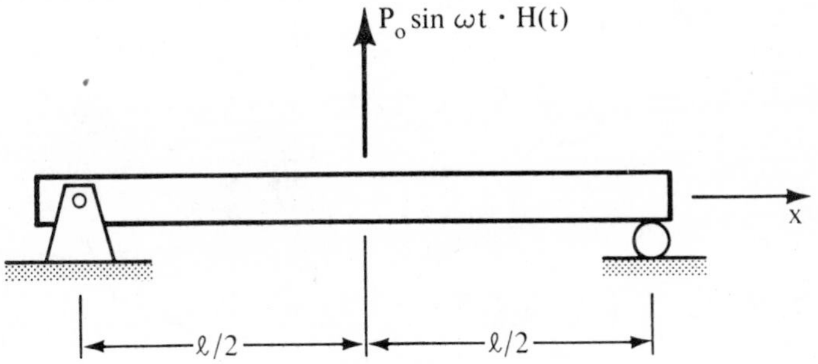

Figure: Exercise 13.10

13.11 A simply supported Timoshenko beam is acted upon by an end moment $M(t)$ as shown in Fig. 13.2. Find the response of the beam if it is at rest in its reference configuration at $t=0$.

13.12 Approximate the improper integral in (13.100) by application of Kelvin's method of stationary phase described in Appendix B. Show that the "approximation", in this case, results in the exact value as given by (13.101).

13.13 An infinite Euler-Bernoulli beam resting on an elastic foundation is acted upon by a concentrated impulse of magnitude J_o at $t=0$, $x=0$. Find the dynamic response of the beam for $t>0$ if it is at rest in its reference configuration at $t=0$.

13.14 An infinite Euler-Bernoulli beam resting on an elastic foundation is subjected to a harmonically oscillating load $F_o \exp i\omega t$. Neglecting transients, find the steady-state response of the beam.

13.15 An infinite Euler-Bernoulli beam is resting on an elastic foundation. A transverse, concentrated load of magnitude F_o acts on the beam and travels with constant speed v in the direction of the positive x-axis (beam axis). Neglecting transients, find the steady-state response of the beam.

13.16 A semi-infinite Timoshenko beam $0 \leq x < \infty$ is subjected to an incoming wave characterized by (13.128). The wave is reflected from the end at $x=0$. Give a complete description of the reflected wave (or waves) if (a) the end at $x=0$ is free, and (b) if the end at $x=0$ is simply supported.

13.17 Consider the forced motion problem characterized by equations (13.18) through (13.21) with $f_1 \equiv f_2 \equiv g_1 \equiv g_2 \equiv 0$. Assume a solution of the form $w(x,t) = \sum_{i=1}^{\infty} W_i(x) \cdot q_i(t)$, where $W_i(x)$ are the eigenfunctions which satisfy the particular homogeneous boundary conditions of the problem. Show that $\ddot{q}_i + \omega_i^2 q_i = P_i(t) = \int_0^{\ell} pW_i dx$, $q_i(0) = \int_0^{\ell} \rho A w_o W_i dA$, $\dot{q}_i(0) = \int_0^{\ell} \rho A \dot{w}_o W_i dx$, and

$$q_i(t) = q_i(0) \cos\omega_i t + \frac{1}{\omega_i} \int_0^t P_i(\tau) \sin\omega_i(t-\tau) d\tau + \frac{1}{\omega_i} \dot{q}_i(0) \sin\omega_i t.$$

13.18 An infinite Euler-Bernoulli beam is acted upon by a concentrated impulsive moment of magnitude M_o applied at x=0 and t=0. The beam is at rest and in its reference configuration at t=0. Show that the differential equation to be solved is

$$EI\frac{\partial^4 w}{\partial x^4} + \rho A\frac{\partial^2 w}{\partial t^2} = \frac{\partial m}{\partial x}$$

where $m(x,t)=M_o\delta(x)\delta(t)$. Show that the solution can be written in the form

$$w = \frac{M_o}{2\lambda^2\rho A}\left[S\left(\frac{x}{\lambda\sqrt{2\pi t}}\right)-C\left(\frac{x}{\lambda\sqrt{2\pi t}}\right)\right]$$

where $\lambda^4 = \frac{EI}{\rho A}$ and $S(z) = \int_0^z \sin\frac{\pi}{2}\xi^2 d\xi$, $C(z) = \int_0^z \cos\frac{\pi}{2}\xi^2 d\xi$ are the Fresnel integrals. Check the result by Kelvin's method of stationary phase.

13.19 A simply supported Euler-Bernoulli beam is at rest in its reference configuration at t=0. A uniformly distributed pressure load of magnitude p_o moves over the beam with speed v as shown in the figure. Give a complete description of the motion of the beam for t>0.

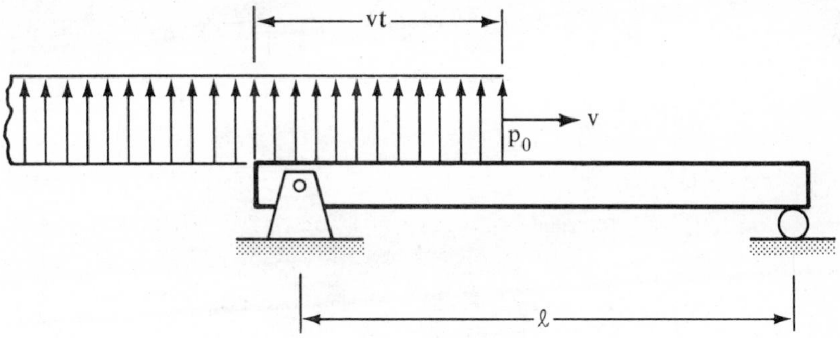

Figure: Exercise 13.19

Chapter

14

DYNAMICS OF ELASTIC PLATES

The laws which govern the small vibrations of elastic plates are contained in the three-dimensional equations of the linear theory of elasticity.

R.D. Mindlin, "An Introduction to the Mathematical Theory of Vibrations of Elastic Plates," 1955.

14.1 PRELIMINARIES

A plate is a load carrying structural member bounded by two parallel planes, called its faces, and a cylindrical surface, called its edge. The generators of the cylindrical surface are perpendicular to the plane faces. The distance between the plane faces is called the thickness of the plate and it is assumed that the thickness is small compared to the characteristic dimensions of the faces (such as length or width of a rectangle, diameter of a circle, etc.). The plane which is parallel to, and equidistant from the faces is called the midplane or median plane of the plate. Its intersection with the cylindrical boundary is assumed to be a piecewise smooth, simple closed curve C. The plane region enclosed by C is called the area of the plate and is denoted by the symbol A. After the region A has been oriented in space then a sense can be assigned to the curve C as follows. The region A always lies to the left of an observer who traverses the curve C in the positive direction on the positive side of A. The plate will be subject to two types of loading, (a) distributed loads which act normal to the faces of the plate, and (b) distributed loads which act on the edges of the plate. Plates which contain holes or cutouts may be included in the preceding definition by permitting the region A to be multiply connected. In these cases the symbol C is understood to represent the sum of all the curves which define the boundary of A. Structural members in the form of plates as defined above are used extensively in the construction of buildings, airplanes, ships, machines, etc., and engineers require methods which facilitate the prediction of their static and dynamic response. Because the solution of plate problems within the framework of three-dimensional elasticity theory is extremely cumbersome, it is customary to utilize (approximate) plate theories, two of which are developed in the present chapter.

Let us consider the "small" deformation of a plate of thickness h. The material plate is referred to an orthogonal Cartesian coordinate system whose x_1x_2 plane coincides with the midplane of the plate as shown in Fig. 14.1. The plate model to be considered in the present chapter is based upon the following displacement assumptions:

$$\begin{aligned} u_\alpha(x_1,x_2,z,t) &= z\psi_\alpha(x_1,x_2,t) \\ u_z(x_1,x_2,z,t) &= w(x_1,x_2,t) \quad . \end{aligned} \tag{14.1}$$

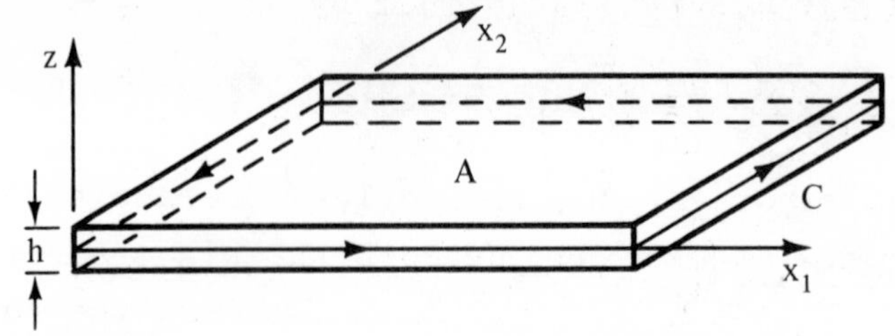

Figure 14.1

In the present chapter, the range of all latin indices (i,j, etc.) is (1,2,3), while the range of all greek indices (α,β, etc.) is (1,2). The displacement assumptions (14.1) imply that

(a) Straight lines which are perpendicular to the midplane of the plate in the undeformed state remain straight during the deformation.

(b) All the points on any straight line which is perpendicular to the midplane before deformation have the same vertical displacement, i.e., there is no "thickness stretch."

(c) There is no stretching of the midplane, i.e., the midplane is the neutral surface of the plate.

By substituting (14.1) into the strain-displacement relations (10.21) we obtain the strain field

$$\left.\begin{aligned} \varepsilon_{\alpha\beta} &= \frac{1}{2} z(\psi_{\alpha,\beta}+\psi_{\beta,\alpha}) \\ \varepsilon_{\alpha z} &= \varepsilon_{z\alpha} = \frac{1}{2}(\psi_{\alpha}+w_{,\alpha}) \\ \varepsilon_{zz} &= 0 \end{aligned}\right\} \tag{14.2}$$

In the subsequent analysis it will be convenient to work with plate stress resultants and plate strain resultants rather than the stresses and strains, respectively. The plate stress resultants are moments $M_{\alpha\beta}$ and shear forces Q_{α} defined by

$$M_{\alpha\beta} = \int_{-\frac{h}{2}}^{\frac{h}{2}} \tau_{\alpha\beta} z dz = \int_{-\frac{h}{2}}^{\frac{h}{2}} \tau_{\beta\alpha} z dz = M_{\beta\alpha} \tag{14.3a}$$

$$Q_{\alpha} = \int_{-\frac{h}{2}}^{\frac{h}{2}} \tau_{z\alpha} dz = \int_{-\frac{h}{2}}^{\frac{h}{2}} \tau_{\alpha z} dz \quad . \tag{14.3b}$$

The plate strain resultants are defined by

$$m_{\alpha\beta}=\frac{12}{h^3}\int_{-\frac{h}{2}}^{\frac{h}{2}}\varepsilon_{\alpha\beta}zdz=\frac{12}{h^3}\int_{-\frac{h}{2}}^{\frac{h}{2}}\varepsilon_{\beta\alpha}zdz=m_{\beta\alpha} \tag{14.4a}$$

$$q_\alpha=\frac{2}{h}\int_{-\frac{h}{2}}^{\frac{h}{2}}\varepsilon_{z\alpha}dz=\frac{2}{h}\int_{-\frac{h}{2}}^{\frac{h}{2}}\varepsilon_{\alpha z}dz \quad . \tag{14.4b}$$

The substitution of (14.2) into (14.4) yields

$$\left.\begin{aligned} m_{\alpha\beta}&=\frac{1}{2}(\psi_{\alpha,\beta}+\psi_{\beta,\alpha})\\ q_\alpha&=\psi_\alpha+w_{,\alpha} \quad . \end{aligned}\right\} \tag{14.5}$$

It can be shown that the quantities ψ_α, Q_α and q_α are vectors while the quantities $M_{\alpha\beta}$ and $m_{\alpha\beta}$ are second order tensors in a two-dimensional space (the x_1x_2 plane). Towards this end, we consider an arbitrary rotation of the x_1x_2 coordinate system about the z-axis as shown in Fig. 14.2.

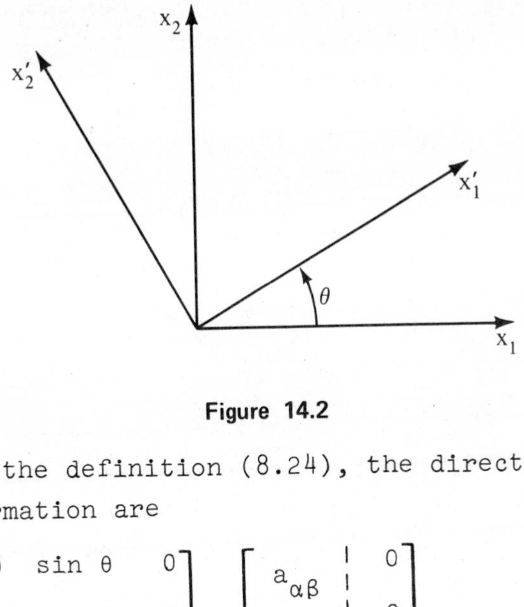

Figure 14.2

In accordance with the definition (8.24), the direction cosines of this coordinate transformation are

$$[a_{ij}]=\begin{bmatrix}\cos\theta & \sin\theta & 0\\ -\sin\theta & \cos\theta & 0\\ 0 & 0 & 1\end{bmatrix}=\left[\begin{array}{cc|c} \multicolumn{2}{c|}{a_{\alpha\beta}} & 0\\ & & 0\\ \hline 0 & 0 & 1\end{array}\right] \quad . \tag{14.6}$$

According to (8.25a) the components of the displacement vector in the primed reference are given by $u_i'=a_{ij}u_j$. In particular, with the aid of (14.6), we conclude that u_1' and u_2' are given by

$$u_\alpha'=a_{\alpha j}u_j=a_{\alpha\beta}u_\beta+a_{\alpha z}u_z=a_{\alpha\beta}u_\beta$$

since $a_{\alpha z}=0$. Consequently, using (14.1) we obtain,

$$z\psi'_{\alpha}=u'_{\alpha}=a_{\alpha\beta}u_{\beta}=a_{\alpha\beta}z\psi_{\beta}$$

or $\psi'_{\alpha}=a_{\alpha\beta}\psi_{\beta}$. Hence the ψ_{α} are the components of a two-dimensional vector. Utilizing (8.29) we find that

$$\tau'_{z\alpha}=a_{zi}a_{\alpha j}\tau_{ij}=a_{z\beta}a_{\alpha j}\tau_{\beta j}+a_{zz}a_{\alpha j}\tau_{zj}$$

$$=a_{\alpha j}\tau_{zj}=a_{\alpha\beta}\tau_{z\beta}+a_{\alpha z}\tau_{zz}=a_{\alpha\beta}\tau_{z\beta}$$

because $a_{z\beta}=a_{\alpha z}=0$ and $a_{zz}=1$. Consequently,

$$Q'_{\alpha}=\int_{-\frac{h}{2}}^{\frac{h}{2}}\tau'_{z\alpha}dz=a_{\alpha\beta}\int_{-\frac{h}{2}}^{\frac{h}{2}}\tau_{z\beta}dz=a_{\alpha\beta}Q_{\beta}\ .$$

Thus, the shear force components transform according to the law for two-dimensional vectors. Utilizing (8.29) once again we find that

$$\tau'_{\alpha\beta}=a_{\alpha i}a_{\beta j}\tau_{ij}=a_{\alpha\gamma}a_{\beta\delta}\tau_{\gamma\delta}+a_{\alpha z}a_{\beta\delta}\tau_{z\delta}$$

$$+a_{\alpha\gamma}a_{\beta z}\tau_{\gamma z}+a_{\alpha z}a_{\beta z}\tau_{zz}=a_{\alpha\gamma}a_{\beta\delta}\tau_{\gamma\delta}$$

since $a_{\alpha z}=a_{\beta z}=0$. Thus,

$$M'_{\alpha\beta}=\int_{-\frac{h}{2}}^{\frac{h}{2}}\tau'_{\alpha\beta}zdz=a_{\alpha\gamma}a_{\beta\delta}\int_{-\frac{h}{2}}^{\frac{h}{2}}\tau_{\gamma\delta}zdz=a_{\alpha\gamma}a_{\beta\delta}M_{\gamma\delta}$$

which is the law of transformation of a two-dimensional tensor of order two. Similar procedures can be used to derive the transformation laws for q_{α} and $m_{\alpha\beta}$ (see Exercise 14.1). In summary,

$$\psi'_{\alpha}=a_{\alpha\beta}\psi_{\beta},\ Q'_{\alpha}=a_{\alpha\beta}Q_{\beta},\ q'_{\alpha}=a_{\alpha\beta}q_{\beta} \tag{14.7}$$

and

$$M'_{\alpha\beta}=a_{\alpha\gamma}a_{\beta\delta}M_{\gamma\delta},\ m'_{\alpha\beta}=a_{\alpha\gamma}a_{\beta\delta}m_{\gamma\delta} \tag{14.8}$$

where

$$\left[a_{\alpha\beta}\right]=\begin{bmatrix}\cos\theta & \sin\theta\\ -\sin\theta & \cos\theta\end{bmatrix}. \tag{14.9}$$

With the aid of (14.1), the kinetic energy of a plate of area A can be expressed as follows:

$$T=\frac{1}{2}\int_V \rho\dot{u}_i\dot{u}_i dV=\frac{1}{2}\int_A\int_{-\frac{h}{2}}^{\frac{h}{2}}\rho[z^2\dot{\psi}_\alpha\dot{\psi}_\alpha+\dot{w}^2]dzdA$$

$$=\frac{1}{2}\int_A(\rho h\dot{w}^2+\frac{\rho h^3}{12}\dot{\psi}_\alpha\dot{\psi}_\alpha)dA \quad . \tag{14.10}$$

The strain energy stored in the plate is given by

$$U=\int_V W^* dV=\int_A W_p^* dA \tag{14.11}$$

where

$$W_p^*=\int_{-\frac{h}{2}}^{\frac{h}{2}}W^* dz$$

is the plate strain energy density (strain energy per unit area of the plate). By substituting (14.11) into (9.101) we obtain

$$\frac{dU}{dt}=\int_A \dot{W}_p^* dA=\int_V \tau_{ij}\dot{\varepsilon}_{ij}dV=\int_A\int_{-\frac{h}{2}}^{\frac{h}{2}}\tau_{ij}\dot{\varepsilon}_{ij}dz\,dA$$

and since A is arbitrary,

$$\dot{W}_p^*=\int_{-\frac{h}{2}}^{\frac{h}{2}}\tau_{ij}\dot{\varepsilon}_{ij}dz \quad .$$

In view of (14.2) and (14.5), we can set

$$\tau_{ij}\dot{\varepsilon}_{ij}=\tau_{\alpha\beta}\dot{\varepsilon}_{\alpha\beta}+\tau_{\alpha z}\dot{\varepsilon}_{\alpha z}+\tau_{z\alpha}\dot{\varepsilon}_{z\alpha}+\tau_{zz}\dot{\varepsilon}_{zz}$$
$$=z\tau_{\alpha\beta}\dot{m}_{\alpha\beta}+\tau_{z\alpha}\dot{q}_\alpha$$

and consequently

$$\dot{W}_p^*=\int_{-\frac{h}{2}}^{\frac{h}{2}}(z\tau_{\alpha\beta}\dot{m}_{\alpha\beta}+\tau_{z\alpha}\dot{q}_\alpha)dz=M_{\alpha\beta}\dot{m}_{\alpha\beta}+Q_\alpha\dot{q}_\alpha \quad . \tag{14.12}$$

However,

$$\dot{W}_p^*=\frac{\partial W_p^*}{\partial m_{\alpha\beta}}\dot{m}_{\alpha\beta}+\frac{\partial W_p^*}{\partial q_\alpha}\dot{q}_\alpha=M_{\alpha\beta}\dot{m}_{\alpha\beta}+Q_\alpha\dot{q}_\alpha$$

and therefore

$$\left(\frac{\partial W_p^*}{\partial m_{\alpha\beta}}-M_{\alpha\beta}\right)\dot{m}_{\alpha\beta}+\left(\frac{\partial W_p^*}{\partial q_\alpha}-Q_\alpha\right)\dot{q}_\alpha=0 \quad .$$

This result must be valid for all processes, i.e., the choice of $\dot{m}_{\alpha\beta}$ and $\dot{q}_\alpha$ is arbitrary. Thus, we conclude that

$$M_{\alpha\beta}=\frac{\partial W_p^*}{\partial m_{\alpha\beta}} \quad \text{and } Q_\alpha=\frac{\partial W_p^*}{\partial q_\alpha} \quad . \tag{14.13}$$

It is evident from (14.13) that the scalar function $W_p^*(m_{11},m_{12},m_{21},m_{22},q_1,q_2)$ completely characterizes the mechanical constitution of a homogeneous plate.

14.2 EQUATIONS OF MOTION

Equations of motion for the plate will now be derived with the aid of Hamilton's principle. With reference to (14.10), the variation of the kinetic energy is

$$\delta T=\int_A(\rho h\dot{w}\delta\dot{w}+\frac{\rho h^3}{12}\dot{\psi}_\alpha\delta\dot{\psi}_\alpha)dA$$

and

$$\int_{t_1}^{t_2}\delta T dt=-\int_{t_1}^{t_2}\int_A(\rho h\ddot{w}\delta w+\frac{\rho h^3}{12}\ddot{\psi}_\alpha\delta\psi_\alpha)dAdt \quad . \tag{14.14}$$

To obtain (14.14) we integrated by parts with respect to t and then invoked the conditions

$$\delta\psi_\alpha=\delta w=0 \text{ at } t=t_1 \text{ and } t=t_2 \quad .$$

With reference to (14.11), (14.13) and (14.5), the variation of the strain energy is found to be

$$\delta U=\int_A \delta W_p^* dA=\int_A\left(\frac{\partial W_p^*}{\partial m_{\alpha\beta}}\delta m_{\alpha\beta}+\frac{\partial W_p^*}{\partial q_\alpha}\delta q_\alpha\right)dA$$

$$=\int_A(M_{\alpha\beta}\delta m_{\alpha\beta}+Q_\alpha\delta q_\alpha)dA$$

$$=\int_A(M_{\alpha\beta}\delta\psi_{\alpha,\beta}+Q_\alpha\delta\psi_\alpha+Q_\alpha\delta w,_\alpha)dA$$

$$=\int_A\left[(M_{\alpha\beta}\delta\psi_\alpha)_{,\beta}+(Q_\alpha\delta w)_{,\alpha}\right]dA$$

$$-\int_A(M_{\alpha\beta,\beta}\delta\psi_\alpha-Q_\alpha\delta\psi_\alpha+Q_{\alpha,\alpha}\delta w)dA$$

By applying Green's theorem to the first integral in the last equation we obtain

$$\delta U=\oint_C(M_{\alpha\beta}\delta\psi_\alpha n_\beta+Q_\alpha\delta w n_\alpha)d\ell$$

$$-\int_A\left[(M_{\alpha\beta,\beta}-Q_\alpha)\delta\psi_\alpha+Q_{\alpha,\alpha}\delta w\right]dA \quad . \tag{14.15}$$

The symbol $\oint_C d\ell$ denotes the line integral taken over the boundary of the plane region A. The symbol $n_\alpha(\ell_\alpha)$ denotes the cosine of the angle included between the unit vector $\hat{n}(\hat{\ell})$, which is perpendicular (tangent) to C, and the x_α axis, as shown in Fig. 14.3.

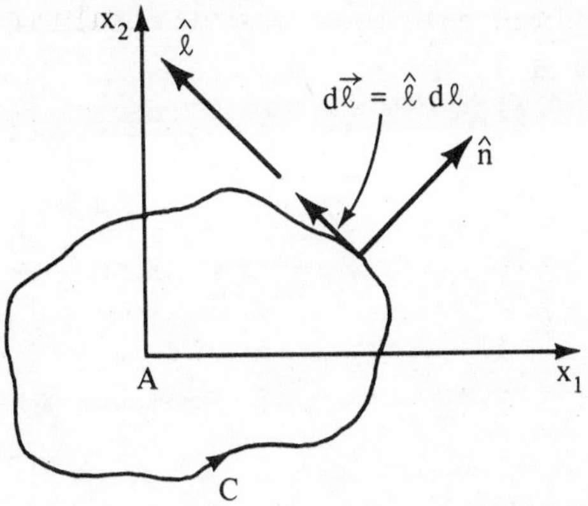

Figure 14.3

On the curve C, the vectors $\vec{\psi}$ and $\vec{Q}$ can be expressed in terms of their components in the $\hat{n}$ and $\hat{\ell}$ directions, i.e.,

$$\vec{\psi}=\psi_n\hat{n}+\psi_\ell\hat{\ell} \text{ and } \vec{Q}=Q_n\hat{n}+Q_\ell\hat{\ell}$$

or, in component form,

$$\psi_\alpha=\psi_n n_\alpha+\psi_\ell \ell_\alpha \text{ and } Q_\alpha=Q_n n_\alpha+Q_\ell \ell_\alpha \tag{14.16a}$$

where

$$\left.\begin{aligned} &\psi_n=\hat{n}\cdot\vec{\psi}=n_\alpha\psi_\alpha, \quad Q_n=\hat{n}\cdot\vec{Q}=n_\alpha Q_\alpha \\ &\text{and} \\ &\psi_\ell=\hat{\ell}\cdot\vec{\psi}=\ell_\alpha\psi_\alpha, \quad Q_\ell=\hat{\ell}\cdot\vec{Q}=\ell_\alpha Q_\alpha. \end{aligned}\right\} \tag{14.16b}$$

Hence, on C,

$$\delta\psi_\alpha=\delta\psi_n n_\alpha+\delta\psi_\ell \ell_\alpha$$

and therefore

$$\begin{aligned} M_{\alpha\beta}\delta\psi_\alpha n_\beta &= M_{\alpha\beta}n_\alpha n_\beta\delta\psi_n+M_{\alpha\beta}\ell_\alpha n_\beta\delta\psi_\ell \\ &= M_{nn}\delta\psi_n+M_{n\ell}\delta\psi_\ell \end{aligned} \tag{14.17a}$$

where

$$\left.\begin{aligned} &M_{nn}=n_\alpha n_\beta M_{\alpha\beta} \qquad \text{(no sum on n)}\\ &\text{and}\\ &M_{n\ell}=M_{\ell n}=\ell_\alpha n_\beta M_{\alpha\beta} \quad . \end{aligned}\right\} \qquad (14.17b)$$

In the preceding equations the summation convention has been suspended for the subscripts n and ℓ. These results also follow from (14.7), (14.8) and (14.9) by letting

$$\left[a_{\alpha\beta}\right] = \begin{bmatrix} n_1 & n_2 \\ \ell_1 & \ell_2 \end{bmatrix} \quad .$$

In view of (14.16) and (14.17), (14.15) can be rewritten as follows:

$$\delta U= \oint_C (M_{nn}\delta\psi_n+M_{n\ell}\delta\psi_\ell+Q_n\delta w)d\ell - \int_A \left[(M_{\alpha\beta,\beta}-Q_\alpha)\delta\psi_\alpha+Q_{\alpha,\alpha}\delta w\right]dA \quad . \qquad (14.18)$$

The virtual work done by the surface tractions acting on the plate is, according to (9.119),

$$\delta W=\int_S (T_\alpha\delta u_\alpha+T_z\delta u_z)dS=\int_S (zT_\alpha\delta\psi_\alpha+T_z\delta w)dS$$

$$=\int_A (zT_\alpha\delta\psi_\alpha+T_z\delta w)\Big|_{z=\pm\frac{h}{2}} dA+ \oint_C \int_{-\frac{h}{2}}^{\frac{h}{2}} (zT_\alpha\delta\psi_\alpha+T_z\delta w)dzd\ell$$

We shall assume that the applied loads act normal to the faces of the plate, i.e., $T_\alpha=0$ at $z=\pm\frac{h}{2}$, and we set $p(x_1,x_2,t)=(T_z)_{z=\frac{h}{2}}+(T_z)_{z=-\frac{h}{2}}$. The load resultants acting on the edges of the plate will be distinguished by superscript asterisks. Thus, (see Exercise 14.2)

$$\left.\begin{aligned} &Q_n^* = \int_{-\frac{h}{2}}^{\frac{h}{2}} T_z dz \\ &M_{nn}^* = \int_{-\frac{h}{2}}^{\frac{h}{2}} zT_n dz, \quad M_{n\ell}^* = \int_{-\frac{h}{2}}^{\frac{h}{2}} zT_\ell dz \end{aligned}\right\} \quad \text{on C.}$$

Since,

$$T_\alpha \delta\psi_\alpha = T_\alpha n_\alpha \delta\psi_n + T_\alpha \ell_\alpha \delta\psi_\ell = T_n \delta\psi_n + T_\ell \delta\psi_\ell$$

on C, we conclude that

$$\delta W = \int_A p\,\delta w\,dA + \oint_C (M^*_{nn}\delta\psi_n + M^*_{n\ell}\delta\psi_\ell + Q^*_n \delta w)\,d\ell \qquad (14.19)$$

We now substitute (14.14), (14.18) and (14.19) into (9.121) and readily obtain

$$\int_{t_1}^{t_2}\int_A \left[(-\rho h\ddot{w} + Q_{\alpha,\alpha} + p)\delta w + (-\frac{\rho h^3}{12}\ddot{\psi}_\alpha + M_{\alpha\beta,\beta} - Q_\alpha)\delta\psi_\alpha\right] dA\,dt$$
$$+ \int_{t_1}^{t_2}\oint_C \left[(M^*_{nn} - M_{nn})\delta\psi_n + (M^*_{n\ell} - M_{n\ell})\delta\psi_\ell + (Q^*_n - Q_n)\delta w\right] d\ell\,dt = 0$$

The time interval $t_2 - t_1$ as well as the area and shape of the plate are arbitrary. Therefore the integrands of the triple and double integrals must vanish separately. Furthermore, δw and $\delta\psi_\alpha$ are independent and arbitrary, thus we conclude that

$$\left.\begin{aligned} \rho h\ddot{w} &= Q_{\alpha,\alpha} + p \\ \frac{\rho h^3}{12}\ddot{\psi}_\alpha &= M_{\alpha\beta,\beta} - Q_\alpha \end{aligned}\right\} \text{ in } A \qquad (14.20)$$

while

either w or Q_n

either ψ_n or M_{nn}

and

either ψ_ℓ or $M_{n\ell}$

are specified on C. (14.21)

Equations (14.20) are the stress equations of (small) motion of a plate, and (14.21) are the associated, admissible boundary conditions.

We shall now derive an expression for energy flux in the plate. With reference to (14.11), (14.12) and (14.5) we have

$$\frac{dU}{dt} = \int_A (M_{\alpha\beta}\dot{m}_{\alpha\beta} + Q_\alpha \dot{q}_\alpha)\,dA$$

$$= \int_A (M_{\alpha\beta}\dot{\psi}_{\alpha,\beta} + Q_\alpha\dot{\psi}_\alpha + Q_\alpha \dot{w}_{,\alpha})\,dA \quad .$$

In view of (14.10) and (14.20) we conclude that

$$\frac{dT}{dt} = \int_A (\rho h\dot{w}\ddot{w} + \frac{\rho h^3}{12}\dot{\psi}_\alpha\ddot{\psi}_\alpha)dA$$

$$= \int_A \left[\dot{w}(Q_{\alpha,\alpha}+p)+\dot{\psi}_\alpha(M_{\alpha\beta,\beta}-Q_\alpha)\right] dA \quad .$$

If we denote the instantaneous energy per unit area of the plate by E*, then

$$\int_A \dot{E}^* dA = \frac{d}{dt}(T+U) = \int_A (p\dot{w}-\vec{\nabla}\cdot\vec{S})dA$$

where $\vec{\nabla}\cdot\vec{S}=S_{\alpha,\alpha}$ is the divergence of $\vec{S}$, the energy flux vector, whose components are $S_\alpha=-Q_\alpha\dot{w}-M_{\beta\alpha}\dot{\psi}_\beta$. Since the area of the plate is arbitrary, we obtain the energy continuity equation (see 9.94)

$$\dot{E}^* = p\dot{w} - \vec{\nabla}\cdot\vec{S} \qquad \text{in A.} \tag{14.22}$$

With the aid of Green's theorem we find that

$$\int_A \vec{\nabla}\cdot\vec{S}dA = \oint_C \vec{S}\cdot\hat{n}d\ell = \oint_C S_n d\ell$$

where

$$S_n=S_\alpha n_\alpha=-Q_\alpha n_\alpha\dot{w}-M_{\beta\alpha}n_\alpha(n_\beta\dot{\psi}_n+\ell_\beta\dot{\psi}_\ell)$$

$$S_n=-Q_n\dot{w}-M_{nn}\dot{\psi}_n-M_{n\ell}\dot{\psi}_\ell$$

is the energy flux (energy per unit length per unit time) leaving the plate across its boundary C. Consequently, if there are no loads acting on the faces of the plate, i.e., if p=0, then

$$\frac{d}{dt}(T+U) = -\oint_C S_n d\ell = \oint_C (Q_n\dot{w}+M_{nn}\dot{\psi}_n+M_{n\ell}\dot{\psi}_\ell)d\ell \quad . \tag{14.23}$$

Thus, the rate at which the total energy of the plate increases is equal to the rate at which energy flows into A across its boundary C. Note that in the case of homogeneous boundary conditions with p=0, $\frac{d}{dt}(T+U)=0$, i.e., the total energy of the plate is constant.

The stress equations of motion (14.20) together with the plate strain-displacement relations (14.5) furnish us with a system of eight partial differential equations relating the thirteen dependent variables w, ψ_α, Q_α, $m_{\alpha\beta}=m_{\beta\alpha}$, q_α, $M_{\alpha\beta}=M_{\beta\alpha}$. In order to complete this system we require five additional relations between these variables.

These are obtained by postulating a mathematical model of the plate's mechanical response, i.e., a constitutive law. We shall assume that the plate is a three-dimensional, linearly elastic, isotropic medium and consequently the stresses are related to the strains through Hooke's law (10.22). Solving (10.22) for the strains in terms of the stresses one obtains, with the aid of (10.19),

$$E\varepsilon_{ij} = (1+\nu)\tau_{ij} - \nu\tau_{kk}\delta_{ij} \quad .$$

In the notation of the present chapter, five of these six constitutive relations can be written as follows

$$E\varepsilon_{\alpha\beta} = (1+\nu)\tau_{\alpha\beta} - \nu\tau_{\gamma\gamma}\delta_{\alpha\beta} - \nu\tau_{zz}\delta_{\alpha\beta} \tag{14.24a}$$

$$\tau_{z\alpha} = 2G\varepsilon_{z\alpha} = Gq_\alpha \tag{14.24b}$$

where $\tau_{\gamma\gamma}=\tau_{11}+\tau_{22}=\tau_{\alpha\alpha}$ and $\delta_{\alpha\beta}$ is the two-dimensional Kronecker delta. For the present plate theory we choose to ignore the sixth constitutive equation $E\varepsilon_{zz}=\tau_{zz}-\nu\tau_{\gamma\gamma}=0$. By setting $\alpha=\beta$ in (14.24a) one obtains

$$E\varepsilon_{\alpha\alpha} = (1-\nu)\tau_{\alpha\alpha} - 2\nu\tau_{zz}$$

or

$$\tau_{\alpha\alpha} = \frac{E}{1-\nu}\,\varepsilon_{\alpha\alpha} + \frac{2\nu}{1-\nu}\,\tau_{zz} \quad .$$

Using the preceding result one can solve (14.24a) for $\tau_{\alpha\beta}$ in terms of $\varepsilon_{\alpha\beta}$ and τ_{zz}. The result is

$$\tau_{\alpha\beta} = \frac{E}{1+\nu}\,\varepsilon_{\alpha\beta} + \frac{\nu E}{1-\nu^2}\,\varepsilon_{\gamma\gamma}\delta_{\alpha\beta} + \frac{\nu}{1-\nu}\,\tau_{zz}\delta_{\alpha\beta}$$

or, in view of (14.2) and (14.5),

$$\tau_{\alpha\beta} = \frac{Ez}{1-\nu^2}\,[(1-\nu)m_{\alpha\beta} + \nu m\delta_{\alpha\beta}] + \frac{\nu}{1-\nu}\,\tau_{zz}\delta_{\alpha\beta} \tag{14.25}$$

where

$$m = m_{\alpha\alpha} = \psi_{\alpha,\alpha} = \vec{\nabla}\cdot\vec{\psi} \quad .$$

Upon substitution of (14.25) into (14.3a) one obtains,

$$M_{\alpha\beta} = D\,[(1-\nu)m_{\alpha\beta} + \nu m\delta_{\alpha\beta}] + \frac{\nu\delta_{\alpha\beta}}{1-\nu}\int_{-\frac{h}{2}}^{\frac{h}{2}} z\tau_{zz}\,dz \tag{14.26a}$$

where $D=\frac{Eh^3}{12(1-\nu^2)}$ is called the plate flexural rigidity. Similarly, the substitution of (14.24b) into (14.3b) yields

$$Q_\alpha = Ghq_\alpha \quad . \tag{14.26b}$$

These results will now be altered in two respects: (a) We drop the terms $\frac{\nu\delta_{\alpha\beta}}{1-\nu}\int_{-\frac{h}{2}}^{\frac{h}{2}} z\tau_{zz}dz$ from (14.26a), and (b) we change G to $\kappa^2 G$ in (14.26b) where κ^2 is a constant which will be determined in Section 14.5. This results in the following plate stress-strain relations:

$$M_{\alpha\beta} = D\left[(1-\nu)m_{\alpha\beta} + \nu m\delta_{\alpha\beta}\right] \tag{14.27a}$$

$$Q_\alpha = \kappa^2 Ghq_\alpha \tag{14.27b}$$

Upon substitution of (14.5) into (14.27), we obtain the plate stress-displacement relations:

$$M_{\alpha\beta} = \frac{D}{2}\left[(1-\nu)(\psi_{\alpha,\beta}+\psi_{\beta,\alpha})+2\nu\psi_{\gamma,\gamma}\delta_{\alpha\beta}\right] \tag{14.28a}$$

$$Q_\alpha = \kappa^2 Gh(\psi_\alpha + w,_\alpha) \text{ or } \vec{Q}=\kappa^2 Gh(\vec{\psi}+\vec{\nabla}w) \tag{14.28b}$$

Further substitution of (14.28) into (14.20) yields the displacement equations of motion

$$\frac{D}{2}\left[(1-\nu)\psi_{\alpha,\beta\beta}+(1+\nu)\psi_{\beta,\beta\alpha}\right]-\kappa^2 Gh(\psi_\alpha + w,_\alpha) = \frac{\rho h^3}{12}\ddot{\psi}_\alpha \tag{14.29a}$$

$$\kappa^2 Gh(\psi_{\alpha,\alpha}+w,_{\alpha\alpha})+p=\rho h\ddot{w}$$

or, in vector notation,

$$\frac{D}{2}\left[(1-\nu)\nabla^2\vec{\psi}+(1+\nu)\vec{\nabla}(\vec{\nabla}\cdot\vec{\psi})\right]-\kappa^2 Gh(\vec{\psi}+\vec{\nabla}w) = \frac{\rho h^3}{12}\ddot{\vec{\psi}} \tag{14.29b}$$

$$\kappa^2 Gh(\vec{\nabla}\cdot\vec{\psi}+\nabla^2 w)+p=\rho h\ddot{w} \quad .$$

Equations (14.29) were first published by R.D. Mindlin, "Influence of Rotatory Inertia and Shear on Flexural Motions of Isotropic, Elastic Plates," Journal of Applied Mechanics, Vol. 18, March 1951, pp. 31-38. The names of E. Reissner and Y.S. Uflyand are also associated with important contributions to this theory.

The plate strain energy density can be expressed in terms of the plate stress and strain resultants as follows. With reference to (14.10) and (10.17b), we have

$$W_p^* = \int_{-\frac{h}{2}}^{\frac{h}{2}} W^* dz = \frac{1}{2}\int_{-\frac{h}{2}}^{\frac{h}{2}} \tau_{ij}\varepsilon_{ij} dz \tag{14.30}$$

However, in view of (14.2) and (14.5),

$$\begin{aligned}\tau_{ij}\varepsilon_{ij} &= \tau_{\alpha\beta}\varepsilon_{\alpha\beta}+\tau_{\alpha z}\varepsilon_{\alpha z}+\tau_{z\alpha}\varepsilon_{z\alpha}+\tau_{zz}\varepsilon_{zz} \\ &= z\tau_{\alpha\beta}m_{\alpha\beta}+\tau_{z\alpha}q_{\alpha} \quad .\end{aligned} \tag{14.31}$$

Substituting (14.31) into (14.30) we obtain, with the aid of (14.3),

$$W_p^* = \frac{1}{2}\left(M_{\alpha\beta}m_{\alpha\beta}+Q_{\alpha}q_{\alpha}\right) \quad . \tag{14.32}$$

The substitution of (14.27) into (14.32) yields

$$W_p^* = \frac{D}{2}\left[(1-\nu)m_{\alpha\beta}m_{\alpha\beta}+\nu m^2\right]+ \frac{1}{2}\kappa^2 Ghq_{\alpha}q_{\alpha} \tag{14.33}$$

In order to express W_p^* in terms of the plate stress resultants we must first invert (14.27). Towards this end, we set $\alpha=\beta$ in (14.27a) to obtain

$$M_{\alpha\alpha}= D(1+\nu)m=M \text{ or } m= \frac{M}{D(1+\nu)} \quad .$$

Utilizing this result we can solve (14.27a) for $m_{\alpha\beta}$ and obtain

$$m_{\alpha\beta}= \frac{1}{(1-\nu^2)D}\left[(1+\nu)M_{\alpha\beta}-\nu M\delta_{\alpha\beta}\right] \quad . \tag{14.34a}$$

From (14.27b) we obtain

$$q_{\alpha}= \frac{Q_{\alpha}}{\kappa^2 Gh} \quad . \tag{14.34b}$$

In view of (14.32) and (14.34) we conclude that

$$W_p^* = \frac{1}{2(1-\nu^2)D}\left[(1+\nu)M_{\alpha\beta}M_{\alpha\beta}-\nu M^2\right] + \frac{Q_{\alpha}Q_{\alpha}}{2\kappa^2 Gh} \quad . \tag{14.35}$$

Equations (14.33) and (14.35) characterize the plate strain energy density as a positive definite quadratic function of the plate strains and stresses, respectively. One can readily verify, with the aid of (3.3) and (14.33) or (14.35), that $W_p^*>0$ implies $D>0$, $-1<\nu<1$ and

$\kappa^2 Gh>0$ which in turn implies $\nu^2<1$, $E>0$ and $\kappa^2>0$. By differentiating (14.33) with respect to the plate strain components we find that

$$\left.\begin{aligned} M_{\alpha\beta} &= \frac{\partial W_p^*}{\partial m_{\alpha\beta}} = \left[D\ (1-\nu)m_{\alpha\beta}+\nu m\delta_{\alpha\beta}\right] \\ &\text{and} \\ Q_\alpha &= \frac{\partial W_p^*}{\partial q_\alpha} = \kappa^2 Ghq_\alpha \quad . \end{aligned}\right\} \tag{14.36}$$

Similarly, by differentiating (14.35) with respect to the plate stress components we find that

$$\left.\begin{aligned} m_{\alpha\beta} &= \frac{\partial W_p^*}{\partial M_{\alpha\beta}} = \frac{1}{D(1-\nu^2)}\left[(1+\nu)M_{\alpha\beta}-\nu M\delta_{\alpha\beta}\right] \\ &\text{and} \\ q_\alpha &= \frac{\partial W_p^*}{\partial Q_\alpha} = \frac{Q_\alpha}{\kappa^2 Gh} \quad . \end{aligned}\right\} \tag{14.37}$$

We next derive the analog for plates of the reciprocal relation of Betti and Rayleigh (see Section 10.3). Let $(m_{\alpha\beta}^{(1)}, q_\alpha^{(1)})$ and $(m_{\alpha\beta}^{(2)}, q_\alpha^{(2)})$ represent two different states of plate strain which a given plate can experience and let $(M_{\alpha\beta}^{(1)}, Q_\alpha^{(1)})$ and $(M_{\alpha\beta}^{(2)}, Q_\alpha^{(2)})$ represent the corresponding plate stresses. In view of (14.36),

$$M_{\alpha\beta}^{(1)} m_{\alpha\beta}^{(2)} = D\left[(1-\nu)m_{\alpha\beta}^{(1)} m_{\alpha\beta}^{(2)} + \nu m^{(1)} m^{(2)}\right] = M_{\alpha\beta}^{(2)} m_{\alpha\beta}^{(1)}$$

$$Q_\alpha^{(1)} q_\alpha^{(2)} = \kappa^2 Ghq_\alpha^{(1)} q_\alpha^{(2)} = Q_\alpha^{(2)} q_\alpha^{(1)} \quad .$$

Consequently, with reference to (14.32),

$$W_p^{*(1,2)} = \frac{1}{2}\left[M_{\alpha\beta}^{(1)} m_{\alpha\beta}^{(2)} + Q_\alpha^{(1)} q_\alpha^{(2)}\right] = \frac{1}{2}\left[M_{\alpha\beta}^{(2)} m_{\alpha\beta}^{(1)} + Q_\alpha^{(2)} q_\alpha^{(1)}\right] = W_p^{*(2,1)}$$

and therefore

$$U^{(1,2)} = \int_A W_p^{*(1,2)} dA = \int_A W_p^{*(2,1)} dA = U^{(2,1)} \tag{14.38}$$

In view of (14.5),

$$U^{(1,2)} = \frac{1}{2}\int_A \left[M_{\alpha\beta}^{(1)} m_{\alpha\beta}^{(2)} + Q_\alpha^{(1)} q_\alpha^{(2)}\right] dA$$

$$= \frac{1}{2}\int_A \left[M_{\alpha\beta}^{(1)} \psi_{\alpha,\beta}^{(2)} + Q_\alpha^{(1)} \psi_\alpha^{(2)} + Q_\alpha^{(1)} w_{,\alpha}^{(2)}\right] dA$$

$$= \frac{1}{2}\int_A \left[(M_{\alpha\beta}^{(1)} \psi_\alpha^{(2)})_{,\beta} + (Q_\alpha^{(1)} w^{(2)})_{,\alpha}\right] dA$$

$$- \frac{1}{2}\int_A \left[(M_{\alpha\beta,\beta}^{(1)} - Q_\alpha^{(1)})\psi_\alpha^{(2)} + Q_{\alpha,\alpha}^{(1)} w^{(2)}\right] dA$$

The application of Green's theorem to the preceding equation yields, with the aid of (14.16) and (14.17),

$$U^{(1,2)} = \frac{1}{2}\oint_C \left[M_{nn}^{(1)} \psi_n^{(2)} + M_{n\ell}^{(1)} \psi_\ell^{(2)} + Q_n^{(1)} w^{(2)}\right] d\ell$$

$$- \frac{1}{2}\int_A \left[(M_{\alpha\beta,\beta}^{(1)} - Q_\alpha^{(1)})\psi_\alpha^{(2)} + Q_{\alpha,\alpha}^{(1)} w^{(2)}\right] dA \quad . \qquad (14.39a)$$

Similarly,

$$U^{(2,1)} = \frac{1}{2}\oint_C \left[M_{nn}^{(2)} \psi_n^{(1)} + M_{n\ell}^{(2)} \psi_\ell^{(1)} + Q_n^{(2)} w^{(1)}\right] d\ell$$

$$- \frac{1}{2}\int_A \left[(M_{\alpha\beta,\beta}^{(2)} - Q_\alpha^{(2)})\psi_\alpha^{(1)} + Q_{\alpha,\alpha}^{(2)} w^{(1)}\right] dA \quad . \qquad (14.39b)$$

By substituting (14.39) into (14.38) we obtain the Betti-Rayleigh reciprocal relation for plates, i.e., $U^{(1,2)}=U^{(2,1)}$. In problems of static equilibrium, $M_{\alpha\beta,\beta}-Q_\alpha=0$ and $Q_{\alpha,\alpha}=-p$ and therefore the Betti-Rayleigh reciprocal relation assumes the form

$$\int_A p^{(1)} w^{(2)} dA + \oint_C (M_{nn}^{(1)} \psi_n^{(2)} + M_{n\ell}^{(1)} \psi_\ell^{(2)} + Q_n^{(1)} w^{(2)}) d\ell$$

$$= \int_A p^{(2)} w^{(1)} dA + \oint_C (M_{nn}^{(2)} \psi_n^{(1)} + M_{n\ell}^{(2)} \psi_\ell^{(1)} + Q_n^{(2)} w^{(1)}) d\ell \qquad (14.40)$$

With reference to (14.40), we can state that if a linearly elastic plate in static equilibrium is subjected to two systems of loads, then the work which would be done by the first system of loads ($p^{(1)}, M_{nn}^{(1)}, M_{n\ell}^{(1)}, Q_n^{(1)}$) in acting through the displacements ($w^{(2)}, \psi_n^{(2)}, \psi_\ell^{(2)}$) produced by the second system of loads is equal to the work which would be done by the second system of loads if they acted through the displacements produced by the first system of loads. A dynamic reciprocal relation similar to (10.37) may be obtained by noting that (14.38) and (14.39) remain valid if each of the quantities appearing in those equations is replaced by its Laplace Transform. The details of this derivation are similar to those followed in arriving at (10.37) and are left as an Exercise for the reader. The result is,

$$U^{*(1,2)} = U^{*(2,1)} \tag{14.41}$$

where

$$U^{*(1,2)} = \int_A p^{(1)} * w^{(2)} dA + \oint_C (M_{nn}^{(1)} * \psi_n^{(2)} + M_{n\ell}^{(1)} * \psi_\ell^{(2)} + Q_n^{(1)} * w^{(2)}) d\ell$$

$$+ \int_A \left[\frac{\rho h^3}{12} (\psi_{\alpha_o}^{(1)} \dot{\psi}_\alpha^{(2)} + \dot{\psi}_{\alpha_o}^{(1)} \psi_\alpha^{(2)}) + \rho h (w_o^{(1)} \dot{w}^{(2)} + \dot{w}_o^{(1)} w^{(2)}) \right] dA \tag{14.42}$$

with

$$w_o = w(x,0), \dot{w}_o = \dot{w}(x,0), \psi_{\alpha_o} = \psi_\alpha(x,0), \dot{\psi}_{\alpha_o} = \dot{\psi}_\alpha(x,0)$$

and

$$f*g \equiv \int_0^t f(\tau) g(t-\tau) d\tau = \int_0^t f(t-\tau) g(\tau) d\tau \quad .$$

To obtain $U^{*(2,1)}$, simply interchange the superscripts (1) and (2) on both sides of (14.42). In the case of zero initial conditions for both states (1) and (2) we have

$$\int_A p^{(1)} * w^{(2)} dA + \oint_C (M_{nn}^{(1)} * \psi_n^{(2)} + M_{n\ell}^{(1)} * \psi_\ell^{(2)} + Q_n^{(1)} * w^{(2)}) d\ell$$

$$= \int_A p^{(2)} * w^{(1)} dA + \oint_C (M_{nn}^{(2)} * \psi_n^{(1)} + M_{n\ell}^{(2)} * \psi_\ell^{(1)} + Q_n^{(2)} * w^{(1)}) d\ell \quad . \tag{14.43}$$

14.3 FORCED MOTION

A major problem in the application of the theory of elastic plates is concerned with the determination of the deformations and stresses in a plate which is subjected to time-dependent, transverse surface loads as well as time-dependent boundary conditions. A well posed forced motion problem can be stated as follows: Find the solution $w=w(x_1,x_2,t)$, $\psi_\alpha=\psi_\alpha(x_1,x_2,t)$ in A (the area bounded by C) of the partial differential equations

$$\rho h\ddot{w} = Q_{\alpha,\alpha}+ p \tag{14.44a}$$

$$\frac{\rho h^3}{12}\ddot{\psi}_\alpha = M_{\alpha\beta,\beta}- Q_\alpha \tag{14.44b}$$

where

$$Q_\alpha= \kappa^2 Gh(\psi_\alpha+w,_\alpha) \tag{14.44c}$$

and

$$M_{\alpha\beta}= \frac{D}{2}\left[(1-\nu)(\psi_{\alpha,\beta}+\psi_{\beta,\alpha})+2\nu\psi_{\gamma,\gamma}\delta_{\alpha\beta}\right] \tag{14.44d}$$

which satisfies the following three boundary conditions on C

$$\text{(1) either } w=f_1(\ell,t) \text{ or } Q_n=Q_\alpha n_\alpha=f_1(\ell,t) \tag{14.45a}$$

$$\text{(2) either } \psi_n=f_2(\ell,t) \text{ or } M_{nn}=M_{\alpha\beta}n_\alpha n_\beta=f_2(\ell,t) \tag{14.45b}$$

$$\text{(3) either } \psi_\ell=f_3(\ell,t) \text{ or } M_{n\ell}=M_{\alpha\beta}\ell_\alpha n_\beta=f_3(\ell,t) \tag{14.45c}$$

and the following initial conditions in A

$$\left.\begin{aligned} &w(x_1,x_2,0)=w^{(o)}(x_1,x_2),\ \dot{w}(x_1,x_2,0)=\dot{w}^{(o)}(x_1,x_2) \\ &\psi_\alpha(x_1,x_2,0)=\psi_\alpha^{(o)}(x_1,x_2),\ \dot{\psi}_\alpha(x_1,x_2,0)=\dot{\psi}_\alpha^{(o)}(x_1,x_2) \quad . \end{aligned}\right\} \tag{14.46}$$

In the preceding equations $p=p(x_1,x_2,t)$ and we may also assume that $\rho=\rho(x_1,x_2)$. As an Exercise, the reader may show that the initial conditions (14.46) are sufficient to insure the uniqueness of the solution of (14.44) and (14.45). To obtain a solution of the problem characterized by equations (14.44) through (14.46) we generalize the technique developed for forced motion problems of rods and beams in Chapters 12 and 13, respectively. As in those cases, we begin by considering the free vibration problem for homogeneous boundary conditions, i.e., we set $p=0$ in (14.44) and $f_i(\ell,t)=0$ in (14.45). Next, we assume a product solution of the form

$$w(x_1,x_2,t)=W^{(i)}(x_1,x_2)\cos\omega_i t \tag{14.47a}$$

$$\psi_\alpha(x_1,x_2,t)=\Psi_\alpha^{(i)}(x_1,x_2)\cos\omega_i t \tag{14.47b}$$

and because of the linearity of the present plate theory we have

$$Q_\alpha(x_1,x_2,t) = Q_\alpha^{(i)}(x_1,x_2)\cos\omega_i t \tag{14.48a}$$

$$M_{\alpha\beta}(x_1,x_2,t) = M_{\alpha\beta}^{(i)}(x_1,x_2)\cos\omega_i t \tag{14.48b}$$

where

$$Q_\alpha^{(i)} = \kappa^2 Gh(\Psi_\alpha^{(i)} + W,_\alpha^{(i)}) \tag{14.49a}$$

and

$$M_{\alpha\beta}^{(i)} = \frac{D}{2}\left[(1-\nu)(\Psi_{\alpha,\beta}^{(i)} + \Psi_{\beta,\alpha}^{(i)}) + 2\nu\Psi_{\gamma,\gamma}^{(i)}\delta_{\alpha\beta}\right] \quad . \tag{14.49b}$$

The superscript (i) is an identification tag. It is not a free index in the sense of Chapter 8. Upon substitution of (14.47), (14.48) and (14.49) into the homogeneous form of (14.44) and (14.45) we obtain

$$\left.\begin{aligned} -\rho h\omega_i^2 W^{(i)} &= Q_{\alpha,\alpha}^{(i)} && (14.50a)\\ -\frac{\rho h^3}{12}\omega_i^2\Psi_\alpha^{(i)} &= M_{\alpha\beta,\beta}^{(i)} - Q_\alpha^{(i)} && (14.50b)\end{aligned}\right\} \text{ in } A$$

and

$$\left.\begin{aligned} &(1) \quad \text{either } W^{(i)}=0 \text{ or } Q_n^{(i)}=0\\ &(2) \quad \text{either } \Psi_n^{(i)}=0 \text{ or } M_{nn}^{(i)}=0\\ &(3) \quad \text{either } \Psi_\ell^{(i)}=0 \text{ or } M_{n\ell}^{(i)}=0\end{aligned}\right\} \text{ on } C \quad . \tag{14.51}$$

Nontrivial solutions of (14.50) and (14.51) exist only for certain values of ω^2 called the eigenvalues ω_i^2, i=1,2,3,... . The solution $(W^{(i)},\Psi_\alpha^{(i)})$ corresponding to the eigenvalue ω_i^2 is called the i'th eigenfunction or normal mode shape function of the plate. As indicated, we are assuming that, for a bounded plate, the set of eigenvalues is denumerable and, with the aid of the reciprocal relation, we will show that the eigenvalues are real and positive. In view of the homogeneous boundary conditions (14.51) imposed on the free vibration problem, (14.39) and (14.38) become

$$\int_A \left[(M_{\alpha\beta,\beta}^{(1)} - Q_\alpha^{(1)})\Psi_\alpha^{(2)} + Q_{\alpha,\alpha}^{(1)}W^{(2)}\right] dA$$

$$= \int_A \left[(M_{\alpha\beta,\beta}^{(2)} - Q_\alpha^{(2)})\Psi_\alpha^{(1)} + Q_{\alpha,\alpha}^{(2)}W^{(1)}\right] dA \tag{14.52}$$

By letting state (1)=state (2)=eigenstate (i) in (14.39) we obtain

$$U^{(i)}=-\frac{1}{2}\int_A\left[(M^{(i)}_{\alpha\beta,\beta}-Q^{(i)}_{\alpha})\Psi^{(i)}_{\alpha}+Q^{(i)}_{\alpha,\alpha}W^{(i)}\right]dA \qquad (14.53a)$$

If, in addition to the trivial solution, we also exclude rigid body motions from the set of eigenfunctions then $W^{*(i)}_p>0$ and therefore

$$U^{(i)}=\int_A W^{*(i)}_p dA>0 \quad . \qquad (14.53b)$$

We will now show that the eigenvalues are real. Let us assume that ω_i^2 and the corresponding eigenfunction $(W^{(i)},\Psi^{(i)}_\alpha)$ are complex. Then, in view of (14.50) and (14.51) one can easily show that $\overset{*}{\omega}{}^2_i$ is also an eigenvalue with corresponding eigenfunction $(\overset{*}{W}{}^{(i)},\overset{*}{\Psi}{}^{(i)}_\alpha)$ where an asterisk denotes the complex conjugate of the indicated quantity. Thus,

$$-\rho h\overset{*}{\omega}{}^2_i\overset{*}{W}{}^{(i)}=\overset{*}{Q}{}^{(i)}_{\alpha,\alpha} \qquad (14.54a)$$

$$-\frac{\rho h^3}{12}\overset{*}{\omega}{}^2_i\overset{*}{\Psi}{}^{(i)}_\alpha=\overset{*}{M}{}^{(i)}_{\alpha\beta,\beta}-\overset{*}{Q}{}^{(i)}_\alpha \quad . \qquad (14.54b)$$

Multiply (14.50a) by $\overset{*}{W}{}^{(i)}$ and (14.54a) by $W^{(i)}$, then subtract the results to obtain

$$(\omega_i^2-\overset{*}{\omega}{}^2_i)\rho hW^{(i)}\overset{*}{W}{}^{(i)}=\overset{*}{Q}{}^{(i)}_{\alpha,\alpha}W^{(i)}-Q^{(i)}_{\alpha,\alpha}\overset{*}{W}{}^{(i)} \qquad (14.55a)$$

Multiply (14.50b) by $\overset{*}{\Psi}{}^{(i)}_\alpha$ and (14.54b) by $\Psi^{(i)}_\alpha$ then subtract the results to obtain

$$(\omega_i^2-\overset{*}{\omega}{}^2_i)\frac{\rho h^3}{12}\Psi^{(i)}_\alpha\overset{*}{\Psi}{}^{(i)}_\alpha=(\overset{*}{M}{}^{(i)}_{\alpha\beta,\beta}-\overset{*}{Q}{}^{(i)}_\alpha)\Psi^{(i)}_\alpha-(M^{(i)}_{\alpha\beta,\beta}-Q^{(i)}_\alpha)\overset{*}{\Psi}{}^{(i)}_\alpha \qquad (14.55b)$$

Add (14.55a) to (14.55b) and then integrate the result over A to obtain, with the aid of (14.52), $(\omega_i^2-\overset{*}{\omega}{}^2_i)\,N_i^2=0$
where

$$N_i^2=\int_A(\rho hW^{(i)}\overset{*}{W}{}^{(i)}+\frac{\rho h^3}{12}\Psi^{(i)}_\alpha\overset{*}{\Psi}{}^{(i)}_\alpha)dA \quad . \qquad (14.56)$$

However, N_i^2 is clearly positive and real and $(\omega_i^2-\overset{*}{\omega}{}^2_i)=2\mathrm{Im}(\omega_i^2)$, therefore, we conclude that $\mathrm{Im}(\omega_i^2)=0$, i.e., ω_i^2 is real. As shown in Section 10.4, the linearity of the present theory together with the reality of the eigenvalues permits us to assume, without any loss in generality, that all the eigenfunctions are real. By substituting (14.50) into (14.53a) we obtain,

$$U^{(i)} = \frac{\omega_i^2}{2} N_i^2$$

or

$$\omega_i^2 = \frac{2U^{(i)}}{N_i^2} > 0 \quad . \tag{14.57}$$

Since $U^{(i)}>0$ and $N_i^2>0$ we conclude that $\omega_i^2>0$, i.e., the eigenvalues are positive. Thus, the natural frequencies ω_i are real. If we now let state (1)=eigenstate (i) and state (2)=eigenstate (j) in (14.52), and subsequently substitute (14.50) into the resulting equation, we obtain

$$(\omega_i^2-\omega_j^2)\int_A (\rho h W^{(i)} W^{(j)} + \frac{\rho h^3}{12} \Psi_\alpha^{(i)} \Psi_\alpha^{(j)}) dA = 0 \quad .$$

If $\omega_i^2 \neq \omega_j^2$, we conclude that

$$\int_A (\rho h W^{(i)} W^{(j)} + \frac{\rho h^3}{12} \Psi_\alpha^{(i)} \Psi_\alpha^{(j)}) dA = 0 \quad . \tag{14.58}$$

This is called the orthogonality relation for the normal modes. Because (14.50) and (14.51) are homogeneous, each normal mode will be specified only to within an indeterminate constant multiplier. Unique expressions for the eigenfunctions will be obtained by imposing the mode normalization condition

$$N_i = \left[\int_A (\rho h W^{(i)} W^{(i)} + \frac{\rho h^3}{12} \Psi_\alpha^{(i)} \Psi_\alpha^{(i)}) dA\right]^{\frac{1}{2}} = 1 \tag{14.59}$$

The form of the normalizing condition was selected because of its convenience in the subsequent solution of forced motion problems. We can combine (14.58) with (14.59) to obtain the orthonormality relation

$$\int_A (\rho h W^{(i)} W^{(j)} + \frac{\rho h^3}{12} \Psi_\alpha^{(i)} \Psi_\alpha^{(j)}) dA = \delta_{ij} \quad . \tag{14.60}$$

The area integral in (14.59) can be transformed into a line integral, and such an expression can be of advantage in certain problems. If we let state (1)=eigenstate (i) and state (2)=eigenstate (j) in (14.38) and (14.39), we obtain, with the aid of (14.50),

$$\int_A (\rho h W^{(i)} W^{(j)} + \frac{\rho h^3}{12} \Psi_\alpha^{(i)} \Psi_\alpha^{(j)}) dA$$

$$= \frac{1}{\omega_i^2-\omega_j^2} \oint_C \Big[M_{nn}^{(j)} \Psi_n^{(i)} - M_{nn}^{(i)} \Psi_n^{(j)} + M_{n\ell}^{(j)} \Psi_\ell^{(i)} - M_{n\ell}^{(i)} \Psi_\ell^{(j)} + Q_n^{(j)} W^{(i)} - Q_n^{(i)} W^{(j)} \Big] d\ell \qquad (14.61)$$

We note that the right-hand side of (14.61) is of the form $\frac{0}{0}$ when i=j. However, this limit can be obtained by applying L'Hospital's rule, i.e., differentiate both numerator and denominator of (14.61) with respect to ω_j and then take the limit as $j\to i$. The result is,

$$\int_A (\rho h W^{(i)} W^{(i)} + \frac{\rho h^3}{12} \Psi_\alpha^{(i)} \Psi_\alpha^{(i)}) dA$$

$$= -\frac{1}{2\omega_i} \oint_C \Big[\frac{dM_{nn}^{(i)}}{d\omega_i} \Psi_n^{(i)} - M_{nn}^{(i)} \frac{d\Psi_n^{(i)}}{d\omega_i} + \frac{dM_{n\ell}^{(i)}}{d\omega_i} \Psi_\ell^{(i)} - M_{n\ell}^{(i)} \frac{d\Psi_\ell^{(i)}}{d\omega_i} + \frac{dQ_n^{(i)}}{d\omega_i} W^{(i)} - Q_n^{(i)} \frac{dW^{(i)}}{d\omega_i} \Big] d\ell \qquad (14.62)$$

For any specified set of admissible boundary conditions, three of the terms in the line integral will vanish and, in many cases, the line integral of the remaining three terms is easier to evaluate than the area integral on the left-hand side of (14.62).

We now proceed to the solution of the forced motion problem as characterized by (14.44) through (14.46). As in the preceding chapters, we shall seek an eigenfunction expansion of the solution in the form

$$w(x_1,x_2,t) = w^{(s)}(x_1,x_2,t) + \sum_{i=1}^{\infty} W^{(i)}(x_1,x_2) g_i(t) \qquad (14.63a)$$

$$\psi_\alpha(x_1,x_2,t) = \psi_\alpha^{(s)}(x_1,x_2,t) + \sum_{i=1}^{\infty} \Psi_\alpha^{(i)}(x_1,x_2) g_i(t) \quad . \qquad (14.63b)$$

In view of (14.44c and d) we will also have

$$Q_\alpha(x_1,x_2,t) = Q_\alpha^{(s)}(x_1,x_2,t) + \sum_{i=1}^{\infty} Q_\alpha^{(i)}(x_1,x_2) g_i(t) \qquad (14.64a)$$

$$M_{\alpha\beta}(x_1,x_2,t) = M_{\alpha\beta}^{(s)}(x_1,x_2,t) + \sum_{i=1}^{\infty} M_{\alpha\beta}^{(i)}(x_1,x_2) g_i(t) \quad . \qquad (14.64b)$$

The functions $w^{(s)}$, $\psi_\alpha^{(s)}$, $Q_\alpha^{(s)}$ and $M_{\alpha\beta}^{(s)}$ represent the solution of the associated quasi-static problem, i.e., they satisfy the stress-displacement relations

$$Q_\alpha^{(s)}=\kappa^2 Gh(\psi_\alpha^{(s)}+w_{,\alpha}^{(s)}) \tag{14.65a}$$

$$M_{\alpha\beta}^{(s)}=\frac{D}{2}[(1-\nu)(\psi_{\alpha,\beta}^{(s)}+\psi_{\beta,\alpha}^{(s)})+2\nu\psi_{\gamma,\gamma}^{(s)}\delta_{\alpha\beta}] \quad , \tag{14.65b}$$

the stress equations of equilibrium

$$\left.\begin{aligned} Q_{\alpha,\alpha}^{(s)}+p&=0\\ Q_\alpha^{(s)}&=M_{\alpha\beta,\beta}^{(s)}\end{aligned}\right\} \quad \text{or } M_{\alpha\beta,\beta\alpha}^{(s)}=-p \text{ in } A \tag{14.66}$$

and the boundary conditions

$$\left.\begin{aligned} &(1)\quad \text{either } w^{(s)}=f_1(\ell,t) \text{ or } Q_n^{(s)}=f_1(\ell,t)\\ &(2)\quad \text{either } \psi_n^{(s)}=f_2(\ell,t) \text{ or } M_{nn}^{(s)}=f_2(\ell,t)\\ &(3)\quad \text{either } \psi_\ell^{(s)}=f_3(\ell,t) \text{ or } M_{n\ell}^{(s)}=f_3(\ell,t) \quad .\end{aligned}\right\} \text{ on } C \tag{14.67}$$

By substituting (14.65b) into (14.66) we obtain the displacement equation for quasi-static equilibrium

$$M_{\alpha\beta,\beta\alpha}^{(s)}=D\nabla^2 m^{(s)}=-p$$

or

$$\nabla^2 m^{(s)}=-\frac{p}{D}, \text{ where } m^{(s)}=\psi_{\alpha,\alpha}^{(s)}=\vec{\nabla}\cdot\vec{\psi}^{(s)} \quad . \tag{14.68a}$$

Next we substitute (14.65a) into the first of (14.66) to obtain

$$-p=Q_{\alpha,\alpha}^{(s)}=\kappa^2 Gh(m^{(s)}+\nabla^2 w^{(s)})$$

or

$$\nabla^2 w^{(s)}=-\frac{p}{\kappa^2 Gh}-m^{(s)} \tag{14.68b}$$

Finally, we substitute (14.65) into the second of (14.66) to obtain

$$Q_\alpha^{(s)}=M_{\alpha\beta,\beta}^{(s)}=\frac{D}{2}[(1-\nu)\nabla^2\psi_\alpha^{(s)}+(1+\nu)m_{,\alpha}^{(s)}]=\kappa^2 Gh(\psi_\alpha^{(s)}+w_{,\alpha}^{(s)}) \tag{14.68c}$$

or

$$(R^2\nabla^2-1)\vec{\psi}^{(s)}=\vec{\nabla}[w^{(s)}-\left(\frac{1+\nu}{1-\nu}\right)R^2 m^{(s)}] \tag{14.68c}$$

where

$$R^2=\frac{D(1-\nu)}{2\kappa^2 Gh}=\frac{h^2}{12\kappa^2} \quad .$$

Equations (14.68) may be manipulated further to obtain

$$\left.\begin{aligned} \nabla^4 w^{(s)} &= \frac{p}{D} - \frac{\nabla^2 p}{\kappa^2 Gh} \\ (R^2\nabla^2 - 1)\nabla^4 \vec{\psi}^{(s)} &= \vec{\nabla}\left[\frac{p}{D} - \frac{(1-\nu)}{2\kappa^2 Gh}\nabla^2 p\right] \end{aligned}\right\} \quad (14.69)$$

However, in many cases $\nabla^2 p$ may not exist because of discontinuities in the load distribution $p(x_1,x_2,t)$. In such cases it is preferable to integrate equations (14.68a), (14.68b) and (14.68c) consecutively to obtain $m^{(s)}, w^{(s)}$ and $\psi_\alpha^{(s)}$ respectively. The eigenfunctions $(W^{(i)}, \Psi_\alpha^{(i)})$ are chosen so that

$$\left.\begin{aligned} &\text{if } w^{(s)} = f_1 \text{ then } W^{(i)} = 0 \\ &\text{if } Q_n^{(s)} = f_1 \text{ then } Q_n^{(i)} = 0 \\ &\text{if } \psi_n^{(s)} = f_2 \text{ then } \Psi_n^{(i)} = 0 \\ &\text{if } M_{nn}^{(s)} = f_2 \text{ then } M_{nn}^{(i)} = 0 \\ &\text{if } \psi_\ell^{(s)} = f_3 \text{ then } \Psi_\ell^{(i)} = 0 \\ &\text{if } M_{n\ell}^{(s)} = f_3 \text{ then } M_{n\ell}^{(i)} = 0 \end{aligned}\right\} \text{ on } C \quad (14.70)$$

where $Q_\alpha^{(i)}$ and $M_{\alpha\beta}^{(i)}$ are determined by (14.49). The normal coordinates of the system $g_i(t)$ are determined as follows. Substitute (14.63) and (14.64) into (14.44) and utilize (14.66) and (14.50) to obtain

$$\sum_{i=1}^{\infty} (\ddot{g}_i + \omega_i^2 g_i)\rho h W^{(i)} = -\rho h \ddot{w}^{(s)} \quad (14.71a)$$

$$\sum_{i=1}^{\infty} (\ddot{g}_i + \omega_i^2 g_i)\frac{\rho h^3}{12}\Psi_\alpha^{(i)} = -\frac{\rho h^3}{12}\ddot{\psi}_\alpha^{(s)} \quad (14.71b)$$

Multiply (14.71a) by $W^{(j)}$ and (14.71b) by $\Psi_\alpha^{(j)}$. Add the resulting equations and then integrate this sum over A. Upon application of the orthonormality relation (14.60) to this result we obtain

$$\ddot{g}_i + \omega_i^2 g_i = \ddot{G}_i(t) \quad (14.72)$$

where

$$G_i(t) = -\int_A \left(\rho h w^{(s)} W^{(i)} + \frac{\rho h^3}{12}\psi_\alpha^{(s)}\Psi_\alpha^{(i)}\right) dA \quad . \quad (14.73)$$

The solution of (14.72) is given by (see Chapter 1)

$$g_i(t)=[g_i(0)-G_i(0)]\cos\omega_i t+\frac{1}{\omega_i}[\dot{g}_i(0)-\dot{G}_i(0)]\sin\omega_i t$$

$$+G_i(t)-\omega_i\int_0^t G_i(\tau)\sin\omega_i(t-\tau)d\tau \quad . \tag{14.74a}$$

or

$$g_i(t)=g_i(0)\cos\omega_i t+\frac{\dot{g}_i(0)}{\omega_i}\sin\omega_i t$$

$$+\frac{1}{\omega_i}\int_0^t \ddot{G}_i(\tau)\sin\omega_i(t-\tau)d\tau \tag{14.74b}$$

A more convenient form of $G_i(t)$ can be obtained as follows. In view of (14.50) and (14.73),

$$\omega_i^2 G_i(t)=\int_A [W^{(s)}Q^{(i)}_{\alpha,\alpha}+\psi^{(s)}_\alpha(M^{(i)}_{\alpha\beta,\beta}-Q^{(i)}_\alpha)]\,dA \tag{14.75}$$

If we let state (1)=eigenstate (i) and state (2)=quasi-static state (s) in (14.39) then by virtue of the reciprocal theorem (14.38) the integral on the right-hand side of (14.75) can be replaced as follows:

$$\omega_i^2 G_i(t)=\int_A [W^{(i)}Q^{(s)}_{\alpha,\alpha}+\psi^{(i)}_\alpha(M^{(s)}_{\alpha\beta,\beta}-Q^{(s)}_\alpha)]\,dA$$

$$+\oint_C [M^{(i)}_{nn}\psi^{(s)}_n-M^{(s)}_{nn}\psi^{(i)}_n+M^{(i)}_{n\ell}\psi^{(s)}_\ell$$

$$-M^{(s)}_{n\ell}\psi^{(i)}_\ell+Q^{(i)}_n W^{(s)}-Q^{(s)}_n W^{(i)}]\,d\ell$$

However, in view of the quasi-static equilibrium equations (14.66) and the boundary conditions (14.67), (14.70) we conclude that

$$\omega_i^2 G_i(t)=-\int_A pW^{(i)}dA+\oint_C [(Q^{(i)}_n-W^{(i)})f_1$$

$$+(M^{(i)}_{nn}-\psi^{(i)}_n)f_2+(M^{(i)}_{n\ell}-\psi^{(i)}_\ell)f_3]\,d\ell \quad . \tag{14.76}$$

All that remains to complete the present solution is the determination of the initial conditions for $g_i(t)$. With reference to (14.46) and (14.63) we have

$$w^{(o)}(x_1,x_2)=w^{(s)}(x_1,x_2,0)+\sum_{i=1}^{\infty} W^{(i)}(x_1,x_2)g_i(0) \tag{14.77a}$$

$$\psi_\alpha^{(o)}(x_1,x_2)=\psi_\alpha^{(s)}(x_1,x_2,0)+\sum_{i=1}^{\infty} \Psi_\alpha^{(i)}(x_1,x_2)g_i(0) \quad . \tag{14.77b}$$

We now multiply (14.77a) by $\rho h W^{(j)}$ and (14.77b) by $\frac{\rho h^3}{12}\Psi_\alpha^{(j)}$ and then add the resulting equations. This sum is then integrated over the area A, making use of the definition (14.73) and the orthonormality relation (14.60), to obtain

$$g_i(0)-G_i(0)= \int_A(\rho h w^{(o)}W^{(i)}+ \frac{\rho h^3}{12}\psi_\alpha^{(o)}\Psi_\alpha^{(i)})dA \tag{14.78a}$$

A completely analogous procedure leads to the result

$$\dot{g}_i(0)-\dot{G}_i(0)= \int_A(\rho h \dot{w}^{(o)}W^{(i)}+ \frac{\rho h^3}{12}\dot{\psi}_\alpha^{(o)}\Psi_\alpha^{(i)})dA \tag{14.78b}$$

Thus, we can conclude that the preceding technique yields a complete resolution of the forced motion problem for plates of bounded extent. This method was first published by H. Reismann, "Forced Motion of Elastic Plates," Transactions of the ASME, Journal of Applied Mechanics, Vol. 35, Series E, No. 3, Sept. 1968, pp. 510-515.

As an example of the foregoing method, we now consider a specific example. A clamped circular plate with radius a is subjected to a suddenly applied load which is uniformly distributed over a circular area with radius $b<a$ which is concentric with respect to the plate boundary. The plate is assumed to be at rest in its reference configuration at $t=0$. Because the plate boundary is a circle, we shall find it convenient to refer all necessary equations to plane, polar coordinates with origin at the center of the plate as shown in Fig. 14.4.

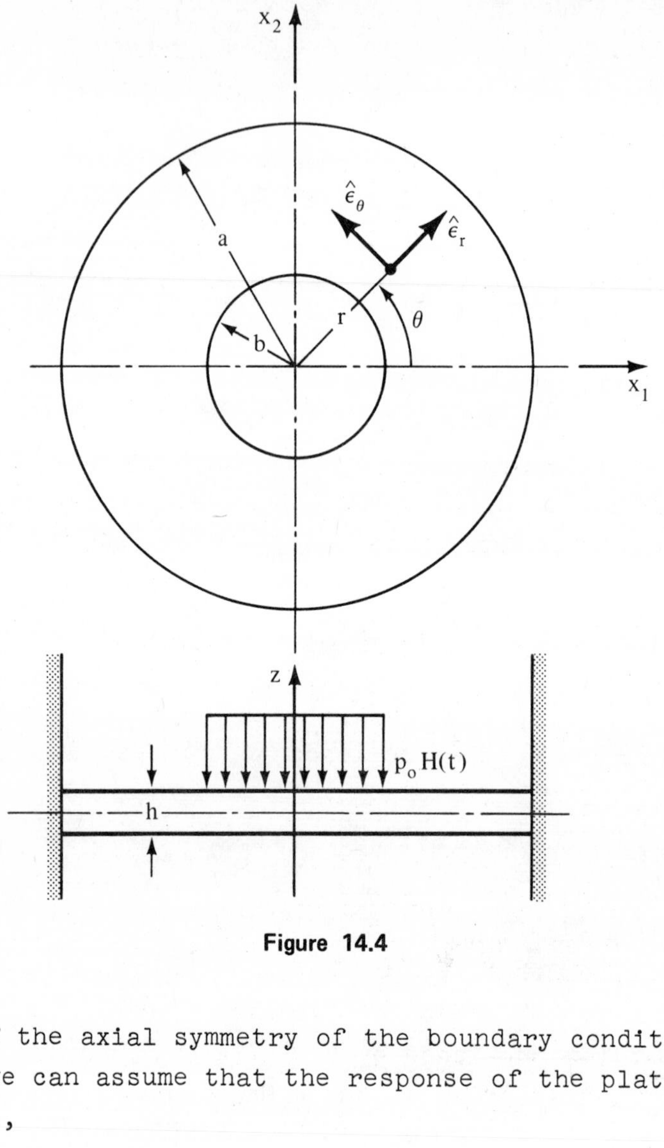

Figure 14.4

In view of the axial symmetry of the boundary conditions and the applied load, we can assume that the response of the plate is axially symmetric, i.e.,

$$w = w(r,t), \ \vec{\psi} = \psi_r(r,t)\hat{\epsilon}_r \quad .$$

The equations of motion (14.29b) become in this case (see Exercise 14.5),

$$\kappa^2 Gh\left(\frac{\partial^2 w}{\partial r^2} + \frac{1}{r}\frac{\partial w}{\partial r} + \frac{\partial \psi_r}{\partial r} + \frac{\psi_r}{r}\right) + p(r,t) = \rho h \frac{\partial^2 w}{\partial t^2}$$
$$D\left(\frac{\partial^2 \psi_r}{\partial r^2} + \frac{1}{r}\frac{\partial \psi_r}{\partial r} - \frac{\psi_r}{r^2}\right) - \kappa^2 Gh\left(\psi_r + \frac{\partial w}{\partial r}\right) = \frac{\rho h^3}{12}\frac{\partial^2 \psi_r}{\partial t^2} \quad . \qquad (14.79)$$

The boundary conditions are

$$w(a,t) = \psi_r(a,t) = 0 \tag{14.80}$$

and the initial conditions are

$$w(r,0)=\dot{w}(r,0)=\psi_r(r,0)=\dot{\psi}_r(r,0)=0 \quad . \tag{14.81}$$

The associated free vibration problem is characterized by the displacement field

$$\begin{aligned} w(r,t)&=W^{(i)}(r)\cos\omega_i t \\ \psi_r(r,t)&=\Psi^{(i)}(r)\cos\omega_i t \quad . \end{aligned} \tag{14.82}$$

Upon substitution of (14.82) into (14.79) with p=0, we obtain

$$(\Delta+\alpha k_i^2)W^{(i)}+\left(\frac{d}{dr}+\frac{1}{r}\right)\Psi^{(i)}= 0 \tag{14.83a}$$

$$-\frac{12}{\alpha h^2}\frac{dW^{(i)}}{dr}+\left(\Delta-\frac{1}{r^2}+k_i^2-\frac{12}{\alpha h^2}\right)\Psi^{(i)}=0 \tag{14.83b}$$

where

$$k_i^2 = \omega_i^2\frac{\rho h^3}{12D} = \frac{\omega_i^2}{v_p^2} \,, \quad v_p=\sqrt{\frac{E}{\rho(1-\nu^2)}} \tag{14.84}$$

$$\alpha = \frac{12}{h^2}\frac{D}{\kappa^2 Gh} = \frac{2}{\kappa^2(1-\nu)}$$

and

$$\Delta \equiv \frac{d^2}{dr^2}+\frac{1}{r}\frac{d}{dr} \quad . \tag{14.85}$$

Equations (14.83a and b) can be uncoupled as follows. Differentiate (14.83a) with respect to r and then subtract (14.83b) from the result to obtain

$$\left(k_i^2-\frac{12}{\alpha h^2}\right)\Psi^{(i)}=\frac{d}{dr}\left(\Delta+\alpha k_i^2+\frac{12}{\alpha h^2}\right)W^{(i)} \quad . \tag{14.86}$$

Now substitute (14.86) into (14.83a) to obtain

$$\left.\begin{aligned}
&\left(\Delta+k_i^{(1)^2}\right)\left(\Delta+k_i^{(2)^2}\right)W^{(i)}=0\\
&\text{where}\\
&k_i^{(1)^2}=k_i^2\delta_1^2,\quad k_i^{(2)^2}=k_i^2\delta_2^2\\
&\text{and}\\
&\begin{Bmatrix}\delta_1^2\\ \\ \delta_2^2\end{Bmatrix}=\frac{1+\alpha}{2}\pm\sqrt{\left(\frac{1-\alpha}{2}\right)^2+\frac{12}{h^2k_i^2}}
\end{aligned}\right\}\tag{14.87}$$

The solution of (14.87) can be written as

$$W^{(i)}=W_1^{(i)}+W_2^{(i)}$$

where

$$\left(\Delta+k_i^{(1)^2}\right)W_1^{(i)}=0\tag{14.88}$$

and

$$\left(\Delta+k_i^{(2)^2}\right)W_2^{(i)}=0\quad .$$

Since $\delta_1^2>0$, $k_i^{(1)^2}>0$ and therefore

$$W_1^{(i)}=C_i^{(1)}J_0(k_i^{(1)}r)+D_i^{(1)}Y_0(k_i^{(1)}r)\quad .\tag{14.89a}$$

If $\frac{12}{h^2k_i^2}<\alpha$, i.e., if $k_i^2>\frac{\kappa^2Gh}{D}$, then $\delta_2^2>0$ which in turn implies $k_i^{(2)^2}>0$ and therefore

$$W_2^{(i)}=C_i^{(2)}J_0(k_i^{(2)}r)+D_i^{(2)}Y_0(k_i^{(2)}r)\quad .\tag{14.89b}$$

However, if $\frac{12}{h^2k_i^2}>\alpha$ or, in other words, if $k_i^2<\frac{\kappa^2Gh}{D}$ then $\delta_2^2<0$ and therefore

$$W_2^{(i)}=C_i^{(2)}I_0(\overline{k}_i^{(2)}r)+D_i^{(2)}K_0(\overline{k}_i^{(2)}r)\tag{14.89c}$$

where $\overline{k}_i^{(2)}=k_i\overline{\delta}_2$ and $\overline{\delta}_2=\sqrt{-\delta_2^2}$. In the preceding equations, J_0 and Y_0 denote the Bessel functions of the first and second kind, respectively, while I_0 and K_0 denote the modified Bessel functions of the first and second kind, respectively. Since the displacement must be bounded at the origin (r=0) we can set $D_i^{(1)}=D_i^{(2)}=0$. In view of (14.88), (14.86) becomes

$$\psi^{(i)}=(\sigma_2-1)\,\frac{dW_1^{(i)}}{dr}+(\sigma_1-1)\,\frac{dW_2^{(i)}}{dr} \tag{14.90}$$

where

$$\sigma_1=\frac{\alpha\delta_1^2}{\alpha-\dfrac{12}{h^2k_i^2}}\,,\quad \sigma_2=\frac{\alpha\delta_2^2}{\alpha-\dfrac{12}{h^2k_i^2}}\ .$$

The combination of (14.88), (14.89) and (14.90) yields the following expressions for the eigenfunctions.

If $k_i^2>\frac{\kappa^2Gh}{D}$, then $\frac{12}{h^2k_i^2}<\alpha$ and

$$W^{(i)}(r)=C_i^{(1)}J_o(k_i^{(1)}r)+C_i^{(2)}J_o(k_i^{(2)}r) \tag{14.91a}$$

$$\psi^{(i)}(r)=k_i^{(1)}(1-\sigma_2)C_i^{(1)}J_1(k_i^{(1)}r)+k_i^{(2)}(1-\sigma_1)C_i^{(2)}J_1(k_i^{(2)}r)\ .$$

If $k_i^2<\frac{\kappa^2Gh}{D}$, then $\frac{12}{h^2k_i^2}>\alpha$ and

$$W^{(i)}(r)=C_i^{(1)}J_o(k_i^{(1)}r)+C_i^{(2)}I_o(\bar{k}_i^{(2)}r) \tag{14.91b}$$

$$\psi^{(i)}(r)=k_i^{(1)}(1-\sigma_2)C_i^{(1)}J_1(k_i^{(1)}r)-\bar{k}_i^{(2)}(1-\sigma_1)C_i^{(2)}I_1(\bar{k}_i^{(2)}r)\ .$$

In view of (14.80) and (14.82), the boundary conditions which the eigenfunctions must satisfy are,

$$W^{(i)}(a)=\psi^{(i)}(a)=0\ . \tag{14.92}$$

The substitution of (14.91) into (14.92) yields the following results.

If $k_i^2>\frac{\kappa^2Gh}{D}$, then

$$-\frac{C_i^{(2)}}{C_i^{(1)}}=\frac{J_o(k_i^{(1)}a)}{J_o(k_i^{(2)}a)}=\frac{\delta_1(1-\sigma_2)J_1(k_i^{(1)}a)}{\delta_2(1-\sigma_1)J_1(k_i^{(2)}a)}\ . \tag{14.93a}$$

If $k_i^2<\frac{\kappa^2Gh}{D}$, then

$$-\frac{C_i^{(2)}}{C_i^{(1)}}=\frac{J_o(k_i^{(1)}a)}{I_o(\bar{k}_i^{(2)}a)}=\frac{\delta_1(1-\sigma_2)J_1(k_i^{(1)}a)}{-\bar{\delta}_2(1-\sigma_1)I_1(\bar{k}_i^{(2)}a)}\ . \tag{14.93b}$$

By introducing the dimensionless frequency

$$\Omega_i = \frac{a\omega_i}{v_p} = ak_i \tag{14.94}$$

and the thickness ratio $H^2 = \frac{1}{12}\left(\frac{h}{a}\right)^2$, the frequency equation (14.93) can be written as follows:

If $\alpha H^2\Omega_i^2 < 1$, then

$$\frac{J_o(\delta_1\Omega_i)}{I_o(\bar{\delta}_2\Omega_i)} \cdot \frac{I_1(\bar{\delta}_2\Omega_i)}{J_1(\delta_1\Omega_i)} = -\frac{\delta_1}{\bar{\delta}_2} \cdot \frac{\alpha H^2\Omega_i^2(1-\delta_2^2)-1}{\alpha H^2\Omega_i^2(1-\delta_1^2)-1} \tag{14.95a}$$

If $\alpha H^2\Omega_i^2 > 1$, then

$$\frac{J_o(\delta_1\Omega_i)}{J_o(\delta_2\Omega_i)} \cdot \frac{J_1(\delta_2\Omega_i)}{J_1(\delta_1\Omega_i)} = \frac{\delta_1}{\delta_2} \cdot \frac{\alpha H^2\Omega_i^2(1-\delta_2^2)-1}{\alpha H^2\Omega_i^2(1-\delta_1^2)-1} \tag{14.95b}$$

where

$$\left\{\begin{matrix}\delta_1^2\\ \\ \delta_2^2\end{matrix}\right\} = \frac{1+\alpha}{2} \pm \sqrt{\left(\frac{1-\alpha}{2}\right)^2 + \frac{1}{H^2\Omega_i^2}}$$

$$\bar{\delta}_2 = \sqrt{-\delta_2^2} \text{ and } \alpha = \frac{2}{\kappa^2(1-\nu)} \quad . \tag{14.96}$$

In the limit as $H \to 0$, (14.96) indicates that

$$\delta_1 \to \sqrt{\frac{1}{H\Omega_i}} \quad , \quad \bar{\delta}_2 \to \sqrt{\frac{1}{H\Omega_i}}$$

and therefore (14.95a) becomes

$$J_o(\lambda_i a)I_1(\lambda_i a) + I_o(\lambda_i a)J_1(\lambda_i a) = 0$$

where $\lambda_i a = \sqrt{\frac{\Omega_i}{H}} = a\sqrt{\omega_i}\sqrt[4]{\frac{\rho h}{D}}$. In the next section we shall demonstrate that this is the frequency equation which one obtains upon application of the classical plate theory to the present problem. The first fourteen roots of (14.95) were computed for $\kappa^2=0.86$, $\nu=0.3$ and are shown as a function of the thickness ratio $\frac{h}{a}$ in Fig. 14.5. After the

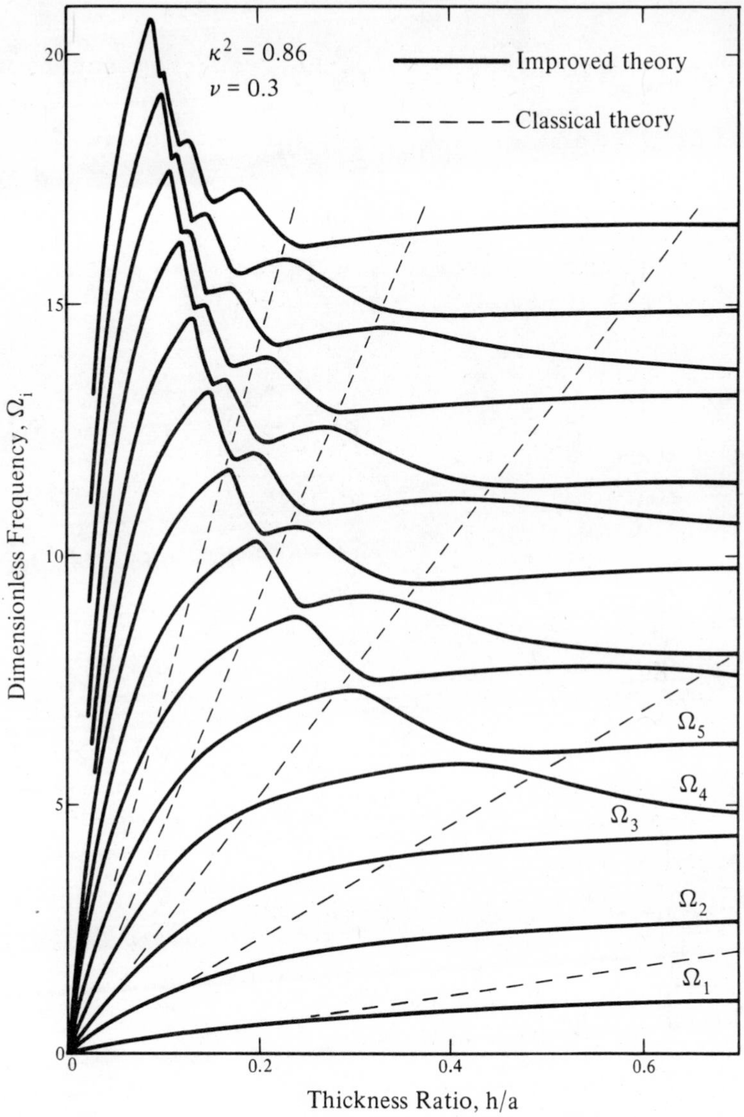

Figure 14.5

natural frequencies have been computed, the ratio of the modal co-efficients $\frac{C_i^{(2)}}{C_i^{(1)}}$ is obtained from (14.93). In order to obtain unique expressions for $C_i^{(1)}$ and $C_i^{(2)}$ we subject the eigenfunctions to the mode normalization condition (14.59). To facilitate this computation we write (14.91) as follows:

$$W^{(i)}=C_i^{(1)}F_i^{(1)}(\Omega_i\frac{r}{a}),\quad \Psi^{(i)}=k_i\delta_1C_i^{(1)}F_i^{(2)}(\Omega_i\frac{r}{a}) \tag{14.97}$$

where,

$$F_i^{(1)}(\Omega_i\frac{r}{a}) = \begin{cases} J_0(\delta_1\Omega_i\frac{r}{a})+\dfrac{C_i^{(2)}}{C_i^{(1)}}J_0(\delta_2\Omega_i\frac{r}{a}), & \text{if } \alpha H^2\Omega_i^2>1 \\ J_0(\delta_1\Omega_i\frac{r}{a})+\dfrac{C_i^{(2)}}{C_i^{(1)}}I_0(\bar{\delta}_2\Omega_i\frac{r}{a}), & \text{if } \alpha H^2\Omega_i^2<1 \end{cases}$$

and

$$F_i^{(2)}(\Omega_i\frac{r}{a}) = \begin{cases} (1-\sigma_2)J_1(\delta_1\Omega_i\frac{r}{a})+\dfrac{\delta_2}{\delta_1}(1-\sigma_1)\dfrac{C_i^{(2)}}{C_i^{(1)}}J_1(\delta_2\Omega_i\frac{r}{a}), & \text{if} \alpha H^2\Omega_i^2>1 \\ (1-\sigma_2)J_1(\delta_1\Omega_i\frac{r}{a})-\dfrac{\bar{\delta}_2}{\delta_1}(1-\sigma_1)\dfrac{C_i^{(2)}}{C_i^{(1)}}I_1(\bar{\delta}_2\Omega_i\frac{r}{a}), & \text{if} \alpha H^2\Omega_i^2<1 \end{cases}$$

where,

$$\sigma_1=\frac{\alpha\delta_1^2H^2\Omega_i^2}{\alpha H^2\Omega_i^2-1} \quad \text{and} \quad \sigma_2=\frac{\alpha\delta_2^2H^2\Omega_i^2}{\alpha H^2\Omega_i^2-1}$$

The subsequent substitution of (14.97) into (14.59) yields the result

$$C_i^{(1)}=\left\{\int_A\rho h\left[(F_i^{(1)})^2+\frac{h^2}{12}k_i^2\delta_1^2(F_i^{(2)})^2\right]dA\right\}^{-\frac{1}{2}}$$

or

$$C_i^{(1)}=\sqrt{\frac{1}{2\pi a^2\rho hP_i^2}} \tag{14.98}$$

where,

$$P_i^2=\int_0^1\left\{\left[F_i^{(1)}(\Omega_ix)\right]^2+H^2\Omega_i^2\delta_1^2\left[F_i^{(2)}(\Omega_ix)\right]^2\right\}xdx \quad .$$

By substituting $F_i^{(1)}$ and $F_i^{(2)}$ into the preceding integral one can obtain an analytical expression for P_i^2. Due to its length this expression will not be reproduced here. However, while performing the computations for a forced motion solution the integral P_i^2 can be evaluated by a numerical quadrature thus yielding the values of $C_i^{(1)}$

rather conveniently. The subsequent substitution of these values into (14.93) yields the numerical values of $C_i^{(2)}$. Having thus obtained the eigenvalues and the normalized eigenfunctions we now turn our attention to the quasi-static problem characterized by the differential equations (see (14.68)),

$$\nabla^2 m^{(s)} = \frac{1}{r}\frac{\partial}{\partial r}\left(r\frac{\partial m^{(s)}}{\partial r}\right) = -\frac{p(r,t)}{D} \tag{14.99a}$$

where,

$$m^{(s)} = \vec{\nabla}\cdot\vec{\psi} = \frac{1}{r}\frac{\partial}{\partial r}(r\psi_r^{(s)}) \tag{14.99b}$$

$$\nabla^2 w^{(s)} = \frac{1}{r}\frac{\partial}{\partial r}\left(r\frac{\partial w^{(s)}}{\partial r}\right) = -\frac{p(r,t)}{\kappa^2 Gh} - m^{(s)} \tag{14.99c}$$

$$\left(R^2\nabla^2 - \frac{R^2}{r^2} - 1\right)\psi_r^{(s)} = \frac{\partial w^{(s)}}{\partial r} - \left(\frac{1+\nu}{1-\nu}\right)R^2\frac{\partial m^{(s)}}{\partial r} \tag{14.99d}$$

and the boundary conditions

$$w^{(s)}(a,t) = \psi_r^{(s)}(a,t) = 0 \tag{14.100}$$

where,

$$p(r,t) = \begin{cases} -p_o H(t) & \text{for } 0\leq r<b \\ 0 & \text{for } b<r<a \end{cases} . \tag{14.101}$$

After substituting (14.101) into (14.99), the resulting equations can be easily integrated to obtain $m^{(s)}$, $\psi_r^{(s)}$ and $w^{(s)}$. These integrals are then subjected to the boundary conditions (14.100), the condition that $w^{(s)}$ and $\psi_r^{(s)}$ remain bounded as r approaches zero, and the condition that $w^{(s)}$, $\psi_r^{(s)}$, $\frac{\partial\psi^{(s)}}{\partial r}$, $\frac{\partial w^{(s)}}{\partial r}$ be continuous at r=b to obtain:

For $0\leq r<b$ (14.102a)

$$\psi_r^{(s)} = \frac{-p_o H(t)}{16D}ab^2\left(B^2 - 4\ \ln B - \frac{r^2}{a^2B^2}\right)\frac{r}{a}$$

$$w^{(s)} = \frac{-p_o H(t)}{64D}a^4\left[\frac{r^4}{a^4} - 2B^4\frac{r^2}{a^2} - 3B^4 + 4B^2 + 16\alpha H^2\left(B^2 - \frac{r^2}{a^2}\right)\right.$$
$$\left. + 4B^2\left(2\frac{r^2}{a^2} + B^2 - 8\alpha H^2\right)\ln B\right]$$

For $b<r<a$ (14.102b)

$$\psi_r^{(s)} = \frac{-p_0 H(t)}{16D}\, ab^2\left(B^2-4\ln\frac{r}{a} - \frac{a^2B^2}{r^2}\right)\frac{r}{a}$$

$$w^{(s)} = \frac{-p_0 H(t)}{32D}\, a^2b^2\left[(B^2+2)\left(1-\frac{r^2}{a^2}\right) + 2\left(2\frac{r^2}{a^2}+B^2-8\alpha H^2\right)\ln\frac{r}{a}\right]$$

where,

$$B = \frac{b}{a}, \quad \alpha = \frac{2}{\kappa^2(1-\nu)}, \quad H^2 = \frac{h^2}{12a^2} .$$

In view of (14.76), (14.100) and (14.101),

$$\omega_i^2 G_i(t) = 2\pi p_0 H(t)\int_0^b W^{(i)} r\,dr .$$

The substitution of (14.97) into the preceding integral yields the result

$$G_i(t) = 2\pi p_0 H(t)\,\frac{ba^3 C_i^{(1)}}{v_P^2\Omega_i^3\delta_1}\, F_i^{(3)}(\Omega_i B) \tag{14.103}$$

where

$$F_i^{(3)}(\Omega_i B) = \begin{cases} J_1(\delta_1\Omega_i B) + \dfrac{\delta_1}{\delta_2}\dfrac{C_i^{(2)}}{C_i^{(1)}} J_1(\delta_2\Omega_i B), & \text{if } \alpha H^2\Omega_i^2>1 \\[2ex] J_1(\delta_1\Omega_i B) + \dfrac{\delta_1}{\bar{\delta}_2}\dfrac{C_i^{(2)}}{C_i^{(1)}} I_1(\bar{\delta}_2\Omega_i B), & \text{if } \alpha H^2\Omega_i^2<1 \end{cases}$$

In view of the quiescent initial conditions (14.81), (14.78) becomes

$$g_i(0) = G_i(0), \quad \dot{g}_i(0) = \dot{G}_i(0)$$

and consequently (14.74) yields, for $t>0$,

$$g_i(t) = G_i(0^+)\cos\omega_i t$$

$$= 2\pi p_0\,\frac{ba^3 C_i^{(1)}}{v_P^2\delta_1\Omega_i^3}\, F_i^{(3)}(\Omega_i B)\cos\Omega_i\tau \tag{14.104}$$

where

$$\tau = \frac{\omega_i}{\Omega_i}\, t = \frac{v_P t}{a} .$$

We now substitute (14.97) and (14.104) into (14.63) to obtain, with the aid of (14.98),

$$w(r,t)=w^{(s)}(r,t)+\frac{p_o ab}{\rho h v_P^2}\sum_{i=1}^{\infty}\frac{F_i^{(1)}(\Omega_i\frac{r}{a})F_i^{(3)}(\Omega_i B)}{\delta_1\Omega_i^3 P_i^2}\cos\Omega_i\tau$$

$$\psi_r(r,t)=\psi_r^{(s)}(r,t)+\frac{p_o b}{\rho h v_P^2}\sum_{i=1}^{\infty}\frac{F_i^{(2)}(\Omega_i\frac{r}{a})F_i^{(3)}(\Omega_i B)}{\Omega_i^2 P_i^2}\cos\Omega_i\tau \qquad (14.105)$$

The stress resultants can be obtained by substituting (14.105) into the stress-displacement relations (see Exercise (14.5)),

$$M_{rr}=D\left[\frac{\partial\psi_r}{\partial r}+\frac{\nu}{r}\psi_r\right],\quad Q_r=\kappa^2 Gh(\psi_r+\frac{\partial w}{\partial r})$$

$$M_{\Theta\Theta}=D\left[\nu\frac{\partial\psi_r}{\partial r}+\frac{1}{r}\psi_r\right] \qquad (14.106)$$

$$M_{r\Theta}=Q_\Theta=0\ .$$

Equations (14.105) and (14.106) characterize the plate displacements and stress resultants as functions of the independent variables $\frac{r}{a}$ and τ and the dimensionless parameters B, H and α. Figures (14.6) through (14.13) show the dimensionless displacement $\tilde{w}(\frac{r}{a},\tau)=\frac{\rho v_P^2 h}{p_o b^2}w$ and the dimensionless bending moment $\tilde{M}_{rr}(\frac{r}{a},\tau)=\frac{M_{rr}}{p_o b^2 H^2}$ plotted versus the dimensionless time $\tau=\frac{v_P t}{a}$ at the center of the plate $\frac{r}{a}=0$ and at the edge of the plate $\frac{r}{a}=1$ for various values of $B=\frac{b}{a}$ and $H^2=\frac{h^2}{12a^2}$ with $\nu=0.3$ and $\kappa^2=0.86$ $\left(\alpha=\frac{2}{\kappa^2(1-\nu)}\cong 3.22\right)$. Superimposed upon the present solution (improved theory) are the results for the same problem obtained from the classical plate theory to be presented in the next section. Additional examples of the present method of solution can be found in a report by H. Reismann and J.E. Green, "Forced Motion of Circular Plates," Air Force Office of Scientific Research No. 67-0565, March 1967 and in H. Reismann and J.E. Green, "Forced, Axi-Symmetric Motion of Circular Plates," Developments in Theoretical and Applied Mechanics, Vol. 4, Proceedings of the Tenth Midwestern Mechanics Conference, Johnson Publishing Co., 1968, pp. 929-947. For forced motion problems of rectangular plates, see H. Reismann and Y.C. Lee, "Forced Motion of Rectangular Plates," Developments in Theoretical and Applied Mechanics, Vol. 8, Pergamon Press, Oxford 1970, pp. 3-18.

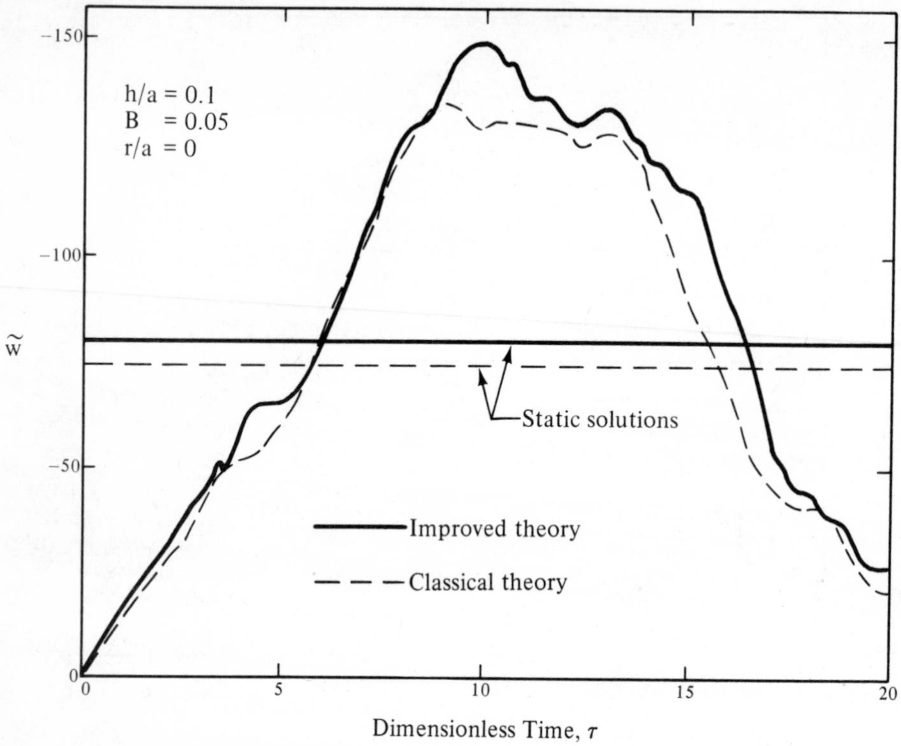
h/a = 0.1
B = 0.05
r/a = 0
-150
-100
-50
0
w̃
Static solutions
Improved theory
Classical theory
0
5
10
15
20
Dimensionless Time, τ

Figure 14.6

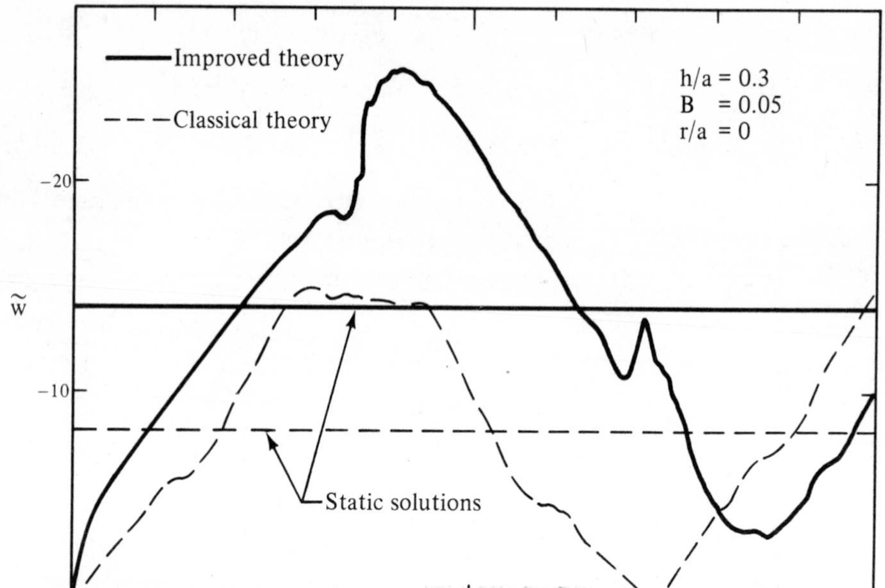
Improved theory
Classical theory
h/a = 0.3
B = 0.05
r/a = 0
-20
-10
0
w̃
Static solutions
0
5
10
Dimensionless Time, τ

Figure 14.7

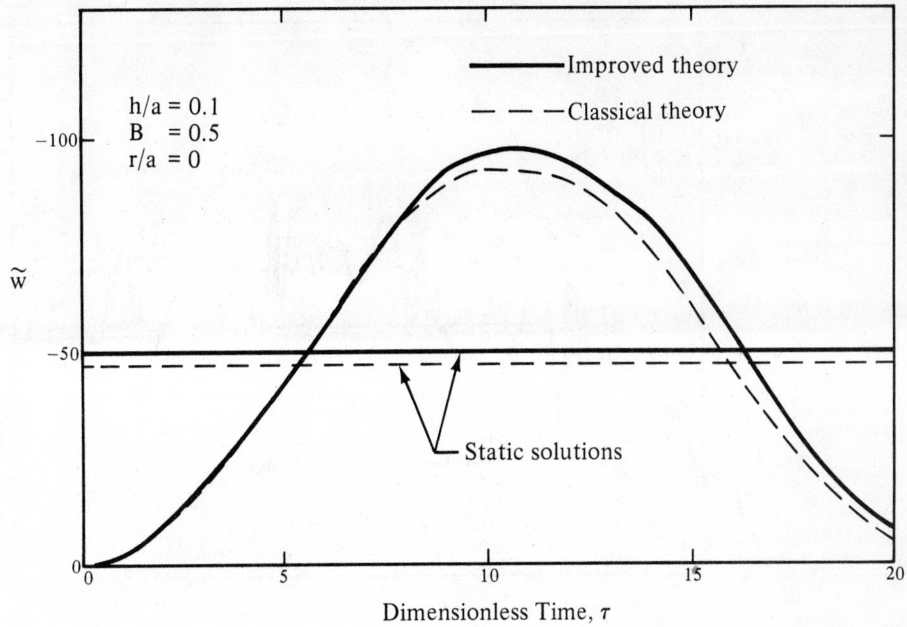
Improved theory
Classical theory
h/a = 0.1
B = 0.5
r/a = 0
−100
−50
0
w̃
Static solutions
0
5
10
15
20
Dimensionless Time, τ

Figure 14.8

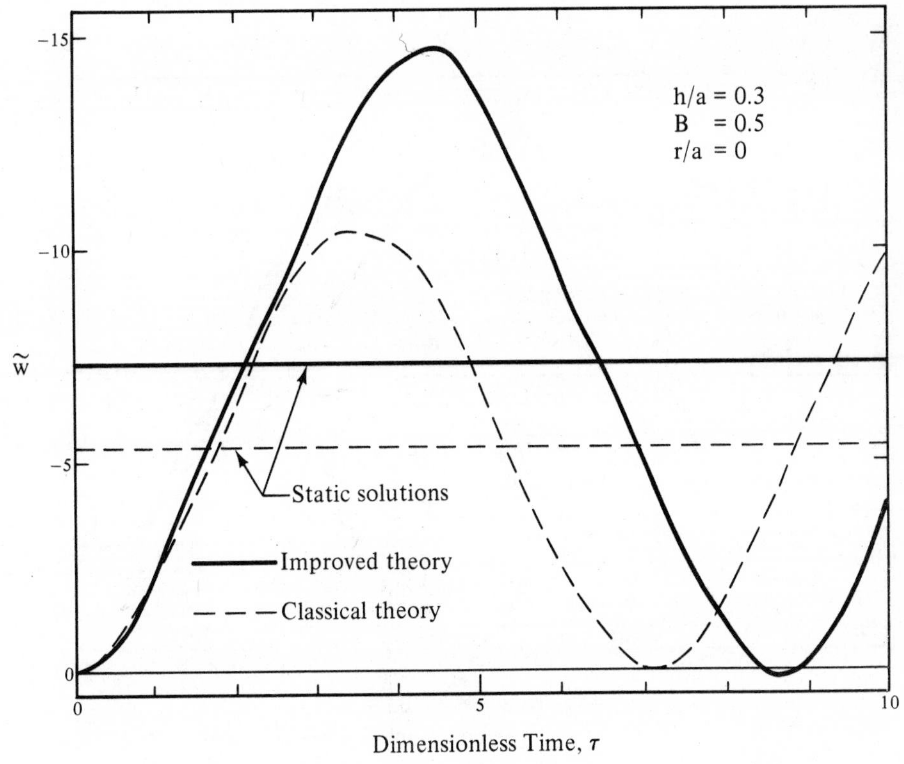
−15
−10
−5
0
w̃
h/a = 0.3
B = 0.5
r/a = 0
Static solutions
Improved theory
Classical theory
0
5
10
Dimensionless Time, τ

Figure 14.9

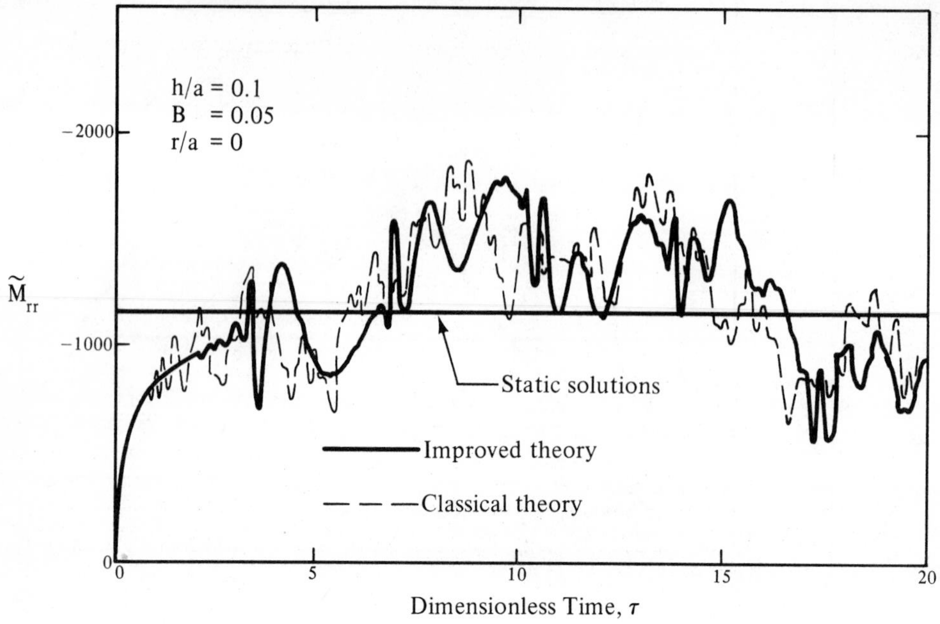

Figure 14.10

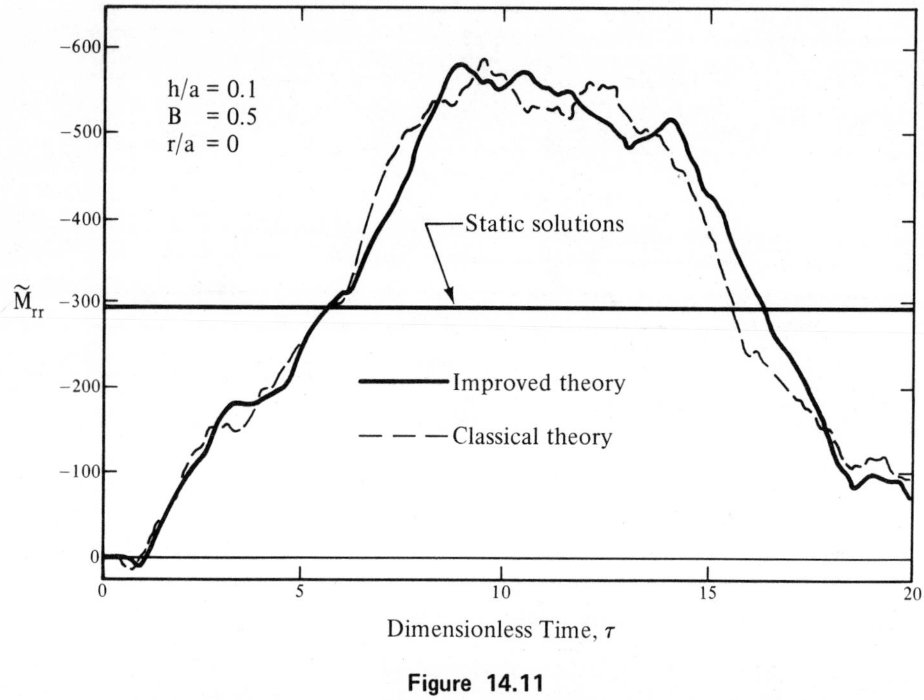

Figure 14.11

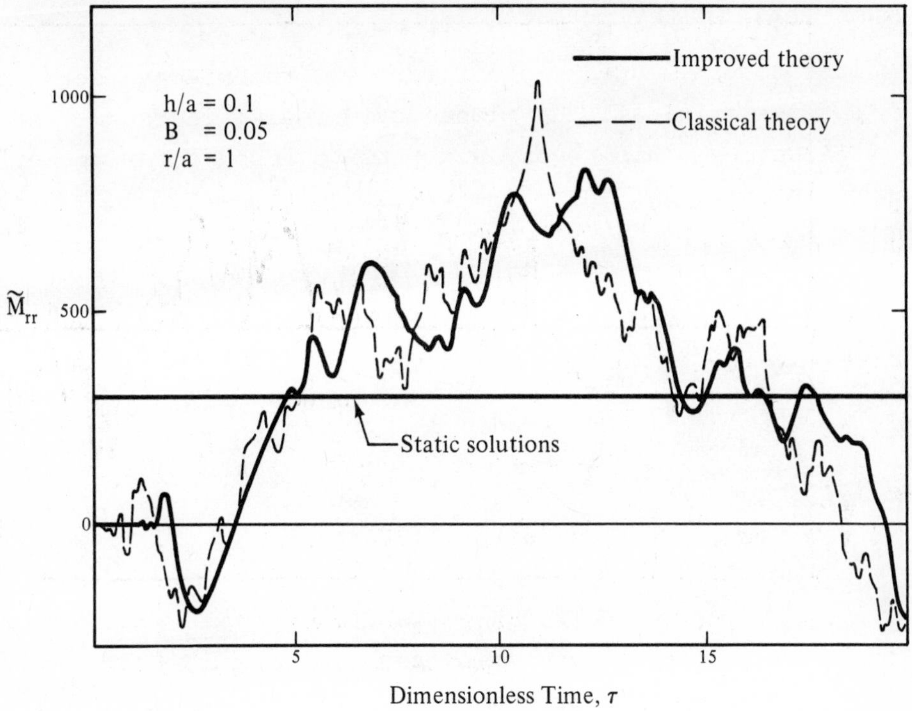

Figure 14.12

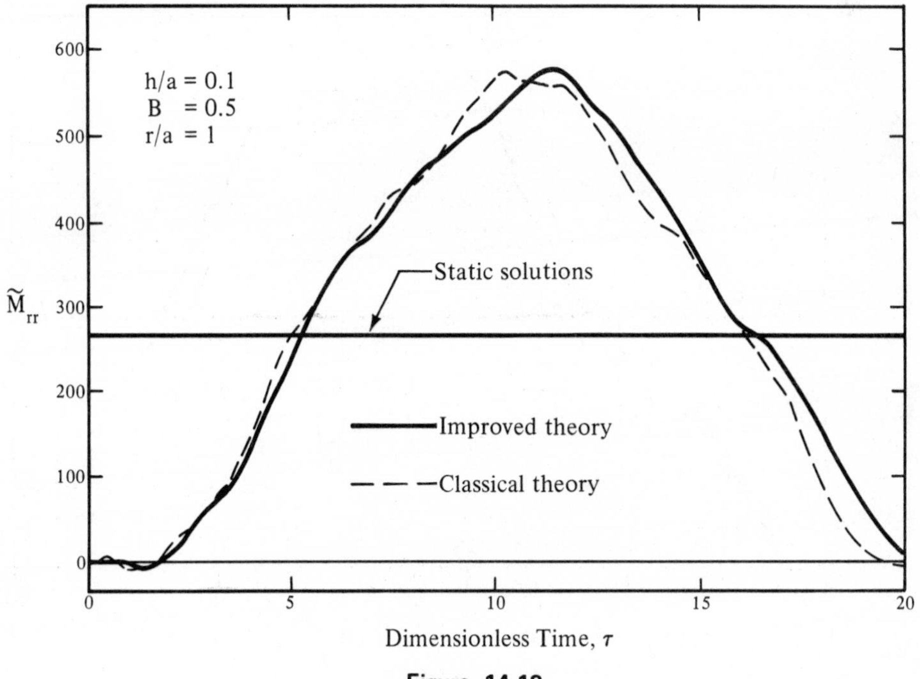

Figure 14.13

14.4 CLASSICAL PLATE THEORY

The most commonly used plate theory in present-day technical applications was originally developed by Germain, Lagrange, and Cauchy in the 19th century. This theory neglects to account for the effects of rotatory inertia, and therefore we can set

$$\frac{\rho h^3}{12}\ddot{\psi}_\alpha = 0 \tag{14.107}$$

in (14.20) to obtain the stress equations of motion for classical plate theory:

$$Q_{\alpha,\alpha} + p = \rho h\ddot{w} \tag{14.108a}$$

$$Q_\alpha = M_{\alpha\beta,\beta} \tag{14.108b}$$

or, after combining (14.108a) with (14.108b),

$$M_{\alpha\beta,\beta\alpha} + p = \rho h\ddot{w} \quad . \tag{14.109}$$

Furthermore, it is assumed that material line elements which were originally straight and normal to the plate's mid-plane in the undeformed (or natural) state remain straight and, in addition, they remain normal to the plate's median surface during the deformation, i.e., there is no transverse shear deformation, therefore,

$$\varepsilon_{\alpha z} = \frac{1}{2}(\psi_\alpha + w,_\alpha) = 0$$

or

$$\psi_\alpha = -w,_\alpha \quad . \tag{14.110}$$

As a result of (14.110), the plate strain-displacement relations (14.5) become

$$m_{\alpha\beta} = -w,_{\alpha\beta},\ q_\alpha = 0 \ . \tag{14.111}$$

By substituting (14.111) into (14.27a) we obtain, with the aid of (14.108b), the stress-displacement relations of classical plate theory:

$$M_{\alpha\beta} = -D\left[(1-\nu)w,_{\alpha\beta} + \nu(\nabla^2 w)\delta_{\alpha\beta}\right] \tag{14.112a}$$

$$Q_\alpha = M_{\alpha\beta,\beta} = -D(\nabla^2 w),_\alpha \tag{14.112b}$$

or, using vector notation,

$$\vec{Q} = -D\vec{\nabla}(\nabla^2 w) \quad . \tag{14.112c}$$

Since the plate is rigid with respect to transverse shear deformations, (14.27b) yields,

$$q_\alpha = \lim_{\kappa^2 G\to\infty} \frac{Q_\alpha}{\kappa^2 Gh} = 0 \quad .$$

By differentiating (14.112b) we find that

$$Q_{\alpha,\alpha} = -D(\nabla^2 w)_{,\alpha\alpha} = -D\nabla^2(\nabla^2 w) = -D\nabla^4 w$$

and therefore (14.108a) may be written as follows:

$$D\nabla^4 w + \rho h\ddot{w} = p \quad . \tag{14.113}$$

This is the displacement equation of motion of classical plate theory. To obtain the admissible boundary conditions appropriate to the present theory, we shall examine the variational derivation of (14.108). In view of (14.107), (14.14) becomes

$$\int_{t_1}^{t_2} \delta T dt = -\int_{t_1}^{t_2}\int_A \rho h\ddot{w}\delta w dA dt \quad . \tag{14.114}$$

After substituting (14.108) into (14.18) we obtain

$$\delta U = \oint_C (M_{nn}\delta\psi_n + M_{n\ell}\delta\psi_\ell + Q_n\delta w)d\ell$$

$$- \int_A (\rho h\ddot{w} - p)\delta w dA$$

where

$$\delta\psi_n = \delta(\psi_\alpha n_\alpha) = -\delta(w_{,\alpha} n_\alpha) = -\delta\left(\frac{\partial w}{\partial n}\right)$$

and

$$\delta\psi_\ell = \delta(\psi_\alpha \ell_\alpha) = -\delta(w_{,\alpha} \ell_\alpha) = -\delta\left(\frac{\partial w}{\partial \ell}\right) \quad .$$

However,

$$\oint_C M_{n\ell}\delta\psi_\ell d\ell = -\oint_C M_{n\ell}\frac{\partial}{\partial\ell}(\delta w)d\ell = \oint_C \frac{\partial M_{n\ell}}{\partial \ell}\delta w d\ell$$

since,

$$0 = \oint_C \frac{\partial}{\partial\ell}(M_{n\ell}\delta w)d\ell = \oint_C \frac{\partial M_{n\ell}}{\partial\ell}\delta w d\ell + \oint_C M_{n\ell}\frac{\partial}{\partial\ell}(\delta w)d\ell$$

provided that $M_{n\ell}\delta w$ is single valued on C.

Thus,

$$\delta U = \oint_C \left[\left(Q_n + \frac{\partial M_{n\ell}}{\partial\ell}\right)\delta w - M_{nn}\delta\left(\frac{\partial w}{\partial n}\right)\right]d\ell$$

$$- \int_A (\rho h\ddot{w} - p)\delta w dA \quad . \tag{14.115}$$

In a similar manner, one can show that (14.19) becomes, in the present case,

$$\delta W= \int_A p\delta w dA+ \oint_C \left[\left(Q_n+ \frac{\partial M_{n\ell}}{\partial \ell}\right)^{*} \delta w-M^{*}_{nn}\delta\left(\frac{\partial w}{\partial n}\right)\right]d\ell \tag{14.116}$$

After substituting (14.114), (14.115) and (14.116) into Hamilton's principle,

$$\int_{t_1}^{t_2}(\delta T-\delta U+\delta W)dt=0,$$

we obtain

$$\int_{t_1}^{t_2}\oint_C\left\{\left[\left(Q_n+ \frac{\partial M_{n\ell}}{\partial \ell}\right)^{*} -\left(Q_n+ \frac{\partial M_{n\ell}}{\partial \ell}\right)\right]\delta w -(M^{*}_{nn}-M_{nn})\delta\left(\frac{\partial w}{\partial n}\right)\right\} d\ell dt=0 \quad .$$

Since the interval (t_1,t_2) and the curve C are both arbitrary, we conclude that

$$\text{either } w \text{ or } \left(Q_n+ \frac{\partial M_{n\ell}}{\partial \ell}\right) \tag{14.117a}$$

and

$$\text{either } \frac{\partial w}{\partial n} \text{ or } M_{nn} \tag{14.117b}$$

must be specified on C. Thus, the classical plate theory requires the specification of only two boundary conditions along the curve C, whereas improved theory requires three (see (14.21)).

In view of (14.32), (14.111) and (14.27a), the plate strain energy density function, in classical plate theory, is given by

$$\begin{aligned} W^{*}_p &= \frac{1}{2} M_{\alpha\beta}m_{\alpha\beta}= \frac{D}{2}[(1-\nu)m_{\alpha\beta}m_{\alpha\beta}+\nu m^2] \\ &= \frac{D}{2}[(1-\nu)w_{,\alpha\beta}w_{,\alpha\beta}+\nu(\nabla^2 w)^2] \end{aligned} \tag{14.118}$$

The Betti-Rayleigh reciprocal relation for classical plate theory can be obtained by substituting $\psi_n=-\frac{\partial w}{\partial n}$, $\psi_\ell=-\frac{\partial w}{\partial \ell}$ and $M_{\alpha\beta,\beta}=Q_\alpha$ into (14.38) and (14.39).

The result is,

$$\begin{aligned} &\int_A Q^{(1)}_{\alpha,\alpha}w^{(2)}dA- \oint_C \left[\left(Q^{(1)}_n+ \frac{\partial M^{(1)}_{n\ell}}{\partial \ell}\right) w^{(2)}-M^{(1)}_{nn}\frac{\partial w^{(2)}}{\partial n}\right]d\ell \\ &=\int_A Q^{(2)}_{\alpha,\alpha}w^{(1)}dA- \oint_C \left[\left(Q^{(2)}_n+ \frac{\partial M^{(2)}_{n\ell}}{\partial \ell}\right) w^{(1)}-M^{(2)}_{nn}\frac{\partial w^{(1)}}{\partial n}\right]d\ell \quad . \end{aligned} \tag{14.119}$$

An energy continuity equation for classical plate theory can be derived with the aid of (14.108) and (14.118) as follows:

$$\frac{dU}{dt} = \int_A \dot{W}^*_p dA = \int_A M_{\alpha\beta}\dot{m}_{\alpha\beta} dA$$

$$= -\int_A M_{\alpha\beta}\dot{w}_{,\alpha\beta} dA$$

$$= \int_A \left[M_{\alpha\beta,\beta}\dot{w}_{,\alpha} - (M_{\alpha\beta}\dot{w}_{,\alpha})_{,\beta}\right] dA$$

$$= \int_A \left[Q_\alpha \dot{w}_{,\alpha} - (M_{\beta\alpha}\dot{w}_{,\beta})_{,\alpha}\right] dA$$

$$\frac{dT}{dt} = \int_A \rho h \ddot{w}\dot{w} dA = \int_A (Q_{\alpha,\alpha} + p)\dot{w} dA$$

Thus,

$$\frac{d}{dt}(T+U) = \int_A \dot{E}^* dA = \int_A [p\dot{w} - \vec{\nabla}\cdot\vec{S}] dA$$

where $\vec{\nabla}\cdot\vec{S} = S_{\alpha,\alpha}$, and

$$S_\alpha = -Q_\alpha \dot{w} + M_{\beta\alpha}\dot{w}_{,\beta} \tag{14.120}$$

is the energy flux vector. Since A is arbitrary we conclude that

$$\dot{E}^* = p\dot{w} - \vec{\nabla}\cdot\vec{S} \quad \text{in } A \quad . \tag{14.121}$$

With the aid of Green's theorem and (14.120), we find that

$$-\int_A \vec{\nabla}\cdot\vec{S} dA = -\oint_C S_\alpha n_\alpha d\ell = \oint_C \left[\left(Q_n + \frac{\partial M_{n\ell}}{\partial \ell}\right)\dot{w} - M_{nn}\frac{\partial \dot{w}}{\partial n}\right] d\ell$$

Thus,

$$\begin{aligned} \frac{d}{dt}(T+U) &= \int_A p\dot{w} dA - \oint_C S_n d\ell \\ &= \int_A p\dot{w} dA + \oint_C \left[\left(Q_n + \frac{\partial M_{n\ell}}{\partial \ell}\right)\dot{w} - M_{nn}\frac{\partial \dot{w}}{\partial n}\right] d\ell \quad . \end{aligned} \tag{14.122}$$

If there are no lateral loads on the plate (p=0) then (14.122) states that the rate of increase of energy in the plate equals the influx of energy across its boundary C. In particular, if the boundary conditions are homogeneous, then the total energy of the plate is a constant.

The central problem of any plate theory is the prediction of the dynamic response of plates subjected to time-dependent loads and/or boundary conditions. We shall now consider a method which can be used to resolve this problem for bounded plates within the framework of the classical plate theory. According to this theory, a properly posed forced motion problem can be stated as follows: Find the solution of the equation of motion

$$Q_{\alpha,\alpha} + p = \rho h\ddot{w} \qquad \text{in } A \tag{14.123a}$$

where,

$$Q_{\alpha} = M_{\alpha\beta,\beta} = -D(\nabla^2 w)_{,\alpha} \tag{14.123b}$$

and

$$M_{\alpha\beta} = -D\left[(1-\nu)w_{,\alpha\beta} + \nu(\nabla^2 w)\delta_{\alpha\beta}\right] \tag{14.123c}$$

subject to the boundary conditions

$$\begin{aligned}&(1)\ \text{either}\quad w=f_1(\ell,t)\ \text{or}\ \left(Q_n + \frac{\partial M_{n\ell}}{\partial \ell}\right)=f_1(\ell,t)\\ &(2)\ \text{either}\quad \frac{\partial w}{\partial n}=f_2(\ell,t)\ \text{or}\ M_{nn}=f_2(\ell,t)\end{aligned} \tag{14.124}$$

on the boundary curve C and satisfying the initial conditions

$$\left.\begin{aligned}w(x_1,x_2,0) &= w_o(x_1,x_2)\\ \dot{w}(x_1,x_2,0) &= \dot{w}_o(x_1,x_2)\end{aligned}\right\} \quad \text{in } A\ . \tag{14.125}$$

In (14.123), ν, D and h are constants, $\rho=\rho(x_1,x_2)$ and $p=p(x_1,x_2,t)$. It can be shown (see Exercise 14.8) that the initial conditions (14.125) are sufficient to insure a unique solution of (14.123) and (14.124). The method we shall employ to resolve this problem is the same as that which was used in Section 14.3. Thus, we first consider the free vibration problem (p=0) of a plate subject to the homogeneous boundary conditions

$$\left.\begin{aligned}&(1)\ \text{either}\quad w=0\ \text{or}\ \left(Q_n + \frac{\partial M_{n\ell}}{\partial \ell}\right)=0\\ &(2)\ \text{either}\quad \frac{\partial w}{\partial n}=0\ \text{or}\ M_{nn}=0\end{aligned}\right\}\ \text{on } C\ . \tag{14.126}$$

We shall assume that the displacement field in a plate executing such free vibrations is of the form

$$w(x_1,x_2,t)=W^{(i)}(x_1,x_2)\cos\omega_i t\ . \tag{14.127}$$

By substituting (14.127) into the homogeneous form of (14.123) we find that

$$Q^{(i)}_{\alpha,\alpha}=-\rho h\omega_i^2 W^{(i)} \tag{14.128a}$$

where

$$Q^{(i)}_{\alpha}=M^{(i)}_{\alpha\beta,\beta}=-D(\nabla^2 W^{(i)})_{,\alpha} \tag{14.128b}$$

$$M^{(i)}_{\alpha\beta}=-D\left[(1-\nu)W^{(i)}{}_{,\alpha\beta}+\nu(\nabla^2 W^{(i)})\delta_{\alpha\beta}\right] \tag{14.128c}$$

and

$$\begin{aligned} Q_\alpha(x_1,x_2,t)&=Q^{(i)}_\alpha(x_1,x_2)\cos\omega_i t \\ M_{\alpha\beta}(x_1,x_2,t)&=M^{(i)}_{\alpha\beta}(x_1,x_2)\cos\omega_i t \quad . \end{aligned} \tag{14.129}$$

Consequently, the eigenfunctions $W^{(i)}(x_1,x_2)$ of the plate are characterized by the partial differential equation

$$D\nabla^4 W^{(i)}=\rho h\omega_i^2 W^{(i)} \qquad \text{in } A \tag{14.130}$$

and the homogeneous boundary conditions

$$\left.\begin{aligned} &(1)\ \text{either } W^{(i)}=0 \text{ or } \left(Q^{(i)}_n+\frac{\partial M^{(i)}_{n\ell}}{\partial \ell}\right)=0 \\ &(2)\ \text{either } \frac{\partial W^{(i)}}{\partial n}=0 \text{ or } M^{(i)}_{nn}=0 \end{aligned}\right\} \text{ on } C \quad . \tag{14.131}$$

Nontrivial solutions of (14.130) which also satisfy (14.131) exist only for certain discrete values of ω_i called the natural frequencies of the plate. There is a denumerable infinity of real natural frequencies ω_i, i=1,2,3,... (see Exercise 14.9). The solution $W^{(i)}(x_1,x_2)$ corresponding to the natural frequency ω_i is called the i'th normal mode (or eigenfunction) of the plate. In view of their homogeneous character, equations (14.130) and (14.131) specify each normal mode only to within an indeterminate, nonzero constant multiplier. By imposing the mode normalization condition

$$N_i=\left[\int_A \rho h {W^{(i)}}^2 dA\right]^{\frac{1}{2}}= 1 \tag{14.132}$$

unique expressions for the eigenfunctions are obtained. This form of the normalization condition was chosen because of its convenience in the subsequent solution of forced motion problems. Consider two different normal modes of vibration and their associated natural frequencies ($W^{(i)}$, ω_i) and ($W^{(j)}$, ω_j). If we let state (1)=eigenstate (i) and state (2)=eigenstate (j) in the reciprocal relation (14.119), then in view of (14.128a) and (14.131) we conclude that

$$(\omega_i^2-\omega_j^2)\int_A \rho h W^{(i)} W^{(j)} dA=0 \quad .$$

Thus, if $\omega_i^2 \neq \omega_j^2$ we conclude that the eigenfunctions $W^{(i)}$ and $W^{(j)}$ are orthogonal, i.e.,

$$\int_A \rho h W^{(i)} W^{(j)} dA=0 \tag{14.133}$$

We can combine (14.132) with (14.133) to obtain the orthonormality relation

$$\int_A \rho h W^{(i)} W^{(j)} dA=\delta_{ij} \quad . \tag{14.134}$$

We will now proceed to the solution of the forced motion problem characterized by (14.123) through (14.125). We shall assume that the solution can be represented by an eigenfunction expansion of the form

$$w(x_1,x_2,t)=w^{(s)}(x_1,x_2,t)+\sum_{i=1}^{\infty} W^{(i)}(x_1,x_2)g_i(t) \quad . \tag{14.134a}$$

Because of the linearity of the present theory, we also have

$$M_{\alpha\beta}(x_1,x_2,t)=M_{\alpha\beta}^{(s)}(x_1,x_2,t)+\sum_{i=1}^{\infty} M_{\alpha\beta}^{(i)}(x_1,x_2)g_i(t) \tag{14.134b}$$

$$Q_\alpha(x_1,x_2,t)=Q_\alpha^{(s)}(x_1,x_2,t)+\sum_{i=1}^{\infty} Q_\alpha^{(i)}(x_1,x_2)g_i(t) \quad . \tag{14.134c}$$

The quasi-static solution $w^{(s)}(x_1,x_2,t)$ satisfies the partial differential equation

$$Q_{\alpha,\alpha}^{(s)}=-D\nabla^4 w^{(s)}=-p \qquad \text{in } A \tag{14.135a}$$

where,

$$Q_\alpha^{(s)}=M_{\alpha\beta,\beta}^{(s)}=-D(\nabla^2 w^{(s)})_{,\alpha} \tag{14.135b}$$

$$M_{\alpha\beta}^{(s)}=-D\left[(1-\nu)w_{,\alpha\beta}^{(s)}+\nu(\nabla^2 w^{(s)})\delta_{\alpha\beta}\right] \tag{14.135c}$$

and the following boundary conditions on C:

$$\begin{aligned}&(1)\ \text{either } w^{(s)}=f_1(\ell,t) \text{ or } \left(Q_n^{(s)}+\frac{\partial M_{n\ell}^{(s)}}{\partial \ell}\right)=f_1(\ell,t)\\ &(2)\ \text{either } \frac{\partial w^{(s)}}{\partial n}=f_2(\ell,t) \text{ or } M_{nn}^{(s)}=f_2(\ell,t) \quad .\end{aligned} \tag{14.136}$$

The boundary conditions to be satisfied by the eigenfunctions $W^{(i)}(x_1,x_2)$ in (14.134) are determined as follows:

$$\left.\begin{aligned} &\text{If } w^{(s)}=f_1 \text{ then } W^{(i)}=0 \\ &\text{If } \left(Q_n^{(s)}+\frac{\partial M_{n\ell}^{(s)}}{\partial \ell}\right)=f_1 \text{ then } \left(Q_n^{(i)}+\frac{\partial M_{n\ell}^{(i)}}{\partial \ell}\right)=0 \\ &\text{If } \frac{\partial w^{(s)}}{\partial n}=f_2 \text{ then } \frac{\partial W^{(i)}}{\partial n}=0 \\ &\text{If } M_{nn}^{(s)}=f_2 \text{ then } M_{nn}^{(i)}=0 \end{aligned}\right\} \text{ on } C \qquad (14.137)$$

We now substitute (14.134) into (14.123) and utilize (14.128) and (14.135) to obtain

$$\sum_{i=1}^{\infty}(\ddot{g}_i+\omega_i^2 g_i)\rho h W^{(i)}=-\rho h\ddot{w}^{(s)} \quad .$$

Both sides of the preceding equation are now multiplied by $W^{(j)}$ and the resulting equation is then integrated over the area A to obtain, with the aid of (14.134),

$$\ddot{g}_i+\omega_i^2 g_i=\ddot{G}_i(t) \qquad (14.138a)$$

where

$$G_i(t)=-\int_A \rho h w^{(s)} W^{(i)} dA \qquad (14.138b)$$

The solution of (14.138a) is given by (14.74). The forcing function $G_i(t)$ can be expressed in terms of the boundary conditions and the applied load as follows. In view of (14.128a) and (14.119), (14.138b) can also be written as

$$\begin{aligned} \omega_i^2 G_i(t) &= \int_A w^{(s)} Q_{\alpha,\alpha}^{(i)} dA \\ &= \int_A W^{(i)} Q_{\alpha,\alpha}^{(s)} dA \\ &\quad + \oint_C \left[\left(Q_n^{(i)}+\frac{\partial M_{n\ell}^{(i)}}{\partial \ell}\right) w^{(s)} - \left(Q_n^{(s)}+\frac{\partial M_{n\ell}^{(s)}}{\partial \ell}\right) W^{(i)} \right. \\ &\qquad \left. + M_{nn}^{(s)}\frac{\partial W^{(i)}}{\partial n} - M_{nn}^{(i)}\frac{\partial w^{(s)}}{\partial n}\right] d\ell \end{aligned}$$

With the aid of (14.135a), (14.136) and (14.137) we obtain the final result

$$\omega_i^2 G_i(t) = -\int_A pW^{(i)} dA + \oint_C \left\{ \left[\left(Q_n^{(i)} + \frac{\partial M_{n\ell}^{(i)}}{\partial \ell} \right) - W^{(i)} \right] f_1(\ell,t) + \left[\frac{\partial W^{(i)}}{\partial n} - M_{nn}^{(i)} \right] f_2(\ell,t) \right\} d\ell \qquad (14.139)$$

The initial conditions on $g_i(t)$ are obtained as follows: At t=0, (14.134a) and (14.125) can be combined to yield

$$w_o(x_1,x_2) = w^{(s)}(x_1,x_2,0) + \sum_{i=1}^{\infty} W^{(i)}(x_1,x_2)g_i(0) \qquad (14.140a)$$

$$\dot{w}_o(x_1,x_2) = \dot{w}^{(s)}(x_1,x_2,0) + \sum_{i=1}^{\infty} W^{(i)}(x_1,x_2)\dot{g}_i(0) \quad . \qquad (14.140b)$$

Multiply both sides of (14.140a) and (14.140b) by $\rho h W^{(j)}$ and then integrate the resulting equations over the area A, making use of (14.134) and (14.138b), to obtain

$$g_i(0) = G_i(0) + \int_A \rho h w_o W^{(i)} dA \qquad (14.141a)$$

$$\dot{g}_i(0) = \dot{G}_i(0) + \int_A \rho h \dot{w}_o W^{(i)} dA \qquad (14.141b)$$

This completes the formal solution of the forced motion problem characterized by (14.123) through (14.125).

We shall now consider two specific forced motion problems to illustrate the method of solution just presented.

(a) Forced Motion of a Simply Supported Rectangular Plate.

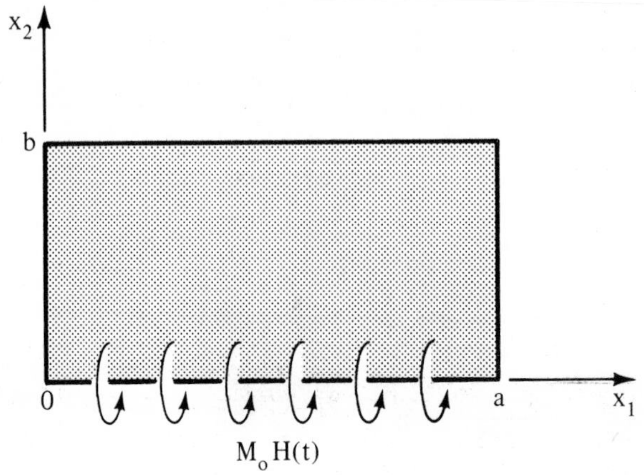

Figure 14.14

The rectangular plate shown in Fig. 14.14 is simply supported along the edges $x_2=0$, $x_2=b$, $x_1=0$ and $x_1=a$. At the instant $t=0$ a uniformly distributed bending moment M_o is suddenly applied to the edge $x_2=0$. Find the response of the plate assuming that it is at rest in its reference configuration at $t=0$. With reference to (14.123) and (14.125) we have, in the present problem,

$$p=0,\ w_o=\dot{w}_o=0 \tag{14.142}$$

and, in accordance with (14.124), the boundary conditions

$$w(0,x_2,t)=M_{11}(0,x_2,t)=0 \tag{14.143a}$$

$$w(a,x_2,t)=M_{11}(a,x_2,t)=0 \tag{14.143b}$$

$$w(x_1,0,t)=0,\ M_{22}(x_1,0,t)=M_oH(t) \tag{14.143c}$$

$$w(x_1,b,t)=M_{22}(x_1,b,t)=0 \tag{14.143d}$$

The eigenfunctions for the present problem are determined by the differential equation

$$\nabla^4W^{(i)}-\lambda_i^4W^{(i)}=0,\ \lambda_i^4=\frac{\rho h}{D}\,\omega_i^2 \tag{14.144a}$$

and the boundary conditions

$$W^{(i)}(0,x_2)=W^{(i)}(a,x_2)=W^{(i)}(x_1,0)=W^{(i)}(x_1,b)=0 \tag{14.144b}$$

$$M_{11}^{(i)}(0,x_2)=M_{11}^{(i)}(a,x_2)=M_{22}^{(i)}(x_1,0)=M_{22}^{(i)}(x_1,b)=0\quad . \tag{14.144c}$$

The solution of (14.144) can easily be shown to be

$$W^{(i)}(x_1,x_2)=A_{mn}\sin m\pi\,\frac{x_1}{a}\,\sin n\pi\,\frac{x_2}{b} \tag{14.145}$$

where $m=1,2,3,\ldots$, $n=1,2,3,\ldots$, A_{mn} is a constant and

$$\lambda_i^4=\left[\left(\frac{m\pi}{a}\right)^2+\left(\frac{n\pi}{b}\right)^2\right]^2$$

or (14.146)

$$\omega_i=\sqrt{\frac{D}{\rho h}}\,\left[\left(\frac{m\pi}{a}\right)^2+\left(\frac{n\pi}{b}\right)^2\right].$$

The correspondence between the integers i, m and n is established by ordering the natural frequencies so that $\omega_1\le\omega_2\le\omega_3\le\ldots$. For example, in the case of a square plate, $a=b$, and we can construct a table of natural frequencies as shown in Table 14.1. An examination of Table

TABLE 14.1

NATURAL FREQUENCIES OF A SIMPLY SUPPORTED SQUARE PLATE

i	m	n	$\sqrt{\frac{\rho h}{D}}\left(\frac{a}{\pi}\right)^2\omega_i$
1	1	1	2
2	1	2	5
3	2	1	5
4	2	2	8
5	1	3	10
6	3	1	10
7	2	3	13
8	3	2	13
9	1	4	17
10	4	1	17

14.1 reveals that some degeneracies occur, i.e., $\omega_2=\omega_3$ but $W^{(2)}\neq W^{(3)}$, $\omega_5=\omega_6$ but $W^{(5)} \neq W^{(6)}$, etc. However, one can easily show that $W^{(2)}$ and $W^{(3)}$ are orthogonal in the sense of (14.133). The same is true for $W^{(5)}$ and $W^{(6)}$ and all the other degenerate cases. Since our derivation of the forced motion solution is dependent upon the orthogonality relation (14.133) it is imperative that we consider only mutually orthogonal eigenfunctions. If there are no degeneracies then (14.133) results from the reciprocal relation as shown earlier. However, when degeneracies occur, the orthogonality of the degenerate eigenfunctions must be established before they can be used in the forced motion solution (see Section 10.4). As we have just seen, in the case of a simply supported square plate, degeneracies occur, however, the eigenfunctions as given by (14.145) for these degenerate cases are mutually orthogonal and can therefore be used, without further modification, in our forced motion solution. Returning to (14.146) we see that degeneracies will occur whenever the ratio $\frac{a}{b}$ is a rational number, however, the eigenfunctions (14.145) remain mutually orthogonal even in these degenerate cases. The constants A_{mn} in

(14.145) are determined by subjecting the eigenfunctions to the normalization condition (14.132) resulting in

$$A_{mn}=\left[\int_0^b\int_0^a \rho h \sin^2 m\pi \frac{x_1}{a} \sin^2 n\pi \frac{x_2}{b}\, dx_1 dx_2\right]^{-\frac{1}{2}}$$

$$= \frac{2}{\sqrt{ab\rho h}} \quad .$$

Thus, the normalized eigenfunctions are given by

$$W^{(i)}= \frac{2}{\sqrt{ab\rho h}} \sin \frac{m\pi x_1}{a} \sin \frac{n\pi x_2}{b} \, . \qquad (14.147)$$

The quasi-static solution of the present problem satisfies the differential equation

$$\nabla^4 w^{(s)} = 0 \qquad (14.148)$$

and the boundary conditions

$$\begin{aligned} &w^{(s)}(0,x_2,t)=w^{(s)}(a,x_2,t)=w^{(s)}(x_1,0,t)=w^{(s)}(x_1,b,t)=0 \\ &M_{11}^{(s)}(0,x_2,t)=M_{11}^{(s)}(a,x_2,t)=0 \\ &M_{22}^{(s)}(x_1,0,t)=M_0 H(t), \; M_{22}^{(s)}(x_1,b,t)=0 \end{aligned} \qquad (14.149)$$

As shown in Exercise 14.10, the solution of (14.148) which satisfies (14.149) is

$$w^{(s)}(x_1,x_2,t)= \frac{2M_0 H(t)}{D} \frac{a^2}{\pi^2} \sum_{m=1,3,5}^{\infty} F_m\left(\frac{x_2}{b}\right) \frac{\sin m\pi \frac{x_1}{a}}{m^2} \qquad (14.150)$$

where

$$F_m\left(\frac{x_2}{b}\right)= \frac{b}{a}\left[\frac{x_2}{b} \frac{\cosh\beta_m(1-\frac{x_2}{b})}{\sinh\beta_m} - \frac{\sinh\beta_m \frac{x_2}{b}}{\sinh^2\beta_m}\right]$$

and, $\beta_m=m\pi \frac{b}{a}$.

By substituting (14.150) into (14.135c) we find that

$$M_{22}^{(s)}(x_1,x_2,t)= \frac{4}{\pi} M_0 H(t) \sum_{m=1,3,5}^{\infty} \left[\frac{\sinh\beta_m(1-\frac{x_2}{b})}{m\sinh\beta_m} - \frac{\pi}{2}(1-\nu)F_m\left(\frac{x_2}{b}\right)\right] \sin m\pi \frac{x_1}{a} \qquad (14.151)$$

Since $F_m(0)=F_m(1)=0$, it is evident that

$$w^{(s)}(x_1,0,t)=w^{(s)}(x_1,b,t)=M_{22}^{(s)}(x_1,b,t)=0,$$

and

$$M_{22}^{(s)}(x_1,0,t)=M_oH(t)\,\frac{4}{\pi}\sum_{m=1,3,5}^{\infty}\frac{1}{m}\sin\frac{m\pi x_1}{a}=M_oH(t)\ \text{for}\ 0<x_1<a\quad.$$

If (14.142), (14.144) and (14.149) are substituted into (14.139) one obtains

$$\omega_i^2G_i(t)=\int_0^a\left(\frac{\partial W^{(i)}}{\partial n}\right)_{x_2=0}M_oH(t)dx_1\quad.$$

However, at $x_2=0$, $n_1=0$ and $n_2=-1$, and therefore

$$\left(\frac{\partial W^{(i)}}{\partial n}\right)_{x_2=0}=-\left(\frac{\partial W^{(i)}}{\partial x_2}\right)_{x_2=0}=-\frac{2}{\sqrt{ab\rho h}}\frac{n\pi}{b}\sin\frac{m\pi x_1}{a}\quad.$$

Consequently,

$$\omega_i^2G_i(t)=-\frac{4M_oH(t)}{\sqrt{ab\rho h}}\frac{na}{mb}\ ,\ m=1,3,5,\ldots$$

$$\omega_i^2G_i(t)=0,\ m=2,4,6,\ \ldots\quad. \tag{14.152}$$

The substitution of (14.142) into (14.141) yields

$$g_i(0)=G_i(0)\ \text{and}\ \dot{g}_i(0)=\dot{G}_i(0). \tag{14.153}$$

The normal coordinates of the plate can now be obtained by substituting (14.152) and (14.153) into (14.74). The result is,

$$g_i(t)=G_i(t)\cos\omega_it=\begin{cases}-\dfrac{4M_oH(t)}{\sqrt{ab\rho h}}\dfrac{an}{bm}\dfrac{\cos\omega_it}{\omega_i^2}, & m=1,3,5,\ldots\\[2ex] 0 & ,\ m=2,4,6,\ldots\end{cases} \tag{14.154}$$

The complete solution is now obtained by substituting (14.146), (14.147), (14.154) and (14.150) into (14.134). In view of (14.154), the summation over all values of i is equivalent to the double summation over all odd values of m and all values of n, i.e.,

$$\sum_{i=1}^{\infty}W^{(i)}g_i=-\frac{8M_oH(t)}{\pi^4D}\frac{a^4}{b^2}\sum_{m=1,3,5}^{\infty}\sum_{n=1}^{\infty}\frac{n}{m}\frac{\sin m\pi\frac{x_1}{a}\sin n\pi\frac{x_2}{b}\cos\Omega_{mn}\tau}{\Omega_{mn}^2} \tag{14.155}$$

where,

$$\Omega_{mn}=m^2+n^2\frac{a^2}{b^2}\ ,\quad \tau=\sqrt{\frac{D}{\rho h}}\left(\frac{\pi}{a}\right)^2 t\ .$$

Thus,

$$w(x_1,x_2,t)=\frac{2M_oH(t)}{D}\frac{a^2}{\pi^2}\sum_{m=1,3,5}^{\infty}\frac{1}{m^2}\sin m\pi\frac{x_1}{a}\left[F_m\left(\frac{x_2}{b}\right)\right.$$

$$\left.-\frac{4m}{\pi^2}\frac{a^2}{b^2}\sum_{n=1}^{\infty}\frac{n}{\Omega_{mn}^2}\sin n\pi\frac{x_2}{b}\cos\Omega_{mn}\tau\right] \qquad (14.156)$$

Figure 14.15a depicts the time history of the dimensionless displacement

$\tilde{w}=\frac{Dw}{M_oa^2}$ at the center $\left(\frac{x_1}{a}=\frac{x_2}{b}=\frac{1}{2}\right)$ of a square plate $\left(\frac{b}{a}=1\right)$.

Figure 14.15b illustrates the variation of the dimensionless displacement $\tilde{w}$ with distance from the loaded edge $\left(\tilde{w}\text{ vs. }\frac{x_2}{b}\text{ for }\frac{x_1}{a}=\frac{1}{2}\right)$ for the square plate at the instant $\tau=\frac{\pi}{\Omega_{11}}=\frac{\pi}{2}$.

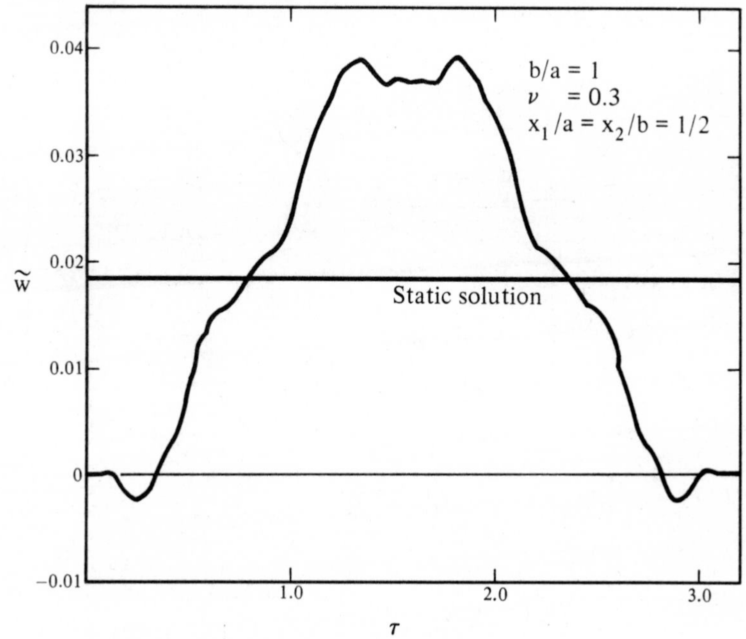

Figure 14.15 (a)

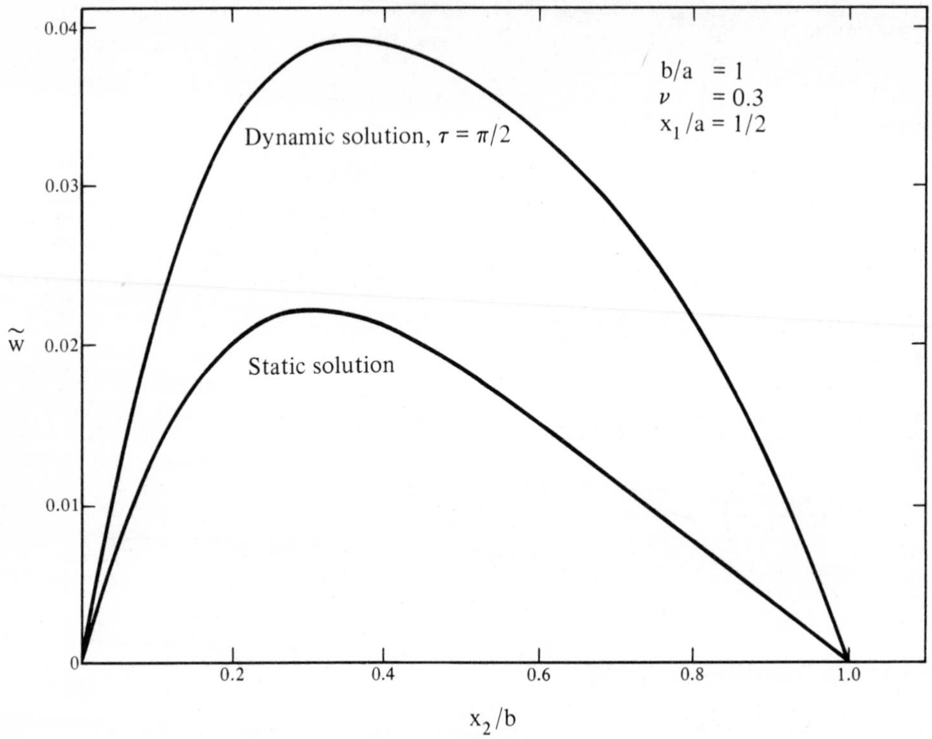

Figure 14.15 (b)

(b) Forced Motion of a Clamped, Circular Plate.

As our second example we shall reconsider, within the framework of classical plate theory, the problem of a clamped, circular plate of radius a, subjected to a suddenly applied load which is uniformly distributed over a circular area with radius $b<a$ which is concentric with respect to the plate boundary (see Fig. 14.4). The present problem is characterized by the load (14.101), the initial conditions

$$w(r,0)=\dot{w}(r,0)=0 \tag{14.157}$$

and the boundary conditions

$$w(a,t) = 0, \ \left(\frac{\partial w}{\partial r}\right)_{r=a} = 0 \ . \tag{14.158}$$

The eigenfunctions satisfy the equation

$$\nabla^4 W^{(i)} - \lambda_i^4 W^{(i)} = 0 \ , \ \lambda_i^4 = \frac{\rho h}{D}\omega_i^2$$

or (14.159)

$$(\nabla^2+\lambda_i^2)(\nabla^2-\lambda_i^2)W^{(i)}=0$$

where, $\nabla^2 = \frac{d^2}{dr^2} + \frac{1}{r}\frac{d}{dr}$. Since $\lambda_i^2 > 0$, the solution of (14.159) can be written as

$$W^{(i)}(r) = A_i J_0(\lambda_i r) + C_i Y_0(\lambda_i r)$$
$$+ B_i I_0(\lambda_i r) + D_i K_0(\lambda_i r) \quad .$$

However, we shall impose the condition that $W^{(i)}$ remain bounded as $r \to 0$. Thus, $C_i = D_i = 0$ and we obtain

$$W^{(i)}(r) = A_i J_0(\lambda_i r) + B_i I_0(\lambda_i r) \tag{14.160}$$

In view of (14.158), the eigenfunctions must satisfy the boundary conditions

$$W^{(i)}(a) = \left(\frac{dW^{(i)}}{dr}\right)_{r=a} = 0 \quad . \tag{14.161}$$

The substitution of (14.160) into (14.161) yields

$$A_i J_0(\lambda_i a) + B_i I_0(\lambda_i a) = 0 \tag{14.162a}$$
$$A_i J_1(\lambda_i a) - B_i I_1(\lambda_i a) = 0$$

or

$$-\frac{B_i}{A_i} = \frac{J_0(\lambda_i a)}{I_0(\lambda_i a)} = -\frac{J_1(\lambda_i a)}{I_1(\lambda_i a)} \quad . \tag{14.162b}$$

In view of (14.162b), the natural frequencies must satisfy the frequency equation

$$J_0(\lambda_i a) I_1(\lambda_i a) + I_0(\lambda_i a) J_1(\lambda_i a) = 0 \quad . \tag{14.162c}$$

Equation (14.162c) has a denumerable infinity of positive real roots, the first few of which are approximately given by

$$\lambda_1 a = 3.190, \ \lambda_2 a = 6.306, \ \lambda_3 a = 9.425, \text{ etc.} \quad .$$

The natural frequencies are related to these roots by the relation, $\omega_i = \frac{1}{a^2}\sqrt{\frac{D}{\rho h}}\,(\lambda_i a)^2$. The eigenfunctions are now normalized by substituting (14.160) into (14.132). The result is,

$$\left[\int_0^a (A_i J_0(\lambda_i a) + B_i I_0(\lambda_i a))^2 r dr\right]^{\frac{1}{2}} = \frac{1}{\sqrt{2\pi\rho h}} \quad .$$

However,

$$\int_0^a rJ_o^2(\lambda_i r)dr = \frac{a^2}{2}\left[J_o^2(\lambda_i a)+J_1^2(\lambda_i a)\right]$$

$$\int_0^a rI_o^2(\lambda_i r)dr = \frac{a^2}{2}\left[I_o^2(\lambda_i a)-I_1^2(\lambda_i a)\right]$$

$$\int_0^a rJ_o(\lambda_i r)I_o(\lambda_i r)dr = \frac{a}{2\lambda_i}\left[J_o(\lambda_i a)I_1(\lambda_i a)+I_o(\lambda_i a)J_1(\lambda_i a)\right] = 0 \quad .$$

Consequently, the normalization condition becomes

$$A_i\left\{\left[J_o^2(\lambda_i a)+J_1^2(\lambda_i a)\right]+\frac{B_i^2}{A_i^2}\left[I_o^2(\lambda_i a)-I_1^2(\lambda_i a)\right]\right\}^{\frac{1}{2}} = \frac{1}{\sqrt{\pi a^2\rho h}} \quad . \qquad (14.163)$$

Combining (14.162b) with (14.163) yields, with the aid of (14.162c),

$$A_i = \frac{1}{a\sqrt{2\pi\rho h}\; J_o(\lambda_i a)}$$

$$B_i = \frac{-1}{a\sqrt{2\pi\rho h}\; I_o(\lambda_i a)} \qquad (14.164)$$

and therefore

$$W^{(i)}(r) = \frac{1}{a\sqrt{2\pi\rho h}}\left[\frac{J_o(\lambda_i r)}{J_o(\lambda_i a)} - \frac{I_o(\lambda_i r)}{I_o(\lambda_i a)}\right] \quad . \qquad (14.165)$$

The quasi-static solution of the present problem is characterized by the equation

$$\nabla^4 w^{(s)} = \frac{1}{r}\frac{d}{dr}\left\{r\frac{d}{dr}\left[\frac{1}{r}\frac{d}{dr}\left(r\frac{dw^{(s)}}{dr}\right)\right]\right\} = \frac{p(r,t)}{D} \qquad (14.166)$$

where

$$p(r,t) = \begin{cases} -p_o H(t), & 0\le r<b \\ 0, & b<r<a \end{cases}$$

and the boundary conditions

$$w^{(s)}(a,t)=0, \left(\frac{\partial w^{(s)}}{\partial r}\right)_{r=a} = 0 \quad . \qquad (14.167)$$

The solution of (14.166) can be expressed as follows:

For $0\le r<b$ (14.168a)

$$w^{(s)} = -\frac{p_o H(t)}{64D} r^4 + C_1 + C_2 r^2 + C_3 r^2 \ln\frac{r}{a} + C_4 \ln\frac{r}{a}$$

For $b<r<a$ (14.168b)

$$w^{(s)}=\bar{C}_1+\bar{C}_2r^2+\bar{C}_3r^2\ln\frac{r}{a}+\bar{C}_4\ln\frac{r}{a}$$

The eight constants C_1, C_2, C_3, C_4, $\bar{C}_1$, $\bar{C}_2$, $\bar{C}_3$ and $\bar{C}_4$ are determined by invoking the boundary conditions (14.67) together with the conditions that $w^{(s)}$, $\frac{\partial w^{(s)}}{\partial r}$, $\frac{\partial^2 w^{(s)}}{\partial r^2}$ and $\frac{\partial^3 w^{(s)}}{\partial r^3}$ remain bounded as $r\to 0$ and be continuous at $r=b$. This results in the following solution.

For $0\le r<b$ (14.169a)

$$w^{(s)}(r,t)=-\frac{p_oH(t)}{64D}a^4\left[\frac{r^4}{a^4}-2B^4\frac{r^2}{a^2}-3B^4+4B^2+4B^2\left(2\frac{r^2}{a^2}+B^2\right)\ln B\right]$$

For $b<r<a$ (14.169b)

$$w^{(s)}(r,t)=-\frac{p_oH(t)}{32D}a^2b^2\left[(B^2+2)\left(1-\frac{r^2}{a^2}\right)+2\left(2\frac{r^2}{a^2}+B^2\right)\ln\frac{r}{a}\right]$$

where, $B=\frac{b}{a}$. A comparison of (14.169) with (14.102) reveals that the static deflection predicted by classical plate theory differs from that predicted by the improved theory and this difference is of the order of magnitude of $\frac{h^2}{a^2}$. Thus, for very thin plates ($\frac{h}{a}<<1$) the two theories are in excellent agreement as far as the static response is concerned. By substituting (14.165), (14.166) and (14.167) into (14.139) we obtain

$$\omega_i^2G_i(t)=2\pi p_oH(t)\int_0^b W^{(i)}(r)rdr$$

$$=\frac{2\pi p_oH(t)b}{\sqrt{2\pi\rho h}\,a\lambda_i}\left[\frac{J_1(\lambda_ib)}{J_0(\lambda_ia)}-\frac{I_1(\lambda_ib)}{I_o(\lambda_ia)}\right]. \quad (14.170)$$

In view of (14.157), (14.141) becomes

$$g_i(0)=G_i(0) \text{ and } \dot{g}_i(0)=\dot{G}_i(0) \quad . \quad (14.171)$$

The substitution of (14.170) and (14.171) into (14.134) yields

$$g_i(t)=G_i(t)\cos\omega_it=$$

$$=\frac{2\pi p_oH(t)b}{\sqrt{2\pi\rho h}\,a\lambda_i}\left[\frac{J_1(\lambda_ib)}{J_0(\lambda_ia)}-\frac{I_1(\lambda_ib)}{I_o(\lambda_ia)}\right]\frac{\cos\omega_it}{\omega_i^2}. \quad (14.172)$$

The final result is now obtained by substituting (14.165), (14.172) and (14.169) into (14.134).

$$w(r,t)=w^{(s)}(r,t)+\frac{p_oH(t)a^3b}{D}\sum_{i=1}^{\infty}\left[\frac{J_1(\lambda_i b)}{J_o(\lambda_i a)}\right.$$

$$\left.-\frac{I_1(\lambda_i b)}{I_o(\lambda_i a)}\right]\left[\frac{J_o(\lambda_i r)}{J_o(\lambda_i a)}-\frac{I_o(\lambda_i r)}{I_o(\lambda_i a)}\right]\frac{\cos\omega_i t}{(\lambda_i a)^5} \qquad (14.173)$$

The corresponding stress resultants are obtained by substituting (14.173) into the stress-displacement relations (see Exercise 14.5)

$$M_{rr}=-D\left[\frac{\partial^2 w}{\partial r^2}+\frac{\nu}{r}\frac{\partial w}{\partial r}\right]$$

$$M_{\Theta\Theta}=-D\left[\frac{1}{r}\frac{\partial w}{\partial r}+\nu\frac{\partial^2 w}{\partial r^2}\right] \qquad (14.174)$$

$$Q_r=-D\frac{\partial}{\partial r}(\nabla^2 w)\ ,\quad Q_\Theta=M_{r\Theta}=0\ .$$

Figures (14.6) through (14.13) show the dimensionless displacement $\tilde{w}(\frac{r}{a},\tau)=\frac{\rho v_P^2 h}{p_o b^2}w$ and the dimensionless bending moment $\tilde{M}_{rr}(\frac{r}{a},\tau)=\frac{M_{rr}}{p_o b^2 H^2}$ plotted versus dimensionless time $\tau=\frac{v_P t}{a}$ at the center of the plate ($\frac{r}{a}=0$) and at the edge of the plate ($\frac{r}{a}=1$) for various values of $B(=\frac{b}{a})$ and $H^2\left(=\frac{h^2}{12a^2}\right)$ with $\nu=0.3$. Figures 14.6 through 14.9 indicate that the displacement at the center of the plate predicted by classical plate theory is in good agreement with that predicted by improved theory for relatively thin plates ($\frac{h}{a}\leq .1$). However, there is a considerable discrepancy between the two theories in the case of a thick plate ($\frac{h}{a}=.3$). Figures 14.10 through 14.13 show that the largest bending moment occurs at the center of the plate.

14.5 WAVE MOTION IN PLATES

Let us return, briefly, to the free vibration problem of a simply supported, rectangular plate. In the preceding section we found that its normal modes of vibration were characterized by the displacement field (14.127) with the eigenvalues and eigenfunctions given by (14.146) and (14.145), respectively. The standing wave solution

(14.127) may be given a somewhat different physical interpretation as follows. By substituting (14.145) into (14.127) we obtain, with the aid of some trigonometric identities,

$$w(x_1,x_2,t)=A_{mn}\sin k_1x_1\sin k_2x_2\cos\omega_i t$$

$$=-\frac{A_{mn}}{4}\Big[\cos(\vec{k}_1\cdot\vec{r}-\omega_i t)+\cos(\vec{k}_1\cdot\vec{r}+\omega_i t)$$

$$-\cos(\vec{k}_2\cdot\vec{r}-\omega_i t)-\cos(\vec{k}_2\cdot\vec{r}+\omega_i t)\Big] \qquad (14.175)$$

where,

$$k_1=\frac{m\pi}{a}\ ,\ k_2=\frac{n\pi}{b}\ ,\ \vec{r}=x_\alpha\hat{\varepsilon}_\alpha=x_1\hat{\varepsilon}_1+x_2\hat{\varepsilon}_2,$$

$$\vec{k}_1=k_\alpha\hat{\varepsilon}_\alpha=k_1\hat{\varepsilon}_1+k_2\hat{\varepsilon}_2=k\hat{k}_1=k(\cos\alpha\,\hat{\varepsilon}_1+\sin\alpha\ \hat{\varepsilon}_2),$$

$$\vec{k}_2=k_1\hat{\varepsilon}_1-k_2\hat{\varepsilon}_2=k\hat{k}_2=k(\cos\alpha\,\hat{\varepsilon}_1-\sin\alpha\ \hat{\varepsilon}_2),$$

$$k^2=|\vec{k}_1|^2=|\vec{k}_2|^2=k_1^2+k_2^2=\left(\frac{m\pi}{a}\right)^2+\left(\frac{n\pi}{b}\right)^2=\lambda_i^2=\omega_i\sqrt{\frac{\rho h}{D}}\ ,$$

and

$$\tan\alpha=\frac{k_2}{k_1}=\frac{an}{bm}\ .$$

The standing wave $w(x_1,x_2,t)$ is thus a superposition of four traveling waves. Take, for example, the expression $\cos(\vec{k}_1\cdot\vec{r}-\omega_i t)$. To an observer moving through space with velocity $\vec{v}_1=\frac{\omega_i}{k}\hat{k}_1$, the phase $\psi_1=\vec{k}_1\cdot\vec{r}-\omega_i t$ appears to be stationary, i.e.,

$$\frac{d\psi_1}{dt}=\vec{k}_1\cdot\frac{d\vec{r}}{dt}-\omega_i=\vec{k}_1\cdot\vec{v}_1-\omega_i=\omega_i-\omega_i=0\ .$$

Thus, a stationary observer would see the wave form $\cos\psi_1$ advancing in the $\hat{k}_1$ direction with speed (phase velocity)

$$v=|\vec{v}_1|=\frac{\omega_i}{k}=k\sqrt{\frac{D}{\rho h}}=\sqrt{\frac{D}{\rho h}}\sqrt{\left(\frac{m\pi}{a}\right)^2+\left(\frac{n\pi}{b}\right)^2}\ . \qquad (14.176)$$

At any fixed instant of time, the crests of this wave are found by setting

$$\cos\psi_1=1$$

or

$$\psi_1=0,\ \pm2\pi,\ \pm4\pi,\ldots\ .$$

Thus, the wave crests are defined by the equations

$$\vec{k}_1 \cdot \vec{r} - \omega_1 t = 2p\pi, \quad p = 0, \pm 1, \pm 2, \ldots$$

or

$$x_1 \cos\alpha + x_2 \sin\alpha = \frac{\omega_1 t + 2p\pi}{k} \quad .$$

This represents a family of equally spaced, parallel, straight lines as shown in Fig. 14.16.

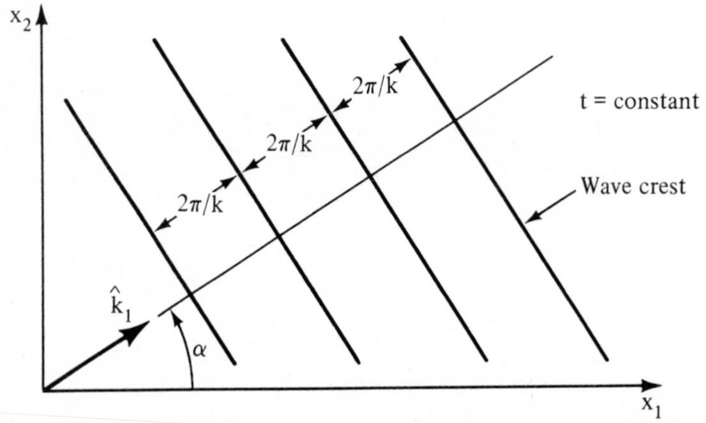

Figure 14.16

Hence, the expression $\cos(\vec{k}_1 \cdot \vec{r} - \omega_1 t)$ represents a straight crested wave. Consequently, the standing wave $w(x_1, x_2, t)$ is the superposition of four straight crested, traveling waves which are advancing with speed v in the directions $\hat{k}_1, -\hat{k}_1, \hat{k}_2$ and $-\hat{k}_2$, respectively. Considerations of this type indicate the close relationship between standing waves and traveling waves. For these as well as other reasons it is often illuminating (and sometimes even necessary) to consider the subject of wave motion in plates of unbounded extent.

(a) Straight Crested Waves in an Unbounded Plate. (Determination of the Shear Coefficient)

In our derivation of the plate displacement equations of motion (14.29), we arbitrarily introduced a shear coefficient κ^2. The value of this coefficient can be determined in a rational manner as follows. Consider a straight crested, harmonic wave propagating through the plate in the x_1 direction. The dispersion relations for such a wave were determined in Section 11.5 within the framework of the three-dimensional theory of elasticity. We will now proceed to determine the dispersion relations for this wave using plate theory. To simplify this calculation we shall eliminate $\vec{\psi}$ from (14.29b) as follows. Taking the divergence of the first of equations (14.29b) yields the result,

$$L(m)=\left(D\nabla^2-\frac{\rho h^3}{12}\frac{\partial^2}{\partial t^2}-\kappa^2 Gh\right)m=\kappa^2 Gh\nabla^2 w$$

where $m=m_{\alpha\alpha}=\psi_{\alpha,\alpha}=\vec{\nabla}\cdot\vec{\psi}$. Next, we let $p=0$ and then operate on the second of equations (14.29b) with L, using the preceding result, to obtain

$$\kappa^2 Gh\nabla^2 w+L(\nabla^2 w)=\frac{\rho}{\kappa^2 G}L(\ddot{w})$$

or

$$\left(D\nabla^2-\frac{\rho h^3}{12}\frac{\partial^2}{\partial t^2}\right)\left(\nabla^2-\frac{\rho}{\kappa^2 G}\frac{\partial^2}{\partial t^2}\right)w=-\rho h\frac{\partial^2 w}{\partial t^2}\quad . \tag{14.177}$$

Now, if we assume that

$$w = A\cos k(x_1-vt) \tag{14.178}$$

then in order to satisfy (14.177) we must have

$$v^4-\left(c_P^2+\kappa^2 c_S^2+\frac{12\kappa^2 c_S^2}{k^2h^2}\right)v^2+\kappa^2 c_S^2 c_P^2=0$$

where, (14.179)

$$c_P^2=\frac{E}{\rho(1-\nu^2)}\ ,\ c_S^2=\frac{G}{\rho}\quad .$$

The roots of this equation are,

$$\begin{pmatrix}v_2^2\\ v_1^2\end{pmatrix}=\frac{c_P^2+\kappa^2 c_S^2}{2}+\frac{6\kappa^2 c_S^2}{k^2h^2}\pm\left[\left(\frac{c_P^2-\kappa^2 c_S^2}{2}\right)^2+(c_P^2+\kappa^2 c_S^2)\left(\frac{6\kappa^2 c_S^2}{k^2h^2}\right)+\left(\frac{6\kappa^2 c_S^2}{k^2h^2}\right)^2\right]^{\frac{1}{2}}. \tag{14.180}$$

In the limit as $kh\to 0$, (14.180) yields

$$v_1^2\to\frac{D}{\rho h}k^2 \text{ and } v_2^2\to\frac{12\kappa^2 c_S^2}{k^2h^2}\quad . \tag{14.181a}$$

On the other hand, as $kh\to\infty$ we obtain

$$v_1^2\to\kappa^2 c_S^2 \text{ and } v_2^2\to c_P^2\quad . \tag{14.181b}$$

The value of the shear coefficient is now determined by matching the dispersion curve of the first mode of propagation $v_1(k)$ with the dispersion curve of the first anti-symmetric mode of propagation obtained from the three-dimensional theory of elasticity. A comparison of (14.181a) with (11.72) reveals that both theories exhibit the same asymptotic behavior as $kh \to 0$. A simple calculation will reveal that $v = \sqrt{\frac{D}{\rho h}}\, k$ is the complete dispersion relation for classical plate theory. Thus, the three theories under discussion agree quite well for very long waves ($kh = 2\pi \frac{h}{\ell} \to 0$). For very short waves ($kh = 2\pi \frac{h}{\ell} \to \infty$), the theory of elasticity predicts that the wave speed will approach that of Rayleigh's surface waves, i.e., (see 11.71)

$$\left(2 - \frac{v^2}{c_S^2}\right)^2 = 4\sqrt{1 - \frac{v^2}{c_S^2}}\;\sqrt{1 - \gamma^2 \frac{v^2}{c_S^2}} \quad . \tag{14.182}$$

According to the present plate theory $\frac{v_1^2}{c_S^2} = \kappa^2$ for very short waves (see (14.181b)). The value of κ^2 which makes these two limits coincide is obtained by substituting $\frac{v^2}{c_S^2} = \kappa^2$ into (14.182) to obtain

$$(2 - \kappa^2)^2 = 4\sqrt{1 - \kappa^2}\;\sqrt{1 - \gamma^2 \kappa^2} \tag{14.183}$$

where, $\gamma^2 = \frac{c_S^2}{c_D^2} = \frac{1 - 2\nu}{2(1 - \nu)}$ and $0 < \kappa^2 < 1$. Thus, κ depends on ν in exactly the same way as the Rayleigh wave speed does. Consequently, Fig. 11.8 can be interpreted as a graph of κ vs. ν. We see from Fig. 11.8 that the variation of κ with ν is nearly linear for $0.874 < \kappa < 0.955$ and $0 < \nu < \frac{1}{2}$. For example, if $\nu = 0.3$ then, $\kappa = 0.9275$ or $\kappa^2 = 0.86$. The complete dispersion relation for the first antisymmetric mode of propagation as predicted by the classical plate theory, the improved plate theory and the three-dimensional theory of elasticity are compared in Fig. 14.17 for $\nu = \frac{1}{2}$. With $\kappa = 0.955$ the difference between the improved and elasticity theories is so small that it cannot be detected on the graph. The present method for the determination of κ is due to R.D. Mindlin and was first presented by him in the paper cited in Section 11.5.

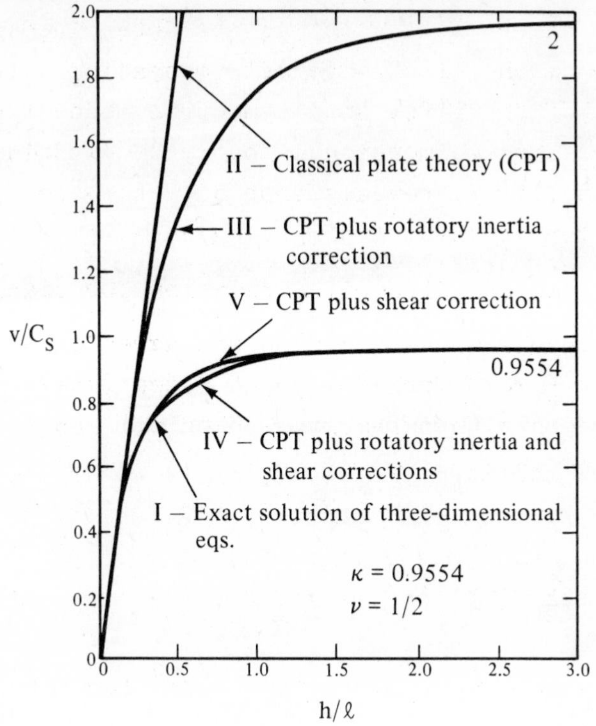

Figure 14.17

We shall conclude our discussion of straight crested waves in plates by examining the energy flux associated with the wave

$$w = A\cos(\vec{k}\cdot\vec{r}-\omega t) \tag{14.184}$$

within the framework of classical plate theory. The phase velocity of this wave is $\vec{v}=\frac{\omega}{k}\hat{k}$ where $\vec{k}=k_\alpha\hat{\varepsilon}_\alpha$ is the propagation vector and $k=|\vec{k}|$ is the wave number. In order to satisfy (14.113) (with p=0) we must set

$$Dk^4 = \rho h\omega^2 \text{ or } v=\frac{\omega}{k}=\sqrt{\frac{D}{\rho h}}\;k\;. \tag{14.185}$$

Hence, the phase velocity is $\vec{v}=\sqrt{\frac{D}{\rho h}}\,\vec{k}$ and since it depends on the wave number k the medium is dispersive. The group velocity is

$$\vec{v}_G=\frac{d\omega}{dk}\hat{k}=2\sqrt{\frac{D}{\rho h}}\;\vec{k}=2\vec{v}\;. \tag{14.186}$$

The energy density in the plate is

$$E^{*}=\frac{\rho h}{2}\dot{w}^2+W_p^{*}=\frac{1}{2}Dk^4A^2=\frac{1}{2}\rho h\omega^2A^2$$

and the energy flux vector has components

$$S_\alpha = -Q_\alpha \dot{w} + M_{\beta\alpha}\dot{w}_{,\beta} = Dk^4A^2v_\alpha = \rho h\omega^2A^2v_\alpha$$

or, in vector form,

$$\vec{S} = Dk^4A^2\vec{v} = 2E^*\vec{v} = E^*\vec{v}_G \quad . \tag{14.187}$$

Consequently, we can interpret the group velocity $\vec{v}_G$ as the velocity with which energy is transported through the plate by the harmonic wave (14.184).

(b) Forced, Time-harmonic Motion of a Plate of Infinite Extent.

We shall now consider the steady state motion of an unbounded plate under the influence of the concentrated, time-harmonic load $Fe^{i\omega t}$ applied at r=0. The analysis will be carried out within the framework of classical plate theory. Thus, we seek a solution of the equation

$$D\nabla^4 w + \rho h\ddot{w} = 0 \text{ for } r>0 \tag{14.188}$$

where $\nabla^2 \equiv \frac{\partial^2}{\partial r^2} + \frac{1}{r}\frac{\partial}{\partial r}$ in polar coordinates. For a steady state response of the form

$$w(r,t) = W(r)e^{i\omega t} \tag{14.189}$$

equation (14.188) becomes

$$\nabla^4 W - k^4 W = 0$$

where

$$k^4 = \frac{\rho h}{D}\omega^2 \quad . \tag{14.190}$$

We note that (14.190) has the same form as (14.159). Thus,

$$W(r) = \bar{A}J_o(kr) + \bar{C}Y_o(kr)$$
$$+ BI_o(kr) + DK_o(kr) \quad . \tag{14.191a}$$

To facilitate the application of boundary conditions at infinity we shall use the Hankel functions

$$H_o^{(1)}(kr) = J_o(kr) + iY_o(kr)$$

$$H_o^{(2)}(kr) = J_o(kr) - iY_o(kr)$$

rather than the Bessel functions J_o and Y_o in our solution. Thus,

$$W(r) = AH_o^{(1)}(kr) + CH_o^{(2)}(kr) + BI_o(kr) + DK_o(kr) \tag{14.191b}$$

where

$$A = \frac{\bar{A}-i\bar{C}}{2} \quad \text{and} \quad C = \frac{\bar{A}+i\bar{C}}{2} \quad .$$

For sufficiently large values of kr, the Hankel functions are closely approximated by the asymptotic formulae

$$\begin{aligned} H_o^{(1)}(kr) &\cong \sqrt{\frac{2}{\pi kr}} \quad e^{i(kr-\frac{\pi}{4})} \\ H_o^{(2)}(kr) &\cong \sqrt{\frac{2}{\pi kr}} \quad e^{-i(kr-\frac{\pi}{4})} \quad . \end{aligned} \tag{14.192}$$

Consequently, the expression $H_o^{(1)}(kr)e^{i\omega t}$ represents an incoming wave, i.e., a wave which is propagating towards the source point r=0. On the other hand, the expression $H_o^{(2)}(kr)e^{i\omega t}$ represents an outgoing wave, i.e., one which is propagating away from the source point r=0. In view of (14.190) the phase velocity of these waves is

$$v = \frac{\omega}{k} = \sqrt{\frac{D}{\rho h}} \quad k \tag{14.193a}$$

and their group velocity is

$$v_G = \frac{d\omega}{dk} = 2\sqrt{\frac{D}{\rho h}} \quad k=2v \quad . \tag{14.193b}$$

As shown in the preceding section, the group velocity is the average velocity of energy transport in a harmonic wave. Since the sign of the group velocity is the same as the sign of the phase velocity we conclude that the wave $H_o^{(1)}(kr)e^{i\omega t}$ transports energy towards the source point r=0 while the wave $H_o^{(2)}(kr)e^{i\omega t}$ transports energy away from the source point r=0. In the present problem the energy input to the plate occurs at the source point and consequently the energy flux must be directed away from the source. Thus, we must set A=0 in (14.191b). For sufficiently large values of kr, the modified Bessel functions $I_o(kr)$ and $K_o(kr)$ are closely approximated by the asymptotic formulae

$$\begin{aligned} I_o(kr) &\cong \frac{e^{kr}}{\sqrt{2\pi kr}} \\ K_o(kr) &\cong \sqrt{\frac{\pi}{2kr}} \quad e^{-kr} \quad . \end{aligned} \tag{14.194}$$

Consequently, if the solution is to remain bounded as $r\to\infty$ we must set $B=0$ in (14.191b). In view of the above discussion our solution now appears as follows:

$$w(r) = CH_0^{(2)}(kr) + DK_0(kr)$$

or (14.195)

$$W(r) = C\left[J_0(kr)-iY_0(kr)\right] + DK_0(kr).$$

In order to determine the constants C and D we must impose some restrictions on the behavior of the solution in the neighborhood of $r=0$. The following technique for obtaining these restrictions is due to A. Kalnins ("On Fundamental Solutions and Green's Functions in the Theory of Elastic Plates," Journal of Appl. Mech., 88, March 1966, pp. 31-38). In the present case, the equation of motion (14.108a) becomes

$$Q_{\alpha,\alpha} = \vec{\nabla}\cdot\vec{Q} = \frac{1}{r}\frac{d}{dr}(rQ_r) = -\rho h\omega^2 W(r) - p(r). \tag{14.196}$$

A function $W^*(r)$ will be called a well-behaved solution of (14.196) corresponding to the load $p^*(r)$ if W^*, $W^{*\prime}$, $W^{*\prime\prime}$ and $W^{*\prime\prime\prime}$ are continuous for $0<r<\infty$, if

$$\lim_{r\to 0^+}\left[W^*, W^{*\prime}, W^{*\prime\prime}, W^{*\prime\prime\prime}\right] = \left[W^*(0), W^{*\prime}(0), W^{*\prime\prime}(0), W^{*\prime\prime\prime}(0)\right] < \infty$$

and if (14.197)

$$\lim_{r\to\infty}\left[rW^*, rW^{*\prime}, rW^{*\prime\prime}, rW^{*\prime\prime\prime}\right] = \left[0,0,0,0\right].$$

Obviously, (14.195) does not characterize a well-behaved solution. The reciprocal relation will now be applied to a circular region of radius R. If we let $w^{(1)}=W(r)$ and $w^{(2)}=W^*(r)$ then, with the aid of (14.196), the reciprocal relation can be written as follows:

$$\int_0^{2\pi}\int_0^R p(r)W^*(r)\,r\,dr\,d\theta = \int_0^{2\pi}\int_0^R p^*(r)W(r)\,r\,dr\,d\theta + I(R) \tag{14.198a}$$

where,

$$I(R) = \int_0^{2\pi}\left[Q_r^*W - Q_rW^* + M_{rr}W^{*\prime} - M_{rr}^*W'\right]_{r=R} R\,d\theta \quad . \tag{14.198b}$$

Within the context of the present problem, the term "concentrated force" will be interpreted as follows. A distribution p(r) will be said to represent a concentrated force of magnitude F if

$$\int_0^{2\pi}\int_0^{\infty} p(r)W^*(r)rdrd\Theta = FW^*(0) \qquad (14.199)$$

for every well-behaved solution of (14.196). If we now let R approach infinity in (14.198) then, by virtue of (14.197) and (14.199), we obtain

$$FW^*(0) = \int_0^{2\pi}\int_0^{\infty} p^*(r)W(r)rdrd\Theta \quad . \qquad (14.200)$$

The characterization of a concentrated force given by (14.199) was not used in our derivation of (14.195). That derivation relied on our intuitive concept of a concentrated force, i.e., we set p(r)=0 for r>0. In order to reconcile these two points of view we shall require that they both yield (14.200) for any well-behaved function W*(r). To accomplish this end, we apply the reciprocal relation (14.119) to the region A bounded by the concentric circles $r=\varepsilon$ and r=R, where $0<\varepsilon<R$. Using our intuitive concept of a concentrated force we set p(r)=0 throughout the region A to obtain

$$0 = \int_0^{2\pi}\int_{\varepsilon}^{R} p^*(r)W(r)rdrd\Theta + I(R) - I(\varepsilon) \quad . \qquad (14.201)$$

In the limit as R approaches infinity and as ε approaches zero, (14.201) should yield (14.200). Since $\lim_{R\to\infty} I(R)=0$, we conclude that

$$\lim_{\varepsilon\to 0} I(\varepsilon) = FW^*(0)$$

or, in view of (14.198b),

$$\lim_{\varepsilon\to 0} \varepsilon W(\varepsilon) = 0 \qquad (14.202a)$$

$$\lim_{\varepsilon\to 0} \varepsilon W'(\varepsilon) = 0 \qquad (14.202b)$$

$$\lim_{\varepsilon\to 0} \varepsilon M_{rr}(\varepsilon) = 0 \qquad (14.202c)$$

$$\lim_{\varepsilon\to 0} \varepsilon Q_r(\varepsilon) = -\frac{F}{2\pi} \quad . \qquad (14.202d)$$

The condition (14.202a) is identically satisfied by the solution (14.195). In order to satisfy (14.202b) we differentiate (14.195) to obtain

$$rW'(r)=C\left[-krJ_1(kr)+ikrY_1(kr)\right]-DkrK_1(kr) \quad .$$

Since $\lim_{x\to 0} xJ_1(x)=0$, $\lim_{x\to 0} xY_1(x)=-\frac{2}{\pi}$ and $\lim_{x\to 0} xK_1(x)=1$, we conclude that

$$\lim_{\varepsilon\to 0} \varepsilon W'(\varepsilon)=-\frac{2i}{\pi}C-D=0 \text{ or } D=-\frac{2i}{\pi}C \quad .$$

Thus,

$$W(r)=C\left[J_0(kr)-iY_0(kr)-\frac{2i}{\pi}K_0(kr)\right] \quad . \qquad (14.203)$$

The condition (14.202c) is also identically satisfied by (14.203). The last condition (14.202d) is now used to determine the constant C. The substitution of (14.203) into (14.174) yields

$$rQ_r(r)=-Dk^2C(kr)\left[J_1(kr)-iY_1(kr)+\frac{2i}{\pi}K_1(kr)\right] \quad .$$

Consequently,

$$\lim_{\varepsilon\to 0} \varepsilon Q_r(\varepsilon)=-Dk^2C\frac{4i}{\pi}=-\frac{F}{2\pi} \text{ or } C=\frac{-iF}{8k^2D}$$

and our solution can now be written as

$$W(r)=\frac{-F}{8k^2D}\left[Y_0(kr)+\frac{2}{\pi}K_0(kr)+iJ_0(kr)\right] \quad . \qquad (14.204)$$

The final result is then

$$\begin{aligned} w(r,t) &= W(r)e^{i\omega t} \\ &=-\frac{F}{8k^2D}\left\{\left[Y_0(kr)+\frac{2}{\pi}K_0(kr)\right]\cos\omega t-J_0(kr)\sin\omega t\right\} \\ &\quad -\frac{iF}{8k^2D}\left\{\left[Y_0(kr)+\frac{2}{\pi}K_0(kr)\right]\sin\omega t+J_0(kr)\cos\omega t\right\} \qquad (14.205) \end{aligned}$$

If the applied force is cosinusoidal then the response is obtained by taking the real part of (14.205). Fig. 14.18 shows $\tilde{w}=\frac{8k^2D}{F}\text{Re}(w)$ plotted versus kr for various values of ωt.

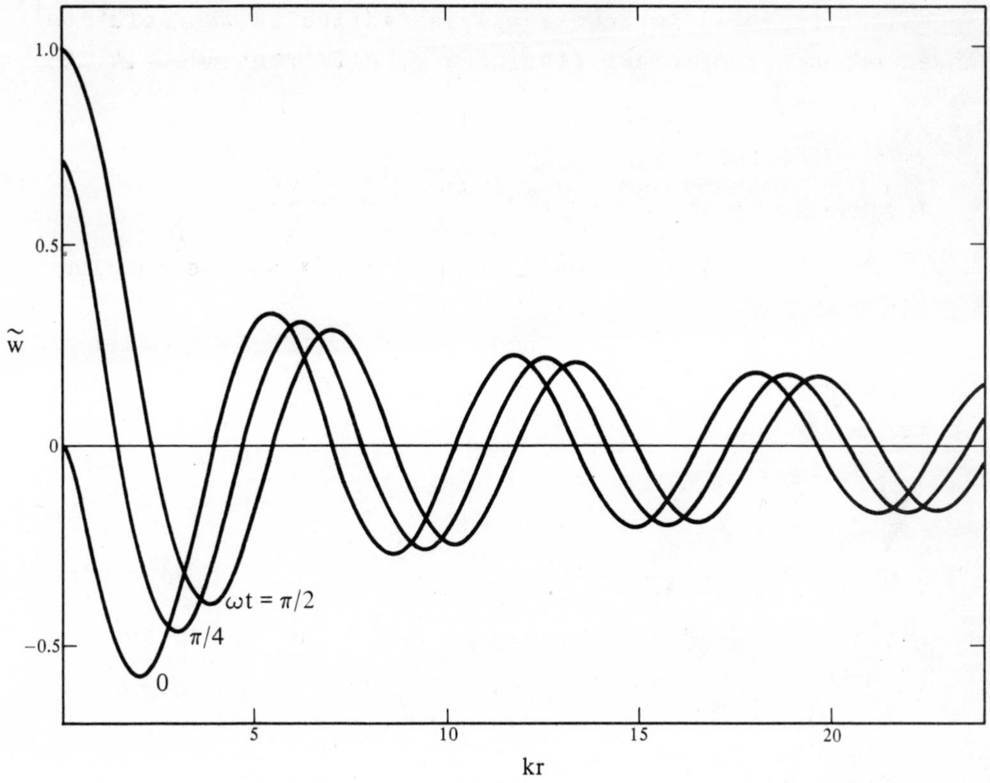

Figure 14.18

(c) The Impulsively Loaded Plate.

Consider an unbounded plate which is subjected to a concentrated, pure impulse of magnitude I at r=0 and t=0. We wish to find the dynamic response of the plate for t>0 assuming that it is at rest and in its equilibrium configuration at t=0. We shall investigate this problem within the framework of classical plate theory using polar coordinates. Thus, we seek a solution of the equation

$$D\nabla^4 w + \rho h\ddot{w} = I\delta(r)\delta(t) \tag{14.206}$$

which satisfies the initial conditions

$$w(r,0) = \dot{w}(r,0) = 0 \quad . \tag{14.207}$$

The symbol δ denotes Dirac's delta function. The resolution of this problem is facilitated by the use of Laplace transforms (see Chapter 1) and Hankel transforms (see, for instance, I.N. Sneddon, <u>Fourier</u>

Transforms, McGraw-Hill Book Co., Inc., New York 1951, pp. 48-70).
The Hankel transform of order zero of w(r,t) is defined by

$$w^*(k,t) = \int_0^\infty rw(r,t)J_0(kr)dr \tag{14.208a}$$

and its inverse is

$$w(r,t) = \int_0^\infty kw^*(k,t)J_0(kr)dk \tag{14.208b}$$

where $J_0(kr)$ denotes the Bessel function of the first kind of order zero. By analogy with (14.199) the Hankel transform of a concentrated impulse is

$$[I\delta(r)]^* = \frac{I}{2\pi}\int_0^{2\pi}\int_0^\infty \delta(r)J_0(kr)rdrd\Theta = \frac{I}{2\pi}J_0(0) = \frac{I}{2\pi} \quad . \tag{14.209}$$

If we assume that rw, $r\frac{\partial w}{\partial r}$, $r\frac{\partial^2 w}{\partial r^2}$ and $r\frac{\partial^3 w}{\partial r^3}$ all vanish as r approaches zero and as r approaches infinity then by repeated integration by parts one can show that

$$(\nabla^4 w)^* = k^4 w^* \quad . \tag{14.210}$$

The equation of motion (14.206) is now transformed by multiplying both sides by $rJ_0(kr)$ and then integrating the result over r between the limits zero and infinity. In view of (14.209) and (14.210) the result is

$$Dk^4 w^* + \rho h\ddot{w}^* = \frac{I}{2\pi}\delta(t). \tag{14.211}$$

We now take the Laplace transform of (14.211) to obtain, with the aid of (14.207),

$$Dk^4\bar{w}^* + \rho hp^2\bar{w}^* = \frac{I}{2\pi} \tag{14.212}$$

where,

$$\bar{w}^* = \int_0^\infty w^*(k,t)\exp(-pt)dt \quad .$$

Thus,

$$\bar{w}^* = \frac{I}{2\pi\sqrt{\rho hD}}\,\frac{1}{k^2}\,\frac{\lambda^2k^2}{p^2+\lambda^4k^4} \tag{14.213}$$

where, $\lambda^4 = \frac{D}{\rho h}$.

With the aid of entry number 11 in Table 1.1 we readily invert (14.213) to obtain

$$w^*(k,t) = \frac{I}{2\pi\sqrt{\rho h D}}\frac{1}{k^2}\sin\lambda^2k^2t$$

and consequently

$$w(r,t)= \frac{I}{2\pi\sqrt{\rho hD}}\int_0^\infty \sin\lambda^2k^2t\, J_0(kr)\,\frac{dk}{k} \quad . \tag{14.214}$$

By differentiating (14.214) with respect to t the following expression is obtained for the velocity:

$$\dot{w}(r,t)= \frac{I\lambda^2}{2\pi\sqrt{\rho hD}}\int_0^\infty k\cos\lambda^2k^2t\, J_0(kr)dk$$

$$= \frac{I}{4\pi\sqrt{\rho hD}\, t}\sin\frac{r^2}{4\lambda^2t} \tag{14.215}$$

(This integral is entry no. 4 of Section 6.728, p. 758, in I.S. Gradshteyn and I.M. Ryzhik, Table of Integrals, Series, and Products, Fourth Edition, Academic Press, New York, 1965.) If we set $\tau=4\lambda^2t$, then (14.215) can be written as

$$\frac{d\tilde{w}}{d\tau} = \frac{2}{\pi}\frac{1}{\tau}\sin(\frac{r^2}{\tau})$$

where $\tilde{w}= \frac{8}{I}\sqrt{\rho hD}\, w$ and since $\tilde{w}=0$ at $\tau=0$,

$$\tilde{w}= \frac{2}{\pi}\int_0^\tau \sin(\frac{r^2}{x})\,\frac{dx}{x} \quad .$$

The change of variable $\xi= \frac{r^2}{x}$ transforms the above integral to

$$\tilde{w}= \frac{2}{\pi}\int_{\frac{r^2}{\tau}}^\infty \frac{\sin\xi}{\xi}\,d\xi= \frac{2}{\pi}\int_0^\infty\frac{\sin\xi}{\xi}\,d\xi- \frac{2}{\pi}\int_0^{\frac{r^2}{\tau}}\frac{\sin\xi}{\xi}\,d\xi$$

or

$$\tilde{w}=1- \frac{2}{\pi}\,Si(\frac{r^2}{\tau})=1- \frac{2}{\pi}\,Si(\frac{R^2}{T}) \tag{14.216}$$

where, $R= \frac{r}{h}$ and $T= \frac{C_pt}{h} =\sqrt{\frac{E}{\rho(1-\nu^2)}}\;\frac{t}{h}$. The function, $Si(x)=\int_0^x \frac{\sin\xi}{\xi}\,d\xi$ is called the sine integral. The dimensionless displacement $\tilde{w}(R,T)$ is shown in Fig. 14.19 plotted versus the dimensionless radius R for various values of the dimensionless time T. For sufficiently large values of its argument $\frac{R^2}{T}$, the sine integral can be approximated by the asymptotic formula (see, for instance, M. Abramowitz and I. Stegun, Handbook of Mathematical Functions, Dover Pub., Inc. New York, 1965, pp. 232-233):

$$Si(\frac{R^2}{T}) \cong \frac{\pi}{2} - \frac{T}{R^2} \cos \frac{R^2}{T}$$

and therefore

$$\tilde{w} \cong \frac{2}{\pi} \frac{T}{R^2} \cos \frac{R^2}{T} \text{ for sufficiently large } \frac{R^2}{T} \quad . \tag{14.217}$$

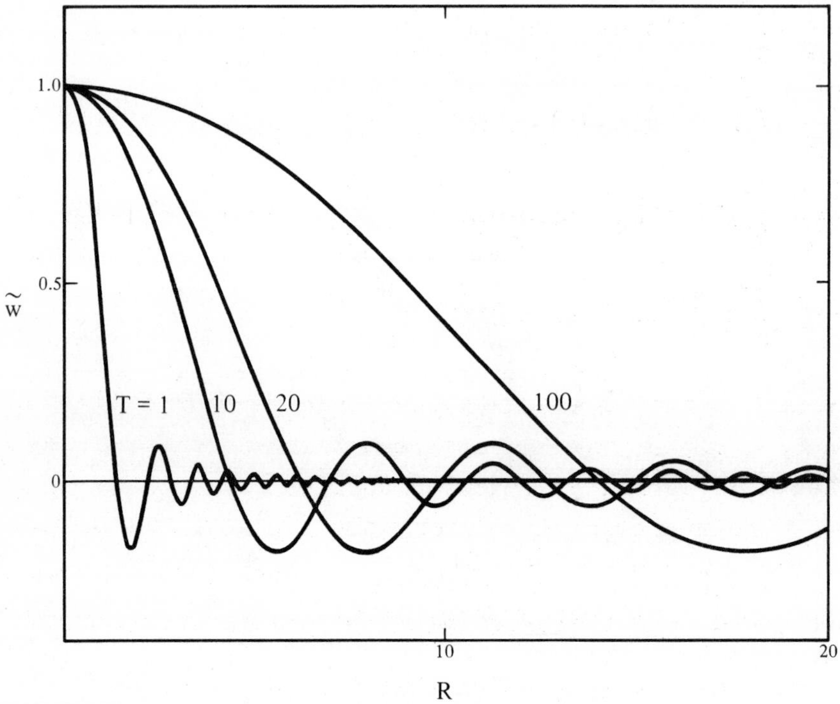

Figure 14.19

Although the present problem has an exact solution (14.216) in terms of known functions, an exact evaluation of inversion integrals of the type (14.208b) is often not possible. In such cases, Kelvin's method of stationary phase can be used to obtain the asymptotic behavior of the solution (see Appendix B). To illustrate the method we shall apply it to the present problem. For large values of kr, the Bessel function $J_0(kr)$ can be approximated by the asymptotic formula

$$J_0(kr) \cong \sqrt{\frac{2}{\pi kr}} \cos (kr- \frac{\pi}{4})$$

and consequently we will approximate (14.214) by the expression

$$\tilde{w} \cong \frac{4}{\pi}\sqrt{\frac{2}{\pi r}} \int_0^\infty k^{-3/2} \sin kvt \cos(kr - \frac{\pi}{4})dk \qquad (14.218)$$

where $v=\lambda^2 k$ is the phase velocity and $v_G=2\lambda^2 k=2v$ is the group velocity of cylindrical waves in the plate (see (14.193)). With the aid of some trigonometric identities, (14.218) can be written as follows:

$$\tilde{w}= \frac{2}{\pi}\sqrt{\frac{2}{r}}\,[\phi_1+\phi_2+\phi_3-\phi_4] \qquad (14.219)$$

where

$$\phi_1= \frac{1}{\sqrt{2\pi}} \int_0^\infty k^{-3/2} \sin k(vt+r)dk$$

$$\phi_2= \frac{1}{\sqrt{2\pi}} \int_0^\infty k^{-3/2} \sin k(vt-r)dk$$

$$\phi_3= \frac{1}{\sqrt{2\pi}} \int_0^\infty k^{-3/2} \cos k(vt-r)dk$$

$$\phi_4= \frac{1}{\sqrt{2\pi}} \int_0^\infty k^{-3/2} \cos k(vt+r)dk \ .$$

With reference to the terminology developed in Appendix B, the expressions ϕ_1, ϕ_2, ϕ_3, and ϕ_4 are recognized as traveling wave packets. Kelvin's method will now be applied to the wave packet ϕ_2. The phase of the function $\sin k(vt-r)=\sin(\lambda^2k^2t-kr)$ is stationary when $\frac{d}{dk}(\lambda^2k^2t-kr)=2\lambda^2kt-r=0$ or $k=\frac{r}{2\lambda^2 t}=\frac{2r}{\tau}$. Thus, the predominant wave number is $k_0=\frac{r}{2\lambda^2 t}$ and therefore $v(k_0)=\lambda^2k_0=\frac{r}{2t}$, $v_G(k_0)=2\lambda^2k_0=\frac{r}{t}$ and $v_G'(k_0)=2\lambda^2$. In the notation of (B.49a) $a(k_0)=k_0^{-3/2}=(\frac{\tau}{2r})^{3/2}$. The substitution of these results into (B.49a) yields

$$\phi_2 \cong (\frac{\tau}{2r})^{3/2} \sqrt{\frac{2}{\tau}}\ \sin(k_0v_0t-k_0r+\frac{\pi}{4})$$

or

$$\phi_2 \cong -\frac{\tau}{2r^{3/2}} \sin\left(\frac{r^2}{\tau}-\frac{\pi}{4}\right) \ .$$

Similarly, it can be shown that

$$\phi_3 \cong \frac{\tau}{2r^{3/2}} \cos\left(\frac{r^2}{\tau}-\frac{\pi}{4}\right) \ .$$

Therefore,

$$\phi_2+\phi_3 \cong \frac{\tau}{r\sqrt{2r}} \cos \frac{r^2}{\tau} \quad . \tag{14.220}$$

If the same technique is applied to ϕ_1 and ϕ_4 we find that their phase is stationary when $k=-\frac{2r}{\tau}$. However, since $r>0$, $\tau>0$ and $k>0$ we conclude that the contribution of ϕ_1 and ϕ_4 to (14.219) is negligible and therefore,

$$\tilde{w} \cong \frac{2}{\pi} \sqrt{\frac{2}{r}} (\phi_2+\phi_3) \cong \frac{2}{\pi} \frac{\tau}{r^2} \cos \frac{r^2}{\tau} \quad . \tag{14.221}$$

Thus, Kelvin's method of stationary phase yields the asymptotic form of the exact solution (14.216) for large values of $\frac{r^2}{\tau}$, i.e., (14.221) is identical with (14.217).

(d) Moving Line Load on an Infinite Plate Strip.

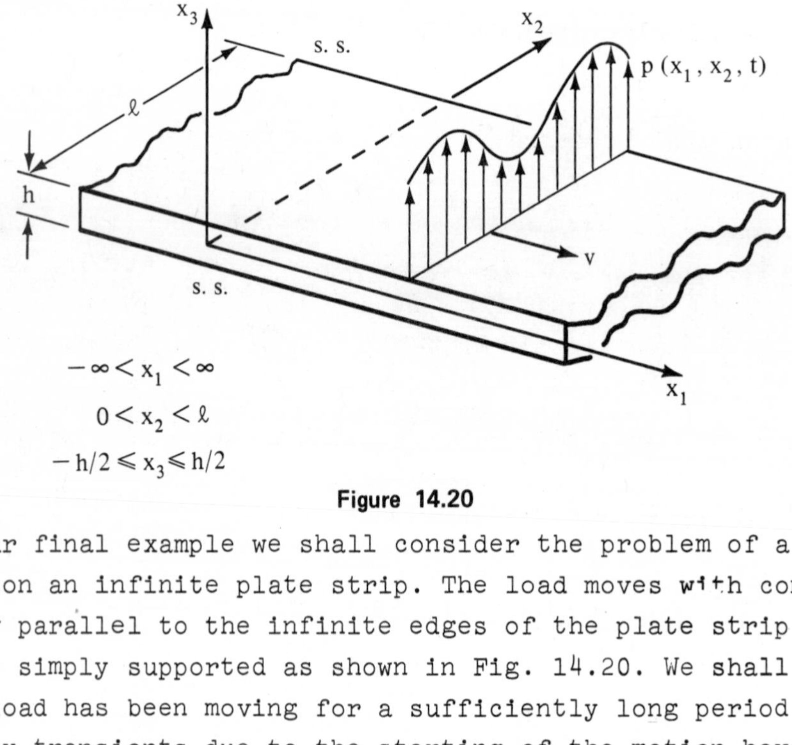

Figure 14.20

As our final example we shall consider the problem of a moving line load on an infinite plate strip. The load moves with constant velocity v parallel to the infinite edges of the plate strip which we assume are simply supported as shown in Fig. 14.20. We shall assume that the load has been moving for a sufficiently long period of time so that any transients due to the starting of the motion have become negligible and we can focus our attention on the steady state motion. To facilitate our investigation of this problem we shall effect the following change of variables:

$$-\infty<x=\frac{\pi}{\ell}(x_1-vt)<\infty, \quad 0<y=\frac{\pi}{\ell}x_2<\pi \tag{14.222}$$

This change of variables can be interpreted as follows: An observer fixed with respect to the x_1x_2 coordinate system sees the load $p(x_1,x_2,t)$ advancing in the positive x_1 direction and to him the defection of any point (x_1,x_2) on the plate will appear to vary with time. However, an observer who is fixed with respect to the xy coordinate system moves with the load and to him the deflection of any point (x,y) will appear stationary, i.e., it will be independent of time. If the load is located at $x_1=0$ when $t=0$ then it can be represented as follows:

$$p(x,y)=\delta(x)f(y)=\delta\left[\frac{\pi}{\ell}(x_1-vt)\right] f\left(\frac{\pi x_2}{\ell}\right) \tag{14.223}$$

If the transformation (14.222) is applied to the equations of classical plate theory (14.113) one obtains

$$\frac{\partial^4 w}{\partial x^4} + 2\frac{\partial^4 w}{\partial x^2 \partial y^2} + \frac{\partial^4 w}{\partial y^4} + \frac{\rho h}{D}\left(\frac{\ell}{\pi}\right)^2 v^2 \frac{\partial^2 w}{\partial x^2} = \frac{\delta(x)f(y)}{D}\left(\frac{\ell}{\pi}\right)^4 \tag{14.224}$$

In view of the simply supported edge conditions we must require that

$$w = M_{22} = 0 \text{ at } y = 0, \pi \quad . \tag{14.225}$$

Thus, it is natural to assume a solution in the form

$$w(x,y) = \sum_{n=1}^{\infty} w_n(x) \sin ny \tag{14.226}$$

and to represent the load distribution by the Fourier sine series (half range expansion)

$$f(y) = \sum_{n=1}^{\infty} a_n \sin ny \tag{14.227}$$

where

$$a_n = \frac{2}{\pi}\int_0^{\pi} f(y)\sin ny\,dy \ .$$

The substitution of (14.226) and (14.227) into (14.224) yields the following equations for w_n:

$$\frac{d^4 w_n}{dx^4} + 2n^2(2\Theta_n^2 - 1)\frac{d^2 w_n}{dx^2} + n^4 w_n = \frac{\ell^4 a_n}{\pi^4 D}\delta(x) \tag{14.228}$$

$$n=1,2,3,\ldots$$

where

$$\Theta_n^2 = \frac{\rho h}{D}\frac{\ell^2 v^2}{4\pi^2 n^2} \quad . \tag{14.229}$$

The quantity Θ_n is a dimensionless load speed parameter. The solution of (14.228) will be obtained with the aid of the Fourier transform defined by

$$w_n^*(k) = \frac{1}{\sqrt{2\pi}} \int_{-\infty}^{\infty} w_n(x)\exp(ikx)dx \tag{14.230a}$$

$$w_n(x) = \frac{1}{\sqrt{2\pi}} \int_{-\infty}^{\infty} w_n^*(k)\exp(-ikx)dk \tag{14.230b}$$

Here, $w_n^*(k)$ is called the Fourier transform of $w_n(x)$. If $w_n^*(k)$ is known then $w_n(x)$ can be determined by applying the inversion formula (14.230b). To apply this transformation to the present problem we multiply both sides of (14.228) by $\frac{\exp(ikx)}{\sqrt{2\pi}}$ and then integrate the result over the range $(-\infty<x<\infty)$ to obtain

$$\left[k^4-2n^2(2\Theta_n^2-1)k^2+n^4\right]w_n^* = \frac{\ell^4}{\pi^4 D}\frac{a_n}{\sqrt{2\pi}} \quad . \tag{14.231}$$

In the derivation of (14.231) we have assumed that w_n, w_n', w_n'' and w_n''' all vanish as $|x|\to\infty$. Furthermore, we have used the fact that

$$\int_{-\infty}^{\infty} \delta(x)\exp(ikx)dx = \exp\left[ik(0)\right] = 1 \quad .$$

The desired solution can now be obtained by solving (14.231) for $w_n^*(k)$ and then substituting this result into (14.230b) to obtain

$$w_n(x) = \frac{\ell^4}{\pi^4 D}\frac{a_n}{2\pi}\int_{-\infty}^{\infty}\frac{\exp(-ikx)dk}{P_n(k)} \tag{14.232a}$$

where,

$$P_n(k) = k^4-2n^2(2\Theta_n^2-1)k^2+n^4 \quad . \tag{14.233a}$$

If we let

$$Q_n(k_n) = \left[k_n^4-2(2\Theta_n^2-1)k_n^2+1\right] \tag{14.233b}$$

where $k_n = \frac{k}{n}$, then (14.232a) may be rewritten as

$$w_n(x_n) = \frac{\ell^4}{\pi^4 D}\frac{a_n}{2\pi n^3}\int_{-\infty}^{\infty}\frac{\exp(-ik_n x_n)dk_n}{Q_n(k_n)} \tag{14.232b}$$

where $x_n = nx$. We note from (14.233b) that $Q_n(k_n)$ may be factored as follows:

$$Q_n(k_n) = (k_n^2 - \alpha_1^2)(k_n^2 - \alpha_2^2)$$

where (14.234)

$$\left\{\begin{matrix}\alpha_2^2 \\ \\ \alpha_1^2\end{matrix}\right\} = (2\Theta_n^2 - 1) \pm 2\Theta_n\sqrt{\Theta_n^2 - 1} \quad .$$

If $\Theta_n^2 > 1$ then it is evident from (14.234) that α_1^2 and α_2^2 are both positive and real. However, if $0 < \Theta_n^2 < 1$ then α_1^2 and α_2^2 are both complex. Let us consider these two cases further.

Case I: $0 < \Theta_n < 1$

For reasons which will later become obvious we call this the sub-critical speed range. In this case (14.234) can be written as

$$\left\{\begin{matrix}\alpha_2^2 \\ \\ \alpha_1^2\end{matrix}\right\} = (2\Theta_n^2 - 1) \pm 2i\Theta_n\sqrt{1 - \Theta_n^2} \quad .$$

If we let

$$\left.\begin{matrix}\Theta_n = \cos\phi_n \\ \text{and} \\ \sqrt{1 - \Theta_n^2} = \sin\phi_n\end{matrix}\right\} \quad 0 < \phi_n < \frac{\pi}{2} \tag{14.235}$$

then,

$$\left\{\begin{matrix}\alpha_2^2 \\ \\ \alpha_1^2\end{matrix}\right\} = \cos 2\phi_n \pm i \sin 2\phi_n = \exp(\pm 2i\phi_n)$$

and consequently $Q_n(k_n)$ may be factored as follows:

$$Q_n(k_n) = (k_n + \alpha_1)(k_n - \alpha_1)(k_n + \alpha_2)(k_n - \alpha_2)$$

where,

$$\left\{\begin{matrix}\alpha_2 \\ \\ \alpha_1\end{matrix}\right\} = \exp(\pm i\phi_n) = \cos\phi_n \pm i \sin\phi_n = \Theta_n \pm i\sqrt{1 - \Theta_n^2}$$

The integral on the right-hand side of (14.232b) will now be evaluated by contour integration. Consider the contour integral

$$I_{C_1} = \oint_{C_1} \frac{\exp(-ix_n z)dz}{Q_n(z)} \tag{14.236}$$

where C_1 is the closed contour shown in Fig. 14.21 and z is a complex variable ($z=k_n+i\kappa$). For any R>1, the residue theorem can be applied to (14.236) to obtain (see for example, G.F. Carrier, M. Krook, C.E. Pearson, Functions of a Complex Variable, McGraw-Hill Book Co., New York, 1966)

$$\begin{aligned} I_{C_1} &= 2\pi i\left[\mathrm{Res}(\alpha_2)+\mathrm{Res}(-\alpha_1)\right] \\ &= \frac{\pi}{2}\,\frac{\exp(x_n\sin\phi_n)\cos(\phi_n+x_n\cos\phi_n)}{\Theta_n\sqrt{1-\Theta_n^2}} \end{aligned} \tag{14.237}$$

where,

$$\mathrm{Res}(\alpha_i) = \lim_{z\to\alpha_i} \frac{(z-\alpha_i)\exp(-ix_n z)}{Q_n(z)}$$

is the residue of the integrand of (14.236) at the simple pole $z=\alpha_i$. We shall now demonstrate that I_{C_1} converges to the integral on the right-hand side of (14.232b) as R approaches infinity. On the real axis, $z=k_n$ and on the semicircle of radius R, $z=R\exp(i\psi)$, $(0<\psi<\pi)$. Thus, (14.236) can be written as

$$I_{C_1} = \int_{-R}^{R} \frac{\exp(-ix_n k_n)dk_n}{Q_n(k_n)} + \int_0^{\pi} \exp(-ix_n R\cos\psi)\exp(x_n R\sin\psi)\,\frac{iR\exp(i\psi)d\psi}{Q_n\left[R\exp(i\psi)\right]} \tag{14.238}$$

If $x_n<0$, then the second integral on the right-hand side of (14.238) vanishes as $R\to\infty$ because of the term $\exp(x_n R\sin\psi)$ $(\sin\psi>0$ for $0<\psi<\pi)$. Consequently, for $x_n<0$

$$\lim_{R\to\infty} I_{C_1} = \int_{-\infty}^{\infty} \frac{\exp(-ix_n k_n)dk_n}{Q_n(k_n)} \tag{14.239}$$

The substitution of (14.239) into (14.232b) yields, in view of (14.237),

$$w_n(x_n) = \frac{\ell^4}{\pi^4 D}\,\frac{a_n}{4n^3}\,\frac{\exp(x_n\sin\phi_n)\cos(\phi_n+x_n\cos\phi_n)}{\Theta_n\sqrt{1-\Theta_n^2}} \tag{14.240}$$

for $x_n<0$ and $0<\Theta_n<1$.

The solution for $x_n>0$ can be obtained in a similar manner by considering the contour integral

$$I_{C_2} = \oint_{C_2} \frac{\exp(-ix_n z)dz}{Q_n(z)} \tag{14.241}$$

where C_2 is the closed contour shown in Fig. 14.21. For any R>1, the residue theorem can be applied to (14.241) to obtain

$$\begin{aligned} I_{C_2} &= -2\pi i\left[\mathrm{Res}(\alpha_1)+\mathrm{Res}(-\alpha_2)\right] \\ &= \frac{\pi}{2}\,\frac{\exp(-x_n\sin\phi_n)\cos(\phi_n - x_n\cos\phi_n)}{\Theta_n\sqrt{1-\Theta_n^2}} \quad . \end{aligned} \tag{14.242}$$

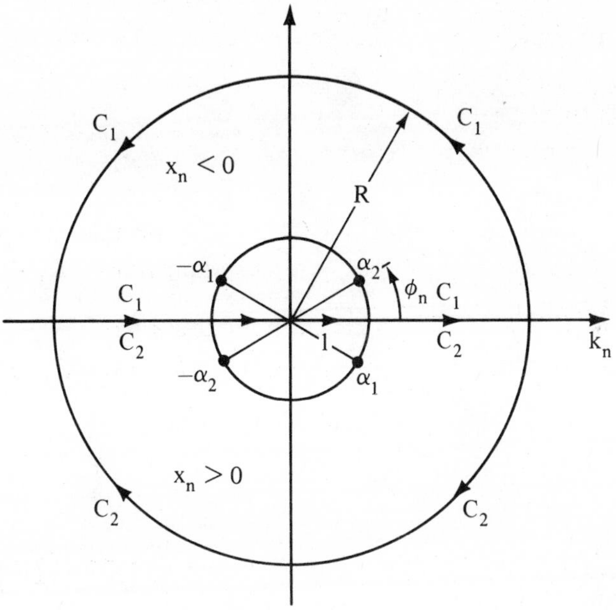

Figure 14.21

Furthermore, one can easily show that, for $x_n>0$,

$$\lim_{R\to\infty} I_{C_2} = \int_{-\infty}^{\infty} \frac{\exp(-ix_n k_n)dk_n}{Q_n(k_n)} \tag{14.243}$$

and therefore

$$w_n(x_n) = \frac{\ell^4}{\pi^4 D}\,\frac{a_n}{4n^3}\,\frac{\exp(-x_n\sin\phi_n)\cos(\phi_n - x_n\cos\phi_n)}{\Theta_n\sqrt{1-\Theta_n^2}} \tag{14.244}$$

for $x_n>0$ and $0<\Theta_n<1$.

Equations (14.240) and (14.244) can be combined into a single expression which is symmetric in x_n. With the aid of (14.235), this result is

$$w_n(x_n) = \frac{\ell^4}{\pi^4 D}\frac{a_n}{4n^3}\frac{\exp(-|x_n|\sqrt{1-\Theta_n^2})\cos(\phi_n-\Theta_n|x_n|)}{\Theta_n\sqrt{1-\Theta_n^2}} \qquad (14.245)$$

for $-\infty<x_n<\infty$ and $0<\Theta_n<1$

where,

$$0<\phi_n=\arc\tan\sqrt{\frac{1}{\Theta_n^2}-1}<\frac{\pi}{2} \quad .$$

In the limit as Θ_n approaches zero we obtain the static solution

$$\Big[w_n(x_n)\Big]_{static} = \frac{\ell^4}{\pi^4 D}\frac{a_n}{4n^3}(1+|x_n|)\exp(-|x_n|) \qquad (14.246)$$

The maximum static displacement occurs at $x_n=0$ and is

$$W_n = \frac{\ell^4}{\pi^4 D}\frac{a_n}{4n^3} \quad . \qquad (14.247)$$

On the other hand, as Θ_n approaches one, $w_n(x_n)$ becomes unbounded. Thus, $\Theta_n=1$ is called the dimensionless critical speed corresponding to the n'th harmonic component of the moving line load.

Case II: $\Theta_n>1$

This is called the supercritical speed range. If we let

$$\Theta_n=\cosh\phi_n$$

and

$$\sqrt{\Theta_n^2-1}=\sinh\phi_n$$

then (14.234) can be written as

$$\begin{Bmatrix}\alpha_2^2\\ \\ \alpha_1^2\end{Bmatrix} = \cosh 2\phi_n \pm \sinh 2\phi_n = \exp(\pm 2\phi_n)$$

and consequently $Q_n(k_n)$ may be factored as follows:

$$Q_n(k_n)=(k_n+\alpha_1)(k_n-\alpha_1)(k_n+\alpha_2)(k_n-\alpha_2) \qquad (14.248)$$

where,

$$\begin{Bmatrix}\alpha_2\\ \\ \alpha_1\end{Bmatrix} = \exp(\pm\phi_n)=\cosh\phi_n\pm\sinh\phi_n=\Theta_n\pm\sqrt{\Theta_n^2-1} \quad .$$

Thus, for $\Theta_n>1$, $0<\alpha_1<1$ and $1<\alpha_2<\infty$. Once again we must evaluate the integral

$$I=\int_{-\infty}^{\infty}\frac{\exp(-ik_nx_n)dk_n}{Q_n(k_n)} \tag{14.249}$$

However, a new difficulty has now arisen. Since $Q_n(k_n)$ has real zeroes, the integrand of I becomes unbounded as k_n approaches ($\pm\alpha_1$ and $\pm\alpha_2$) and therefore I has no meaning in the purely mathematical sense. In spite of this difficulty, a physically and mathematically acceptable meaning can be given to I as follows. Consider the polynomial

$$Q_n(k_n;\varepsilon_1,\varepsilon_2,\varepsilon_3,\varepsilon_4)=(k_n-\bar{\alpha}_1)(k_n-\bar{\alpha}_2)(k_n-\bar{\alpha}_3)(k_n-\bar{\alpha}_4)$$

whose zeroes are located at $\bar{\alpha}_1=\alpha_1\pm i\varepsilon_1$, $\bar{\alpha}_2=\alpha_2\pm i\varepsilon_2$, $\bar{\alpha}_3=-\alpha_1\pm i\varepsilon_3$, $\bar{\alpha}_4=-\alpha_2\pm i\varepsilon_4$ where $\varepsilon_1,\varepsilon_2,\varepsilon_3$ and ε_4 are small but finite positive real numbers. The integral

$$I(\varepsilon_1,\varepsilon_2,\varepsilon_3,\varepsilon_4)=\int_{-\infty}^{\infty}\frac{\exp(-ik_nx_n)dk_n}{Q_n(k_n;\varepsilon_1,\varepsilon_2,\varepsilon_3,\varepsilon_4)}$$

can be evaluated by contour integration since it has no poles on the real axis. We can then define

$$I=\lim_{(\varepsilon_1,\varepsilon_2,\varepsilon_3,\varepsilon_4)\to 0} I(\varepsilon_1,\varepsilon_2,\varepsilon_3,\varepsilon_4) \quad .$$

However, since each pole can be removed from the real axis in one of two ways, i.e., it can be moved up ($\bar{\alpha}_1=\alpha_1+i\varepsilon_1$) or it can be moved down ($\bar{\alpha}_1=\alpha_1-i\varepsilon_1$), there are a total of $2^4=16$ different values which can be assigned to I by this procedure. Each of these values is mathematically acceptable, however, only one of them is physically acceptable. To determine which one that is, either of the following two methods can be used. Suppose we were to admit a viscous damping force ($-\varepsilon\dot{w}$) in addition to the applied force $p(x_1,x_2,t)$. The differential equation of motion would then become

$$D\nabla^4 w + \rho h\ddot{w} + \varepsilon\dot{w} = p(x_1,x_2,t) \tag{14.250}$$

and consequently we would obtain, in place of (14.233b),

$$Q_n(k_n)=k_n^4-2(2\Theta_n^2-1)k_n^2+8i\Theta_n\varepsilon_nk_n+1 \tag{14.251a}$$

where,

$$\varepsilon_n=\left(\frac{\ell}{\pi}\right)^2\frac{\varepsilon}{4n^2\sqrt{\rho hD}}>0 \quad .$$

In this case,

$$Q_n(k_n)=(k_n-\bar{\alpha}_1)(k_n-\bar{\alpha}_2)(k_n-\bar{\alpha}_3)(k_n-\bar{\alpha}_4) \tag{14.251b}$$

where $\bar{\alpha}_1,\bar{\alpha}_2,\bar{\alpha}_3$ and $\bar{\alpha}_4$ are complex. Thus, by including viscous damping we have removed the zeroes of $Q_n(k_n)$ from the real axis and the integral I can now be evaluated by contour integration as in the preceding subcritical case. We now require that the solution of our undamped problem be the same as the solution of the damped case as ε_n approaches zero. Thus, we first solve the problem with damping and subsequent to the evaluation of I we let $\varepsilon_n \to 0$. Since the roots of (14.251a) are continuous functions of ε_n, they may be approximated, for sufficiently small damping ($\varepsilon_n \ll 1$), by the expressions

$$\begin{aligned}
\bar{\alpha}_1 &\cong \alpha_1+i\beta_1\varepsilon_n \\
\bar{\alpha}_2 &\cong \alpha_2+i\beta_2\varepsilon_n \\
\bar{\alpha}_3 &\cong -\alpha_1+i\beta_3\varepsilon_n \\
\bar{\alpha}_4 &\cong -\alpha_2+i\beta_4\varepsilon_n
\end{aligned} \tag{14.252}$$

where $\pm\alpha_1$ and $\pm\alpha_2$ are the roots of $Q_n(k_n)=0$ when $\varepsilon_n=0$ (see (14.248)). To determine the constants β_1 through β_4 we substitute (14.252) into (14.251b) to obtain, after neglecting terms of order $\varepsilon_n^2, \varepsilon_n^3$ and ε_n^4,

$$k_n^4 - Ik_n^3 + IIk_n^2 - IIIk_n + IV = 0$$

where, (14.253)

$$I = i\varepsilon_n(\beta_1+\beta_2+\beta_3+\beta_4)$$

$$II = -2(2\Theta_n^2-1)+i\varepsilon_n\left[\alpha_1(\beta_3-\beta_1)+\alpha_2(\beta_4-\beta_2)\right]$$

$$III = -i\varepsilon_n\left[\alpha_1^2(\beta_2+\beta_4)+\alpha_2^2(\beta_1+\beta_3)\right]$$

$$IV = 1+i\varepsilon_n\left[\alpha_1^2\alpha_2(\beta_2-\beta_4)+\alpha_1\alpha_2^2(\beta_1-\beta_3)\right] .$$

A comparison of (14.253) with (14.251a) reveals that

$$\beta_1=\beta_3=-\beta_2=-\beta_4=\frac{1}{\sqrt{\Theta_n^2-1}} .$$

Thus, for sufficiently small damping,

$$\bar{\alpha}_1 \cong \alpha_1 + \frac{i\varepsilon_n}{\sqrt{\Theta_n^2-1}}\ , \quad \bar{\alpha}_2 \cong \alpha_2 - \frac{i\varepsilon_n}{\sqrt{\Theta_n^2-1}}$$

$$\bar{\alpha}_3 \cong -\alpha_1 + \frac{i\varepsilon_n}{\sqrt{\Theta_n^2-1}}\ , \quad \bar{\alpha}_4 \cong -\alpha_2 - \frac{i\varepsilon_n}{\sqrt{\Theta_n^2-1}} \qquad (14.254)$$

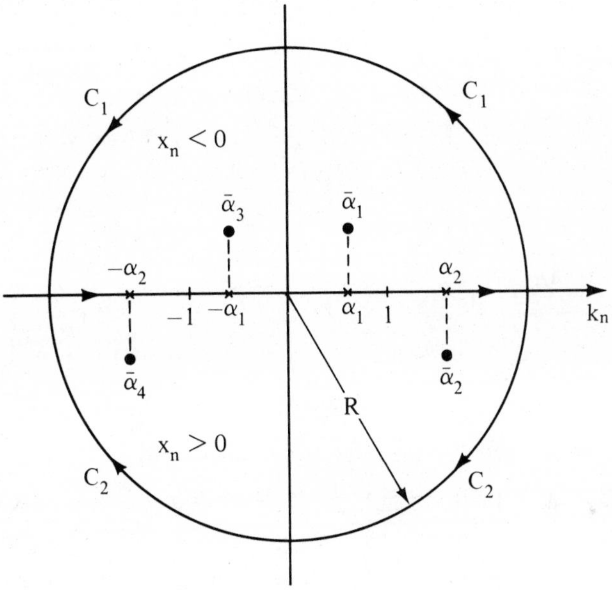

Figure 14.22

The location of these "perturbed" poles is illustrated in Fig. 14.22. In a manner similar to that used in the subcritical case, the integral I can now be evaluated by contour integration around the closed contours C_1 and C_2 shown in Fig. 14.22. The result is:

for $x_n<0$,

$$I = \lim_{\varepsilon_n\to 0}\left[\lim_{R\to\infty} I_{C_1}\right] = \lim_{\varepsilon_n\to 0} 2\pi i\left[\mathrm{Res}(\bar{\alpha}_3)+\mathrm{Res}(\bar{\alpha}_1)\right]$$

$$= 2\pi i\left[\mathrm{Res}(-\alpha_1)+\mathrm{Res}(\alpha_1)\right]$$

$$= -\frac{\pi}{2}\,\frac{\sin\alpha_1 x_n}{\alpha_1\Theta_n\sqrt{\Theta_n^2-1}} \qquad (14.255a)$$

for $x_n>0$,

$$I = \lim_{\varepsilon_n\to 0}\left[\lim_{R\to\infty} I_{C_2}\right] = -\lim_{\varepsilon_n\to 0} 2\pi i\left[\mathrm{Res}(\bar{\alpha}_4)+\mathrm{Res}(\bar{\alpha}_2)\right]$$
$$= -2\pi i\left[\mathrm{Res}(-\alpha_2)+\mathrm{Res}(\alpha_2)\right]$$
$$= -\frac{\pi}{2}\,\frac{\sin\alpha_2 x_n}{\alpha_2\Theta_n\sqrt{\Theta_n^2-1}} \quad . \tag{14.255b}$$

Consequently, the desired solution is

$$w_n(x_n) = -\frac{\ell^4}{\pi^4 D}\,\frac{a_n}{4n^3}\,\frac{\sin\left[x_n\left(\Theta_n-\sqrt{\Theta_n^2-1}\right)\right]}{\Theta_n\sqrt{\Theta_n^2-1}\,\left(\Theta_n-\sqrt{\Theta_n^2-1}\,\right)},\quad x_n<0$$
$$w_n(x_n) = -\frac{\ell^4}{\pi^4 D}\,\frac{a_n}{4n^3}\,\frac{\sin\left[x_n\left(\Theta_n+\sqrt{\Theta_n^2-1}\right)\right]}{\Theta_n\sqrt{\Theta_n^2-1}\,\left(\Theta_n+\sqrt{\Theta_n^2-1}\,\right)},\quad x_n>0 \tag{14.256}$$

for $\Theta_n>1$.

The source of the difficulty in evaluating the integral I for the case $\Theta_n>1$ is evident from an inspection of (14.256), i.e., $w_n(x_n)$ and its derivatives do not vanish as $|x_n|$ approaches infinity. Thus, strictly speaking, the Fourier transform technique is not applicable to this problem. However, the damped solution and its derivatives vanish for $|x_n|\to\infty$ as a result of the term $\exp\left(-\frac{\varepsilon_n|x_n|}{\sqrt{\Theta_n^2-1}}\right)$ which arises from the residues at the poles $\bar{\alpha}_1$ and $\bar{\alpha}_3$ for $x_n<0$ and $\bar{\alpha}_2$ and $\bar{\alpha}_4$ for $x_n>0$. Consequently, in the limit as $\varepsilon_n\to 0$ the physically appropriate solution for the undamped case is obtained. We conclude from the preceding discussion that the integral I for $\Theta_n>1$ (supercritical case) characterizes the physically appropriate solution provided it is interpreted as a limit. To apply this limiting procedure an additional condition must be imposed upon the solution in order to determine how the poles are to be removed from the real axis. For example, in the preceding discussion we required the solution of our problem to be identical with the solution of the damped problem when the damping constant goes to zero. Another condition, which is frequently used in physics, is obtained by invoking the principle of causality, viz., an effect cannot precede its cause. This is done as follows. The dispersion relation for a freely vibrating plate strip is obtained by setting $Q_n(k_n)=0$. With the aid of (14.233b) this dispersion relation can be written as $\Theta_n^2=\left(\frac{k_n^2+1}{2k_n}\right)^2$ or $\Theta_n=\pm\left(\frac{k_n^2+1}{2k_n}\right)$. The corresponding group velocity G_n can be obtained from the formula (see Appendix B) $G_n=\Theta_n+k_n\frac{d\Theta_n}{dk_n}=\pm k_n$. These dispersion curves are shown in Fig. 14.23.

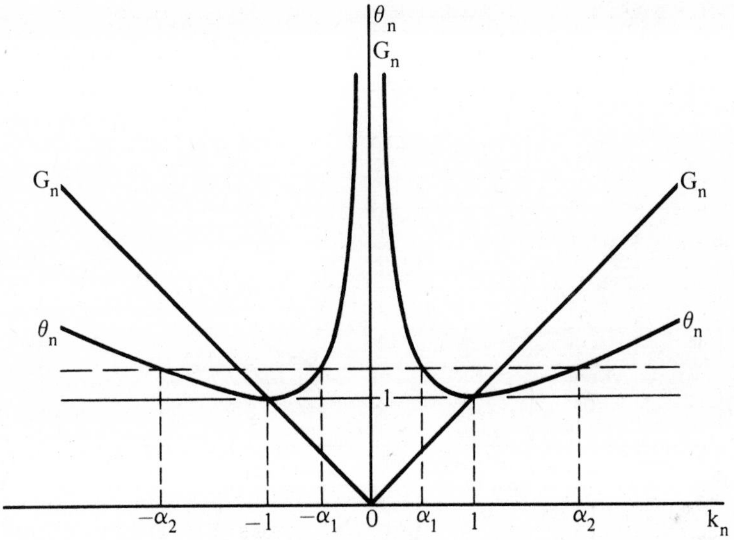

Figure 14.23

It is clear from this figure that there are four wave numbers $(\pm\alpha_1,\pm\alpha_2)$ corresponding to any fixed value of $\Theta_n>1$. For the wave numbers $k_n=\pm\alpha_1$ the phase velocity Θ_n is greater than the group velocity G_n. Since the group velocity is the average velocity of energy transport in the plate and since Θ_n corresponds to the load speed in our problem we conclude that the load is moving faster than the speed of energy propagation in the plate. Therefore, no energy can be transmitted ahead of the line load at these wave numbers. Consequently, the poles at $k_n=\pm\alpha_1$ can only contribute to the solution behind the line load $(x_n<0)$. Similarly, for the wave numbers $k_n=\pm\alpha_2$ the phase velocity Θ_n is less than the group velocity G_n. Therefore, at these wave numbers the speed of energy propagation through the plate is greater than the speed of the line load and consequently energy is transmitted ahead of the load. We conclude that the poles at $k_n=\pm\alpha_2$ can only contribute to the solution ahead of the line load $(x_n>0)$. With reference to Fig. 14.22 we see that in order for the poles $k_n=\pm\alpha_1$ to contribute only to the solution $x_n<0$ they must be replaced by the poles $k_n=\pm\alpha_1+i\varepsilon$ where ε is an arbitrarily small but finite positive real number. Similarly, in order that the poles $k_n=\pm\alpha_2$ only contribute to the solution $x_n>0$ they must be replaced by the poles $k_n=\pm\alpha_2-i\varepsilon$. The integral I can then be evaluated by contour integration after which the limit is taken as ε approaches zero. This procedure yields the same result (14.256) as was obtained by the inclusion of damping which was subsequently set equal to zero. It should also be pointed out that (14.256) is not the Cauchy principal value of I. Fig. 14.24 shows the dimensionless deflection profile $\tilde{w}_n(x_n)=\dfrac{w_n(x_n)}{W_n}$ plotted

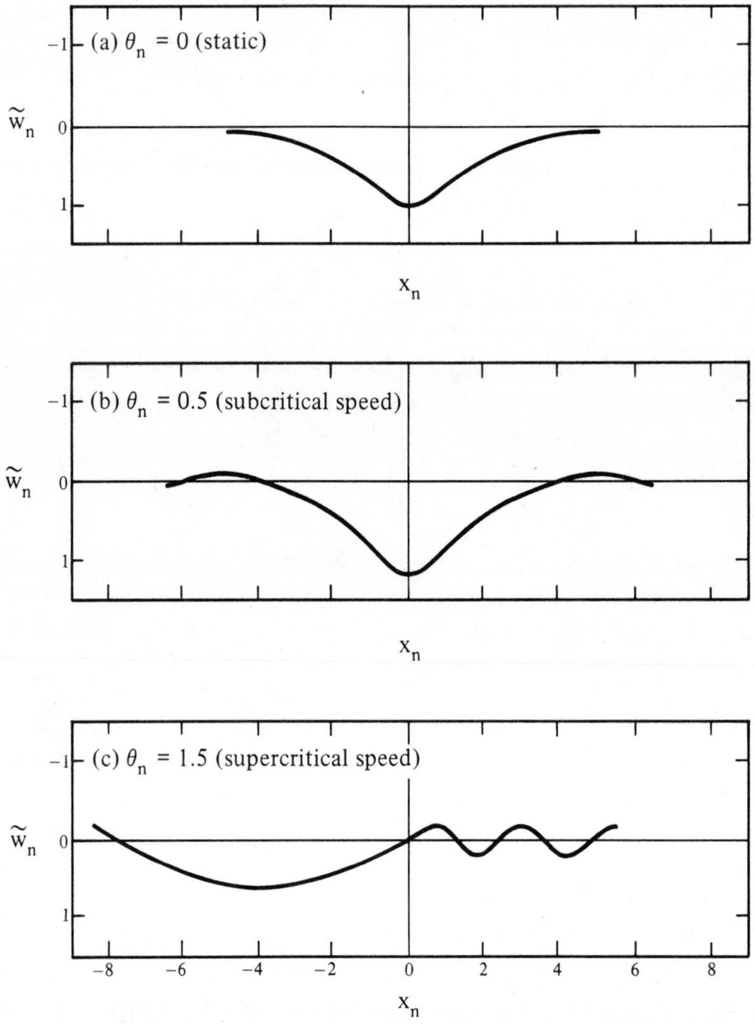

Figure 14.24

versus the dimensionless moving coordinate $x_n = \frac{n\pi}{\ell}(x_1 - vt)$ for the static, subcritical and supercritical cases. The dynamic amplification factor $A_n = \text{Max}|\tilde{w}_n|$ is shown in Fig. 14.25 plotted versus the dimensionless load speed Θ_n. Studies of this problem for the case of arbitrary viscous damping or prestress may be found in the papers by H. Reismann, "Dynamic Response of an Elastic Plate Strip to a Moving Line Load," AIAA Journal, Vol. 1, No. 2, pp. 354-360, 1963 and, "Response of a Prestressed Elastic Plate Strip to a Moving Pressure Load," Journal of The Franklin Institute, Vol. 277, No. 7, pp. 8-19, 1964.

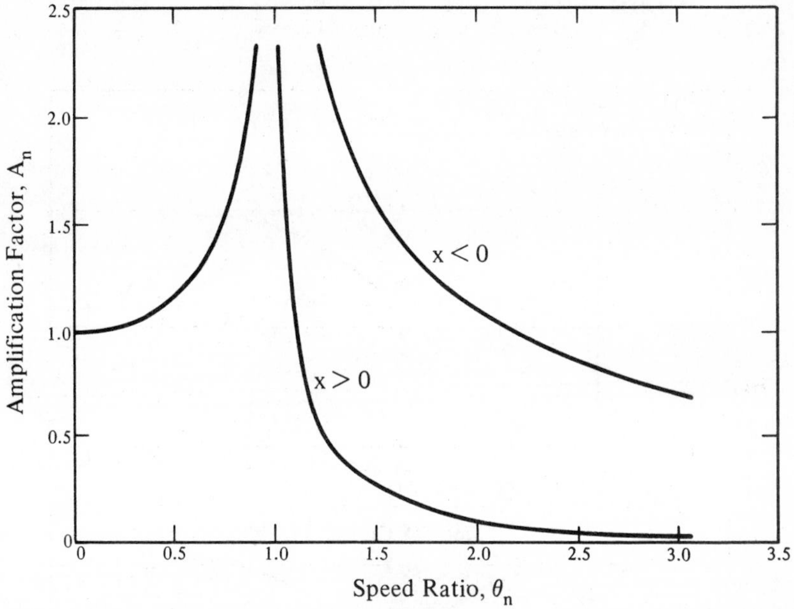

Figure 14.25

EXERCISES

14.1 Show that q_α are the components of a vector and that $m_{\alpha\beta}$ are the components of a second order tensor in two-dimensions.

14.2 Show that

$$Q_n = \int_{-\frac{h}{2}}^{\frac{h}{2}} T_z dz$$

$$M_{nn} = \int_{-\frac{h}{2}}^{\frac{h}{2}} zT_n dz \quad \text{and} \quad M_{n\ell} = \int_{-\frac{h}{2}}^{\frac{h}{2}} zT_\ell dz$$

where $\hat{n}$ and $\hat{\ell}$ are the unit normal and tangent vectors, respectively to a curve C in A and $T_i = \tau_{ij} n_j$ is the stress vector acting on the cylindrical surface whose intersection with the midplane is C,

14.3 Derive the dynamic reciprocal relation for plates (14.41).

14.4 Show that the forced motion problem characterized by (14.44) through (14.46) has a unique solution. Follow the method of Section 10.2.

14.5 Rederive all the major results of Sections 14.1 and 14.2 in plane polar coordinates (r,Θ), where $x_1=r\cos\Theta$ and $x_2=r\sin\Theta$. Begin with the displacement assumptions $u_r=z\psi_r(r,\Theta,t)$, $u_\Theta=z\psi_\Theta(r,\Theta,t)$, $u_z=w(r,\Theta,t)$ and define plate stress resultants $(M_{rr},M_{\Theta\Theta},M_{r\Theta},Q_r,Q_\Theta)$ and plate strain resultants $(m_{rr},m_{\Theta\Theta},m_{r\Theta},q_r,q_\Theta)$ in accordance with (14.3) and (14.4), i.e.,

$$M_{rr}=\int_{-\frac{h}{2}}^{\frac{h}{2}}\tau_{rr}z\,dz,\quad m_{rr}=\frac{12}{h^3}\int_{-\frac{h}{2}}^{\frac{h}{2}}\varepsilon_{rr}z\,dz,\ \text{etc.}$$

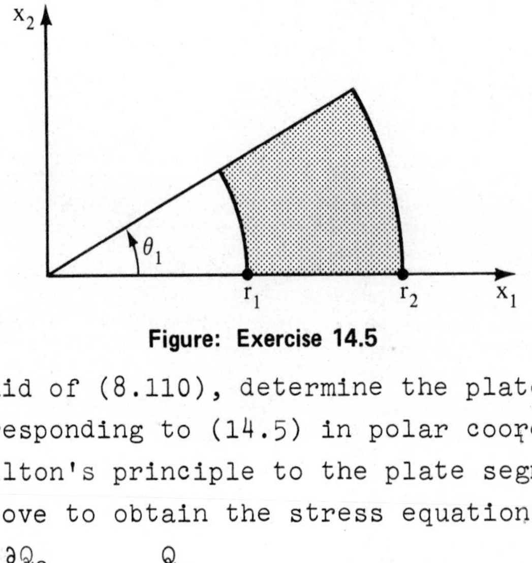

Figure: Exercise 14.5

(a) With the aid of (8.110), determine the plate strain-displacement relations corresponding to (14.5) in polar coordinates.

(b) Apply Hamilton's principle to the plate segment $(0\le r_1<r<r_2,\ 0<\Theta<\Theta_1\le 2\pi)$ shown above to obtain the stress equations of motion

$$-\rho h\ddot{w}+\frac{\partial Q_r}{\partial r}+\frac{1}{r}\frac{\partial Q_\Theta}{\partial \Theta}+p+\frac{Q_r}{r}=0$$

$$-\frac{\rho h^3}{12}\ddot{\psi}_r+\frac{\partial M_{rr}}{\partial r}+\frac{1}{r}\frac{\partial M_{r\Theta}}{\partial\Theta}+\frac{M_{rr}-M_{\Theta\Theta}}{r}-Q_r=0$$

$$-\frac{\rho h^3}{12}\ddot{\psi}_\Theta+\frac{\partial M_{r\Theta}}{\partial r}+\frac{1}{r}\frac{\partial M_{\Theta\Theta}}{\partial\Theta}+\frac{2M_{r\Theta}}{r}-Q_\Theta=0$$

and the admissible boundary conditions: at $r=r_1$ and $r=r_2$ one member of each of the following pairs must be specified (w,Q_r), (ψ_r,M_{rr}), $(\psi_\Theta,M_{r\Theta})$ and at $\Theta=0$ and $\Theta=\Theta_1$ one member of each of the following pairs must be specified (w,Q_Θ), $(\psi_r,M_{r\Theta})$, $(\psi_\Theta,M_{\Theta\Theta})$.

(c) Determine the plate stress-strain law in polar coordinates and consequently obtain the stress-displacement relations

$$M_{rr}=D\left[\frac{\partial\psi_r}{\partial r}+\frac{\nu}{r}\left(\psi_r+\frac{\partial\psi_\Theta}{\partial\Theta}\right)\right]$$

$$M_{\Theta\Theta}=D\left[\nu\frac{\partial\psi_r}{\partial r}+\frac{1}{r}\left(\psi_r+\frac{\partial\psi_\Theta}{\partial\Theta}\right)\right]$$

$$M_{r\Theta}=M_{\Theta r}=\frac{D}{2}(1-\nu)\left[\frac{\partial\psi_\Theta}{\partial r}+\frac{1}{r}\frac{\partial\psi_r}{\partial\Theta}-\frac{1}{r}\psi_\Theta\right]$$

$$Q_r=\kappa^2 Gh\left(\psi_r+\frac{\partial w}{\partial r}\right)$$

$$Q_\Theta=\kappa^2 Gh\left(\psi_\Theta+\frac{1}{r}\frac{\partial w}{\partial\Theta}\right)$$

(d) Substitute the stress-displacement relations into the stress equations of motion to obtain the displacement equations of motion. Check your result by expressing (14.29b) in polar coordinates. The results of Exercise 8.18 will be helpful for this second approach.

(e) Specialize all the preceding results to the classical plate theory by neglecting the rotatory inertia($\frac{\rho h^3}{12}\ddot{\vec{\psi}}=0$) and the transverse shear deformation ($\vec{\psi}=-\vec{\nabla}w$).

14.6 Consider the mathematical model of a "shear plate," i.e., a plate which deforms in shear only, but remains rigid with respect to other types of deformation. Assume (and justify) the displacement field $u_1=0$, $u_2=0$, $u_3=w(x_1,x_2,t)$ and derive the stress equations of motion and admissible boundary conditions using Hamilton's principle. They are:

$$-\rho h\ddot{w}+\vec{\nabla}\cdot\vec{Q}+p=0 \quad \text{in } A$$

and w or Q_n must be specified on C. Show that the stress-displacement relations are given by

$$Q_\alpha=\int_{-\frac{h}{2}}^{\frac{h}{2}}\tau_{z\alpha}dz=\kappa^2 Ghw_{,\alpha}$$

where κ^2 is the shear coefficient. Formulate a well posed forced motion problem and discuss its solution. Derive an expression for energy flux in the plate.

14.7 Derive a mathematical model for the sandwich plate shown in the figure. Assume that the facings are monolithically bonded to the core of the plate. Include the effects of shear deformation and rotatory inertia.

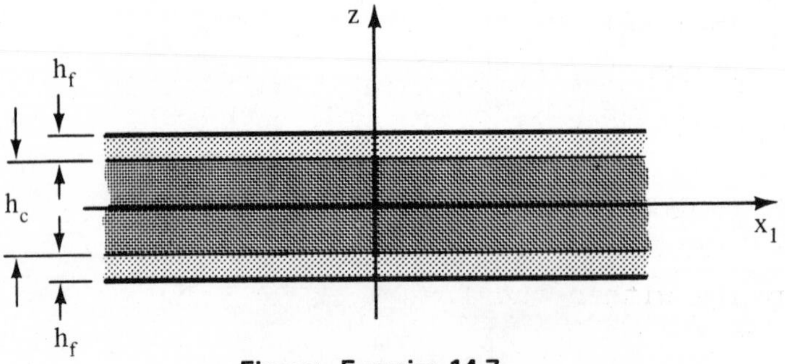

Figure: Exercise 14.7

14.8 Show that the forced motion problem characterized by (14.123) through (14.125) has a unique solution. Follow the method of Section 10.2.

14.9 Show that the natural frequencies associated with the classical plate theory are real.

14.10 Using the separation of variables technique, obtain the solution of the quasi-static problem characterized by (14.148) and (14.149). Begin by assuming that

$$w^{(s)}(x_1,x_2,t)= \sin \frac{m\pi x_1}{a} f(x_2)H(t) \quad .$$

14.11 Obtain the response of a clamped, circular plate to a concentrated force which is suddenly applied at its center by letting $B= \frac{b}{a}$ tend to zero as $\pi b^2 p_0$ tends to F in the solution of example (b) of Section 14.4.

14.12 A simply supported, circular plate is subjected to a uniformly distributed, suddenly applied pressure of magnitude p_0. The plate is at rest in its reference configuration at t=0 when the load is applied. Find the dynamic response within the framework of classical plate theory.

14.13 A simply supported, rectangular plate is subjected to a uniformly distributed, suddenly applied pressure of magnitude p_0. The plate is at rest in its reference configuration at t=0 when the pressure is applied. Find the dynamic response within the framework of classical plate theory.

14.14 Consider the static deformation of a simply supported, rectangular plate subjected to a uniformly distributed load of intensity p_o. Assume a solution in the form

$$w(x_1,x_2) = \sum_{m=1}^{\infty} \sum_{n=1}^{\infty} C_{mn} W_{mn}(x_1,x_2)$$

where $W_{mn}(x_1,x_2)$ are the normalized eigenfunctions given by (14.147) for classical plate theory. Find the constants C_{mn}, $m=1,2,3,\ldots$, $n=1,2,3,\ldots$.

14.15 Determine the eigenfunctions for a simply supported, rectangular plate within the framework of improved plate theory, i.e., assume that

$$w(x_1,x_2,t)=W^{(m,n)}(x_1,x_2)\cos\omega_{mn}t$$

$$\psi_1(x_1,x_2,t)=\Psi_1^{(m,n)}(x_1,x_2)\cos\omega_{mn}t$$

$$\psi_2(x_1,x_2,t)=\Psi_2^{(m,n)}(x_1,x_2)\cos\omega_{mn}t$$

and show that

$$W^{(m,n)}=A_{mn}\sin\frac{m\pi x_1}{a}\sin\frac{n\pi x_2}{b}$$

$$\Psi_1^{(m,n)}=B_{mn}\cos\frac{m\pi x_1}{a}\sin\frac{n\pi x_2}{b}$$

$$\Psi_2^{(m,n)}=C_{mn}\sin\frac{m\pi x_1}{a}\cos\frac{n\pi x_2}{b}$$

Normalize the eigenfunctions and find an expression for the natural frequencies ω_{mn}. (See the paper by H. Reismann and Y.C. Lee, "Forced Motion of Rectangular Plates," Developments in Theoretical and Applied Mechanics, Vol. 4, Pergamon Press, Oxford, 1970, pp. 3-18).

14.16 An infinite elastic plate is subjected to a suddenly applied concentrated load of magnitude F applied at r=0, t=0. The plate is at rest in its reference configuration at t=0. Using classical plate theory, show that the solution is

$$w(r,t)=\frac{F\tau}{16\pi D}\left[\frac{\pi}{2}-\mathrm{Si}\left(\frac{r^2}{\tau}\right)+\frac{r^2}{\tau}\,\mathrm{Ci}\left(\frac{r^2}{\tau}\right)-\sin\frac{r^2}{\tau}\right]$$

where $\tau=4\lambda^2 t$, $\lambda^4=\frac{D}{\rho h}$, and $\mathrm{Si}(x)=\int_0^x \frac{\sin\xi}{\xi}\,d\xi$ and $\mathrm{Ci}(x)=\int_0^x \frac{\cos\xi}{\xi}\,d\xi$ are the sine and cosine integral functions, respectively. Show that for sufficiently large values of $\frac{r^2}{\tau}$, the solution can be approximated by the asymptotic formula

$$w(r,t) \cong -\frac{F\tau}{16\pi D}\left(\frac{\tau}{r^2}\right)^2 \sin\left(\frac{r^2}{\tau}\right) \quad .$$

14.17 Consider the forced motion problem characterized by (14.23) through (14.25) with $f_1 \equiv f_2 \equiv 0$. Assume a solution of the form

$$w(x_1,x_2,t) = \sum_{i=1}^{\infty} W_i(x_1,x_2)q_i(t)$$

where $W_i(x_1,x_2)$ are the eigenfunctions satisfying the particular homogeneous boundary conditions of the problem. Show that

$$\ddot{q}_i + \omega_i^2 q_i = P_i(t) = \int_A pW_i dA$$

$$q_i(0) = \int_A \rho h w_o W_i da,$$

$$\dot{q}_i(0) = \int_A \rho h \dot{w}_o W_i dA$$

and

$$q_i(t) = q_i(0)\cos\omega_i t + \frac{\dot{q}_i(0)}{\omega_i}\sin\omega_i t + \frac{1}{\omega_i}\int_0^t P_i(\tau)\sin\omega_i(t-\tau)d\tau \quad .$$

REFERENCES: PART II

CONTINUUM MECHANICS

Y.C. Fung, Foundations of Solid Mechanics, Prentice Hall, Inc., Englewood Cliffs, N.J., 1965.

W. Jaunzemis, Continuum Mechanics, The Macmillan Co., New York, 1967.

L.E. Malvern, Introduction to the Mechanics of a Continuous Medium, Prentice Hall, Englewood Cliffs, N.J., 1969.

W. Prager, Introduction to Mechanics of Continua, Ginn and Co., Boston, 1961.

C. Truesdell and R.A. Toupin, "The Classical Field Theories," Handbuch der Physik, Vol. III/1, Springer Verlag 1960, pp. 226-790.

ELASTICITY THEORY

L.D. Landau and E.M. Lifshitz, Theory of Elasticity, Second Edition, Pergamon Press, London, 1970.

H. Leipholz, Einführung in die Elastizitätstheorie, G. Braun, Karlsruhe, 1968.

A.E.H. Love, A Treatise on the Mathematical Theory of Elasticity, Fourth Edition, Cambridge University Press, 1927.

N.I. Muskhelishvili, Some Basic Problems of the Mathematical Theory of Elasticity, Third Edition, P. Noordhoff, Groningen, 1953.

V.V. Novozhilov, Theory of Elasticity, Pergamon Press, New York, 1961.

C.E. Pearson, Theoretical Elasticity, Harvard University Press, Cambridge, Mass., 1959.

I.N. Sneddon and D.S. Berry, "The Classical Theory of Elasticity," Handbuch der Physik, Vol. VI, Springer Verlag, Berlin 1958, pp. 1-126.

I.S. Sokolnikoff, Mathematical Theory of Elasticity, Second Edition, McGraw-Hill Book Co., Inc., New York, 1956.

S.P. Timoshenko and J.N. Goodier, Theory of Elasticity, Third Edition, McGraw-Hill Book Co., Inc., New York, 1970.

THEORY OF PLATES

K. Girkmann, Flächentragwerke, 5th Edition, Springer Verlag, Vienna, 1959.

A.W. Leissa, Vibration of Plates, NASA SP-160, U.S. Government Printing Office, Washington, 1969.

E.H. Mansfield, The Bending and Stretching of Plates, The Macmillan Co., New York, 1964.

R.D. Mindlin, An Introduction to the Mathematical Theory of Vibrations of Elastic Plates, U.S. Army Signal Corps Engineering Laboratory, Fort Monmouth, N.J., 1955, Signal Corps Contract DA-36-039 SC-56777.

A. Nádai, Die Elastischen Platten, Springer Verlag, Berlin, 1925.

S. Timoshenko and S. Woinowsky-Krieger, Theory of Plates and Shells, Second Edition, McGraw-Hill Book Co., Inc., New York, 1959.

WAVE MOTION

C.A. Coulson, Waves, Oliver and Boyd Ltd., Edinburgh, 1965.

W.C. Elmore and M.A. Heald, The Physics of Waves, McGraw-Hill Book Co., New York, 1969.

W.M. Ewing, W.S. Jardetzky and F. Press, Elastic Waves in Layered Media, McGraw-Hill Book Co., New York, 1957.

H. Kolsky, Stress Waves in Solids, Dover Publications Inc., 1963.

R.B. Lindsay, Mechanical Radiation, McGraw-Hill Book Co., New York, 1960.

J.M. Pearson, A Theory of Waves, Allyn and Bacon, Inc., Boston, 1966.

VIBRATION THEORY

Y. Chen, Vibrations: Theoretical Methods, Addison-Wesley Publishing Co., Inc., Reading, Mass., 1966.

A.P. Filipov, Vibrations in Elastic Systems, Academy of Sciences, Ukrainian SSR, Kiev, 1956.

L. Meirovitch, Analytical Methods in Vibrations, The Macmillan Co., New York, 1967.

W. Nowacki, Dynamics of Elastic Systems, John Wiley, New York, 1963.

MATHEMATICAL METHODS

A.I. Borisenko and I.E. Tarapov, Vector and Tensor Analysis with Applications, Prentice Hall, Inc., Englewood Cliffs, N.J., 1968.

L. Brand, Vector and Tensor Analysis, John Wiley and Sons, Inc., New York, 1947

L. Brillouin, Tensors in Mechanics and Elasticity, Academic Press, New York, 1964.

G.F. Carrier, M. Krook, C.E. Pearson, Functions of a Complex Variable, McGraw-Hill Book Co., New York, 1966.

H. Jeffreys, Cartesian Tensors, Cambridge University Press, London, 1969.

A.J. McConnell, Applications of Tensor Analysis, Dover Publications, Inc., New York, 1957.

S.G. Mikhlin, The Problem of The Minimum of a Quadratic Functional, Holden-Day, Inc., San Francisco, 1965.

S.G. Mikhlin, Variational Methods in Mathematical Physics, Pergamon Press, New York, 1964.

V.I. Smirnov, A Course of Higher Mathematics, Vols. I through V, Pergamon Press, Oxford, 1964.

I.N. Sneddon, Fourier Transforms, McGraw-Hill Book Co., Inc., New York, 1951.

I.S. Sokolnikoff, Tensor Analysis, John Wiley & Sons, Inc., New York, 1951.

Appendices

A

THE CALCULUS OF VARIATIONS

The Great Architect seems to be a mathematician.

Sir James Jeans

In this appendix we shall be concerned with the connection between Hamilton's principle and Lagrange's equations, as discussed in Chapter 2, with certain aspects of the calculus of variations.

Let us consider a conservative, holonomic, rheonomous, mechanical system characterized by its Lagrangian function

$$L = L(q_1,q_2,\ldots,q_n,\dot{q}_1,\dot{q}_2,\ldots,\dot{q}_n;t) \equiv L(q_i,\dot{q}_i;t) \qquad i = 1,2,\ldots,n \tag{A.1}$$

where the $q_i(t)$ are the generalized coordinates. We now form the integral

$$I = \int_{t_1}^{t_2} L(q_i,\dot{q}_i;t)dt \tag{A.2}$$

where t_1 and t_2 denote two different but otherwise arbitrary instants. It will be useful as well as enlightening to pose and subsequently answer the following question: What are the (necessary) conditions which the function $L(q_i,\dot{q}_i;t)$ must satisfy in order that the integral I be an extremum (maximum or minimum)?

To answer this question, we consider the functions

$$q_i(t) = f_i(t) \tag{A.3}$$

$$q_i(\varepsilon,t) = f_i(t) + \varepsilon\eta_i(t) \equiv f_i + \varepsilon\eta_i \tag{A.4}$$

defined in the interval $t_1 \le t \le t_2$. By definition, (A.3) is that function which will render I an extremum. The function defined by (A.4) is equal to the function in (A.3) when $\varepsilon=0$. In addition, we select the functions $\eta_i(t)$ in such a manner as to insure that the extremizing functions (A.3) and their associated comparison functions (A.4) are equal at the instant t_1 and t_2, resulting in the condition

$$\eta_i(t_1) = \eta_i(t_2) = 0 \tag{A.5}$$

It is noted that ε is a parameter which can assume any real value and the $\eta_i(t)$, $i=1,2,\ldots,n$, are sufficiently smooth and mutually independent functions of time. We now define

$$I(\varepsilon) = \int_{t_1}^{t_2} L(f_i+\varepsilon\eta_i,\ \dot{f}_i+\varepsilon\dot{\eta}_i;t)dt \tag{A.6}$$

and according to the formal rule of finding an extremum, we set

$$\left[\frac{dI(\varepsilon)}{d\varepsilon}\right]_{\varepsilon=0} = 0 \tag{A.7}$$

Upon substitution of (A.6) into (A.7), and subsequent utilization of (A.3), we obtain

$$\int_{t_1}^{t_2}\left\{\sum_{i=1}^{n}\left(\frac{\partial L}{\partial q_i}\eta_i + \frac{\partial L}{\partial \dot{q}_i}\dot{\eta}_i\right)\right\}dt = 0 \tag{A.8}$$

Integrating by parts,

$$\int_{t_1}^{t_2}\frac{\partial L}{\partial \dot{q}_i}\dot{\eta}_i dt = \left(\frac{\partial L}{\partial \dot{q}_i}\eta_i\right)_{t=t_1}^{t=t_2} - \int_{t_1}^{t_2}\eta_i\frac{d}{dt}\left(\frac{\partial L}{\partial \dot{q}_i}\right)dt \tag{A.9}$$

and in view of (A.5), the first term on the right-hand side of (A.9) vanishes. Consequently

$$\int_{t_1}^{t_2}\left\{\sum_{i=1}^{n}\left[\frac{d}{dt}\left(\frac{\partial L}{\partial \dot{q}_i}\right) - \frac{\partial L}{\partial q_i}\right]\eta_i\right\}dt = 0 \tag{A.10}$$

Since t_2-t_1 is an arbitrary time interval, the integrand in (A.10) vanishes and

$$\sum_{i=1}^{n}\left[\frac{d}{dt}\left(\frac{\partial L}{\partial \dot{q}_i}\right) - \frac{\partial L}{\partial q_i}\right]\eta_i = 0 \tag{A.11}$$

The functions η_i are independent, hence

$$\frac{d}{dt}\left(\frac{\partial L}{\partial \dot{q}_i}\right) - \frac{\partial L}{\partial q_i} = 0 \qquad i=1,2,\ldots,n \tag{A.12}$$

Equations (A.12) are the necessary conditions which must be imposed on the Lagrangian function L to render the integral I an extremum. We also note that (A.12) are Lagrange's equations of motion for a conservative, holonomic system as shown in Chapter 2.

We shall now establish the connection between Lagrange's equations and Hamilton's principle. In the spirit of (A.6), we expand $L(\varepsilon)$ in a Taylor series about $L(0)$:

$$L(\varepsilon) = L(0) + \left(\frac{\partial L}{\partial \varepsilon}\right)_{\varepsilon=0} \varepsilon + \tfrac{1}{2} \left(\frac{\partial^2 L}{\partial \varepsilon^2}\right)_{\varepsilon=0} \varepsilon^2 + O(\varepsilon^3)$$

or

$$L(\varepsilon) - L(0) = \varepsilon \left(\frac{\partial L}{\partial \varepsilon}\right)_{\varepsilon=0} + O(\varepsilon^2) \tag{A.13}$$

We have

$$\left(\frac{\partial L}{\partial \varepsilon}\right)_{\varepsilon=0} = \sum_{i=1}^{n} \left(\frac{\partial L}{\partial q_i} \eta_i + \frac{\partial L}{\partial \dot{q}_i} \dot{\eta}_i\right) \tag{A.14}$$

and with the aid of (A.6), (A.2), (A.13), and (A.14) we obtain the "variation" of the integral I:

$$\delta I = I(\varepsilon) - I(0) = \varepsilon I_1 + \varepsilon^2 I_2 + O(\varepsilon^3) \tag{A.15}$$

where

$$I_1 = \int_{t_1}^{t_2} \left\{ \sum_{i=1}^{n} \left(\frac{\partial L}{\partial q_i} \eta_i + \frac{\partial L}{\partial \dot{q}_i} \dot{\eta}_i\right) \right\} dt \tag{A.16}$$

and the terms εI_1 and $\varepsilon^2 I_2$ are the first and second variation of I, respectively. In order that I be a maximum, $\delta I < 0$ for all sufficiently small ε. Therefore a sufficient condition which insures that I is a maximum is provided by $I_1 = 0$ and $I_2 < 0$. In order that I be a minimum, $\delta I > 0$ for all sufficiently small ε. Hence a sufficient condition which insures that I is a minimum is provided by $I_1 = 0$, $I_2 > 0$. Thus the requirement that I be an extremum results in the necessary and sufficient condition that the first variation vanishes, i.e., $I_1 = 0$. With reference to (A.16) together with the arguments leading from (A.8) to (A.12), we see that this requirement results in Lagrange's equations Within the context of Chapter 2, Hamilton's principle for a conservative, holonomic mechanical systems is stated as

$$\delta I = \delta \int_{t_1}^{t_2} L(q_i, \dot{q}_i; t)\, dt = 0 \tag{A.17}$$

to the first order in ε, and with reference to (A.15) this agrees with the present findings obtained in an entirely different manner.

B

KINEMATICS OF WAVE MOTION

From a physical point of view the group-velocity is perhaps even more important and significant than the wave-velocity. The latter may be greater or less than the former, and it is even possible to imagine mechanical media in which it would have the opposite direction; i.e., a disturbance might be propagated outwards from a centre in the form of a group, whilst the individual waves composing the group were themselves travelling backwards, coming into existence at the front, and dying out as they approach the rear.

Sir Horace Lamb, Hydrodynamics, 6th Ed., 1932.

B.1 PHASE VELOCITY

We consider a straight crested wave propagating in the positive x-direction, i.e., the wave profile is expressed as a function of x and t, and it is independent of y and z. The most commonly used example is a harmonic disturbance propagating in the positive x-direction characterized by

$$\phi(x,t) = A \cos 2\pi \left(\frac{t}{T_o} - \frac{x}{\lambda_o}\right) \tag{B.1}$$

where A is the amplitude, λ_o the wave length, and T_o is the period of the wave (see Fig. B.1). It is convenient to set

$$k_o = \frac{2\pi}{\lambda_o}, \quad \omega_o = \frac{2\pi}{T_o}, \quad v_o = \frac{\omega_o}{k_o} \tag{B.2}$$

so that (B.1) can be written

$$\begin{aligned} \phi(x,t) &= A \cos(\omega_o t - k_o x) = A \cos k_o(v_o t - x) \\ &= A \operatorname{Re}\left\{\exp ik_o(v_o t - x)\right\} \end{aligned} \tag{B.3}$$

where Re stands for "real part of". In equations (B.3), k_o is the wave number, ω_o is the frequency, and v_o is the phase velocity.

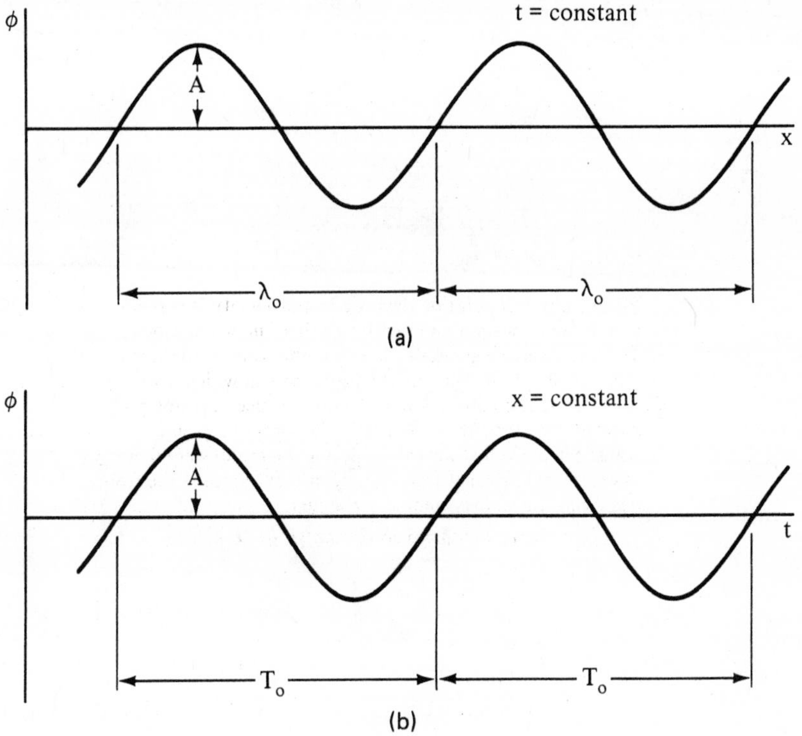

Figure B.1

The argument of the trigonometric function in (B.3) is called the phase ψ of the wave, and we have

$$\psi = \omega_0 t - k_0 x \tag{B.4}$$

If the phase remains constant in time, then the shape of the wave profile does not change as the wave propagates. Such a condition defines a phase velocity v_0, i.e., v_0 is the velocity of propagation of the wave form. To insure ψ=constant with respect to time, we set

$$\frac{d\psi}{dt} = \omega_0 - k_0 \frac{dx}{dt} = 0 \tag{B.5}$$

so that

$$v_0 = \frac{dx}{dt} = \frac{\omega_0}{k_0} \tag{B.6}$$

If we view the propagating wave from a position which moves with speed v_0 parallel to the x-axis, we shall see a stationary wave.

The phase velocity v_0 is the velocity of propagation of a harmonic (monochromatic) wave with frequency ω_0 and wave number k_0. For waves propagating through a medium, the values of frequency and wave number can not be independently assigned. They are related by a dispersion relation of the form $\omega=\omega(k)$ which describes the characteristics of the medium with respect to wave propagation. A medium for which the phase velocity v is constant for all frequencies (or wave numbers) is called a non-dispersive medium, whereas a medium for which $v=v(\omega)$(or $v=v(k)$) is called a dispersive medium.

As an illustrative example, let us consider the torsional motion of a rod characterized by the wave equation (4.78a):

$$\rho \frac{\partial^2 \phi}{\partial t^2} - G \frac{\partial^2 \phi}{\partial x^2} = 0 \tag{B.7}$$

Assuming a torsional wave propagating in the positive x-direction characterized by

$$\phi(x,t) = A \cos(\omega_0 t - k_0 x) \tag{B.8}$$

and substituting (B.8) into (B.7), we obtain

$$\rho\omega_0^2 = Gk_0^2, \text{ or } v_0 = \frac{\omega_0}{k_0} = \sqrt{\frac{G}{\rho}} \tag{B.9}$$

In this case the phase velocity of torsional waves is constant, depending only upon the material properties of the rod. Since v_o is independent of the frequency (or wave number), the medium characterized by (B.7) is non-dispersive.

Let us next consider the case of torsional waves in a rod embedded in a restraining medium. If the torsional restraint is linearly proportional to local rod rotation ϕ, then the rod motion is characterized by

$$J\rho \frac{\partial^2 \phi}{\partial t^2} - JG \frac{\partial^2 \phi}{\partial x^2} + c\phi = 0 \tag{B.10}$$

where $c>0$ is the restraining constant. Assuming torsional waves characterized by (B.8), and substituting (B.8) into (B.10), we obtain a relationship between frequency and wave number

$$\frac{J\rho}{c}\, \omega_o^2 = \frac{JG}{c}\, k_o^2 + 1$$

or

$$\omega_o = \sqrt{\frac{G}{\rho}\, k_o^2 + \frac{c}{J\rho}} \tag{B.11}$$

The corresponding phase velocity is given by

$$\sqrt{\frac{\rho}{G}}\, v_o = \sqrt{\frac{\rho}{G}}\, \frac{\omega_o}{k_o} = \sqrt{1 + \frac{1}{\left(\frac{JG}{c}\right) k_o^2}} \tag{B.12a}$$

or

$$\sqrt{\frac{\rho}{G}}\, v_o = \frac{\sqrt{\left(\frac{J\rho}{c}\right)}\, \omega_o}{\sqrt{\left(\frac{J\rho}{c}\right) \omega_o^2 - 1}} \tag{B.12b}$$

Dimensionless plots of (B.12) are shown in Fig. B.2. It is obvious that the medium is dispersive for $c>0$. For the present case it can be shown that traveling waves are possible only for $\omega_o > \sqrt{c/J\rho}$ (see Chapter 12). When $0<\omega_o<\sqrt{c/J\rho}$, the phase velocity is imaginary (complex, with real part equal to zero), and this implies the existence of standing waves. The frequency $\omega_o=\sqrt{c/J\rho}$ which separates the spectrum of traveling waves from that of standing waves is called the cut-off frequency.

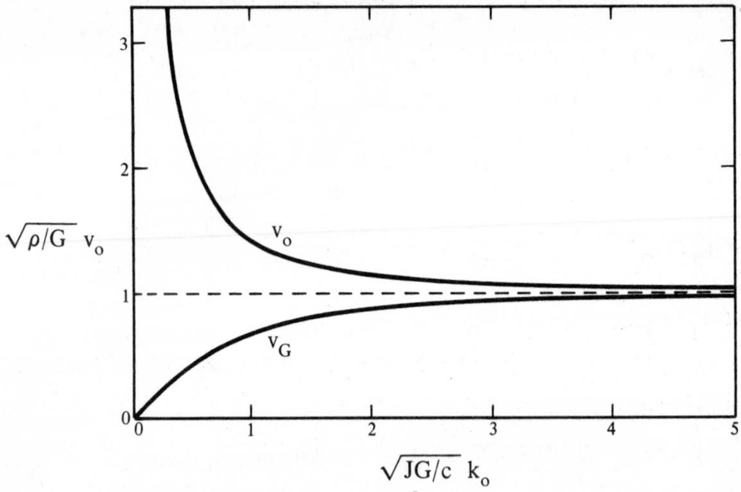

(a)

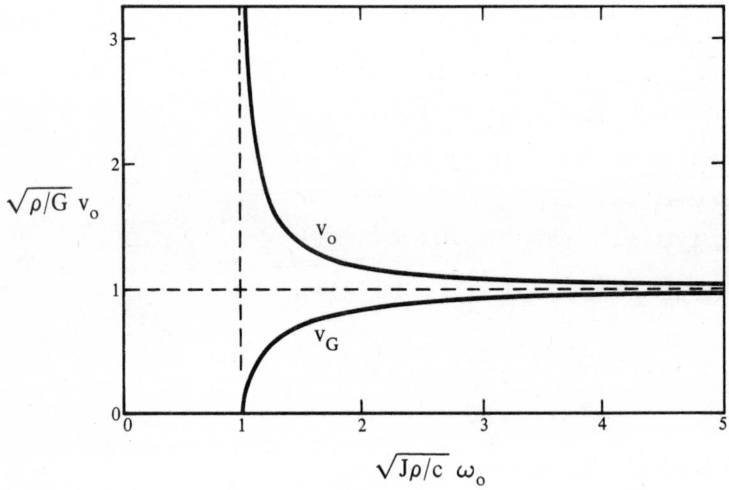

(b)

Figure B.2

B.2 GROUP VELOCITY: THE GAUSSIAN WAVE PACKET

It is instructive to consider the wave resulting from the superposition of two different harmonic, traveling waves with equal amplitudes:

$$\begin{aligned}\phi(x,t) &= A\ \mathrm{Re}\left[\exp i(\omega_1 t-k_1 x) + \exp i(\omega_2 t-k_2 x)\right]\\ &= A\cos(\omega_1 t-k_1 x) + A\cos(\omega_2 t-k_2 x)\end{aligned} \tag{B.13}$$

With the aid of trigonometric indentities, (B.13) may be written as

$$\phi(x,t) = A_{mod}(x,t)\cdot\cos(\omega_o t - k_o x) \tag{B.14}$$

where

$$A_{mod} = 2A\cos(\omega_{mod}t - k_{mod}x) \tag{B.15}$$

and

$$\omega_{mod} = \tfrac{1}{2}(\omega_1-\omega_2), \qquad k_{mod} = \tfrac{1}{2}(k_1-k_2)$$
$$\omega_o = \tfrac{1}{2}(\omega_1+\omega_2), \qquad k_o = \tfrac{1}{2}(k_1+k_2) \tag{B.16}$$

Inspection of (B.14) reveals that $\phi(x,t)$ is a modulated harmonic wave propagating with velocity $v_o=\omega_o/k_o = (\omega_1+\omega_2)/(k_1+k_2)$. Its amplitude modulation is characterized by (B.15). The velocity of the modulation curve (or envelope) is $v_{mod} = \omega_{mod}/k_{mod}=(\omega_1-\omega_2)/(k_1-k_2)$. The situation is shown in Fig. B.3. In general, the modulated wave will have a velocity which differs from the velocity of the wave group (or envelope). The modulated wave shown in Fig. B.3 advances through the profile, gradually increasing and then decreasing its amplitude. If it is now assumed that the medium through which the wave (B.13) propagates is a dispersive one, the frequency will be a known function of the wave number, i.e., $\omega=\omega(k)$. Expanding the frequencies of the two component waves in Taylor series about the average wave number k_o, we obtain

$$\omega(k_1) = \omega(k_o) + (k_1-k_o)v_G + O[(k_1-k_o)^2] \tag{B.17a}$$

$$\omega(k_2) = \omega(k_o) + (k_2-k_o)v_G + O[(k_2-k_o)^2] \tag{B.17b}$$

where

$$v_G = \left(\frac{d\omega}{dk}\right)_{k=k_o} \tag{B.18}$$

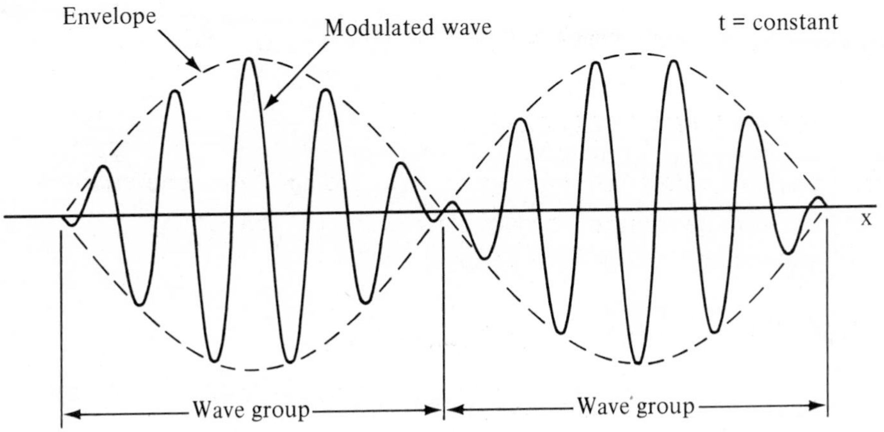

Figure B.3

If it is now assumed that the wave numbers k_1 and k_2 are located in a sufficiently small band of the wave number spectrum, then we may justify the neglect of $O[(k_1-k_0)^2]$ and $O[(k_2-k_0)^2]$ in (B.17). In this case we obtain, with the help of (B.17) and (B.18),

$$v_G = \left(\frac{d\omega}{dk}\right)_{k=k_0} = \frac{\omega(k_2)-\omega(k_1)}{k_2-k_1} = \frac{\omega_2-\omega_1}{k_2-k_1} = v_{mod} \tag{B.19}$$

Thus, if the wave numbers of the two component waves are sufficiently close to one another, the speed of the modulation curve (or wave group) is given by

$$v_G = \left(\frac{d\omega}{dk}\right)_{k=k_0} = \left[\frac{d(kv)}{dk}\right]_{k=k_0} = k_0 \left(\frac{dv}{dk}\right)_{k=k_0} + v_0 \tag{B.20}$$

For this reason the quantity v_G is called the group velocity.

The present development encompassing only two component waves is readily extended to the case where there is a continuous distribution of component waves. It is now assumed that the wave profile consists of a continuous distribution of harmonic waves of the type $\exp i(\omega t-kx)$ where the wave number k can assume all possible values. It is also assumed that the wave propagates through a dispersive medium with a known dispersion relation $\omega=\omega(k)$. If the amplitudes of the component waves are characterized by the wave number spectrum function $(1/\sqrt{2\pi})a(k)$, then the propagating wave is characterized by

$$\phi(x,t) = \frac{1}{\sqrt{2\pi}} \int_{-\infty}^{\infty} a(k)\cdot\exp i(\omega t-kx)\cdot dk \tag{B.21}$$

where it is understood that only the real (or imaginary) part of (B.21) is of physical interest. If we set t=0 in (B.21),

$$\phi(x,0) = \frac{1}{\sqrt{2\pi}} \int_{-\infty}^{\infty} a(k)\cdot\exp(-ikx)\cdot dk \tag{B.22}$$

By means of the method of Fourier transforms, it is possible to "invert" (B.22), i.e.,

$$a(k) = \frac{1}{\sqrt{2\pi}} \int_{-\infty}^{\infty} \phi(x,0)\cdot\exp ikx\cdot dx \tag{B.23}$$

To be specific, we consider the Gaussian wave packet, the spectrum function of which is given by

$$a(k) = \frac{\sigma}{\sqrt{2}} \exp\left[-(k-k_0)^2 \frac{\sigma^2}{4}\right] \tag{B.24}$$

where σ and k_0 are constants. A graph of (B.24) is shown in Fig. B.4a.

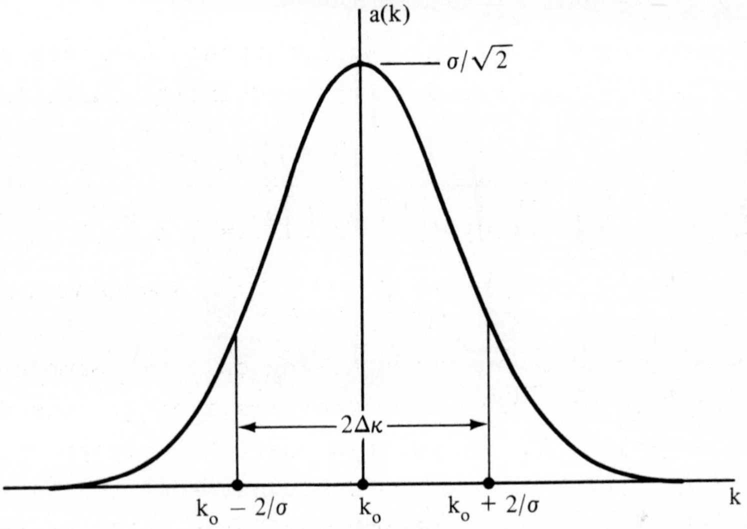

Figure B.4 (a)

We shall first obtain an explicit expression for the wave at t=0. Substituting (B.24) into (B.22), we obtain

$$\phi(x,0) = \frac{1}{\sqrt{2\pi}} \int_{-\infty}^{\infty} \frac{\sigma}{\sqrt{2}} \cdot \exp\left[-ikx-(k-k_0)^2 \frac{\sigma^2}{4}\right] dk \tag{B.25}$$

If we set

$$k=k_0+\kappa \tag{B.26}$$

then (B.25) is readily transformed to

$$\phi(x,0) = \frac{\sigma}{2\sqrt{\pi}} \exp(-ik_0 x) \int_{-\infty}^{\infty} \exp(-ix\kappa- \frac{\sigma^2}{4} \kappa^2)\cdot d\kappa \tag{B.27}$$

With the aid of the known integral

$$\int_{-\infty}^{\infty} \exp(au-bu^2)du = \sqrt{\frac{\pi}{b}} \exp \frac{a^2}{4b} \tag{B.28}$$

equation (B.27) is reduced to

$$\phi(x,0) = \exp(-ik_0 x)\cdot\exp\left(- \frac{x^2}{\sigma^2}\right) \tag{B.29}$$

It is now clear, from (B.29), that $\exp(-ik_o x)$ represents a harmonic wave with wave length $\lambda_o=2\pi/k_o$, while $\exp(-x^2/\sigma^2)$ is the envelope or amplitude modulation. A graph of the real part of (B.29) is shown in Fig. B.4b. Our wave packet consists of a large pulse containing several

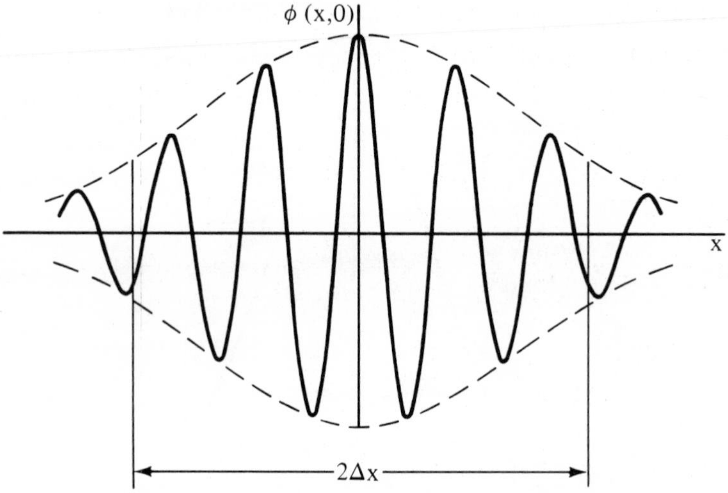

Figure B.4 (b)

oscillations at t=0. If we define the half-length of the pulse as the value of x which reduces the ordinate of the envelope to e^{-1} times its maximum value, then the half-length of the Gaussian wave packet at t=0 is equal to σ. By definition, the effective pulse length 2Δx is twice the half-length, and for the present case Δx=σ. In a similar manner, we can speak of the half-width of the spectrum function a(κ). For the Gaussian wave packet, the half-width of the spectrum function (B.22) is 2/σ, and its effective spectrum width is 2Δk=4/σ. Consequently, Δx·Δk=2, i.e., there is an inverse relationship between effective pulse length and effective wave number spectrum width. It is also interesting to note that we can recover the limiting case of a monochromatic wave by taking the limit of the Gaussian wave packet as σ→∞. In this case $\phi(x,0) = \exp(-ik_o x)$, and with the aid of (B.23),

$$a(k) = \frac{1}{\sqrt{2\pi}} \int_{-\infty}^{\infty} \exp i(k-k_o)x \cdot dx = \delta(k_o)$$

i.e., the spectrum function approaches a line spectrum consisting of a single line and the wave packet approaches a monochromatic wave with wave length $\lambda_o=2\pi/k_o$.

We now proceed to the case $t>0$. It now becomes necessary to integrate (B.21), when $a(k)$ is given by (B.24). It is possible to obtain a useful approximation of this integral by observing that because of the small effective width $2\Delta\kappa=2\Delta(k-k_o)$ of the spectrum function, the major contribution to the value of the integral in (B.21) comes from those wave numbers which are near the predominant wave number k_o (see Fig. B.4a). As a consequence of this observation, we expand the frequency $\omega=\omega(k)$ into a Taylor series about the predominant wave number k_o:

$$\omega(k) = \omega_o+v_G(k-k_o) + \tfrac{1}{2}v'_G(k-k_o)^2 + O[(k-k_o)^3] \tag{B.30}$$

where

$$\omega_o = \omega(k_o),\quad v_G = \left(\frac{d\omega}{dk}\right)_{k=k_o},\quad v'_G = \left(\frac{d^2\omega}{dk^2}\right)_{k=k_o}$$

As a first approximation, we retain the first two terms of the series expansion in (B.30), i.e., we drop terms $O[(k-k_o)^2]$. In this case

$$\omega t-kx = (\omega_o t-k_o x) + \kappa(v_G t-x)$$

and (B.21) in conjunction with (B.24) and (B.26) becomes

$$\phi(x,t) = \frac{\sigma}{2\sqrt{\pi}}\cdot\exp i(\omega_o t-k_o x)\cdot\int_{-\infty}^{\infty}\exp[i(v_G t-x)\kappa-\frac{\sigma^2}{4}\kappa^2]d\kappa \tag{B.31}$$

If we utilize (B.28) to evaluate the integral in (B.31), we obtain

$$\phi(x,t) = \exp\left[-\frac{(v_G t-x)^2}{\sigma^2}\right]\cdot\exp i(\omega_o t-k_o x) \tag{B.32}$$

Inspection of (B.32) reveals that the modulated wave travels with velocity $v_o=\omega_o/k_o$, but its envelope travels without suffering distortion with velocity $v_G=\left(\frac{d\omega}{dk}\right)_{k=k_o}$. Thus the wave packet (or group) moves with the velocity v_G (group velocity). We also note that upon setting $t=0$ in (B.32), we obtain (B.29), providing a check on the present computation.

A better approximation of the value of the integral (B.21) can be obtained by taking the first three terms of the series expansion (B.30) and neglecting terms $O[(k-k_o)^3]$. In this case

$$\omega t-kx = (\omega_o t-k_o x) + \kappa(v_G t-x) + \tfrac{1}{2}\kappa^2 v'_G t$$

and (B.21) in conjunction with (B.24) and (B.26) becomes

$$\phi(x,t) = \frac{\sigma}{2\sqrt{\pi}} \cdot \exp i(\omega_0 t - k_0 x) \int_{-\infty}^{\infty} \exp\left[i(v_G t - x)\kappa - \left(\frac{\sigma^2}{4} - i\tfrac{1}{2}v_G' t\right)\kappa^2\right] d\kappa \tag{B.33}$$

The integral in (B.33) can be evaluated with the aid of (B.28). The result is

$$\phi(x,t) = A(x,t)\cdot \exp i(\omega_0 t - k_0 x) \tag{B.34}$$

where

$$A(x,t) = \frac{\sigma}{\sqrt{\sigma^2 - i2v_G' t}} \exp\left[-\frac{(x-v_G t)^2}{\sigma^2 - i2v_G' t}\right]$$

A straightforward computation reveals that $A(x,t)$ can be written as

$$A(x,t) = B(x,t)e^{i\theta(x,t)} \tag{B.35a}$$

where

$$B(x,t) = \frac{\sigma}{\sqrt[4]{\sigma^4 + 4v_G'^2 t^2}} \exp\left[-\frac{\sigma^2(x-v_G t)^2}{\sigma^4 + 4v_G'^2 t^2}\right] \tag{B.35b}$$

$$\theta(x,t) = -\frac{2v_G' t(x-v_G t)^2}{\sigma^4 + 4v_G'^2 t^2} + \tfrac{1}{2}\tan^{-1}\left(\frac{2v_G' t}{\sigma^2}\right) \tag{B.35c}$$

An inspection of (B.35) reveals the following: the effective pulse length is now

$$L(t) = \frac{2}{\sigma}\sqrt{\sigma^4 + 4v_G'^2 t^2} \tag{B.36}$$

and a time-dependent phase angle θ must be added to the term $(\omega_0 t - k_0 x)$. The wave packet (envelope) moves with the group velocity

$$v_G = \left(\frac{d\omega}{dk}\right)_{k=k_0}.$$

However, the envelope now spreads in a manner such that its effective width at t is given by (B.36). Since $L(0)=2\Delta(x)=2\sigma$, we have

$$\frac{L(t)}{L(0)} = \sqrt{1 + \frac{4v_G'^2 t^2}{\sigma^4}} = \sqrt{1 + \frac{64v_G'^2 t^2}{[L(0)]^4}} \tag{B.37}$$

for the fractional increase in pulse length. With reference to (B.37), we note that the smaller the pulse length at $t=0$, the greater will be its subsequent rate of deformation.

The foregoing observations should provide a more fundamental understanding of the phenomenon of the dispersion of a propagating wave packet in a dispersive medium. Each component harmonic wave of the moving wave packet travels with the phase velocity corresponding to its particular wave number. Since the pulse consists of an ensemble of trigonometric waves, each moving with a different velocity, the shape of the pulse will change as it moves along. It is important to note, however, that a wave packet characterized by a sufficiently narrow spectrum function will move with group velocity v_G and, to a first approximation, will not change its shape as it propagates.

B.3 GROUP VELOCITY: GENERAL

The previous section is now generalized to encompass traveling wave packets (or groups of waves) with arbitrary envelope functions, except for the following restrictions: (a) the spectrum function corresponding to the packet envelope has a predominant wave number and (b) the effective width of the spectrum function must be sufficiently small.

As motivated in Section 2, a wave packet can be characterized by

$$\phi(x,t) = \frac{1}{\sqrt{2\pi}} \int_{-\infty}^{\infty} a(\kappa) \exp i[\omega t-(k_o+\kappa)x]d\kappa \tag{B.38}$$

If we set t=0,

$$\phi(x,0) = \frac{1}{\sqrt{2\pi}} \int_{-\infty}^{\infty} a(\kappa) \exp[-i(k_o+\kappa)x]d\kappa$$

or

$$\phi(x,0) = \phi_1(x) \exp(-ik_o x) \tag{B.39}$$

where

$$\phi_1(x) = \frac{1}{\sqrt{2\pi}} \int_{-\infty}^{\infty} a(\kappa) \exp(-i\kappa x)d\kappa \tag{B.40}$$

Equation (B.39) characterizes the modulated wave at t=0 with amplitude modulation $\phi_1(x)$. Assuming propagation in a dispersive medium, we have $\omega=\omega(k)$, and upon expanding ω in a Taylor series about the predominant wave number $k=k_o$, we obtain

$$\begin{aligned}\omega(k) &= \omega(k_o) + (k-k_o)v_G + O[(k-k_o)^2] \\ &= \omega_o + v_G\kappa + O(\kappa^2)\end{aligned} \tag{B.41}$$

where

$$\omega_o = \omega(k_o), \quad v_G = \left(\frac{d\omega}{dk}\right)_{k=k_o} = \left(\frac{d\omega}{dk}\right)_{\kappa=0}$$

$$\kappa = k-k_o.$$

Assuming that our spectrum function $a(\kappa)$ has a sufficiently narrow effective width $\Delta\kappa$, we neglect $O(\kappa^2)$ in (B.41). Thus (B.38) assumes the approximate value

$$\phi(x,t) = \frac{1}{\sqrt{2\pi}} \int_{-\infty}^{\infty} a(\kappa) \exp i[(\omega_o t-k_o x) + \kappa(v_G t-x)]d\kappa$$

$$= \phi_1(v_G t-x)\cdot\exp i\, k_o(v_o t-x) \tag{B.42}$$

Equation (B.42) characterizes a modulated wave traveling with the phase velocity which corresponds to the predominant wave number k_o of the ensemble. However, the modulation envelope in (B.42) travels with the group velocity $v_G = \left(\frac{d\omega}{dk}\right)_{k=k_o}$.

The concept of group velocity is important because it can be shown to be equal to the velocity of energy transport in most situations where dispersion occurs. For several demonstrations of that fact, the reader is urged to consult Chapters 7, 12, and 13.

To illustrate these results, we now consider two examples.

(a) Torsional Waves in an Elastic Rod. With reference to (B.9), we note that the phase velocity of harmonic torsional waves $v=\sqrt{G/\rho}$ is independent of the wave number. There is no dispersion, and all harmonic waves travel with constant speed. Consequently a wave packet will also travel with the speed $\sqrt{G/\rho}$.

(b) Torsional Waves in an Elastically Restrained Rod. The motion of the rod is characterized by (B.10). Its dispersion relation is given by (B.11). With the aid of the latter

$$\sqrt{\frac{\rho}{G}}\, v_G = \sqrt{\frac{\rho}{G}} \left(\frac{d\omega}{dk}\right)_{k=k_o} = \sqrt{1 - \frac{1}{\left(\frac{J\rho}{c}\right)\omega_o^2}} = \frac{\sqrt{\frac{JG}{c}}\, k_o}{\sqrt{1+\frac{JG}{c}k_o^2}} \tag{B.43}$$

In the present case we have a dispersive medium for $c>0$, and the group velocity is always smaller than the phase velocity of the harmonic wave with predominant wave number k_o as shown in Fig. B.2.

B.4 THE METHOD OF STATIONARY PHASE

In the analysis of transient wave propagation it often becomes necessary to evaluate integrals of the type

$$\phi(x,t) = \frac{1}{\sqrt{2\pi}} \int_{-\infty}^{\infty} a(k)\cdot \exp\, i\, k(vt-x)\cdot dk \tag{B.44}$$

where $v=v(k)$. In most instances, an exact evaluation is either not possible or extremely cumbersome, and it is helpful to have an approximate method available for this task. A method for the approximate evaluation of (B.44) valid for large t and x, which utilizes the concepts developed in Sections 1 through 3 and which has considerable heuristic appeal is due to Lord Kelvin. Kelvin asserted that in the case of the integral of a rapidly oscillating function, the major contribution to the integral comes from that part of the range of integration near which the phase of the trigonometric function involved is stationary. Subsequent investigators have shown that Kelvin's assertion can be rigorously proved for a large class of oscillatory functions.

Let us assume that the phase $k(vt-x)=\omega t-kx$ has an extremum at the phase number $k=k_o$ and nowhere else. Moreover, let us assume that the spectrum function $a(k)$ is approximately constant in a small neighborhood of the point $k=k_o$. Then according to Kelvin's conjecture, the approximate value of the integral (B.44) is given by

$$\phi(x,t) \simeq \frac{a(k_o)}{\sqrt{2\pi}} \int_{k_o-\varepsilon}^{k_o+\varepsilon} \exp\, i(\omega t-kx)\cdot dk \tag{B.45}$$

where $\varepsilon<\Delta k$ is a sufficiently small positive quantity and k_o is a root of the equation

$$\frac{d}{dk}(\omega t-kx) = 0$$

i.e.,

$$x = v_G t, \text{ where } v_G = \left(\frac{d\omega}{dk}\right)_{k=k_o} \tag{B.46}$$

If we expand $\omega(k)$ in a Taylor series about the point $k=k_o$, we obtain

$$\omega(k) = \omega_o + v_G\kappa + \tfrac{1}{2}v_G'\kappa^2 + \frac{1}{6}v_G''\kappa^3 + O(\kappa^4) \tag{B.47}$$

where

$$\kappa = k-k_o, \quad \omega_o = \omega(k_o), \quad v_G' = \left(\frac{d^2\omega}{dk^2}\right)_{k=k_o}, \quad v_G'' = \left(\frac{d^3\omega}{dk^3}\right)_{k=k_o}$$

With the aid of (B.46) and (B.47),

$$k(vt-x) = \omega t-kx = k_o(v_ot-x) + \tfrac{1}{2} v_G' t\kappa^2 + \frac{1}{6} v_G'' t\kappa^3 + O(\kappa^4) \tag{B.48}$$

If it is now assumed that $|\kappa|\leq\varepsilon$ is sufficiently small so that we can neglect $O(\kappa^3)$ in (B.48), then we obtain the approximation

$$k(vt-x) = \omega t-kx \simeq k_o(v_ot-x) + \mu^2\kappa^2$$

where $\mu^2 = \tfrac{1}{2}v_G't>0$ if $v_G'>0$. Consequently, the integral (B.45) is now approximated by

$$\phi(x,t) \simeq \frac{a(k_o)}{\sqrt{2\pi}} \exp i\, k_o(v_ot-x) \cdot \int_{-\varepsilon}^{\varepsilon} \exp i\mu^2\kappa^2\cdot d\kappa$$

But

$$\int_{-\varepsilon}^{\varepsilon} \exp(i\mu^2\kappa^2)d\kappa = 2\int_0^{\varepsilon} \exp(i\mu^2\kappa^2)\cdot d\kappa = \frac{1}{\mu}\int_0^{\mu^2\varepsilon^2} (z)^{-\frac{1}{2}}\exp i\, z\cdot dz$$

where we let $z = \mu^2\kappa^2$. We now assume that $t\to\infty$, and therefore $\mu^2\varepsilon^2\to\infty$ provided $\varepsilon^2>0$. Moreover,

$$\int_0^{\infty} (z)^{-\frac{1}{2}}\cdot\exp i\, z\cdot dz = \int_0^{\infty} \frac{\cos z}{\sqrt{z}}\, dz + i\int_0^{\infty} \frac{\sin z}{\sqrt{z}}\, dz$$

$$= \sqrt{\frac{\pi}{2}} + i\sqrt{\frac{\pi}{2}} = \sqrt{\pi}\exp\left(i\,\frac{\pi}{4}\right).$$

Consequently, for sufficiently large t and x

$$\phi(x,t) \simeq \frac{a(k_o)}{\sqrt{v_G't}} \exp i\left[k_o(v_ot-x) + \frac{\pi}{4}\right], \tag{B.49a}$$

where $v_G'>0$. An entirely analogous analysis reveals that

$$\phi(x,t) \simeq \frac{a(k_o)}{\sqrt{-v_G't}} \exp i\left[k_o(v_ot-x)-\frac{\pi}{4}\right] \tag{B.49b}$$

where $v_G'<0$.

The approximations (B.49) break down if $v_G'=0$. In this case we must take an additional term in (B.48). Thus

$$k(vt-x) = \omega t-kx = k_o(v_ot-x) + \eta^3\kappa^3 \tag{B.50}$$

where $\eta^3 = \frac{1}{6}v_G''t>0$ if $v_G''>0$. Substitution of (B.50) into (B.45) results in

$$\phi(x,t) = \frac{a(k_o)}{\sqrt{2\pi}} \exp i\, k_o(v_ot-x)\cdot\int_{-\varepsilon}^{\varepsilon} \exp i\,\eta^3\kappa^3 d\kappa$$

But $\int_{-\varepsilon}^{\varepsilon} \exp i\,\eta^3\kappa^3 d\kappa = 2\int_0^{\varepsilon} \cos \eta^3\kappa^3 d\kappa = \frac{2}{3\eta}\int_0^{\eta^3\varepsilon^3} (z)^{-2/3} \cos z dz$

where we let $z = \eta^3\kappa^3$. For $t\to\infty$ and $\varepsilon^3>0$, $\mu^3\varepsilon^3\to\infty$. We also note the result

$$\int_0^{\infty} (z)^{-\frac{2}{3}} \cos z\, dz = \frac{\sqrt{3}}{2}\,\Gamma(\tfrac{1}{3})$$

where $\Gamma(1/3)\simeq 2.67893$ is the Gamma function with argument 1/3. Consequently, for sufficiently large t and x, we obtain the approximation

$$\phi(x,t) \simeq \frac{\sqrt[3]{6}}{\sqrt{6}}\,\frac{\Gamma(\frac{1}{3})}{\sqrt{\pi}}\,\frac{a(k_o)}{\sqrt[3]{v_G''t}} \exp i\, k_o(v_ot-x)$$

$$\text{where } v_G''>0 \text{ and } v_G'=0 \tag{B.51a}$$

An entirely analogous analysis reveals that

$$\phi(x,t) \simeq \frac{\sqrt[3]{6}}{\sqrt{6}}\,\frac{\Gamma(\frac{1}{3})}{\sqrt{\pi}}\,\frac{a(k_o)}{\sqrt[3]{-v_G''t}} \exp i\, k_o(v_ot-x)$$

$$\text{where } v_G''<0 \text{ and } v_G'=0. \tag{B.51b}$$

To illustrate an application of the method of the stationary phase, we consider the transient wave motion characterized by (see (12.101)).

$$\phi(x,t) = \frac{1}{\sqrt{2\pi}}\int_{-\infty}^{\infty} \frac{\sin kvt}{kv}\cdot \exp ixk\cdot dk$$

where (B.52)

$$v = \frac{\sqrt{1+k^2}}{k}$$

This integral has its origin in the study of transient wave motion in an elastic rod. In this case it is possible to obtain an exact evaluation with the aid of the theory of residues and contour integration, resulting in

$$\phi(x,t) = \begin{cases} 0 \text{ for } |x|>t \\ \sqrt{\frac{\pi}{2}}\, J_o\left(\sqrt{t^2-x^2}\right) \text{ for } |x|<t \end{cases} \tag{B.53}$$

where J_o denotes the Bessel function of the first kind, of order zero. At this point we note that a well-known asymptotic representation for the Bessel function is given by

$$\sqrt{\frac{\pi}{2}}\, J_o\left(\sqrt{t^2-x^2}\right) \simeq \frac{1}{\sqrt[4]{t^2-x^2}} \cos\left(\sqrt{t^2-x^2} - \frac{\pi}{4}\right) \tag{B.54}$$

valid for sufficiently large argument $\sqrt{t^2-x^2}$.

We now proceed to an approximate evaluation of (B.52) by the method of the stationary phase. Equation (B.52) can be written

$$\phi(x,t) = \frac{2}{\sqrt{2\pi}} \int_0^\infty \frac{\sin kvt}{kv} \cdot \cos xk \cdot dk = \phi_1+\phi_2 \tag{B.55}$$

where

$$\phi_1 = \frac{1}{\sqrt{2\pi}} \int_0^\infty \frac{\sin k(vt+x)}{\sqrt{1+k^2}}\, dk \tag{B.56a}$$

$$\phi_2 = \frac{1}{\sqrt{2\pi}} \int_0^\infty \frac{\sin k(vt-x)}{\sqrt{1+k^2}}\, dk \tag{B.56b}$$

It is clear that ϕ_2 characterizes a wave which travels in the direction of the positive x-axis, while ϕ_1 is a wave which travels in the opposite direction. Applying the method of the stationary phase to evaluate ϕ_2, we readily compute:

$$v_G = \frac{x}{t} = \frac{k_o}{\sqrt{1+k_o^2}}; \qquad k_o = \frac{x}{\sqrt{t^2-x^2}} > 0$$

$$v_G' = (1+k_o^2)^{-\frac{3}{2}} = \frac{(t^2-x^2)^{\frac{3}{2}}}{t^3} > 0 \text{ if } t>x.$$

Upon substitution in (B.49a), we obtain the approximation

$$\phi(x,t) \simeq \frac{1}{\sqrt[4]{t^2-x^2}} \sin(\sqrt{t^2-x^2} + \frac{\pi}{4})$$

$$= \frac{1}{\sqrt[4]{t^2-x^2}} \cos(\sqrt{t^2-x^2} - \frac{\pi}{4}) \qquad \text{(B.57)}$$

valid for sufficiently large t and x, and $t>x>>0$. Comparison of (B.57) with (B.54) reveals that in this particular instance, the approximate solution obtained by the stationary phase method results in the asymptotic representation of the exact solution.

INDEX

Author Index

A

E.B. Adams, 83
E. Almansi, 218
M. Abramowitz, 528

B

D. Bancroft, 354
H. Bateman, 360
E. Betti, 275
L. Boltzmann, 68
L. Brand, 181

C

G.F. Carrier, 534
C. Chree, 353
A. Cauchy, 231
G.R. Cowper, 407

D

B.M. Davies, 358
P. Duhem, 317

E

D.M. Eidus, 278
A. Einstein, 157

F

B. Finzi, 259
K. Friedrichs, 278

G

G. Galilei, 2
D. Graffi, 276
J. Green, 491
I.S. Gradshteyn, 527

H

W. Hamilton, 25, 288
O. Heaviside, 12
R. Hooke, 203
G.E. Hudson, 357

I

C. Inglis, 93

J

J. Jeans, 554

K

A. Kalnins, 522
Lord Kelvin (J.J. Thompson), 124, 571
H. Kolsky, 343
M. Krook, 534

L

J.L. Lagrange, 20, 53, 289
H. Lamb, 315, 340, 558
G. Lamé, 268, 317
P.S. Laplace, 5, 24
Y.C. Lee, 313, 491, 547
A.E.H. Love, 399

M

J.C. Maxwell, 66, 259
S.G. Mikhlin, 278, 281
R.D. Mindlin, 346, 456, 468, 515
O. Mohr, 73

N

C.L.M.H. Navier, 317
F. Neumann, 271
I. Newton, 3, 25

P

P.S. Pawlik, 313
C.E. Pearson, 534
J.S. Pistiner, 76
L. Pochhammer, 353
H. Poincaré, v
S.D. Poisson, 268

R

Lord Rayleigh (J.W. Strutt), 108, 110, 275, 281, 335, 340
H. Reismann, 76, 288, 481, 491, 543, 547
E. Reissner, 468
I.M. Ryzhik, 527

S

I.N. Sneddon, 439, 523
I. Stegun, 528
E. Sternberg, 317

T

C. Truesdell, 258, 260

U

Y.S. Uflyand, 468

Y

T. Young, 268

Subject Index

A

Action integral, 247
Affine deformation, 257
Almansi's strain tensor, 218
Angle of incidence, 328
Angle of twist, 362

B

Band pass filter, 142
Base vectors, 168
Beams, 400
  on elastic foundation, 440
Beats, 17
Bending moment, 402
Bodies, 284
Body force, 228

C

Christoffel symbols, 190
Classical plate theory, 496
Concentrated force, 523
Configuration space, 26
Conservation body force, 258
Constitutive relation, 266
Constraints, 102
Continuum, 204
Contraction, 174
Cauchy's lemma, 231
Cauchy's theorem on stress, 231
Critical speed, 536
Cross product, 161
Curl, 180
Curvilinear coordinates, 182
Cut-off frequency, 130, 384, 394, 561
Cylindrical coordinates, 43, 194, 249

D

Damping
   in a one degree of freedom
      system, 3, 7, 8, 9
   in a two degree of freedom
      system, 114
   n degrees of freedom, 112
   in rods, 381
Degrees of freedom, 36
Difference equations, 81
Dilatation, 317
Dilatational waves, 327
Divergence, 180, 182
Direction cosine, 168
Dispersion relations, 384
   in an iterated structure, 129
   for plane irrotational waves,321
   in a rod, 570
   in an Euler Bernoulli Beam,
      431, 441
   in a Timoshenko Beam, 447
   in a Plate (classical), 541
   in a Plate (improved), 517
Dissipation function, 109
Dot product, 160
Dynamic amplification 543
   for one degree of freedom
      system, 12, 17
   for two degrees of freedom
      system, 76

E

Eigenfunctions, 279, 370, 374,
   474
Eigenvalues, 278, 370, 474
Elastic foundations moduli, 271
Energy continuity equation
   in an iterated structure, 127
   for solids, 237, 365
   in a rod, 365
   in a beam, 405
   in a plate, 466
Energy equation of motion
   for a system of particles, 94
Energy flux vector, 387, 391,
   404, 499, 520
   in a solid, 237
   in a rod, 364, 367
   in an iterated structure, 126
   in a plate, 322, 326, 327, 465
Energy transport, 127
Equivoluminal waves, 324
Euler-Bernoulli beam, 408
Eulerian coordinates, 207
Eulerian strain tensor, 218

F

Flexural rigidity, 468
Force fields, 228
Forced motion
   one degree of freedom, 12, 13,
      14
   two degrees of freedom, 75
   n degrees of freedom, 64
   of plates, 472, 477, 502
   of bounded elastic bodies, 284
   of spherical shell, 292
   of rods, 371
   of Euler-Bernoulli beams, 408
   of simply supported rectangular
      plates, 504
Free vibrations
   one degree of freedom, 7
   two degrees of freedom, 72
   n degrees of freedom, 56
   of plates, 473
   of bounded elastic bodies, 277
   of rods, 369
Frequency, 9, 10, 11, 588
Frequency spectrum, 82
Friction, 110

G

Gauss' theorem, 181
Generalized coordinates, 36
Gaussian Wave Packet, 565
Green's deformation tensor, 221
Group velocity, 431, 441, 444,
   447, 562, 569

H

Hamilton's principle, 246, 462
   for discrete systems, 30
Harmonic forcing function
   one degree of freedom, 14
   two degrees of freedom, 77
   n degrees of freedom, 117
Holonomic constraints, 36

I

Index notation, 158
Influence coefficient, 66
Invariance, 173
Invariants
   of the strain tensor, 241
Irrotational waves, 321
Isotropic tensors, 178
Iterated (periodic, lattice)
   structure, 79, 125

J

Jacobian determinant, 182

K

Kinetic energy, 243
particle, 4
particle system, 27, 37
solid, 236
in a plate, 461
Kronecker delta symbol, 162

L

Logarithmic decrement, 11
Lagrange's equations, 42, 289
Lagrange's theorem, 99
Lagrangian, 246
coordinates, 205, 265
strain tensor, 213
metric tensor, 221
Laplace transform, 5
table of transformations, 19
Laplacian operator, 180
Linear theory of elasticity, 269
Longitudinal motion
of rods, 366, 367

M

Material derivative, 208, 223
Mechanical filter, 128
Modal vector, 57, 64
Mohr circle, 74
Momentum continuity equation
in an iterated structure, 128
in a solid, 237
Momentum flux
in an iterated structure, 126
in a rod, 365, 367
Momentum flux tensor
for a solid, 237

N

Natural frequencies
stationary properties, 97
bounds, 100
of elastic body, 281
of beams, 409
of spherical shell, 299
of plates, 476, 501, 506
Natural state, 235
Newton's laws, 227
Normal coordinates, 63, 98
Normal mode
discrete undamped system, 57, 64
discrete, damped system, 120
Normalization, 284
Normalization condition, 370, 410, 420, 476, 501
n degrees of freedom, 58

O

Orthogonal curvilinear coordinates, 187
Orthogonal transformations, 172
Orthogonality, 283
Orthogonality condition, 371, 410, 420
Orthogonality relation, 476, 502
n degrees of freedom, 61
rods, 371
beams, 410, 420
plates, 476, 502
Orthonormality equations, 61, 284

P

Pendulum, 33
Period, 9, 11, 559
Permutation tensor, 163
Phase angle, 16
Phase velocity, 384, 431, 441, 444, 447, 559
Piola-Kirchhoff stress tensor, 240
Plate strain resultants, 459
Plate stress resultants, 458
Plates, 458
Potential energy, 247
particle system, 55
Potential function, 28

Q

Quasi-static solution, 284
Quotient law of tensor, 175

R

Rate of deformation tensor, 223
Rayleigh-Lamb Problem, 340
Rayleigh Quotient, 281
Rayleigh Surface Wave, 335
Rayleigh's beam, 452
Rayleigh's principle, 100
Reciprocal base vectors, 186
Reciprocal relation, 274, 471, 498
Reciprocal theorem
for discrete systems, 95
Rate of deformation tensor, 223

Resonant frequency
one degree of freedom, 15, 17
two degrees of freedom, 78
n degrees of freedom, 119
iterated structure, 86
Reynold's Transport theorem, 257
Rigid body, 36, 38, 40
Rigid body motion, 209
Rods
torsional motion of, 361
of finite length, 368
of unbounded length, 381
Rotation, 317

S

Scalar triple product, 165
Scleronomous constraints, 36
Shear beam, 451
Shear coefficient, 406, 518
Shear force, 402
Shear plate, 545
Shear wave, 325
Simple shearing deformation, 256
Simply supported Euler-Bernoulli beams, 413
Simply supported Timoshenko beams, 423
Small deformation assumption, 261
Small displacement assumption,262
Spherical coordinates, 196
Stability, 54, 241
Stationary phase, 439, 528, 571
Steady state solution
one degree of freedom, 15
two degrees of freedom, 77
n degrees of freedom, 119
Straight crested waves, 516
Strain
normal, 213
shear, 216
Lagrangian, 213
Eulerian, 218
Strain energy, 236, 244, 266,469
in a rod, 362
in a beam, 402
in a plate, 461
Stress equations of motion
for a solid, 234
in spherical coordinates, 259
Stress functions, 258
Stress tensor, 232
Piola-Kirchhoff, 240
Subcritical speed range, 533
Supercritical speed range, 536
Surface traction, 228

T

Tensile test, 255
Tensor, 173
Timoshenko beams, 418
Transient wave motion
in a simple periodic structure, 137
in a rod, 394
in a beam, 437
in a plate, 525
Transmission coefficient, 388, 391
Transmission and reflection of waves
in an iterated structure, 134
in a solid, 328
in a rod, 385
in a beam, 433
Twisting moment, 363
Two degrees of freedom system, 71

U

Uniform dilation, 255
Uniqueness theorem, 271
for particle systems, 94

V

Variational equation of motion, 248, 265
Velocity of energy transport
in a simple periodic structure, 131
in a beam, 444
in a plate, 541
Virtual displacement, 29
Virtual work, 40, 246, 363
Volume element, 193

W

Wave length, 559
Wave motion
in beams, 431
in two bonded semi-infinite rods, 385
in plates, 515
Wave number, 321, 559
Wave propagation
in an iterated structure, 125
Wave reflection, 328
Waves in cylinders, 347
longitudinal, 352
torsional, 351
flexural, 355
Waves in unbounded media, 316

†